PREFACE

1. Scope

The **Department of Defense Dictionary of Military and Associated Terms** (short title: Joint Pub 1-02 or JP 1-02) sets forth standard US military and associated terminology to encompass the joint activity of the Armed Forces of the United States in both US joint and allied joint operations, as well as to encompass the Department of Defense (DOD) as a whole. These military and associated terms, together with their definitions, constitute approved DOD terminology for general use by all components of the Department of Defense. The Secretary of Defense, by DOD Directive 5025.12, 23 August 1989, *Standardization of Military and Associated Terminology*, has directed the use of JP 1-02 throughout the Department of Defense to ensure standardization of military and associated terminology.

2. Purpose

This publication supplements standard English-language dictionaries with standard terminology for military and associated use. However, it is not the intent of this publication to restrict the authority of the joint force commander (JFC) from organizing the force and executing the mission in a manner the JFC deems most appropriate to ensure unity of effort in the accomplishment of the overall mission.

3. Application — DOD and NATO Activities

JP 1-02 is promulgated for mandatory use by the Office of the Secretary of Defense, Military Departments, Joint Staff, combatant commands, Defense agencies, and any other DOD components. DOD terminology herein is to be used without alteration unless a distinctly different context or application is intended. To provide a common interpretation of terminology at home and abroad, US officials, when participating in the North Atlantic Treaty Organization (NATO) or dealing with NATO matters, will use NATO terminology. When a NATO standard for a term or definition does not exist, applicable DOD terminology (if any) may be used.

Note concerning DOD-NATO Standardization: The United States is a signatory to NATO Standardization Agreement (STANAG) 3680, which ratifies the **NATO Glossary of Terms and Definitions (English and French)** (short title: AAP-6). Under the provisions of STANAG 3680, AAP-6 is established as the primary glossary for NATO. The United States carries out its obligation to implement STANAG 3680 in the following manner: (a) English-language entries approved for AAP-6 may be proposed by DOD elements for inclusion in JP 1-02 as DOD-NATO entries. The purpose of such proposals is to increase multinational standardization. After DOD-wide staffing by the US NATO Military Terminology Group (USNMTG), terminology so approved for inclusion in JP 1-02 and DOD-wide use will appear, along with DOD-only entries, in JP 1-02 with an asterisk in parentheses after the term to denote DOD-NATO standardization of terminology, referred to as "alignment" in NATO. (b) As stated in paragraph 3, US officials will adhere to NATO terminology when engaged in NATO matters, provided that applicable terminology exists. (c) An electronic copy of AAP-6 is provided under "Other Publications" at the internet address cited in paragraph 7.

4. Criteria for Terms

The following criteria are used to determine the acceptability of terminology for inclusion in JP 1-02:

a. Inadequate coverage in a standard, commonly accepted dictionary, e.g., by Merriam-Webster.

b. Terminology should be of general military or associated significance. Technical or highly specialized terms may be included if they can be defined in easily understood language and if their inclusion is of general military or associated significance.

c. Terms for weaponry are limited to generic weapon systems.

d. Unless there are special reasons to the contrary, terms and definitions are not to consist of or contain abbreviations or other shortened forms, e.g., acronyms.

e. Only UNCLASSIFIED terminology will be included.

f. Dictionary entries will not be provided for prowords, code words, brevity words, or NATO-only terms.

g. Dictionary entries will not be Service-specific or functionality-specific unless they are commonly employed by US joint forces as a whole.

h. Dictionary entries will not consist of components or sub-components contained in missiles, aircraft, equipment, weapons, etc.

5. Other DOD Dictionaries

Other dictionaries or glossaries for DOD use will be published ONLY AFTER coordination with the USNMTG and approval by the Director for Operational Plans and Joint Force Development (J-7), Joint Staff.

6. Publication Format

This edition of JP 1-02 has been published in two basic parts:

a. **Main Body.** This part of the dictionary contains all terms and definitions approved for use within the Department of Defense, to include those terms and definitions that are approved for both DOD and NATO use. Each entry approved for both DOD and NATO appears with an asterisk in parentheses, i.e., (*), after the term to denote DOD-NATO acceptance.

Note: In rare instances, a term may have a combination of DOD-only definitions and DOD-NATO definitions. In these instances, though an asterisk will appear after the term to denote DOD-NATO standardization, DOD-only definitions will be preceded by "DOD only" in parentheses.

b. **Appendix A.** Appendix A contains a listing of current abbreviations and acronyms in common use within the Department of Defense. This is by no means a complete list of DOD abbreviations and acronyms. Rather, it serves as a guide to current DOD usage in abbreviations and acronyms.

7. JP 1-02 on the Internet

JP 1-02 is accessible on-line at the following internet address:

http://www.dtic.mil/doctrine/jel/doddict

As changes are approved for JP 1-02, they are added to the internet version, making the internet version of JP 1-02 more up-to-date than any printed edition. The internet version thus provides the latest changes worldwide between regular printed editions.

For the Chairman of the Joint Chiefs of Staff:

S. A. FRY
Vice Admiral, U.S. Navy
Director, Joint Staff

Intentionally Blank

TABLE OF CONTENTS

Intentionally Blank

A

abort — **(*)** 1. To terminate a mission for any reason other than enemy action. It may occur at any point after the beginning of the mission and prior to its completion. 2. To discontinue aircraft takeoff or missile launch.

above-the-line publications — The upper level publications in the hierarchy of joint publications which includes capstone, keystone, and other key joint doctrine publications that the Chairman of the Joint Chiefs of Staff signs and are intended to be used by combatant commanders, subunified commanders, joint task force commanders, Service Chiefs, and Joint Staff directors. See also **below-the-line publications; capstone publication; joint publication; keystone publications.** (JP 1-01)

absolute altimeter — **(*)** A type of altimeter which measures vertical distance to the surface below, using radio, radar, sonic, laser, or capacitive technology.

absolute dud — A nuclear weapon which, when launched at or emplaced on a target, fails to explode.

absolute filter — **(*)** A filter capable of cutting off 100% by weight of solid particles greater than a stated micron size.

absolute height — **(*)** The height of an aircraft directly above the surface or terrain over which it is flying. See also **altitude.**

absorbed dose — **(*)** The amount of energy imparted by nuclear (or ionizing) radiation to unit mass of absorbing material. The unit is the rad.

acceptability — Operation plan review criterion. The determination as to whether the contemplated course of action is worth the cost in manpower, materiel, and time involved; is consistent with the law of war; and is militarily and politically supportable. See also **adequacy; feasibility.**

access to classified information — The ability and opportunity to obtain knowledge of classified information. Persons have access to classified information if they are permitted to gain knowledge of the information or if they are in a place where they would be expected to gain such knowledge. Persons do not have access to classified information by being in a place where classified information is kept if security measures prevent them from gaining knowledge of the information.

accidental attack — An unintended attack which occurs without deliberate national design as a direct result of a random event, such as a mechanical failure, a simple human error, or an unauthorized action by a subordinate.

accompanying supplies — Unit supplies that deploy with forces.

accountability — The obligation imposed by law or lawful order or regulation on an officer or other person for keeping accurate record of property, documents, or funds. The person having this obligation may or may not have actual possession of the property, documents, or funds. Accountability is concerned primarily with records, while responsibility is concerned primarily with custody, care, and safekeeping. See also **responsibility.**

accounting line designator — A five-character code, consisting of the target desired ground zero designator and the striking command suffix, to indicate a specific nuclear strike by a specified weapon delivery system on a target objective to the operation plan. Also called **ALD.**

accuracy of fire — (*) The precision of fire expressed by the closeness of a grouping of shots at and around the center of the target.

accuracy of information — See **evaluation.**

acoustical surveillance — Employment of electronic devices, including sound-recording, -receiving, or -transmitting equipment, for the collection of information.

acoustic circuit — A mine circuit which responds to the acoustic field of a target. See also **mine.**

acoustic intelligence — (*) Intelligence derived from the collection and processing of acoustic phenomena. Also called **ACINT.**

acoustic jamming — The deliberate radiation or reradiation of mechanical or electroacoustic signals with the objectives of obliterating or obscuring signals that the enemy is attempting to receive and of disrupting enemy weapons systems. See also **barrage jamming; electronic warfare; jamming; spot jamming.**

acoustic mine — (*) A mine with an acoustic circuit which responds to the acoustic field of a ship or sweep. See also **mine.**

acoustic minehunting — (*) The use of a sonar to detect mines or mine-like objects which may be on or protruding from the seabed, or buried.

acoustic warfare — (*) Action involving the use of underwater acoustic energy to determine, exploit, reduce, or prevent hostile use of the underwater acoustic spectrum and actions which retain friendly use of the underwater acoustic spectrum. Also called **AW.** There are three divisions within acoustic warfare. 1. **acoustic warfare support measures.** That aspect of acoustic warfare involving actions to search for, intercept, locate, record, and analyze radiated acoustic energy in water for the purpose of exploiting such radiations. The use of acoustic warfare support measures involves no intentional underwater acoustic emission and is generally not detectable by the enemy. Also called **AWSM.** 2. **acoustic warfare countermeasures.** That aspect of acoustic warfare involving actions taken to prevent or reduce an enemy's effective use of the underwater acoustic spectrum. Acoustic warfare countermeasures involve intentional underwater acoustic emissions for deception and jamming. Also called **AWCM.** 3. **acoustic warfare counter-countermeasures.** That aspect of acoustic warfare involving actions taken to ensure friendly effective use of the underwater acoustic spectrum despite the enemy's use of underwater acoustic warfare. Acoustic warfare counter-countermeasures involve anti-acoustic warfare support measures and anti-acoustic warfare countermeasures, and may not involve underwater acoustic emissions. Also called **AWCCM.**

acoustic warfare counter-countermeasures — See **acoustic warfare Part 3.**

acoustic warfare countermeasures — See **acoustic warfare Part 2.**

acoustic warfare support measures — See **acoustic warfare Part 1.**

acquire — 1. When applied to acquisition radars, the process of detecting the presence and location of a target in sufficient detail to permit identification. 2. When applied to tracking radars, the process of positioning a radar beam so that a target is in that beam to permit the effective employment of weapons. See also **target acquisition.**

acquire (radar) — See **acquire.**

acquisition — See **collection (acquisition).**

acquisition and cross-servicing agreement — Agreements negotiated on a bilateral basis with US allies or coalition partners that allow US forces to exchange most common types of support, including food, fuel, transportation, ammunition, and equipment. Authority to negotiate these agreements is usually delegated to the combatant commander by the Secretary of Defense. Authority to execute these agreements lies with the Secretary of Defense, and may or may not be delegated. Governed by legal guidelines, these agreements are used for contingencies, peacekeeping operations, unforeseen emergencies, or exercises to correct logistic deficiencies that cannot be adequately corrected by national means. The support received or given is reimbursed under the conditions of the acquisition and cross-servicing agreement. Also called **ACSA.** See also **cross-servicing; servicing.** (JP 4-07)

action agent — In intelligence usage, one who has access to, and performs actions against, the target.

action deferred — Tactical action on a specific track is being withheld for better tactical advantage. Weapons are available and commitment is pending.

action information center — See **air defense control center; combat information center.**

action phase — In an amphibious operation, the period of time between the arrival of the landing forces of the amphibious force in the operational area and the accomplishment of their mission. See also **amphibious force; amphibious operation; landing force; mission.** (JP 3-02)

activation — Order to active duty (other than for training) in the Federal service. See also **active duty; federal service.** (JP 4-05)

activation detector — **(*)** A device used to determine neutron flux or density by virtue of the radioactivity induced in it as a result of neutron capture.

active air defense — Direct defensive action taken to destroy, nullify, or reduce the effectiveness of hostile air and missile threats against friendly forces and assets. It includes the use of aircraft, air defense weapons, electronic warfare, and other available weapons. See also **air defense.** (JP 3-01)

active communications satellite — See **communications satellite.**

active defense — The employment of limited offensive action and counterattacks to deny a contested area or position to the enemy. See also **passive defense.**

active duty — Full-time duty in the active military service of the United States. This includes members of the Reserve Components serving on active duty or full-time training duty, but does not include full-time National Guard duty. Also called **AD.** See also **active duty for training; inactive duty training.**

active duty for special work — A tour of active duty for reserve personnel authorized from military and reserve personnel appropriations for work on active or reserve component programs. This includes annual screening, training camp operations, training ship operations, and unit conversion to new weapon systems when such duties are essential. Active duty for special work may also be authorized to support study groups, training sites and exercises, short-term projects, and doing administrative or support functions. By policy, active duty for special work tours are normally limited to 179 days or less in one fiscal year. Tours exceeding 180 days are accountable against active duty end strength.

active duty for training — A tour of active duty which is used for training members of the Reserve Components to provide trained units and qualified persons to fill the needs of the Armed Forces in time of war or national emergency and such other times as the national security requires. The member is under orders that provide for return to non-active status when the period of active duty for training is completed. This includes annual training, special tours of active duty for training, school tours, and the initial duty for training performed by nonprior service enlistees. Also called **ADT.**

Active Guard and Reserve — National Guard and Reserve members who are on voluntary active duty providing full-time support to National Guard, Reserve, and Active Component organizations for the purpose of organizing, administering, recruiting, instructing, or training the Reserve Components. Also called **AGR.** (JP 1-03.17)

active homing guidance — (*) A system of homing guidance wherein both the source for illuminating the target and the receiver for detecting the energy reflected from the target as the result of the illumination are carried within the missile.

active material — (*) Material, such as plutonium and certain isotopes of uranium, which is capable of supporting a fission chain reaction.

active mine — (*) A mine actuated by the reflection from a target of a signal emitted by the mine.

active public affairs policy — Open dissemination of information to inform the news media and public about an issue or activity. An active approach is characterized by announcing the event or addressing the issue through news media advisories, news releases, personal contacts, news conferences, or other forms of public presentation. Such a policy encourages and supports news media coverage. See also **public affairs.** (JP 3-61)

active sealift forces — Military Sealift Command active, common-user sealift and the afloat pre-positioning force, including the required cargo handling and delivery systems as well as necessary operating personnel. See also **afloat pre-positioning force; common-user sealift; Military Sealift Command.** (JP 4-01.2)

active status — Status of all Reserves except those on an inactive status list or in the Retired Reserve. Reservists in an active status may train for points and/or pay and may be considered for promotion.

activity — 1. A unit, organization, or installation performing a function or mission, e.g., reception center, redistribution center, naval station, naval shipyard. 2. A function, mission, action, or collection of actions. Also called **ACT.** See also **establishment.**

act of mercy — In evasion and recovery operations, assistance rendered to evaders by an individual or elements of the local population who sympathize or empathize with the evaders' cause or plight. See also **evader; evasion; evasion and recovery; recovery; recovery operations.** (JP 3-50.3)

actual ground zero — (*) The point on the surface of the Earth at, or vertically below or above, the center of an actual nuclear detonation. See also **desired ground zero; ground zero.**

actuate — **(*)** To operate a mine-firing mechanism by an influence or a series of influences in such a way that all the requirements of the mechanism for firing, or for registering a target count, are met.

acute radiation dose — **(*)** Total ionizing radiation dose received at one time and over a period so short that biological recovery cannot occur.

adequacy — Operation plan review criterion. The determination as to whether the scope and concept of a planned operation are sufficient to accomplish the task assigned. See also **acceptability; feasibility.**

adjust — An order to the observer or spotter to initiate an adjustment on a designated target.

administrative airlift service — The airlift service normally provided by specifically identifiable aircraft assigned to organizations or commands for internal administration.

administrative control — Direction or exercise of authority over subordinate or other organizations in respect to administration and support, including organization of Service forces, control of resources and equipment, personnel management, unit logistics, individual and unit training, readiness, mobilization, demobilization, discipline, and other matters not included in the operational missions of the subordinate or other organizations. Also called **ADCON.** (JP 0-2)

administrative escort — A warship or merchant ship under naval control, carrying a convoy commodore and staff, and serving as a platform for simultaneous communication with an operational control authority and a coastal convoy.

administrative landing — An unopposed landing involving debarkation from vessels that have been administratively loaded. See also **administrative loading; administrative movement; logistics over-the-shore operations.**

administrative lead time — The interval between initiation of procurement action and letting of contract or placing of order. See also **procurement lead time.**

administrative loading — **(*)** A loading system which gives primary consideration to achieving maximum utilization of troop and cargo space without regard to tactical considerations. Equipment and supplies must be unloaded and sorted before they can be used. Also called **commercial loading.** See also **loading.**

administrative map — A map that contains graphically recorded information pertaining to administrative matters, such as supply and evacuation installations, personnel installations, medical facilities, collecting points for stragglers and enemy prisoners of war, train bivouacs, service and maintenance areas, main supply roads, traffic circulation, boundaries, and other details necessary to show the administrative situation. See also **map.**

administrative movement — **(*)** A movement in which troops and vehicles are arranged to expedite their movement and conserve time and energy when no enemy interference, except by air, is anticipated.

administrative order — **(*)** An order covering traffic, supplies, maintenance, evacuation, personnel, and other administrative details.

administrative shipping — Support shipping that is capable of transporting troops and cargo from origin to destination, but that cannot be loaded or unloaded without non-organic personnel and/or equipment (e.g., cargo handling personnel, stevedores, piers, barges, cranes, materials handling equipment, vessels, etc.). See also **administrative loading; administrative movement.**

advanced base — A base located in or near an operational area whose primary mission is to support military operations.

advanced operations base — In special operations, a small temporary base established near or within a joint special operations area to command, control, and/or support training or tactical operations. Facilities are normally austere. The base may be ashore or afloat. If ashore, it may include an airfield or unimproved airstrip, a pier, or an anchorage. An advanced operations base is normally controlled and/or supported by a main operations base or a forward operations base. Also called **AOB.** See also **forward operations base; main operations base.** (JP 3-05.3)

advance force — **(*)** A temporary organization within the amphibious task force which precedes the main body to the objective area. Its function is to participate in preparing the objective for the main assault by conducting such operations as reconnaissance, seizure of supporting positions, minesweeping, preliminary bombardment, underwater demolitions, and air support.

advance guard — Detachment sent ahead of the main force to ensure its uninterrupted advance; to protect the main body against surprise; to facilitate the advance by removing obstacles and repairing roads and bridges; and to cover the deployment of the main body if it is committed to action.

advance guard reserve — Second of the two main parts of an advance guard, the other being the advance guard support. It protects the main force and is itself protected by the advance guard support. Small advance guards do not have reserves.

advance guard support — First of the two main parts of an advance guard, the other being the advance guard reserve. It is made up of three smaller elements, in order from front to rear, the advance guard point, the advance party, and the support proper. The advance guard support protects the advance guard reserve.

adverse weather — Weather in which military operations are generally restricted or impeded. See also **marginal weather.**

adverse weather aerial delivery system — The precise delivery of personnel, equipment, and supplies during adverse weather, using a self-contained aircraft instrumentation system without artificial ground assistance or the use of ground navigational aids. Also called **AWADS.** (JP 3-17)

advisory area — **(*)** A designated area within a flight information region where air traffic advisory service is available.

aerial picket — See **air picket.**

aerial port — An airfield that has been designated for the sustained air movement of personnel and materiel as well as an authorized port for entrance into or departure from the country where located. Also called **APORT.** See also **port of debarkation; port of embarkation.**

aerial port control center — The agency responsible for the management and control of all aerial port resources and for the receipt and dissemination of all airlift requirements received from the airlift coordination cell as the joint force commander's agent. Also called **APCC.** (JP 3-17)

aerial port squadron — An Air Force organization that operates and provides the functions assigned to aerial ports, including processing personnel and cargo, rigging for airdrop, packing parachutes, loading equipment, preparing air cargo and load plans, loading and securing aircraft, ejecting cargo for inflight delivery, and supervising units engaged in aircraft loading and unloading operations.

aerodynamic missile — (*) A missile which uses aerodynamic forces to maintain its flight path. See also **ballistic missile; guided missile.**

aeromedical evacuation — The movement of patients under medical supervision to and between medical treatment facilities by air transportation. Also called **AE.**

aeromedical evacuation control officer — An officer of the air transport force or air command controlling the flow of patients by air.

aeromedical evacuation coordination center — A coordination center within the joint air operations center's airlift coordination cell that monitors all activities related to aeromedical evacuation (AE) operations execution. It manages the medical aspects of the AE mission and serves as the net control station for AE communications. It coordinates medical requirements with airlift capability, assigns medical missions to the appropriate AE elements, and monitors patient movement activities. Also called **AECC.** See also **aeromedical evacuation; aeromedical evacuation system; aeromedical evacuation unit.** (JP 4-01.1)

aeromedical evacuation system — A system that provides: a. control of patient movement by air transport; b. specialized medical aircrew, medical crew augmentees, and specialty medical attendants and equipment for inflight medical care; c. facilities on or in the vicinity of air strips and air bases for the limited medical care of intransit patients entering, en route via, or leaving the system; and d. communication with originating, destination, and en route medical facilities concerning patient transportation. Also called **AES.** See also **aeromedical evacuation.** (JP 4-02.2)

aeromedical evacuation unit — An operational medical organization concerned primarily with the management and control of patients being transported via an aeromedical evacuation system or system echelon. See also **forward aeromedical evacuation.**

aeronautical chart — A specialized representation of mapped features of the Earth, or some part of it, produced to show selected terrain, cultural and hydrographic features, and supplemental information required for air navigation, pilotage, or for planning air operations.

aeronautical information overprint — (*) Additional information which is printed or stamped on a map or chart for the specific purpose of air navigation.

aeronautical plotting chart — (*) A chart designed for the graphical processes of navigation.

aerosol — A liquid or solid composed of finely divided particles suspended in a gaseous medium. Examples of common aerosols are mist, fog, and smoke. (JP 3-11)

aerospace — Of, or pertaining to, Earth's envelope of atmosphere and the space above it; two separate entities considered as a single realm for activity in launching, guidance, and control of vehicles that will travel in both entities.

aerospace defense — 1. All defensive measures designed to destroy or nullify attacking enemy aircraft and missiles and also negate hostile space systems. 2. An inclusive term encompassing air defense, ballistic missile defense, and space defense. See also **air defense**; **space defense.** (JP 3-01.1)

affiliation training — Military training based on allied and/or coalition, joint, and/or Service doctrine or tactics, techniques, and procedures, as applicable, to prepare personnel or units for multinational operations. Usually conducted between US and non-US forces. May also be referred to as multinational training. See also **command post exercise; exercise; field training exercise; maneuver.**

afloat pre-positioning force — Shipping maintained in full operational status to afloat pre-position military equipment and supplies in support of combatant commanders' operation plans. The afloat pre-positioning force consists of the three maritime pre-positioning ships squadrons and the afloat pre-positioning ships. Also called **APF.** See also **afloat pre-positioning ships; maritime pre-positioning ships.** (JP 4-01.2)

afloat pre-positioning operations — Pre-positioning of ships, preloaded with equipment and supplies (including ammunition and petroleum) that provides for an alternative to land-based programs. This concept provides for ships and onboard force support equipment and supplies positioned near potential crisis areas that can be delivered rapidly to joint airlifted forces in the operational area. Afloat pre-positioning in forward areas enhances a force's capability to respond to a crisis, resulting in faster reaction time. See also **operation.** (JP 4-01.6)

afloat pre-positioning ships — Forward deployed merchant ships loaded with tactical equipment and supplies to support the initial deployment of military forces. Also called **APS.** See also **merchant ship.** (JP 4-01.2)

afloat support — **(*)** A form of logistic support outside the confines of a harbor in which fuel, ammunition, and supplies are provided for operating forces either underway or at anchor. See also **floating base support.**

afterwinds — Wind currents set up in the vicinity of a nuclear explosion directed toward the burst center, resulting from the updraft accompanying the rise of the fireball.

agency — **(*)** In intelligence usage, an organization or individual engaged in collecting and/or processing information. Also called **collection agency.** See also **agent**; **intelligence cycle**; **source.**

agent — In intelligence usage, one who is authorized or instructed to obtain or to assist in obtaining information for intelligence or counterintelligence purposes.

agent authentication — The technical support task of providing an agent with personal documents, accoutrements, and equipment which have the appearance of authenticity as to claimed origin and which support and are consistent with the agent's cover story.

agent net — An organization for clandestine purposes that operates under the direction of a principal agent.

aggressor forces — 1. Forces engaged in aggressive military action. 2. In the context of training exercises, the "enemy" created to add realism in training maneuvers and exercises.

air — (*) In artillery and naval gunfire support, a spotting, or an observation, by a spotter or an observer to indicate that a burst or group of bursts occurred before impact.

air alert — See **airborne alert; air defense warning conditions; alert; ground alert.**

air and space expeditionary task force — A deployed numbered air force (NAF) or command echelon immediately subordinate to a NAF provided as the US Air Force component command committed to a joint operation. Also called **AETF.** See also **air expeditionary force; air expeditionary wing.** (JP 3-33)

air apportionment — See **apportionment (air).** (JP 3-56.1)

air assault — The movement of friendly assault forces (combat, combat support, and combat service support) by rotary-wing aircraft to engage and destroy enemy forces or to seize and hold key terrain. See also **assault.** (JP 3-18)

air attack — 1. **coordinated** — A combination of two or more types of air attack (dive, glide, low-level) in one strike, using one or more types of aircraft. 2. **deferred** — A procedure in which attack groups rendezvous as a single unit. It is used when attack groups are launched from more than one station with their departure on the mission being delayed pending further orders. 3. **divided** — A method of delivering a coordinated air attack which consists of holding the units in close tactical concentration up to a point, then splitting them to attack an objective from different directions.

airborne — 1. In relation to personnel, troops especially trained to effect, following transport by air, an assault debarkation, either by parachuting or touchdown. 2. In relation to equipment, pieces of equipment that have been especially designed for use by airborne troops during or after an assault debarkation. It also designates some aeronautical equipment used to accomplish a particular mission. 3. When applied to materiel, items that form an integral part of the aircraft. 4. The state of an aircraft, from the instant it becomes entirely sustained by air until it ceases to be so sustained. A lighter-than-air aircraft is not considered to be airborne when it is attached to the ground, except that moored balloons are airborne whenever sent aloft. Also called **ABN.** See also **air transportable unit.**

airborne alert — (*) A state of aircraft readiness wherein combat-equipped aircraft are airborne and ready for immediate action. See also **fighter cover. (DOD only)** It is designed to reduce reaction time and to increase survivability. See also **combat air patrol; fighter cover; ground alert.**

airborne assault — See **assault phase, Part 2.**

airborne assault weapon — An unarmored, mobile, full-tracked gun providing a mobile antitank capability for airborne troops. Can be airdropped.

airborne battlefield command and control center — A United States Air Force aircraft equipped with communications, data link, and display equipment; it may be employed as an airborne command post or a communications and intelligence relay facility. Also called **ABCCC.**

airborne command post — (*) A suitably equipped aircraft used by the commander for the control of his or her forces.

airborne early warning — The detection of enemy air or surface units by radar or other equipment carried in an airborne vehicle, and the transmitting of a warning to friendly units. Also called **AEW.**

airborne early warning and control — **(*)** Air surveillance and control provided by airborne early warning aircraft which are equipped with search and height-finding radar and communications equipment for controlling weapon systems. Also called **AEW & C.** See also **air picket.**

airborne force — **(*)** A force composed primarily of ground and air units organized, equipped, and trained for airborne operations. See also **force(s).**

airborne interception equipment — **(*)** A fire control system, including radar equipment, installed in interceptor aircraft used to effect air interception.

airborne lift — The total capacities expressed in terms of personnel and cargo that are, or can be, carried by available aircraft in one trip.

airborne mission commander — The commander serves as an airborne extension of the executing component's rescue coordination center (RCC) and coordinates the combat search and rescue (CSAR) effort between the combat search and rescue task force (CSARTF) and the RCC (or joint search and rescue center) by monitoring the status of all CSARTF elements, requesting additional assets when needed, and ensuring the recovery and supporting forces arrive at their designated areas to accomplish the CSAR mission. The airborne mission commander (AMC) may be designated by the component RCC or higher authority. The AMC appoints, as necessary, an on-scene commander. Also called **AMC.** See also **combat search and rescue; combat search and rescue task force; rescue coordination center.** (JP 3-50.21)

airborne operation — An operation involving the air movement into an objective area of combat forces and their logistic support for execution of a tactical, operational, or strategic mission. The means employed may be any combination of airborne units, air transportable units, and types of transport aircraft, depending on the mission and the overall situation. See also **assault; assault phase.**

airborne order — A command and authorization for flight when a predetermined time greater than five minutes is established for aircraft to become airborne.

airborne radio relay — Airborne equipment used to relay radio transmission from selected originating transmitters.

airborne sensor operator — An individual trained to operate sensor equipment aboard aircraft and to perform limited interpretations of collected information produced in flight.

airborne troops — Those ground units whose primary mission is to make assault landings from the air. See also **troops.**

air-breathing missile — A missile with an engine requiring the intake of air for combustion of its fuel, as in a ramjet or turbojet. To be contrasted with the rocket missile, which carries its own oxidizer and can operate beyond the atmosphere.

airburst — **(*)** An explosion of a bomb or projectile above the surface as distinguished from an explosion on contact with the surface or after penetration. See also **types of burst.**

air-capable ship — All ships other than aircraft carriers; aircraft carriers, nuclear;

amphibious assault ships, landing platform helicopter; general purpose amphibious assault ships; or general purpose amphibious assault ships (with internal dock) from which aircraft can take off, be recovered, or routinely receive and transfer logistic support. See also **aviation ship.** (JP 3-04.1)

air cargo — (*) Stores, equipment or vehicles, which do not form part of the aircraft, and are either part or all of its payload.

Air Carrier Initiative Program — Mutual assistance program with signatory commercial air carriers to assist in illegal drug detection and detection of internal conspiracies. (JP 3-07.4)

air cartographic camera — (*) A camera having the accuracy and other characteristics essential for air survey or cartographic photography. Also called **mapping camera.**

air cartographic photography — (*) The taking and processing of air photographs for mapping and charting purposes.

air control operations — The employment of air forces, supported by ground and naval forces, as appropriate, to achieve military objectives in vital airspace areas. Such operations include destruction of enemy air and surface-to-air forces, interdiction of enemy air operations, protection of vital air lines of communication, and the establishment of local military superiority in areas of air operations. See also **operation.** (JP 3-18)

air corridor — (*) A restricted air route of travel specified for use by friendly aircraft and established for the purpose of preventing friendly aircraft from being fired on by friendly forces.

aircraft — See **inactive aircraft inventory; program aircraft; reserve aircraft; supporting aircraft; unit aircraft.**

aircraft arresting barrier — (*) A device, not dependent on an aircraft arresting hook, used to stop an aircraft by absorbing its forward momentum in an emergency landing or an aborted takeoff. Also called **barricade; emergency barrier.** See also **aircraft arresting system.**

aircraft arresting cable — (*) That portion of an aircraft arresting system which spans the runway surface or flight deck landing area and is engaged by the aircraft arresting hook. Also called **aircraft arresting wire.**

aircraft arresting gear — (*) A device used to engage hook-equipped aircraft to absorb the forward momentum of a routine or emergency landing or aborted takeoff. See also **aircraft arresting system.**

aircraft arresting hook — (*) A device fitted to an aircraft to engage arresting gear. Also called **tail hook.** See also **aircraft arresting system.**

aircraft arresting system — (*) A series of components used to stop an aircraft by absorbing its momentum in a routine or emergency landing or aborted takeoff. See also **aircraft arresting barrier; aircraft arresting gear; aircraft arresting hook.**

aircraft arresting wire — See **aircraft arresting cable.** See also **aircraft arresting system.**

aircraft arrestment — (*) Controlled stopping of an aircraft by external means.

aircraft block speed — True airspeed in knots under zero wind conditions adjusted in relation to length of sortie to compensate for takeoff, climbout, letdown, instrument approach, and landing.

aircraft captain — See **aircraft commander.**

aircraft carrier — A warship designed to support and operate aircraft, engage in attacks on targets afloat or ashore, and engage in sustained operations in support of other forces. Designated as **CV or CVN.** CVN is nuclear powered.

aircraft commander — **(*)** The aircrew member designated by competent authority as being in command of an aircraft and responsible for its safe operation and accomplishment of the assigned mission. Also called **AC.**

aircraft control and warning system — A system established to control and report the movement of aircraft. It consists of observation facilities (radar, passive electronic, visual, or other means), control center, and necessary communications.

aircraft cross-servicing — **(*)** Services performed on an aircraft by an organization other than that to which the aircraft is assigned, according to an established operational aircraft cross-servicing requirement, and for which there may be a charge. Aircraft cross-servicing has been divided into two categories: a. **Stage A cross-servicing**: The servicing of an aircraft on an airfield/ship which enables the aircraft to be flown to another airfield/ship. b. **Stage B cross-servicing**: The servicing of an aircraft on an airfield/ship which enables the aircraft to be flown on an operational mission. See also **aircraft transient servicing.**

aircraft loading table — A data sheet used by the airlift commander containing information as to the load that actually goes into each aircraft.

aircraft mission equipment — **(*)** Equipment that must be fitted to an aircraft to enable it to fulfill a particular mission or task. Also called **aircraft role equipment.**

aircraft modification — **(*)** A change in the physical characteristics of aircraft, accomplished either by a change in production specifications or by alteration of items already produced.

aircraft monitoring and control — That equipment installed in aircraft to permit monitoring and control of safing, arming, and fuzing functions of nuclear weapons or nuclear weapon systems.

aircraft piracy — Any seizure or exercise of control, by force or violence, or threat of force or violence or by any other form of intimidation and with wrongful intent, of an aircraft within the special aircraft jurisdiction of the United States. (JP 3-07.2)

aircraft role equipment — See **aircraft mission equipment.**

aircraft scrambling — **(*)** Directing the immediate takeoff of aircraft from a ground alert condition of readiness.

aircraft store — **(*)** Any device intended for internal or external carriage and mounted on aircraft suspension and release equipment, whether or not the item is intended to be separated in flight from the aircraft. Aircraft stores are classified in two categories as follows. a. **expendable store** — An aircraft store normally separated from the aircraft in flight such as a missile, rocket, bomb, nuclear weapon, mine, torpedo, pyrotechnic device, sonobuoy, signal underwater sound device, or other similar items. b. **nonexpendable store** — An aircraft store which is not normally separated from the aircraft in flight such as a tank (fuel and spray), line-source disseminator, pod (refueling, thrust augmentation, gun, electronic attack, data link, etc.), multiple rack, target, cargo drop container, drone, or other similar items. See also **payload.**

aircraft tiedown — Securing aircraft when parked in the open to restrain movement due to the weather or condition of the parking area.

aircraft transient servicing — (*) Services performed on an aircraft by an organization other than that to which the aircraft is assigned and for which there may be a financial charge. This activity is separate from the established aircraft cross-servicing program and requires that the transient aircrew supervise the correct application of ground crew procedures. See also **aircraft cross-servicing.**

aircraft utilization — Average numbers of hours during each 24-hour period that an aircraft is actually in flight.

aircraft vectoring — (*) The directional control of in-flight aircraft through transmission of azimuth headings.

air cushion vehicle — A vehicle capable of being operated so that its weight, including its payload, is wholly or significantly supported on a continuously generated cushion or "bubble" of air at higher than ambient pressure. Also called **ACV.** (Note: NATO uses the term "ground effect machine.")

air defense — All defensive measures designed to destroy attacking enemy aircraft or missiles in the Earth's envelope of atmosphere, or to nullify or reduce the effectiveness of such attack. Also called **AD.** See also **active air defense; aerospace defense; passive air defense.**

air defense action area — (*) An area and the airspace above it within which friendly aircraft or surface-to-air weapons are normally given precedence in operations except under specified conditions. See also **air defense operations area.**

air defense area — 1. **overseas** — A specifically defined airspace for which air defense must be planned and provided. 2. **United States** — Airspace of defined dimensions designated by the appropriate agency within which the ready control of airborne vehicles is required in the interest of national security during an air defense emergency.

air defense artillery — Weapons and equipment for actively combatting air targets from the ground. Also called **ADA.**

air defense battle zone — A volume of airspace surrounding an air defense fire unit or defended area, extending to a specified altitude and range, in which the fire unit commander will engage and destroy targets not identified as friendly under criteria established by higher headquarters.

air defense control center — (*) The principal information, communications, and operations center from which all aircraft, antiaircraft operations, air defense artillery, guided missiles, and air warning functions of a specific area of air defense responsibility are supervised and coordinated. Also called **air defense operations center.** See also **combat information center.**

air defense direction center — An installation having the capability of performing air surveillance, interception, control, and direction of allocated air defense weapons within an assigned sector of responsibility. It may also have an identification capability.

air defense division — A geographic subdivision of an air defense region. See also **air defense sector.**

air defense early warning — See **early warning.**

air defense emergency — An emergency condition, declared by the Commander in Chief, North American Air Defense Command, that exists when attack upon the continental United States, Alaska, Canada, or United States installations in Greenland by hostile aircraft or missiles is considered probable, is imminent, or is taking place. Also called **ADE.**

air defense ground environment — (*) The network of ground radar sites and command and control centers within a specific theater of operations which are used for the tactical control of air defense operations.

air defense identification zone — Airspace of defined dimensions within which the ready identification, location, and control of airborne vehicles are required. Also called **ADIZ.** See also **air defense operations area.**

air defense operations area — An area and the airspace above it within which procedures are established to minimize mutual interference between air defense and other operations. It may include designation of one or more of the following: air defense action area, air defense area; air defense identification zone, and/or firepower umbrella. See also **air defense action area; air defense identification zone; positive identification and radar advisory zone.** (JP 3-52)

air defense operations center — See **air defense control center.**

air defense operations team — A team of United States Air Force ground environment personnel assigned to certain allied air defense control and warning units/elements.

air defense readiness — An operational status requiring air defense forces to maintain higher than ordinary preparedness for a short period of time.

air defense region — (*) A geographical subdivision of an air defense area.

air defense sector — (*) A geographical subdivision of an air defense region. See also **air defense division.**

air defense suppression — In air operations, actions taken to degrade fixed and mobile surface-based components of enemy air defense systems so that offensive air forces may effectively attack a target.

air defense warning conditions — A degree of air raid probability according to the following code. The term air defense division/sector referred to herein may include forces and units afloat and/or deployed to forward areas, as applicable. **Air defense warning yellow** — attack by hostile aircraft and/or missiles is probable. This means that hostile aircraft and/or missiles are en route toward an air defense division/sector, or unknown aircraft and/or missiles suspected to be hostile are en route toward or are within an air defense division/sector. **Air defense warning red** — attack by hostile aircraft and/or missiles is imminent or is in progress. This means that hostile aircraft and/or missiles are within an air defense division/sector or are in the immediate vicinity of an air defense division/sector with high probability of entering the division/sector. **Air defense warning white** — attack by hostile aircraft and/or missiles is improbable. May be called either before or after air defense warning yellow or red. The initial declaration of air defense emergency will automatically establish a condition of air defense warning other than white for purposes of security control of air traffic.

air delivery — See **airdrop; air landed; air movement; air supply.**

air delivery container — A sling, bag, or roll, usually of canvas or webbing, designed to hold supplies and equipment for air delivery.

air delivery equipment — Special items of equipment (such as parachutes, air delivery containers, platforms, tie downs, and related items) used in air delivery of personnel, supplies, and equipment.

air direct delivery — The strategic air movement of cargo or personnel from an airlift point of embarkation to a point as close as practicable to the user's specified final destination, thereby minimizing transshipment requirements. Air direct delivery eliminates the traditional Air Force two step strategic and theater airlift transshipment mission mix.

airdrop — The unloading of personnel or materiel from aircraft in flight. See also **airdrop platform; air movement; free drop; free fall; high velocity drop; low velocity drop.**

airdrop platform — A base upon which vehicles, cargo, or equipment are loaded for airdrop. See also **airdrop.**

air employment/allocation plan — The means by which subordinate commanders advise the joint force commander of planned employment/allocation of organic or assigned assets, of any expected excess sorties, or of any additional air support requirements.

air expeditionary force — Deployed US Air Force wings, groups, and squadrons committed to a joint operation. Also called **AEF.** See also **air and space expeditionary task force.** (JP 3-33)

air expeditionary wing — A wing or wing slice placed under the administrative control of an air and space expeditionary task force or air and space task force by Department of the Air Force orders for a joint operation. Also called **AEW.** See also **air and space expeditionary task force.** (JP 3-33)

air facility — An installation from which air operations may be or are being conducted. See also **facility.**

airfield — An area prepared for the accommodation (including any buildings, installations, and equipment), landing, and takeoff of aircraft. See also **alternate airfield; departure airfield; landing area; landing point; landing site; main airfield; redeployment airfield.** (DOD Note: In all entries involving "airfield" or "aerodrome," the US uses "airfield," and NATO uses "aerodrome." The terms are synonymous.)

airfield traffic — **(*)** All traffic on the maneuvering area of an airfield and all aircraft flying in the vicinity of an airfield.

air fire plan — A plan for integrating and coordinating tactical air support of ground forces with other fire support.

Air Force Component Headquarters — The field headquarters facility of the Air Force commander charged with the overall conduct of Air Force operations. It is composed of the command section and appropriate staff elements.

Air Force special operations base — A base, airstrip, or other appropriate facility that provides physical support to Air Force special operations forces (AFSOF). The facility may be used solely to support AFSOF or may be a portion of a larger base supporting other operations. As a supporting facility, it is distinct from the forces operating from or being supported by it. Also called **AFSOB.** (JP 3-05)

move about the battlefield by aircraft to engage in ground combat.

airmobility — **(*)** A capability of airmobile forces which permits them to move by air while retaining the ability to engage in ground combat.

Air Mobility Command — The Air Force component command of the US Transportation Command. Also called **AMC.**

air mobility division — The division within the air operations center responsible for planning, coordinating, tasking, and management of air mobility missions. Also called **AMD.** See also **air operations center; division.** (JP 4-01.8)

air mobility element — The air mobility element is an extension of the Air Mobility Command tanker airlift control center deployed to a theater when requested by the geographic combatant commander. It coordinates strategic airlift operations with the theater airlift management system and collocates with the air operations center whenever possible. Also called **AME.** See also **air operations center; tanker airlift control center.** (JP 3-17)

air movement — Air transport of aircraft, units, personnel, supplies, equipment, and materiel. See also **airdrop; air landed; free drop; high velocity drop; low velocity drop.**

air movement column — In airborne operations, the lead formation and the serials following, proceeding over the same flight path at the same altitude.

air movement table — **(*)** A table prepared by a ground force commander in coordination with an air force commander. This form, issued as an annex to the operation order: a. indicates the allocation of aircraft space to elements of the ground units to be airlifted; b. designates the number and type of aircraft in each serial; c. specifies the departure area, time of loading, and takeoff.

air observation — See **air observer.**

air observation post — See **observation post.**

air observer — **(*)** An individual whose primary mission is to observe or take photographs from an aircraft in order to adjust artillery fire or obtain military information.

air observer adjustment — The correcting of gunfire from an aircraft. See also **spot.**

air offensive — Sustained operations by strategic and/or tactical air weapon systems against hostile air forces or surface targets.

air operations center — The principal air operations installation from which aircraft and air warning functions of combat air operations are directed, controlled, and executed. It is the senior agency of the Air Force Component Commander from which command and control of air operations are coordinated with other components and Services. Also called **AOC.** (JP 3-56.1)

air photographic reconnaissance — **(*)** The obtaining of information by air photography, divided into three types: a. Strategic photographic reconnaissance; b. Tactical photographic reconnaissance; and c. Survey/cartographic photography-air photography taken for survey/ cartographical purposes and to survey/ cartographic standards of accuracy. It may be strategic or tactical.

air picket — **(*)** An airborne early warning aircraft positioned primarily to detect, report, and track approaching enemy aircraft or missiles and to control intercepts.

Also called **aerial picket.** See also **airborne early warning and control.**

air plot — (*) 1. A continuous plot used in air navigation of a graphic representation of true headings steered and air distances flown. 2. A continuous plot of the position of an airborne object represented graphically to show true headings steered and air distances flown. 3. Within ships, a display that shows the positions and movements of an airborne object relative to the plotting ship.

airport — See **airfield.**

air portable — (*) Denotes materiel which is suitable for transport by an aircraft loaded internally or externally, with no more than minor dismantling and reassembling within the capabilities of user units. This term must be qualified to show the extent of air portability. See also **load.**

airport surface detection equipment — Short-range radar displaying the airport surface. Aircraft and vehicular traffic operating on runways, taxiways, and ramps, moving or stationary, may be observed with a high degree of resolution.

airport surveillance radar — Radar displaying range and azimuth that is normally employed in a terminal area as an aid to approach- and departure-control.

airport traffic area — Unless otherwise specifically designated, that airspace within a horizontal radius of five statute miles from the geographic center of any airport at which a control tower is operating, extending from the surface up to, but not including, an altitude of 3,000 feet above the elevation of the airport. Also called **ATA.**

air position — (*) The calculated position of an aircraft assuming no wind effect.

air priorities committee — (*) A committee set up to determine the priorities of passengers and cargo.

air raid reporting control ship — (*) A ship to which the air defense ship has delegated the duties of controlling air warning radar and air raid reporting.

air reconnaissance — The acquisition of information by employing visual observation and/or sensors in air vehicles.

air reconnaissance liaison officer — An Army officer especially trained in air reconnaissance and imagery interpretation matters who is attached to a tactical air reconnaissance unit. This officer assists and advises the air commander and staff on matters concerning ground operations and informs the supported ground commander on the status of air reconnaissance requests.

air refueling — The capability to refuel aircraft in flight, which extends presence, increases range, and serves as a force multiplier. Also called **AR.**

air refueling control point — During refueling operations, the geographic point where the receiver arrives in the observation or precontact position with respect to the tanker. Also called **ARCP.**

air refueling control time — During refueling operations, the time the receiver and tanker arrive at the air refueling control point. Also called **ARCT.**

air refueling initiation point — During refueling operations, a point located upstream from the air refueling control point (inbound to the air refueling control point) where the receiver aircraft initiates the rendezvous. Also called **ARIP.**

air request net — A high frequency, single sideband, nonsecure net monitored by all tactical air control parties (TACPs) and the air support operations center (ASOC) that allows immediate requests to be transmitted from a TACP at any Army echelon directly to the ASOC for rapid response. (JP 3-01.4)

air route — **(*)** The navigable airspace between two points, identified to the extent necessary for the application of flight rules.

air route traffic control center — The principal facility exercising en route control of aircraft operating under instrument flight rules within its area of jurisdiction. Approximately 26 such centers cover the United States and its possessions. Each has a communication capability to adjacent centers.

air smuggling event — In counterdrug operations, the departure of a suspected drug smuggling aircraft, an airdrop of drugs, or the arrival of a suspected drug smuggling aircraft. (JP 3-07.4)

air sovereignty — A nation's inherent right to exercise absolute control and authority over the airspace above its territory. See also **air sovereignty mission.**

air sovereignty mission — The integrated tasks of surveillance and control, the execution of which enforces a nation's authority over its territorial airspace. See also **air sovereignty.**

airspace control — See **airspace control in the combat zone.** (JP 3-52)

airspace control area — Airspace that is laterally defined by the boundaries of the operational area. The airspace control area may be subdivided into airspace control sectors.

airspace control authority — **(*)** The commander designated to assume overall responsibility for the operation of the airspace control system in the airspace control area. Also called **ACA.** See also **airspace control; airspace control area; airspace control system; control; operation.**

airspace control boundary — **(*)** The lateral limits of an airspace control area, airspace control sub-area, high density airspace control zone, or airspace restricted area.

airspace control center — The airspace control authority's primary airspace control facility, including assigned Service component, host-nation, and/or allied personnel and equipment. (JP 3-52)

airspace control facility — Any of the several Service component, host nation, or allied facilities that provide airspace control in the combat zone. (JP 3-52)

airspace control in the combat zone — A process used to increase combat effectiveness by promoting the safe, efficient, and flexible use of airspace. Airspace control is provided in order to prevent fratricide, enhance air defense operations, and permit greater flexibility of operations. Airspace control does not infringe on the authority vested in commanders to approve, disapprove, or deny combat operations. Also called **airspace control; combat airspace control.** (JP 3-52)

airspace control order — An order implementing the airspace control plan that provides the details of the approved requests for airspace control measures. It is published either as part of the air tasking order or as a separate document. Also called **ACO.** (JP 3-52)

airspace control plan — The document approved by the joint force commander that provides specific planning guidance and procedures for the airspace control system for the joint force area of responsibility and/or joint operations area. Also called **ACP.** See also **airspace control system; area of responsibility; joint force commander; joint operations area.** (JP 3-52)

airspace control sector — A subelement of the airspace control area, established to facilitate the control of the overall area. Airspace control sector boundaries normally coincide with air defense organization subdivision boundaries. Airspace control sectors are designated in accordance with procedures and guidance contained in the airspace control plan in consideration of Service component, host-nation, and allied airspace control capabilities and requirements. See also **airspace control area.** (JP 3-52)

airspace control system — **(*)** An arrangement of those organizations, personnel, policies, procedures, and facilities required to perform airspace control functions. Also called **ACS.**

airspace coordination area — A three-dimensional block of airspace in a target area, established by the appropriate ground commander, in which friendly aircraft are reasonably safe from friendly surface fires. The airspace coordination area may be formal or informal. Also called **ACA.** (JP 3-09.3)

airspace management — The coordination, integration, and regulation of the use of airspace of defined dimensions.

airspace reservation — The airspace located above an area on the surface of the land or water, designated and set apart by Executive Order of the President or by a state, commonwealth, or territory, over which the flight of aircraft is prohibited or restricted for the purpose of national defense or for other governmental purposes.

airspace restrictions — **(*)** Special restrictive measures applied to segments of airspace of defined dimensions.

air space warning area — See **danger area.**

airspeed — The speed of an aircraft relative to its surrounding air mass. The unqualified term "airspeed" can mean any one of the following. a. **calibrated airspeed** — Indicated airspeed corrected for instrument installation error. b. **equivalent airspeed** — Calibrated airspeed corrected for compressibility error. c. **indicated airspeed** — The airspeed shown by an airspeed indicator. d. **true airspeed** — Equivalent airspeed corrected for error due to air density (altitude and temperature).

airspeed indicator — **(*)** An instrument which displays the indicated airspeed of the aircraft derived from inputs of pitot and static pressures.

air staging unit — **(*)** A unit situated at an airfield and concerned with reception, handling, servicing, and preparation for departure of aircraft and control of personnel and cargo.

air station — **(*)** In photogrammetry, the point in space occupied by the camera lens at the moment of exposure.

air strike — An attack on specific objectives by fighter, bomber, or attack aircraft on an offensive mission. May consist of several air organizations under a single command in the air.

air strike coordinator — The air representative of the force commander in a target area, who is responsible for directing all aircraft in the target area and coordinating their efforts to achieve the most effective use of air striking power.

air strip — **(*)** An unimproved surface which has been adapted for takeoff or landing of aircraft, usually having minimum facilities. See also **airfield.**

air superiority — **(*)** That degree of dominance in the air battle of one force over another which permits the conduct of operations by the former and its related land, sea, and air forces at a given time and place without prohibitive interference by the opposing force.

air supply — **(*)** The delivery of cargo by airdrop or air landing.

air support — **(*)** All forms of support given by air forces on land or sea. See also **close air support; immediate air support; preplanned air support; tactical air support.**

air support operations center — **(*)** An agency of a tactical air control system collocated with a corps headquarters or an appropriate land force headquarters, which coordinates and directs close air support and other tactical air support. Also called **ASOC.** See also **air support; close air support; operation; tactical air control center.** (JP 4-01.8)

air support request — A means to request preplanned and immediate close air support, air interdiction, air reconnaissance, surveillance, escort, helicopter airlift, and other aircraft missions. Also called **AIRSUPREQ.** (JP 3-56.1)

air supremacy — **(*)** That degree of air superiority wherein the opposing air force is incapable of effective interference.

air surface zone — **(*)** A restricted area established for the purpose of preventing friendly surface vessels and aircraft from being fired upon by friendly forces and for permitting antisubmarine operations, unrestricted by the operation of friendly submarines. See also **restricted area.**

air surveillance — **(*)** The systematic observation of airspace by electronic, visual or other means, primarily for the purpose of identifying and determining the movements of aircraft and missiles, friendly and enemy, in the airspace under observation. See also **satellite and missile surveillance; surveillance.**

air surveillance officer — **(*)** An individual responsible for coordinating and maintaining an accurate, current picture of the air situation within an assigned airspace area.

air survey camera — See **air cartographic camera.**

air survey photography — See **air cartographic photography.**

air target chart — A display of pertinent air target intelligence on a specialized graphic base. It is designed primarily to support operations against designated air targets by various weapon systems. Also called **ATC.**

Air Target Materials Program — A Department of Defense program under the management control of the National Imagery and Mapping Agency established for and limited to the production of medium- and large-scale map, chart, and geodetic products, that supports worldwide targeting requirements of the unified and specified commands, the Military Departments, and allied participants. It encompasses the determination of production and coverage requirements, standardization of products, establishment of production priorities and schedules, and the production, distribution, storage, and release/exchange of products included under it.

air target mosaic — A large-scale mosaic providing photographic coverage of an area and permitting comprehensive portrayal of pertinent target detail. These mosaics are used for intelligence study and in planning and briefing for air operations.

air tasking order — A method used to task and disseminate to components, subordinate units, and command and control agencies projected sorties, capabilities and/or forces to targets and specific missions. Normally provides specific instructions to include call signs, targets, controlling agencies, etc., as well as general instructions. Also called **ATO.** (JP 3-56.1)

air tasking order/confirmation — A message used to task joint force components; to inform the requesting command and the tasking authority of the action being taken; and/or to provide additional information about the mission. The message is used only for preplanned missions and is transmitted on a daily basis, normally 12 hours prior to the start of the air tasking day or in accordance with established operation plans for the operational area. Also called **ATOCONF.** (JP 3-56.1)

air terminal — A facility on an airfield that functions as an air transportation hub and accommodates the loading and unloading of airlift aircraft and the intransit processing of traffic. The airfield may or may not be designated an aerial port.

air-to-air guided missile — **(*)** An air-launched guided missile for use against air targets. See also **guided missile.**

air-to-surface guided missile — **(*)** An air-launched guided missile for use against surface targets. See also **guided missile.**

air traffic control and landing system — Department of Defense facilities, personnel, and equipment (fixed, mobile, and seaborne) with associated avionics to provide safe, orderly, and expeditious aerospace vehicle movements worldwide. Also called **ATCALS.**

air traffic control center — **(*)** A unit combining the functions of an area control center and a flight information center. Also called **ATCC.** See also **area control center; flight information region.**

air traffic control clearance — **(*)** Authorization by an air traffic control authority for an aircraft to proceed under specified conditions.

air traffic control facility — Any of the component airspace control facilities primarily responsible for providing air traffic control services and, as required, limited tactical control services. (JP 3-52)

air traffic controller — An air controller especially trained for and assigned to the duty of airspace management and traffic control of airborne objects.

air traffic control service — **(*)** A service provided for the purpose of: a. preventing collisions: (1) between aircraft; and (2) on the maneuvering area between aircraft and obstructions; and b. expediting and maintaining an orderly flow of air traffic.

air traffic identification — The use of electronic devices, operational procedures, visual observation, and/or flight plan correlation for the purpose of identifying and locating aircraft flying within the airspace control area.

air traffic section — The link between the staging post and the local air priority committee. It is the key to the efficient handling of passengers and cargo at a staging post. It must include load control (including Customs and Immigrations facilities), freight, and mail sections.

air transportable unit — **(*)** A unit, other than airborne, whose equipment is adapted for air movement. See also **airborne; airborne operation.**

air transported operations — The movement by aircraft of troops and their equipment for an operation.

air transport group — A task organization of transport aircraft units that provides air transport for landing force elements or provides logistic support. (JP 3-02)

airway — **(*)** A control area or portion thereof established in the form of a corridor marked with radio navigational aids.

airways station — A ground communication installation established, manned, and equipped to communicate with aircraft in flight, as well as with other designated airways installations, for the purpose of expeditious and safe movements of aircraft. These stations may or may not be located on designated airways.

air weapons controller — An individual especially trained for and assigned to the duty of employing and controlling air weapon systems against airborne and surface objects.

alert — **(*)** 1. Readiness for action, defense or protection. 2. A warning signal of a real or threatened danger, such as an air attack. 3. The period of time during which troops stand by in response to an alarm. 4. To forewarn; to prepare for action. See also **airborne alert.** 5. **(DOD only)** A warning received by a unit or a headquarters which forewarns of an impending operational mission. 6. **(DOD only)** In aviation, an aircraft and aircrew that are placed in an increased state of readiness so that they may be airborne in a specified period of time after a launch order is received. See also **air defense warning conditions; ground alert; warning order.**

alert force — Specified forces maintained in a special degree of readiness.

alerting service — **(*)** A service provided to notify appropriate organizations regarding aircraft in need of search and rescue aid, and assist such organizations as required.

alert order — 1. A crisis action planning directive from the Secretary of Defense, issued by the Chairman of the Joint Chiefs of Staff, that provides essential guidance for planning and directs the initiation of execution planning for the selected course of action authorized by the Secretary of Defense. 2. A planning directive that provides essential planning guidance and directs the initiation of execution planning after the directing authority approves a military course of action. An alert order does not authorize execution of the approved course of action. See also **course of action; crisis action planning; execution planning.** (JP 5-0)

all appropriate action — Action taken in self-defense that is reasonable in intensity, duration, and magnitude, based on all the facts known to the commander at the time.

alliance — An alliance is the result of formal agreements (i.e., treaties) between two or more nations for broad, long-term objectives that further the common interests of the members. See also **coalition; multinational.** (JP 5-0)

allocation — In a general sense, distribution of limited resources among competing requirements for employment. Specific allocations (e.g., air sorties, nuclear weapons, forces, and transportation) are described as allocation of air sorties, nuclear weapons, etc. See also **allocation (air); allocation (nuclear); allocation (transportation); apportionment.**

allocation (air) — The translation of the air apportionment decision into total numbers of sorties by aircraft type available for each operation or task. See also **allocation.** (JP 3-17)

allocation (nuclear) — The apportionment of specific numbers and types of nuclear weapons to a commander for a stated time period as a planning factor for use in the development of war plans. (Additional authority is required for the actual deployment of allocated weapons to locations desired by the commander to support the war plans. Expenditures of these weapons are not authorized until released by proper authority.)

allocation request — A message used to provide an estimate of the total air effort, to identify any excess and joint force general support aircraft sorties, and to identify unfilled air requirements. This message is used only for preplanned missions and is transmitted on a daily basis, normally 24 hours prior to the start of the next air tasking day. Also called **ALLOREQ.** (JP 3-56.1)

allocation (transportation) — Apportionment by designated authority of available transport capability to users.

allotment — The temporary change of assignment of tactical air forces between subordinate commands. The authority to allot is vested in the commander having combatant command (command authority). See also **combatant command (command authority).**

allowable cabin load — The maximum payload that can be carried on an individual sortie. Also called **ACL.** (JP 3-17)

allowable load — **(*)** The total load that an aircraft can transport over a given distance, taking into account weight and volume. See also **airlift capability; airlift requirement; load; payload.**

allowable stacking weight — The amount of weight that can be stacked on corner posts of a container when subjected to 1.8 times the force of gravity. (JP 4-01.7)

all-source intelligence — 1. Intelligence products and/or organizations and activities that incorporate all sources of information, most frequently including human resources intelligence, imagery intelligence, measurement and signature intelligence, signals intelligence, and open-source data in the production of finished intelligence. 2. In intelligence collection, a phrase that indicates that in the satisfaction of intelligence requirements, all collection, processing, exploitation, and reporting systems and resources are identified for possible use and those most capable are tasked. See also **intelligence.** (JP 2-0)

all-weather air defense fighter — **(*)** A fighter aircraft with equipment and weapons which enable it to engage airborne targets in all weather conditions, day and night.

alongside replenishment — The transfer at sea of personnel and/or supplies by rigs between two or more ships proceeding side by side.

alphabet code — See **phonetic alphabet.**

alternate airfield — **(*)** An airfield specified in the flight plan to which a flight may proceed when it becomes inadvisable to land at the airfield of intended landing. An alternate airfield may be the airfield of departure.

alternate command authority — One or more predesignated officers empowered by the commander through predelegation of authority to act under stipulated emergency conditions in the accomplishment of previously defined functions.

alternate command post — Any location designated by a commander to assume command post functions in the event the command post becomes inoperative. It may be partially or fully equipped and manned or it may be the command post of a subordinate unit.

alternate headquarters — An existing headquarters of a component or subordinate command that is predesignated to assume the responsibilities and functions of another headquarters under prescribed emergency conditions.

alternative — See **variant.**

altitude — **(*)** The vertical distance of a level, a point or an object considered as a point, measured from mean sea level. See also **density altitude; drop altitude; elevation; minimum safe altitude; pressure altitude; transition altitude; true altitude.**

altitude acclimatization — **(*)** A slow physiological adaptation resulting from prolonged exposure to significantly reduced atmospheric pressure.

altitude chamber — See **hypobaric chamber.**

altitude datum — **(*)** The arbitrary level from which vertical displacement is measured. The datum for height measurement is the terrain directly below the aircraft or some specified datum; for pressure altitude, the level at which the atmospheric pressure is 29.92 inches of mercury (1013.2 m.bs); and for true altitude, mean sea level. See also **altitude.**

altitude delay — **(*)** Synchronization delay introduced between the time of transmission of the radar pulse and the start of the trace on the indicator, for the purpose of eliminating the altitude hole on the plan position indicator-type display.

altitude height — See **altitude datum.**

altitude hole — **(*)** The blank area at the origin of a radial display, on a radar tube presentation, the center of the periphery of which represents the point on the ground immediately below the aircraft. In side-looking airborne radar, this is known as the altitude slot.

altitude separation — See **vertical separation.**

altitude slot — See **altitude hole.**

ambient temperature — Outside temperature at any given altitude, preferably expressed in degrees centigrade. (JP 3-04.1)

ambulance exchange point — A location where a patient is transferred from one ambulance to another en route to a medical treatment facility. This may be an established point in an ambulance shuttle or it may be designated independently. Also called **AXP.** See also **medical treatment facility.** (JP 4-02.2)

American National Standards Institute — The United States standards organization that establishes procedures for the development and coordination of voluntary American national standards.

ammunition — See **munition.**

ammunition and toxic material open space — **(*)** An area especially prepared for storage of explosive ammunition and toxic material. For reporting purposes, it does not include the surrounding area restricted for storage because of safety distance factors. It includes barricades and improvised coverings. See also **storage.**

ammunition controlled supply rate — In Army usage, the amount of ammunition estimated to be available to sustain operations of a designated force for a specified time if expenditures are controlled at that rate. It is expressed in terms of rounds per weapon per day for ammunition items fired by weapons, and in terms of units of measure per organization per day for bulk allotment ammunition items. Tactical commanders use this rate to control expenditures of ammunition during tactical operations at planned intervals. It is issued through command channels at each level. It is determined based on consideration of the required supply rates submitted by subordinate commanders and ammunition assets available.

ammunition lot — **(*)** A quantity of homogeneous ammunition, identified by a unique lot number, which is manufactured, assembled, or renovated by one producer under uniform conditions and which is expected to function in a uniform manner.

ammunition supply point — See **distribution point.**

amphibian — A small craft, propelled by propellers and wheels or by air cushions for the purpose of moving on both land and water. (JP 4-01.6)

amphibious assault — The principal type of amphibious operation that involves establishing a force on a hostile or potentially hostile shore. See also **assault; assault phase.** (JP 3-02)

amphibious assault area — See **landing area.**

amphibious assault bulk fuel system — The petroleum, oils, and lubricants discharge system used to support US Marine Corps amphibious assaults and maritime pre-positioning force operations. It consists of 5,000 or 10,000 feet of buoyant 6-inch hose deployed from a landing ship, tank in amphibious assaults, or a maritime pre-positioning ship in maritime pre-positioning force operations. See also **amphibious assault; petroleum, oils, and lubricants.** (JP 4-01.6)

amphibious assault landing — See **amphibious operation, Part e.**

amphibious assault ship (general purpose) — A naval ship designed to embark, deploy, and land elements of a landing force in an assault by helicopters, landing craft, amphibious vehicles, and by combinations of these methods. Designated as **"LHA" or with internal dock as "LHD."**

amphibious aviation assault ship — An amphibious assault ship, landing platform helicopter; general purpose amphibious assault ship; or general purpose amphibious assault ship (with internal dock). (JP 3-04.1)

amphibious chart — **(*)** A special naval chart designed to meet special requirements for landing operations and passive coastal defense, at a scale of 1:25,000 or larger, and showing foreshore and coastal information in greater detail than a combat chart.

amphibious command ship — **(*)** A naval ship from which a commander exercises control in amphibious operations. Designated as **LCC.**

amphibious construction battalion — A permanently commissioned naval unit, subordinate to the Commander, Naval

Beach Group, designed to provide an administrative unit from which personnel and equipment are formed in tactical elements and made available to appropriate commanders to operate pontoon causeways, transfer barges, warping tugs, and assault bulk fuel systems, and to meet salvage requirements of the naval beach party. Also called **PHIBCB.** (JP 3-02)

amphibious control group — (*) Personnel, ships, and craft designated to control the waterborne ship-to-shore movement in an amphibious operation.

amphibious demonstration — (*) A type of amphibious operation conducted for the purpose of deceiving the enemy by a show of force with the expectation of deluding the enemy into a course of action unfavorable to him.

amphibious force — An amphibious task force and a landing force together with other forces that are trained, organized, and equipped for amphibious operations. Also called **AF.** See also **amphibious operation; amphibious task force; landing force.** (JP 3-02)

amphibious group — A command within the amphibious force, consisting of the commander and staff, designed to exercise operational control of assigned units in executing all phases of a division-size amphibious operation. (JP 3-02.2)

amphibious lift — (*) The total capacity of assault shipping utilized in an amphibious operation, expressed in terms of personnel, vehicles, and measurement or weight tons of supplies.

amphibious objective area — A geographical area (delineated for command and control purposes in the order initiating the amphibious operation) within which is located the objective(s) to be secured by the amphibious force. This area must be of sufficient size to ensure accomplishment of the amphibious force's mission and must provide sufficient area for conducting necessary sea, air, and land operations. Also called **AOA.** See also **amphibious force; mission.** (JP 3-02)

amphibious objective study — A study designed to provide basic intelligence data of a permanent or semipermanent nature required for planning amphibious operations. Each study deals with a specific area, the selection of which is based on strategic location, susceptibility to seizure by amphibious means, and other considerations.

amphibious operation — A military operation launched from the sea by an amphibious force, embarked in ships or craft with the primary purpose of introducing a landing force ashore to accomplish the assigned mission. See also **amphibious force; landing force; mission; operation.** (JP 3-02)

amphibious planning — The process of planning for an amphibious operation, distinguished by the necessity for concurrent, parallel, and detailed planning by all participating forces. The planning pattern is cyclical in nature, composed of a series of analyses and judgments of operational situations, each stemming from those that have preceded. (JP 3-02.2)

amphibious raid — (*) A type of amphibious operation involving swift incursion into or temporary occupation of an objective followed by a planned withdrawal. See also **amphibious operation.**

amphibious reconnaissance — (*) An amphibious landing conducted by minor elements, normally involving stealth rather than force of arms, for the purpose of securing information, and usually followed by a planned withdrawal.

amphibious reconnaissance unit — A unit organized, equipped, and trained to conduct and support amphibious reconnaissance missions. An amphibious reconnaissance unit is made up of a number of amphibious reconnaissance teams.

amphibious shipping — Organic Navy ships specifically designed to transport, land, and support landing forces in amphibious assault operations and capable of being loaded or unloaded by naval personnel without external assistance in the amphibious objective area.

amphibious squadron — **(*)** A tactical and administrative organization composed of amphibious assault shipping to transport troops and their equipment for an amphibious assault operation. Also called **PHIBRON.**

amphibious striking forces — Forces capable of projecting military power from the sea upon adjacent land areas for initiating and/or conducting operations in the face of enemy opposition.

amphibious task force — A Navy task organization formed to conduct amphibious operations. The amphibious task force, together with the landing force and other forces, constitutes the amphibious force. Also called **ATF.** See also **amphibious force; amphibious operation; landing force.** (JP 3-02)

amphibious tractor — See **amphibious vehicle.**

amphibious transport dock — A ship designed to transport and land troops, equipment, and supplies by means of embarked landing craft, amphibious vehicles, and helicopters. Designated as **LPD.**

amphibious transport group — A subdivision of an amphibious task force composed primarily of transport ships. The size of the transport group will depend upon the scope of the operation. Ships of the transport group will be combat-loaded to support the landing force scheme of maneuver ashore. A transport unit will usually be formed to embark troops and equipment to be landed over a designated beach or to embark all helicopter-borne troops and equipment. (JP 3-02.2)

amphibious vehicle — **(*)** A wheeled or tracked vehicle capable of operating on both land and water. See also **landing craft.**

amphibious vehicle availability table — A tabulation of the type and number of amphibious vehicles available primarily for assault landings and for support of other elements of the operation.

amphibious vehicle employment plan — A plan showing in tabular form the planned employment of amphibious vehicles in landing operations, including their employment after the initial movement to the beach.

amphibious vehicle launching area — **(*)** An area, in the vicinity of and to seaward of the line of departure, to which landing ships proceed and launch amphibious vehicles.

amphibious withdrawal — A type of amphibious operation involving the extraction of forces by sea in ships or craft from a hostile or potentially hostile shore. See also **amphibious operation.** (JP 3-02)

analysis and production — See **intelligence cycle.** (JP 2-0)

anchorage — A specified location for anchoring or mooring a vessel in-stream or offshore. (JP 4-01.6)

anchor cable — **(*)** In air transport, a cable in an aircraft to which the parachute static lines or strops are attached.

anchor line extension kit — **(*)** A device fitted to an aircraft equipped with removable clamshell doors to enable paratroopers to exit from the rear.

annex — A document appended to an operation order or other document to make it clearer or to give further details.

annotated print — **(*)** A photograph on which interpretation details are indicated by words or symbols.

annotation — **(*)** A marking placed on imagery or drawings for explanatory purposes or to indicate items or areas of special importance.

annual screening — One day of active duty for training required each year for Individual Ready Reserve members so the Services can keep current on each member's physical condition, dependency status, military qualifications, civilian occupational skills, availability for service, and other information.

annual training — The minimal period of training reserve members must perform each year to satisfy the training requirements associated with their Reserve Component assignment. Also called **AT.**

antemortem identification media — Records, samples, and photographs taken prior to death. These include (but are not limited to) fingerprints, dental x-rays, body tissue samples, photographs of tattoos, or other identifying marks. These "predeath" records would be compared against records completed after death to help establish a positive identification of a remains. See also **mortuary affairs.** (JP 4-06)

antenna mine — **(*)** In naval mine warfare, a contact mine fitted with antennae which, when touched by a steel ship, sets up galvanic action to fire the mine. See also **mine.**

antiarmor helicopter — **(*)** A helicopter armed primarily for use in the destruction of armored targets. Also called **antitank helicopter.**

anticountermining device — **(*)** A device fitted in an influence mine designed to prevent its actuation by shock.

antideficiency violations — The incurring of obligations or the making of expenditure (outlays) in excess of amounts available in appropriations or funds. (JP 1-06)

anti-G suit — A device worn by aircrew to counteract the effects on the human body of positive acceleration.

antilift device — A device arranged to detonate the mine to which it is attached, or to detonate another mine or charge nearby, if the mine is disturbed.

antimateriel agent — **(*)** A living organism or chemical used to cause deterioration of, or damage to, selected materiel.

antimateriel operation — **(*)** The employment of antimateriel weapons or agents in military operations.

antipersonnel mine (land mine warfare) — A mine designed to cause casualties to personnel. See also **mine.**

antiradiation missile — **(*)** A missile which homes passively on a radiation source. Also called **ARM.** See also **guided missile.**

antirecovery device — **(*)** In naval mine warfare, any device in a mine designed to prevent an enemy discovering details of the working of the mine mechanism.

antisubmarine action — An operation by one or more antisubmarine-capable ships, submarines, or aircraft (or a combination thereof) against a particular enemy submarine.

antisubmarine air distant support — Antisubmarine air support at a distance from, but directly related to, specific convoys or forces.

antisubmarine air search attack unit — The designation given to one or more aircraft separately organized as a tactical unit to search for and destroy submarines.

antisubmarine barrier — **(*)** The line formed by a series of static devices or mobile units arranged for the purpose of detecting, denying passage to, or destroying hostile submarines. See also **antisubmarine patrol.**

antisubmarine close air support — Air operations for the antisubmarine warfare protection of a supported force.

antisubmarine operation — Operation contributing to the conduct of antisubmarine warfare.

antisubmarine patrol — **(*)** The systematic and continuing investigation of an area or along a line to detect or hamper submarines, used when the direction of submarine movement can be established. See also **antisubmarine barrier.**

antisubmarine screen — **(*)** An arrangement of ships and/or aircraft for the protection of a screened unit against attack by a submarine.

antisubmarine search — **(*)** Systematic investigation of a particular area for the purpose of locating a submarine known or suspected to be somewhere in the area. Some types of search are also used in locating the position of a distress incident.

antisubmarine support operation — **(*)** An operation conducted by an antisubmarine force in the area around a force or convoy, in areas through which the force or convoy is passing, or in defense of geographic areas. Support operations may be completely coordinated with those of the force or convoy, or they may be independent operations coordinated only to the extent of providing operational intelligence and information.

antisubmarine warfare — **(*)** Operations conducted with the intention of denying the enemy the effective use of submarines. Also called **ASW.**

antisubmarine warfare forces — Forces organized primarily for antisubmarine action. May be composed of surface ships, aircraft, submarines, or any combination of these, and their supporting systems.

antisurface air operation — **(*)** An air operation conducted in an air/sea environment against enemy surface forces.

antisweep device — **(*)** Any device incorporated in the mooring of a mine or obstructor, or in the mine circuits to make the sweeping of the mine more difficult.

antisweeper mine — **(*)** A mine which is laid or whose mechanism is designed or adjusted with the specific object of damaging mine countermeasures vessels. See also **mine.**

antitank helicopter — See **antiarmor helicopter.**

antitank mine — **(*)** A mine designed to immobilize or destroy a tank. See also **mine.**

antiterrorism — Defensive measures used to reduce the vulnerability of individuals and property to terrorist acts, to include limited response and containment by local military forces. Also called **AT.** See also **antiterrorism awareness; counterterrorism; proactive measures; terrorism.** (JP 3-07.2)

antiterrorism awareness — Fundamental knowledge of the terrorist threat and measures to reduce personal vulnerability to terrorism. See also **antiterrorism.**

antiwatching device — A device fitted in a moored mine which causes it to sink should it show on the surface, so as to prevent the position of the mine or minefield being disclosed. See also **watching mine.**

any Service member mail — Mail sent by the general public to an unspecified Service member deployed on a contingency operation, as an expression of patriotic support. (JP 1-0)

apogee — The point at which a missile trajectory or a satellite orbit is farthest from the center of the gravitational field of the controlling body or bodies.

apparent horizon — **(*)** The visible line of demarcation between land/sea and sky.

apparent precession — **(*)** The apparent deflection of the gyro axis, relative to the Earth, due to the rotating effect of the Earth and not due to any applied force. Also called **apparent wander.**

appendix — A document appended to an annex of an operation order, operation plan, or other document to clarify or to give further details.

applicable materiel assets — That portion of the total acceptable materiel assets that meets the military or other characteristics as defined by the responsible Military Service and that is in the right condition and location to satisfy a specific military requirement.

application — 1. The system or problem to which a computer is applied. Reference is often made to an application as being either of the computational type (arithmetic computations predominate) or of the data processing type (data handling operations predominate). 2. In the intelligence context, the direct extraction and tailoring of information from an existing foundation of intelligence and near real time reporting. It is focused on and meets specific, narrow requirements, normally on demand. (JP 2-0)

apportionment — In the general sense, distribution for planning of limited resources among competing requirements. Specific apportionments (e.g., air sorties and forces for planning) are described as apportionment of air sorties and forces for planning, etc. See also **allocation; apportionment (air).**

apportionment (air) — The determination and assignment of the total expected effort by percentage and/or by priority that should be devoted to the various air operations for a given period of time. Also called **air apportionment.** See also **apportionment.** (JP 3-0)

approach clearance — Authorization for a pilot conducting flight in accordance with instrument flight rules to commence an approach to an airport.

approach control — A control station in an air operations control center, helicopter direction center, or carrier air traffic control center, that is responsible for controlling air traffic from marshal until hand-off to final control. See also **helicopter direction center; marshal.** (JP 3-04.1)

approach end of runway — **(*)** That end of the runway nearest to the direction from which the final approach is made.

approach lane — An extension of a boat lane from the line of departure toward the transport area.

approach march — **(*)** Advance of a combat unit when direct contact with the enemy is imminent. Troops are fully or partially deployed. The approach march ends when ground contact with the enemy is made or when the attack position is occupied.

approach schedule — The schedule that indicates, for each scheduled wave, the time of departure from the rendezvous area, from the line of departure, and from other control points and the time of arrival at the beach.

approach sequence — **(*)** The order in which two or more aircraft are cleared for an approach.

approach time — The time at which an aircraft is expected to commence approach procedure.

approval authority — A representative (person or organization) of the Commandant, US Coast Guard, authorized to approve containers within terms of the International Conference for Safe Containers. See also **International Convention for Safe Containers.** (JP 4-01.7)

apron — A defined area on an airfield intended to accommodate aircraft for purposes of loading or unloading passengers or cargo, refueling, parking, or maintenance.

archipelagic sea lanes passage — The nonsuspendable right of continuous and expeditious transit through archipelagic waters in the normal mode through and over routes normally used for navigation and overflight.

architecture — A framework or structure that portrays relationships among all the elements of the subject force, system, or activity. (JP 3-05)

archive — When used in the context of deliberate planning, the directed command will remove the referenced operation plan, operation plan in concept format, and any associated Joint Operation Planning and Execution System automated data processing files from its library of active plans. All material will be prepared for shipment to appropriate archive facilities in accordance with appropriate command directives. See also **maintain; retain.**

area air defense commander — Within a unified command, subordinate unified command, or joint task force, the commander will assign overall responsibility for air defense to a single commander. Normally, this will be the component commander with the preponderance of air defense capability and the command, control, and communications capability to plan and execute integrated air defense operations. Representation from the other components involved will be provided, as appropriate, to the area air defense commander's headquarters. Also called **AADC.** (JP 3-52)

area assessment — The commander's prescribed collection of specific information that commences upon employment and is a continuous operation. It confirms, corrects, refutes, or adds to previous intelligence acquired from area studies and other sources prior to employment. (JP 3-05)

area bombing — **(*)** Bombing of a target which is in effect a general area rather than a small or pinpoint target.

area command — **(*)** A command which is composed of those organized elements of one or more of the Armed Services, designated to operate in a specific geographical area, which are placed under a single commander. See also **command.**

area control center — **(*)** A unit established to provide air traffic control service to controlled flights in control areas under its jurisdiction. See also **air traffic control center; flight information region.**

area damage control — **(*)** Measures taken before, during, or after hostile action or natural or manmade disasters, to reduce the probability of damage and minimize its effects. See also **damage control; disaster control.**

area of influence — **(*)** A geographical area wherein a commander is directly capable of influencing operations by maneuver or fire support systems normally under the commander's command or control.

area of intelligence responsibility — An area allocated to a commander in which the commander is responsible for the provision of intelligence within the means at the commander's disposal. See also **area of interest; area of responsibility.**

area of interest — That area of concern to the commander, including the area of influence, areas adjacent thereto, and extending into enemy territory to the objectives of current or planned operations. This area also includes areas occupied by enemy forces who could jeopardize the accomplishment of the mission. Also called **AOI.** See also **area of influence.** (JP 2-03)

area of limitation — A defined area where specific limitations apply to the strength and fortifications of disputing or belligerent forces. Normally, upper limits are established for the number and type of formations, tanks, antiaircraft weapons, artillery, and other weapons systems in the area of limitation. Also called **AOL.** See also **line of demarcation; peace operations.** (JP 3-07.3)

area of militarily significant fallout — **(*)** Area in which radioactive fallout affects the ability of military units to carry out their normal mission.

area of northern operations — A region of variable width in the Northern Hemisphere that lies north of the 50 degrees isotherm — a line along which the average temperature of the warmest 4-month period of the year does not exceed 50 degrees Fahrenheit. Mountain regions located outside of this area are included in this category of operations provided these same temperature conditions exist.

area of operations — An operational area defined by the joint force commander for land and naval forces. Areas of operation do not typically encompass the entire operational area of the joint force commander, but should be large enough for component commanders to accomplish their missions and protect their forces. Also called **AO.** See also **area of responsibility; joint operations area; joint special operations area.** (JP 5-0)

area of responsibility — The geographical area associated with a combatant command within which a combatant commander has authority to plan and conduct operations. Also called **AOR.** See also **combatant command.** (JP 3-0)

area of separation — See **buffer zone.** Also called **AOS.** See also **peace operations.** (JP 3-07.3)

area operations — **(*)** In maritime usage, operations conducted in a geographical area and not related to the protection of a specific force.

area oriented — Personnel or units whose organizations, mission, training, and equipping are based on projected operational deployment to a specific geographic or demographic area. (JP 3-05)

area radar prediction analysis — Radar target intelligence study designed to provide radar-significant data for use in the preparation of radar target predictions.

area search — Visual reconnaissance of limited or defined areas.

area target — **(*)** A target consisting of an area rather than a single point.

armament delivery recording — Motion picture, still photography, and video recordings showing the delivery and impact of ordnance. This differs from reconnaissance imagery in that it records the act of delivery and impact and normally is done by the weapon system delivering the ordnance. Armament delivery recording is used primarily for evaluating strike effectiveness and for combat crew training. It is also one of the principal sources of over-the-target documentation in force employments, and may be used for public affairs purposes. Also called **ADR.**

armed forces — The military forces of a nation or a group of nations. See also **force(s).**

armed forces censorship — The examination and control of personal communications to or from persons in the Armed Forces of the United States and persons accompanying or serving with the Armed Forces of the United States. See also **censorship.**

armed forces courier — An officer or enlisted member in the grade of E-7 or above, of the US Armed Forces, assigned to perform Armed Forces Courier Service duties and identified by possession of an Armed Forces Courier Service Identification Card (ARF-COS Form 9). See also **courier.**

Armed Forces Courier Service — A joint service of the Departments of the Army, the Navy, and the Air Force, with the Chief of Staff, US Army, as Executive Agent. The courier service provides one of the available methods for the secure and expeditious transmission of material requiring protected handling by military courier.

armed forces courier station — An Army, Navy, or Air Force activity, approved by the respective military department and officially designated by Headquarters, Armed Forces Courier Service, for the acceptance, processing, and dispatching of Armed Forces Courier Service material.

Armed Forces of the United States — A term used to denote collectively all components of the Army, Navy, Air Force, Marine Corps, and Coast Guard. See also **United States Armed Forces.**

Armed Forces Radio and Television Service — A worldwide radio and television broadcasting organization that provides US military commanders overseas and at sea with sufficient electronic media resources to effectively communicate theater, local, Department of Defense, and Service-unique command information to their personnel and family members. Also called **AFRTS.** (JP 3-61)

armed helicopter — (*) A helicopter fitted with weapons or weapon systems.

armed mine — (*) A mine from which all safety devices have been withdrawn and, after laying, all automatic safety features and/or arming delay devices have operated. Such a mine is ready to be actuated after receipt of a target signal, influence, or contact.

armed reconnaissance — A mission with the primary purpose of locating and attacking targets of opportunity, i.e., enemy materiel, personnel, and facilities, in assigned general areas or along assigned ground communications routes, and not for the purpose of attacking specific briefed targets.

armed sweep — (*) A sweep fitted with cutters or other devices to increase its ability to cut mine moorings.

arming — As applied to explosives, weapons, and ammunition, the changing from a safe condition to a state of readiness for initiation.

arming delay device — A device fitted in a mine to prevent it being actuated for a preset time after laying.

arming lanyard — See **arming wire.**

arming pin — (*) A safety device inserted in a munition, which until its removal, prevents the unintentional action of the arming cycle. Also called **safety pin.** See also **safety device.**

arming system — That portion of a weapon that serves to ready (arm), safe, or re-safe (disarm) the firing system and fuzing system and that may actuate devices in the nuclear system.

arming wire — (*) A cable, wire or lanyard routed from the aircraft to an expendable aircraft store in order to initiate the arming sequence for the store upon release from the aircraft, when the armed release condition has been selected; it also prevents arming initiation prior to store release and during safe jettison. Also called **arming lanyard.** See also **safety wire.**

armistice — In international law, a suspension or temporary cessation of hostilities by agreement between belligerent powers. (JP 3-07.3)

armistice demarcation line — A geographically defined line from which disputing or belligerent forces disengage and withdraw to their respective sides following a truce or cease fire agreement. Also called cease fire line in some United Nations operations. Also called **ADL.** See also **armistice; cease fire; cease fire line; peace operations.** (JP 3-07.3)

arm or de-arm — Applies to those procedures in the arming or de-arming section of the applicable aircraft loading manual or checklist that places the ordnance or explosive device in a ready or safe condition i.e., rocket launchers, guided missiles, guns — internal and pods, paraflares — (external and SUU-44/25 dispenser). (NOTE: The removal or installation of pylon or bomb rack safety pins from a nonordnance-loaded station is considered a function requiring certification within the purview of this publication.) See also **arming; de-arming; ordnance.** (JP 3-04.1)

armored personnel carrier — A lightly armored, highly mobile, full-tracked vehicle, amphibious and air-droppable, used primarily for transporting personnel and their individual equipment during tactical operations. Production modifications or application of special kits permit use as a mortar carrier, command post, flame thrower, antiaircraft artillery chassis, or limited recovery vehicle. Also called **APC.**

arms control — A concept that connotes: a. any plan, arrangement, or process, resting upon explicit or implicit international agreement, governing any aspect of the following: the numbers, types, and performance characteristics of weapon systems (including the command and control, logistics support arrangements, and any related intelligence-gathering mechanism); and the numerical strength, organization, equipment, deployment, or employment of the Armed Forces retained by the parties (it encompasses disarmament); and b. on some occasions, those measures taken for the purpose of reducing instability in the military environment.

arms control agreement — The written or unwritten embodiment of the acceptance of one or more arms control measures by two or more nations.

arms control agreement verification — A concept that entails the collection, processing, and reporting of data indicating testing or employment of proscribed weapon systems, including country of origin and location, weapon and payload identification, and event type.

arms control measure — Any specific arms control course of action.

Army Air Defense Command Post — The tactical headquarters of an Army air defense commander.

Army air-ground system — The Army system which provides for interface between Army and tactical air support agencies of other Services in the planning, evaluating, processing, and coordinating of air support requirements and operations. It is composed of appropriate staff members, including G-2 air and G-3 air personnel, and necessary communication equipment. Also called **AAGS.**

Army and Air Force Exchange Service imprest fund activity — A military-operated retail activity, usually in remote or forward sites, when regular direct operations exchanges cannot be provided. It is a satellite activity of an Army and Air Force Exchange Service (AAFES) direct operation. The supported unit appoints the officer in charge of an imprest fund activity, who is issued an initial fund by AAFES to purchase beginning inventory. Money generated from sales is used to replenish the merchandise stock. See also **imprest fund.** (JP 1-0)

Army base — A base or group of installations for which a local commander is responsible, consisting of facilities necessary for support of Army activities including security, internal lines of communications, utilities, plants and systems, and real property for which the Army has operating responsibility. See also **base complex.**

Army corps — A tactical unit larger than a division and smaller than a field army. A corps usually consists of two or more divisions together with auxiliary arms and services. See also **field army.**

Army service area — The territory between the corps rear boundary and the combat zone rear boundary. Most of the Army administrative establishment and service troops are usually located in this area. See also **rear area.**

Army special operations component — The Army component of a joint force special operations component. Also called **ARSOC.** See also **Air Force special operations component; Navy special operations component.** (JP 3-05.3)

assumed azimuth — The assumption of azimuth origins as a field expedient until the required data are available.

assumed grid — A grid constructed using an arbitrary scale superimposed on a map, chart, or photograph for use in point designation without regard to actual geographic location. See also **grid.**

assumption — A supposition on the current situation or a presupposition on the future course of events, either or both assumed to be true in the absence of positive proof, necessary to enable the commander in the process of planning to complete an estimate of the situation and make a decision on the course of action.

astern fueling — **(*)** The transfer of fuel at sea during which the receiving ship(s) keep(s) station astern of the delivering ship.

asymmetrical sweep — **(*)** A sweep whose swept path under conditions of no wind or cross-tide is not equally spaced either side of the sweeper's track.

atmospheric environment — The envelope of air surrounding the Earth, including its interfaces and interactions with the Earth's solid or liquid surface.

at my command — **(*)** In artillery and naval gunfire support, the command used when it is desired to control the exact time of delivery of fire.

atomic air burst — See **airburst.**

atomic defense — See **nuclear defense.**

atomic demolition munition — A nuclear device designed to be detonated on or below the ground surface, or under water as a demolition munition against material-type targets to block, deny, and/or canalize the enemy.

atomic underground burst — See **nuclear underground burst.**

atomic underwater burst — See **nuclear underwater burst.**

atomic warfare — See **nuclear warfare.**

atomic weapon — See **nuclear weapon.**

at priority call — **(*)** A precedence applied to the task of an artillery unit to provide fire to a formation/unit on a guaranteed basis. Normally observer, communications, and liaison are not provided. An artillery unit in "direct support" or "in support" may simultaneously be placed "at priority call" to another unit or agency for a particular task and/or for a specific period of time.

at sea — Includes the following maritime areas: foreign internal waters, archipelagic waters, and territorial seas; foreign contiguous zones; foreign exclusive economic zones; the high seas; and US-exclusive economic zone, territorial sea, and internal waters.

attach — 1. The placement of units or personnel in an organization where such placement is relatively temporary. 2. The detailing of individuals to specific functions where such functions are secondary or relatively temporary, e.g., attached for quarters and rations; attached for flying duty. See also **assign.**

attachment — See **attach.**

attack assessment — An evaluation of information to determine the potential or actual nature and objectives of an attack for the purpose of providing information for timely decisions. See also **damage estimation.**

attack cargo ship — A naval ship designed or converted to transport combat-loaded cargo in an assault landing. Capabilities as to carrying landing craft, speed of ship, armament, and size of hatches and booms are greater than those of comparable cargo ship types. Designated as **LKA.**

attack group — (*) A subordinate task organization of the navy forces of an amphibious task force. It is composed of assault shipping and supporting naval units designated to transport, protect, land, and initially support a landing group.

attack heading — 1. The interceptor heading during the attack phase that will achieve the desired track-crossing angle. 2. The assigned magnetic compass heading to be flown by aircraft during the delivery phase of an air strike.

attack helicopter — (*) A helicopter specifically designed to employ various weapons to attack and destroy enemy targets.

attack origin — 1. The location or source from which an attack was initiated. 2. The nation initiating an attack. See also **attack assessment.**

attack pattern — The type and distribution of targets under attack. Also called **target pattern.** See also **attack assessment.**

attack position — The last position occupied by the assault echelon before crossing the line of departure.

attack timing — The predicted or actual time of bursts, impacts, or arrival of weapons at their intended targets.

attenuation — (*) 1. Decrease in intensity of a signal, beam, or wave as a result of absorption of energy and of scattering out of the path of a detector, but not including the reduction due to geometric spreading, i.e., the inverse square of distance effect. 2. In mine warfare, the reduction in intensity of an influence as distance from the source increases. 3. In camouflage and concealment, the process of making an object or surface less conspicuous by reducing its contrast to the surroundings and/or background. Also called **tone down.**

attenuation factor — (*) The ratio of the incident radiation dose or dose rate to the radiation dose or dose rate transmitted through a shielding material. This is the reciprocal of the transmission factor.

attitude — (*) The position of a body as determined by the inclination of the axes to some frame of reference. If not otherwise specified, this frame of reference is fixed to the Earth.

attitude indicator — (*) An instrument which displays the attitude of the aircraft by reference to sources of information which may be contained within the instrument or be external to it. When the sources of information are self-contained, the instrument may be referred to as an artificial horizon.

attrition — (*) The reduction of the effectiveness of a force caused by loss of personnel and materiel.

attrition minefield — (*) In naval mine warfare, a field intended primarily to cause damage to enemy ships. See also **minefield.**

attrition rate — (*) A factor, normally expressed as a percentage, reflecting the degree of losses of personnel or materiel due to various causes within a specified period of time.

attrition reserve aircraft — Aircraft procured for the specific purpose of replacing the anticipated losses of aircraft because of peacetime and/or wartime attrition.

attrition sweeping — **(*)** The continuous sweeping of minefields to keep the risk of mines to all ships as low as possible.

augmentation forces — Forces to be transferred from a supporting commander to the combatant command (command authority) or operational control of a supported commander during the execution of an operation order approved by the National Command Authorities. (JP 5-0)

authenticate — A challenge given by voice or electrical means to attest to the authenticity of a message or transmission.

authentication — 1. A security measure designed to protect a communications system against acceptance of a fraudulent transmission or simulation by establishing the validity of a transmission, message, or originator. 2. A means of identifying individuals and verifying their eligibility to receive specific categories of information. 3. Evidence by proper signature or seal that a document is genuine and official. 4. In evasion and recovery operations, the process whereby the identity of an evader is confirmed. See also **evader; evasion; evasion and recovery; recovery operations; security.** (JP 3-50.3)

authenticator — A symbol or group of symbols, or a series of bits, selected or derived in a prearranged manner and usually inserted at a predetermined point within a message or transmission for the purpose of attesting to the validity of the message or transmission.

autocode format — An abbreviated and formatted message header used in conjunction with the mobile cryptologic support facility (MCSF) to energize the automatic communications relay functions of the MCSF, providing rapid exchange of data through the system.

automated data handling — See **automatic data handling.**

automated identification technology — A suite of tools for facilitating total asset visibility (TAV) source data capture and transfer. Automated identification technology (AIT) includes a variety of devices, such as bar codes, magnetic strips, optical memory cards, and radio frequency tags for marking or "tagging" individual items, multi-packs, equipment, air pallets, or containers, along with the hardware and software required to create the devices, read the information on them, and integrate that information with other logistic information. AIT integration with logistic information systems is key to the Department of Defense's TAV efforts. Also called **AIT.** See also **total asset visibility**. (JP 4-01.8)

automatic approach and landing — A control mode in which the aircraft's speed and flight path are automatically controlled for approach, flare-out, and landing. See also **ground-controlled approach procedure.**

automatic data handling — **(*)** A generalization of automatic data processing to include the aspect of data transfer.

automatic data processing — 1. Data processing largely performed by automatic means. 2. That branch of science and technology concerned with methods and techniques relating to data processing largely performed by automatic means.

automatic flight control system — **(*)** A system which includes all equipment to control automatically the flight of an aircraft or missile to a path or attitude described by references internal or external to the aircraft or missile. Also called **AFCS.**

automatic message processing system — Any organized assembly of resources and methods used to collect, process, and distribute messages largely by automatic means.

automatic resupply — A resupply mission fully planned before insertion of a special operations team into the operations area that occurs at a prearranged time and location, unless changed by the operating team after insertion. See also **emergency resupply; on-call resupply.** (JP 3-50.3)

automatic search jammer — **(*)** An intercept receiver and jamming transmitter system which searches for and jams signals automatically which have specific radiation characteristics.

Automatic Secure Voice Communications Network — A worldwide, switched, secure voice network developed to fulfill DOD long-haul, secure voice requirements. Also called **AUTOSEVOCOM.**

automatic supply — A system by which certain supply requirements are automatically shipped or issued for a predetermined period of time without requisition by the using unit. It is based upon estimated or experience-usage factors.

automation network — The automation network combines all of the information collection devices, automatic identification technologies, and the automated information systems that either support or facilitate the joint reception, staging, onward movement, and integration process. See also **automated identification technology; joint reception, staging, onward movement, and integration.** (JP 4-01.8)

autonomous operation — In air defense, the mode of operation assumed by a unit after it has lost all communications with higher echelons. The unit commander assumes full responsibility for control of weapons and engagement of hostile targets.

availability date — The date after notification of mobilization by which forces will be marshalled at their home station or mobilization station and available for deployment. See also **home station; mobilization; mobilization station.** (JP 4-05)

available payload — The passenger and/or cargo capacity expressed in weight and/or space available to the user.

available-to-load date — A date specified for each unit in a time-phased force and deployment data indicating when that unit will be ready to load at the point of embarkation. Also called **ALD.**

avenue of approach — An air or ground route of an attacking force of a given size leading to its objective or to key terrain in its path. Also called **AA.**

average speed — **(*)** The average distance traveled per hour, calculated over the whole journey, excluding specifically ordered halts.

aviation combat element — The core element of a Marine air-ground task force (MAGTF) that is task-organized to conduct aviation operations. The aviation combat element (ACE) provides all or a portion of the six functions of Marine aviation necessary to accomplish the MAGTF's mission. These functions are antiair warfare, offensive air support, assault support, electronic warfare, air reconnaissance, and control of aircraft and missiles. The ACE is usually composed of an aviation unit headquarters and various other aviation units or their detachments.

It can vary in size from a small aviation detachment of specifically required aircraft to one or more Marine aircraft wings. The ACE itself is not a formal command. Also called **ACE.** See also **combat service support element; command element; ground combat element; Marine air-ground task force; Marine expeditionary force; Marine expeditionary force (forward); Marine expeditionary unit; special purpose Marine air-ground task force; task force.**

aviation life support equipment — See **life support equipment.**

aviation medicine — **(*)** The special field of medicine which is related to the biological and psychological problems of flight.

aviation ship — An aircraft carrier. See also **air-capable ship; aircraft; amphibious aviation assault ship.** (JP 3-04.1)

avoidance — Individual and/or unit measures taken to avoid or minimize nuclear, biological, and chemical (NBC) attacks and reduce the effects of NBC hazards. (JP 3-11)

axial route — A route running through the rear area and into the forward area. See also **route.**

axis of advance — A line of advance assigned for purposes of control; often a road or a group of roads, or a designated series of locations, extending in the direction of the enemy.

azimuth — Quantities may be expressed in positive quantities increasing in a clockwise direction, or in X, Y coordinates where south and west are negative. They may be referenced to true north or magnetic north depending on the particular weapon system used.

azimuth angle — **(*)** An angle measured clockwise in the horizontal plane between a reference direction and any other line.

azimuth guidance — **(*)** Information which will enable the pilot or autopilot of an aircraft to follow the required track.

azimuth resolution — **(*)** The ability of radar equipment to separate two reflectors at similar ranges but different bearings from a reference point. Normally the minimum separation distance between the reflectors is quoted and expressed as the angle subtended by the reflectors at the reference point.

B

backfill — Reserve Component units and individuals recalled to replace deploying active units and/or individuals in the continental United States and outside the continental United States. See also **Reserve Components.** (JP 4-05.1)

background count — The evidence or effect on a detector of radiation caused by background radiation. In connection with health protection, the background count includes but is not limited to radiations produced by naturally occurring radioactivity and cosmic rays.

background radiation — **(*)** Nuclear (or ionizing) radiations arising from within the body and from the surroundings to which individuals are always exposed.

back-haul airlift — The rearward movement of personnel and materiel from an air terminal in forward deployed areas back to a staging base (either in-theater or out) after the normal forward delivery. See also **staging base.** (JP 3-17)

backscatter — Refers to a portion of the laser energy that is scattered back in the direction of the seeker by an obscurant. See also **laser.** (JP 3-09.1)

back-scattering — Radio wave propagation in which the direction of the incident and scattered waves, resolved along a reference direction (usually horizontal), are oppositely directed. A signal received by back-scattering is often referred to as "back-scatter."

backshore — The area of a beach extending from the limit of high water foam lines to dunes or extreme inland limit of the beach. (JP 4-01.6)

back tell — **(*)** The transfer of information from a higher to a lower echelon of command. See also **track telling.**

back-up — **(*)** In cartography, an image printed on the reverse side of a map sheet already printed on one side. Also the printing of such images.

backwash — An even layer of water that moves along the sea floor from the beach through the surf zone and caused by the pile-up of water on the beach from incoming breakers. (JP 4-01.6)

balance — A concept as applied to an arms control measure that connotes: a. adjustments of armed forces and armaments in such a manner that one state does not obtain military advantage over other states agreeing to the measure; and b. internal adjustments by one state of its forces in such manner as to enable it to cope with all aspects of remaining threats to its security in a post arms control agreement era.

balanced stock(s) — 1. That condition of supply when availability and requirements are in equilibrium for specific items. 2. An accumulation of supplies in quantities determined necessary to meet requirements for a fixed period.

balance station zero — See **reference datum.**

bale cubic capacity — **(*)** The space available for cargo measured in cubic feet to the inside of the cargo battens, on the frames, and to the underside of the beams. In a general cargo of mixed commodities, the bale cubic applies. The stowage of the mixed cargo comes in contact with the cargo battens and as a general rule does not extend to the skin of the ship.

balisage — **(*)** The marking of a route by a system of dim beacon lights enabling vehicles to be driven at near day-time speed, under blackout conditions.

ballistic missile — **(*)** Any missile which does not rely upon aerodynamic surfaces to produce lift and consequently follows a ballistic trajectory when thrust is terminated. See also **aerodynamic missile; guided missile.**

ballistic missile early warning system — An electronic system for providing detection and early warning of attack by enemy intercontinental ballistic missiles. Also called **BMEWS.**

ballistics — **(*)** The science or art that deals with the motion, behavior, appearance, or modification of missiles or other vehicles acted upon by propellants, wind, gravity, temperature, or any other modifying substance, condition, or force.

ballistic trajectory — **(*)** The trajectory traced after the propulsive force is terminated and the body is acted upon only by gravity and aerodynamic drag.

ballistic wind — That constant wind that would have the same effect upon the trajectory of a bomb or projectile as the wind encountered in flight.

balloon barrage — See **barrage, Part 2.**

balloon reflector — In electronic warfare, a balloon-supported confusion reflector to produce fraudulent radar echoes.

bandwidth — The difference between the limiting frequencies of a continuous frequency band expressed in hertz (cycles per second). The term bandwidth is also loosely used to refer to the rate at which data can be transmitted over a given communications circuit. In the latter usage, bandwidth is usually expressed in either kilobits per second or megabits per second.

bank angle — **(*)** The angle between the aircraft's normal axis and the Earth's vertical plane containing the aircraft's longitudinal axis.

bar — A submerged or emerged embankment of sand, gravel, or mud created on the sea floor in shallow water by waves and currents. A bar may be composed of mollusk shells. (JP 4-01.6)

bare base — A base having minimum essential facilities to house, sustain, and support operations to include, if required, a stabilized runway, taxiways, and aircraft parking areas. A bare base must have a source of water that can be made potable. Other requirements to operate under bare base conditions form a necessary part of the force package deployed to the bare base. See also **base.** (JP 3-05.3)

barge — A flat-bed, shallow-draft vessel with no superstructure that is used for the transport of cargo and ships' stores or for general utility purposes. See also **watercraft.** (JP 4-01.6)

barometric altitude — **(*)** The altitude determined by a barometric altimeter by reference to a pressure level and calculated according to the standard atmosphere laws. See also **altitude.**

barrage — 1. A prearranged barrier of fires, except that delivered by small arms, designed to protect friendly troops and installations by impeding enemy movements across defensive lines or areas. 2. A protective screen of balloons that is moored to the ground and kept at given heights to prevent or hinder operations by enemy aircraft. This meaning also called **balloon barrage.** 3. A type of electronic attack intended for simultaneous jamming over a wide area of frequency spectrum. See also **barrage jamming; electronic warfare; fires.**

barrage fire — (*) Fire which is designed to fill a volume of space or area rather than aimed specifically at a given target. See also **fire.**

barrage jamming — Simultaneous electromagnetic jamming over a broad band of frequencies. See also **jamming.**

barricade — See **aircraft arresting barrier.**

barrier — A coordinated series of obstacles designed or employed to channel, direct, restrict, delay, or stop the movement of an opposing force and to impose additional losses in personnel, time, and equipment on the opposing force. Barriers can exist naturally, be manmade, or a combination of both. (JP 3-15)

barrier combat air patrol — One or more divisions or elements of fighter aircraft employed between a force and an objective area as a barrier across the probable direction of enemy attack. It is used as far from the force as control conditions permit, giving added protection against raids that use the most direct routes of approach. See also **combat air patrol.**

barrier forces — Air, surface, and submarine units and their supporting systems positioned across the likely courses of expected enemy transit for early detection and providing rapid warning, blocking, and destruction of the enemy.

barrier, obstacle, and mine warfare plan — A comprehensive, coordinated plan that includes responsibilities; general location of unspecified and specific barriers, obstacles, and minefields; special instructions; limitations; coordination; and completion times. The plan may designate locations of obstacle zones or belts. It is normally prepared as an annex to a campaign plan, operation plan, or operation order. (JP 3-15)

bar scale — See **graphic scale; scale.**

base — (*) 1. A locality from which operations are projected or supported. 2. An area or locality containing installations which provide logistic or other support. See also **establishment.** 3. **(DOD only)** Home airfield or home carrier. See also **base of operations; facility.**

base cluster — In base defense operations, a collection of bases, geographically grouped for mutual protection and ease of command and control. (JP 3-10)

base cluster commander — In base defense operations, the senior officer in the base cluster (excluding medical officers, chaplains, and commanders of transient units), with responsibility for coordinating the defense of bases within the base cluster and for integrating defense plans of bases into a base cluster defense plan. (JP 3-10)

base cluster operations center — A command and control facility that serves as the base cluster commander's focal point for defense and security of the base cluster. Also called **BCOC.** (JP 3-10.1)

base command — An area containing a military base or group of such bases organized under one commander. See also **command.**

base commander — In base defense operations, the officer assigned to command a base. (JP 3-10)

base complex — See **Army base; installation complex; Marine base; naval**

base; naval or Marine (air) base. See also **noncontiguous facility.**

base defense — The local military measures, both normal and emergency, required to nullify or reduce the effectiveness of enemy attacks on, or sabotage of, a base, to ensure that the maximum capacity of its facilities is available to US forces.

base defense forces — Troops assigned or attached to a base for the primary purpose of base defense and security as well as augmentees and selectively armed personnel available to the base commander for base defense from units performing primary missions other than base defense. (JP 3-10.1)

base defense operations center — A command and control facility established by the base commander to serve as the focal point for base security and defense. It plans, directs, integrates, coordinates, and controls all base defense efforts and coordinates and integrates into area security operations with the rear area operations center/rear tactical operations center. Also called **BDOC.** (JP 3-10.1)

base defense zone — An air defense zone established around an air base and limited to the engagement envelope of short-range air defense weapons systems defending that base. Base defense zones have specific entry, exit, and identification, friend or foe procedures established. Also called **BDZ.** (JP 3-10.1)

base development (less force beddown) — The acquisition, development, expansion, improvement, and construction and/or replacement of the facilities and resources of an area or location to support forces employed in military operations or deployed in accordance with strategic plans. (JP 4-04)

base development plan — A plan for the facilities, installations, and bases required to support military operations.

base element — See **base unit.**

base line — 1. **(surveying)** A surveyed line established with more than usual care, to which surveys are referred for coordination and correlation. 2. **(photogrammetry)** The line between the principal points of two consecutive vertical air photographs. It is usually measured on one photograph after the principal point of the other has been transferred. 3. **(radio navigation systems)** The shorter arc of the great circle joining two radio transmitting stations of a navigation system. 4. **(triangulation)** The side of one of a series of coordinated triangles the length of which is measured with prescribed accuracy and precision and from which lengths of the other triangle sides are obtained by computation.

baseline environmental survey — A multi-disciplinary site survey conducted prior to or in the initial stage of a joint operational deployment. The survey documents existing deployment-area environmental conditions, determines the potential for present and past site contamination (e.g., hazardous substances, petroleum products, and derivatives), and identifies potential vulnerabilities (to include occupational and environmental health risks). Surveys accomplished in conjunction with joint operational deployments that do not involve training or exercises (e.g., contingency operations) should be completed to the extent practicable consistent with operational requirements. See also **civil engineering; survey.** (JP 4-04)

base map — **(*)** A map or chart showing certain fundamental information, used as a base upon which additional data of specialized nature are compiled or overprinted. Also, a map containing all the information from which maps showing specialized information can be prepared. See also **chart base; map.**

base of operations — An area or facility from which a military force begins its offensive operations, to which it falls back in case of reverse, and in which supply facilities are organized.

base period — That period of time for which factors were determined for use in current planning and programming.

base section — An area within the communications zone in an operational area organized to provide logistic support to forward areas.

base surge — **(*)** A cloud which rolls out from the bottom of the column produced by a subsurface burst of a nuclear weapon. For underwater bursts the surge is, in effect, a cloud of liquid droplets which has the property of flowing almost as if it were a homogeneous fluid. For subsurface land bursts the surge is made up of small solid particles but still behaves like a fluid.

base unit — Unit of organization in a tactical operation around which a movement or maneuver is planned and performed.

baseline costs — The continuing annual costs of military operations funded by the operations and maintenance and military personnel appropriations. (JP 1-06)

basic cover — Coverage of any installation or area of a permanent nature with which later coverage can be compared to discover any changes that have taken place.

basic encyclopedia — A compilation of identified installations and physical areas of potential significance as objectives for attack. Also called **BE.**

basic intelligence — Fundamental intelligence concerning the general situation, resources, capabilities, and vulnerabilities of foreign countries or areas which may be used as reference material in the planning of operations at any level and in evaluating subsequent information relating to the same subject.

basic load — **(*)** The quantity of supplies required to be on hand within, and which can be moved by, a unit or formation. It is expressed according to the wartime organization of the unit or formation and maintained at the prescribed levels.

basic military route network — **(*)** Axial, lateral, and connecting routes designated in peacetime by the host nation to meet the anticipated military movements and transport requirements, both Allied and national.

basic psychological operations study — A document that describes succinctly the characteristics of a country, geographical area, or region which are most pertinent to psychological operations, and which can serve as an immediate reference for the planning and conduct of psychological operations. Also called **BPS.**

basic research — Research directed toward the increase of knowledge, the primary aim being a greater knowledge or understanding of the subject under study. See also **research.**

basic stocks — **(*)** Stocks to support the execution of approved operational plans for an initial predetermined period. See also **sustaining stocks.**

basic stopping power — **(*)** The probability, expressed as a percentage, of a single vehicle being stopped by mines while attempting to cross a minefield.

basic tactical organization — The conventional organization of landing force units for combat, involving combinations of infantry, supporting ground arms, and aviation for accomplishment of missions ashore. This organizational form is employed as soon as possible following the landing of the various assault components of the landing force.

basic undertakings — The essential things, expressed in broad terms, that must be done in order to implement the commander's concept successfully. These may include military, diplomatic, economic, informational, and other measures. See also **strategic concept.**

basis of issue — Authority that prescribes the number of items to be issued to an individual, a unit, a military organization, or for a unit piece of equipment.

bathymetric contour — See **depth contour.**

battalion landing team — In an amphibious operation, an infantry battalion normally reinforced by necessary combat and service elements; the basic unit for planning an assault landing. Also called **BLT.**

battery — **(*)** 1. Tactical and administrative artillery unit or subunit corresponding to a company or similar unit in other branches of the Army. 2. All guns, torpedo tubes, searchlights, or missile launchers of the same size or caliber or used for the same purpose, either installed in one ship or otherwise operating as an entity.

battery center — **(*)** A point on the ground, the coordinates of which are used as a reference indicating the location of the battery in the production of firing data. Also called **chart location of the battery.**

battery (troop) left (right) — A method of fire in which weapons are discharged from the left (right), one after the other, at five second intervals.

battle damage assessment — The timely and accurate estimate of damage resulting from the application of military force, either lethal or non-lethal, against a predetermined objective. Battle damage assessment can be applied to the employment of all types of weapon systems (air, ground, naval, and special forces weapon systems) throughout the range of military operations. Battle damage assessment is primarily an intelligence responsibility with required inputs and coordination from the operators. Battle damage assessment is composed of physical damage assessment, functional damage assessment, and target system assessment. Also called **BDA.** See also **combat assessment.** (JP 2-0)

battle damage repair — **(*)** Essential repair, which may be improvised, carried out rapidly in a battle environment in order to return damaged or disabled equipment to temporary service. Also called **BDR.**

battlefield coordination detachment — An Army liaison provided by the Army component or force commander to the air operations center (AOC) and/or to the component designated by the joint force commander to plan, coordinate, and deconflict air operations. The battlefield coordination detachment processes Army requests for air support, monitors and interprets the land battle situation for the AOC, and provides the necessary interface for exchange of current intelligence and operational data. Also called **BCD.** See also **air operations center; liaison.** (JP 3-01.4)

battlefield illumination — **(*)** The lighting of the battle area by artificial light, either visible or invisible to the naked eye.

battlefield psychological activities — **(*)** Planned psychological activities conducted as an integral part of combat operations and designed to bring psychological pressure to bear on enemy forces and civilians under enemy control in the battle area, to assist in achievement of operational and tactical objectives.

battlefield surveillance — **(*)** Systematic observation of the battle area for the purpose of providing timely information and combat intelligence. See also **surveillance.**

battle force — A standing operational naval task force organization of carriers, surface combatants, and submarines assigned to numbered fleets. A battle force is subdivided into battle groups.

battle reserves — Reserve supplies accumulated by an army, detached corps, or detached division in the vicinity of the battlefield, in addition to unit and individual reserves. See also **reserve supplies.**

battlespace — The environment, factors, and conditions that must be understood to successfully apply combat power, protect the force, or complete the mission. This includes the air, land, sea, space, and the included enemy and friendly forces; facilities; weather; terrain; the electromagnetic spectrum; and the information environment within the operational areas and areas of interest. See also **electromagnetic spectrum; information environment; joint intelligence preparation of the battlespace.**

beach — 1. The area extending from the shoreline inland to a marked change in physiographic form or material, or to the line of permanent vegetation (coastline). 2. In amphibious operations, that portion of the shoreline designated for landing of a tactical organization.

beach capacity — **(*)** An estimate, expressed in terms of measurement tons, or weight tons, of cargo that may be unloaded over a designated strip of shore per day. See also **clearance capacity; port capacity.**

beach group — See **naval beach group; shore party.**

beachhead — A designated area on a hostile or potentially hostile shore that, when seized and held, ensures the continuous landing of troops and materiel, and provides maneuver space requisite for subsequent projected operations ashore. (JP 3-02)

beach landing site — A geographic location selected for across-the-beach infiltration, exfiltration, or resupply operations. Also called **BLS.** (JP 3-05)

beach marker — A sign or device used to identify a beach or certain activities thereon for incoming waterborne traffic. Markers may be panels, lights, buoys, or electronic devices.

beachmaster — The naval officer in command of the beachmaster unit of the naval beach group. Also called **BM.**

beachmaster unit — A commissioned naval unit of the naval beach group designed to provide to the shore party a Navy component known as a beach party, which is capable of supporting the amphibious landing of one division (reinforced). Also called **BMU.** See also **beach party; naval beach group; shore party.** (JP 4-01.6)

beach minefield — **(*)** A minefield in the shallow water approaches to a possible amphibious landing beach. See also **minefield.**

beach organization — In an amphibious operation, the planned arrangement of personnel and facilities to effect movement, supply, and evacuation across beaches and in the beach area for support of a landing force.

beach party — The naval component of the shore party. See also **beachmaster unit; shore party.**

beach party commander — The naval officer in command of the naval component of the shore party.

beach photography — Vertical, oblique, ground, and periscope coverage at varying scales to provide information of offshore, shore, and inland areas. It covers terrain that provides observation of the beaches and is primarily concerned with the geological and tactical aspects of the beach.

beach reserves — **(*)** In an amphibious operation, an accumulation of supplies of all classes established in dumps in beachhead areas. See also **reserve supplies.**

beach support area — In amphibious operations, the area to the rear of a landing force or elements thereof, established and operated by shore party units, which contains the facilities for the unloading of troops and materiel and the support of the forces ashore; it includes facilities for the evacuation of wounded, enemy prisoners of war, and captured materiel. Also called **BSA.**

beach survey — The collection of data describing the physical characteristics of a beach; that is, an area whose boundaries are a shoreline, a coastline, and two natural or arbitrary assigned flanks.

beach width — The horizontal dimensions of the beach measured at right angles to the shoreline from the line of extreme low water inland to the landward limit of the beach (the coastline).

beam rider — A missile guided by an electronic beam.

beam width — The angle between the directions, on either side of the axis, at which the intensity of the radio frequency field drops to one-half the value it has on the axis.

bearing — The horizontal angle at a given point measured clockwise from a specific datum point to a second point. See also **grid bearing; relative bearing; true bearing.**

beaten zone — The area on the ground upon which the cone of fire falls.

begin morning civil twilight — The period of time at which the sun is halfway between beginning morning and nautical twilight and sunrise, when there is enough light to see objects clearly with the unaided eye. At this time, light intensification devices are no longer effective, and the sun is six degrees below the eastern horizon. Also called **BMCT.**

begin morning nautical twilight — The start of that period where, in good conditions and in the absence of other illumination, enough light is available to identify the general outlines of ground objects and conduct limited military operations. Light intensification devices are still effective and may have enhanced capabilities. At this time, the sun is 12 degrees below the eastern horizon. Also called **BMNT.**

beleaguered — See **missing.**

below-the-line publications — The lower level publications in the hierarchy of joint publications which includes supporting joint doctrine and joint tactics, techniques, and procedures publications that are signed by the Director, Joint Staff and contain specific mission-area guidance for the joint community. Included in this level are reference publications and those describing joint personnel, intelligence support, operations, logistic support, planning, and command, control, communications, and computer systems support. See also **above-the-line publications; capstone publication; joint publication; keystone publications.** (JP 1-01)

berm, natural — The nearly horizontal portion of a beach or backshore having an abrupt fall and formed by deposition of material by wave action. A berm marks the limit of ordinary high tide. For air cushion vehicles, berms (constructed) are required to protect materials handling equipment operations. See also **backshore.** (JP 4-01.6)

besieged — See **missing.**

bight — A bend in a coast forming an open bay or an open bay formed by such a bend. (JP 4-01.6)

bilateral infrastructure — (*) Infrastructure which concerns only two NATO members and is financed by mutual agreement between them (e.g., facilities required for the use of forces of one NATO member in the territory of another). See also **infrastructure.**

bill — A ship's publication listing operational or administrative procedures. (JP 3-04.1)

billet — 1. Shelter for troops. 2. To quarter troops. 3. A personnel position or assignment that may be filled by one person.

binary chemical munition — (*) A munition in which chemical substances, held in separate containers, react when mixed or combined as a result of being fired, launched, or otherwise initiated to produce a chemical agent. See also **munition.**

binding — (*) The fastening or securing of items to a movable platform called a pallet. See also **palletized unit load.**

bin storage — Storage of items of supplies and equipment in an individual compartment or subdivision of a storage unit in less than bulk quantities. See also **bulk storage; storage.**

biographical intelligence — That component of intelligence that deals with individual foreign personalities of actual or potential importance.

biological agent — A microorganism that causes disease in personnel, plants, or animals or causes the deterioration of materiel. See also **biological operation; biological weapon; chemical agent.**

biological ammunition — (*) A type of ammunition, the filler of which is primarily a biological agent.

biological defense — (*) The methods, plans, and procedures involved in establishing and executing defensive measures against attacks using biological agents.

biological environment — (*) Conditions found in an area resulting from direct or persisting effects of biological weapons.

biological half-time — See **half-life.**

biological operation — Employment of biological agents to produce casualties in personnel or animals or damage to plants. See also **biological agent; biological threat.** (JP 3-11)

biological threat — A threat that consists of biological material planned to be deployed to produce casualties in personnel or animals or damage plants. See also **biological agent; biological ammunition;**

biological defense; biological environment; chemical, biological, and radiological operation; contamination; contamination control. (JP 3-11)

biological warfare — See **biological operation.**

biological weapon — **(*)** An item of materiel which projects, disperses, or disseminates a biological agent including arthropod vectors.

black — In intelligence handling, a term used in certain phrases (e.g., living black, black border crossing) to indicate reliance on illegal concealment rather than on cover.

black list — An official counterintelligence listing of actual or potential enemy collaborators, sympathizers, intelligence suspects, and other persons whose presence menaces the security of friendly forces.

black propaganda — Propaganda that purports to emanate from a source other than the true one. See also **propaganda.**

blast effect — Destruction of or damage to structures and personnel by the force of an explosion on or above the surface of the ground. Blast effect may be contrasted with the cratering and ground-shock effects of a projectile or charge that goes off beneath the surface.

blast line — A horizontal radial line on the surface of the Earth originating at ground zero on which measurements of blast from an explosion are taken.

blast wave — A sharply defined wave of increased pressure rapidly propagated through a surrounding medium from a center of detonation or similar disturbance.

blast wave diffraction — **(*)** The passage around and envelopment of a structure by the nuclear blast wave.

bleeding edge — **(*)** That edge of a map or chart on which cartographic detail is extended to the edge of the sheet.

blind transmission — Any transmission of information that is made without expectation of acknowledgement. (JP 3-05)

blister agent — **(*)** A chemical agent which injures the eyes and lungs, and burns or blisters the skin. Also called **vesicant agent.**

blocking and chocking — **(*)** The use of wedges or chocks to prevent the inadvertent shifting of cargo in transit.

blocking position — A defensive position so sited as to deny the enemy access to a given area or to prevent the enemy's advance in a given direction.

block shipment — A method of shipment of supplies to overseas areas to provide balanced stocks or an arbitrary balanced force for a specific number of days, e.g., shipment of 30 days' supply for an average force of 10,000 individuals.

block stowage loading — **(*)** A method of loading whereby all cargo for a specific destination is stowed together. The purpose is to facilitate rapid off-loading at the destination, with the least possible disturbance of cargo intended for other points. See also **loading.**

blood agent — **(*)** A chemical compound, including the cyanide group, that affects bodily functions by preventing the normal utilization of oxygen by body tissues.

blood chit — A small sheet of material depicting an American flag and a statement in several languages to the effect that anyone assisting the bearer to safety will be rewarded. See also **evasion aid.** (JP 3-50.3)

blood chit (intelligence) — See **blood chit.**

blowback — **(*)** 1. Escape, to the rear and under pressure, of gases formed during the firing of the weapon. Blowback may be caused by a defective breech mechanism, a ruptured cartridge case, or a faulty primer. 2. Type of weapon operation in which the force of expanding gases acting to the rear against the face of the bolt furnishes all the energy required to initiate the complete cycle of operation. A weapon which employs this method of operation is characterized by the absence of any breech-lock or bolt-lock mechanism.

Blue Bark — US military personnel, US citizen civilian employees of the Department of Defense, and the dependents of both categories who travel in connection with the death of an immediate family member. It also applies to designated escorts for dependents of deceased military members. Furthermore, the term is used to designate the personal property shipment of a deceased member.

boat diagram — In the assault phase of an amphibious operation, a diagram showing the positions of individuals and equipment in each boat.

boat group — The basic organization of landing craft. One boat group is organized for each battalion landing team (or equivalent) to be landed in the first trip of landing craft or amphibious vehicles.

boat lane — **(*)** A lane for amphibious assault landing craft, which extends seaward from the landing beaches to the line of departure. The width of a boat lane is determined by the length of the corresponding beach.

boat space — The space and weight factor used to determine the capacity of boats, landing craft, and amphibious vehicles. With respect to landing craft and amphibious vehicles, it is based on the requirements of one person with individual equipment. The person is assumed to weigh 224 pounds and to occupy 13.5 cubic feet of space. See also **man space.**

boattail — **(*)** The conical section of a ballistic body that progressively decreases in diameter toward the tail to reduce overall aerodynamic drag.

boat wave — See **wave.**

bomb disposal unit — See **explosive ordnance disposal unit.**

bomber — See **intermediate-range bomber aircraft; long-range bomber aircraft; medium-range bomber aircraft.**

bomb impact plot — A graphic representation of the target area, usually a pre-strike air photograph, on which prominent dots are plotted to mark the impact or detonation points of bombs dropped on a specific bombing attack.

bombing angle — **(*)** The angle between the vertical and a line joining the aircraft to what would be the point of impact of a bomb released from it at that instant.

bombing run — **(*)** In air bombing, that part of the flight that begins, normally from an initial point, with the approach to the target, includes target acquisition, and ends normally at the weapon release point.

bomb release line — **(*)** An imaginary line around a defended area or objective over which an aircraft should release its bomb in order to obtain a hit or hits on an area or objective.

bomb release point — **(*)** The point in space at which bombs must be released to reach the desired point of detonation.

bona fides — Good faith. In evasion and recovery operations, the use of verbal or visual communication by individuals who are unknown to one another to establish their authenticity, sincerity, honesty, and truthfulness. See also **evasion; evasion and recovery; recovery; recovery operations.** (JP 3-50.3)

bonding — **(*)** In electrical engineering, the process of connecting together metal parts so that they make low resistance electrical contact for direct current and lower frequency alternating currents. See also **earthing.**

booby trap — **(*)** An explosive or nonexplosive device or other material, deliberately placed to cause casualties when an apparently harmless object is disturbed or a normally safe act is performed.

booster — **(*)** 1. A high-explosive element sufficiently sensitive so as to be actuated by small explosive elements in a fuze or primer and powerful enough to cause detonation of the main explosive filling. 2. An auxiliary or initial propulsion system which travels with a missile or aircraft and which may or may not separate from the parent craft when its impulse has been delivered. A booster system may contain, or consist of, one or more units.

boost phase — That portion of the flight of a ballistic missile or space vehicle during which the booster and sustainer engines operate. See also **midcourse phase; reentry phase; terminal phase.**

border — **(*)** In cartography, the area of a map or chart lying between the neatline and the surrounding framework.

border break — **(*)** A cartographic technique used when it is required to extend a portion of the cartographic detail of a map or chart beyond the sheetlines into the margin.

border crosser — **(*)** An individual, living close to a frontier, who normally has to cross the frontier frequently for legitimate purposes.

boresafe fuze — **(*)** Type of fuze having an interrupter in the explosive train that prevents a projectile from exploding until after it has cleared the muzzle of a weapon.

bottom mine — **(*)** A mine with negative buoyancy which remains on the seabed. Also called **ground mine.** See also **mine.**

bound — **(*)** 1. In land warfare, a single movement, usually from cover to cover, made by troops often under enemy fire. 2. **(DOD only)** Distance covered in one movement by a unit that is advancing by bounds.

boundary — A line that delineates surface areas for the purpose of facilitating coordination and deconfliction of operations between adjacent units, formations, or areas. See also **airspace control boundary.** (JP 3-0)

bouquet mine — **(*)** In naval mine warfare, a mine in which a number of buoyant mine cases are attached to the same sinker, so that when the mooring of one mine case is cut, another mine rises from the sinker to its set depth. See also **mine.**

bracketing — **(*)** A method of adjusting fire in which a bracket is established by obtaining an over and a short along the spotting line, and then successively splitting the bracket in half until a target hit or desired bracket is obtained.

branch — 1. A subdivision of any organization. 2. A geographically separate unit of an activity which performs all or part of the primary functions of the parent activity on a smaller scale. Unlike an annex, a branch is not merely an overflow addition. 3. An arm or service of the Army. 4. The contingency options built into the basic plan. A branch is used for changing the mission, orientation, or direction of movement of a force to aid success of the operation based on anticipated events, opportunities, or disruptions caused by enemy actions and reactions. See also **sequel.** (JP 3-0)

breakaway — **(*)** 1. The onset of a condition in which the shock front moves away from the exterior of the expanding fireball produced by the explosion of a nuclear weapon. 2. **(DOD only)** After completion of attack, turn to heading as directed.

breakbulk cargo — Any commodity that, because of its weight, dimensions, or incompatibility with other cargo, must be shipped by mode other than military van or SEAVAN. See also **breakbulk ship.** (JP 4-01.7)

breakbulk ship — A ship with conventional holds for stowage of breakbulk cargo, below or above deck, and equipped with cargo-handling gear. Ships also may be capable of carrying a limited number of containers, above or below deck. See also **breakbulk cargo.** (JP 4-01.7)

breaker — A wave in the process of losing energy where offshore energy loss is caused by wind action and nearshore energy loss is caused by the impact of the sea floor as the wave enters shallow (shoaling) water. Breakers either plunge, spill, or surge. See also **breaker angle.** (JP 4-01.6)

breaker angle — The angle a breaker makes with the beach. See also **breaker.** (JP 4-01.6)

breakoff position — **(*)** The position at which a leaver or leaver section breaks off from the main convoy to proceed to a different destination.

break-up — **(*)** 1. In detection by radar, the separation of one solid return into a number of individual returns which correspond to the various objects or structure groupings. This separation is contingent upon a number of factors including range, beam width, gain setting, object size and distance between objects. 2. In imagery interpretation, the result of magnification or enlargement which causes the imaged item to lose its identity and the resultant presentation to become a random series of tonal impressions. Also called **split-up.**

brevity code — **(*)** A code which provides no security but which has as its sole purpose the shortening of messages rather than the concealment of their content.

bridgehead — An area of ground held or to be gained on the enemy's side of an obstacle. See also **airhead; beachhead.**

bridgehead line — **(*)** The limit of the objective area in the development of the bridgehead. See also **objective area.**

briefing — **(*)** The act of giving in advance specific instructions or information.

brigade — A unit usually smaller than a division to which are attached groups and/or battalions and smaller units tailored to meet anticipated requirements. Also called **BDE.**

broach — When a water craft is thrown broadside to the wind and waves, against a bar, or against the shoreline. (JP 4-01.6)

buffer distance — **(*)** In nuclear warfare: 1. The horizontal distance which, when added to the radius of safety, will give the desired assurance that the specified degree of risk will not be exceeded. The buffer distance is normally expressed quantitatively in multiples of the delivery error. 2. The vertical distance which is added to the fallout safe-height of burst in order to determine a desired height of burst which will provide the desired assurance that militarily significant fallout will not occur. It is normally expressed quantitatively in multiples of the vertical error.

buffer zone — 1. A defined area controlled by a peace operations force from which disputing or belligerent forces have been excluded. A buffer zone is formed to create an area of separation between disputing or belligerent forces and reduce the risk of renewed conflict. Also called **area of separation** in some United Nations operations. Also called **BZ.** See also **area of separation; line of demarcation; peace operations.** 2. A conical volume centered on the laser's line of sight with its apex at the aperture of the laser, within which the beam will be contained with a high degree of certainty. It is determined by the buffer angle. See also **laser.** (JP 3-07.3)

bug — 1. A concealed microphone or listening device or other audiosurveillance device. 2. To install means for audiosurveillance.

bugged — Room or object that contains a concealed listening device.

building systems — Structures assembled from manufactured components designed to provide specific building configurations (e.g., large steel arch structures, large span tension fabric structures, panelized buildings, and pre-engineered buildings). See also **civil engineering.** (JP 4-04)

buildup — **(*)** The process of attaining prescribed strength of units and prescribed levels of vehicles, equipment, stores, and supplies. Also may be applied to the means of accomplishing this process.

bulk cargo — That which is generally shipped in volume where the transportation conveyance is the only external container; such as liquids, ore, or grain.

bulk petroleum product — **(*)** A liquid petroleum product transported by various means and stored in tanks or containers having an individual fill capacity greater than 250 liters.

bulk storage — 1. Storage in a warehouse of supplies and equipment in large quantities, usually in original containers, as distinguished from bin storage. 2. Storage of liquids, such as petroleum products in tanks, as distinguished from drum or packaged storage. See also **bin storage; storage.**

bullseye — An established reference point from which the position of an object can be referenced. See also **reference point.** (JP 3-60)

burn notice — An official statement by one intelligence agency to other agencies, domestic or foreign, that an individual or group is unreliable for any of a variety of reasons.

burnout — **(*)** The point in time or in the missile trajectory when combustion of fuels in the rocket engine is terminated by other than programmed cutoff.

burnout velocity — (*) The velocity attained by a missile at the point of burnout.

burn-through range — The distance at which a specific radar can discern targets through the external interference being received.

Intentionally Blank

C

cache — In evasion and recovery operations, source of subsistence and supplies, typically containing items such as food, water, medical items, and/or communications equipment, packaged to prevent damage from exposure and hidden in isolated locations by such methods as burial, concealment, and/or submersion, to support evaders in current or future operations. See also **concealment; evader; evasion; evasion and recovery; recovery; recovery operations.** (JP 3-50.3)

calibrated focal length — (*) An adjusted value of the equivalent focal length, so computed as to equalize the positive and negative values of distortion over the entire field used in a camera.

call fire — Fire delivered on a specific target in response to a request from the supported unit. See also **fire.**

call for fire — (*) A request for fire containing data necessary for obtaining the required fire on a target.

call sign — (*) Any combination of characters or pronounceable words, which identifies a communication facility, a command, an authority, an activity, or a unit; used primarily for establishing and maintaining communications. Also called **CS.** See also **collective call sign; indefinite call sign; international call sign; net call sign; tactical call sign; visual call sign; voice call sign.**

camera axis — (*) An imaginary line through the optical center of the lens perpendicular to the negative photo plane.

camera axis direction — (*) Direction on the horizontal plane of the optical axis of the camera at the time of exposure. This direction is defined by its azimuth expressed in degrees in relation to true/magnetic north.

camera calibration — (*) The determination of the calibrated focal length, the location of the principal point with respect to the fiducial marks and the lens distortion effective in the focal plane of the camera referred to the particular calibrated focal length.

camera cycling rate — (*) The frequency with which camera frames are exposed, expressed as cycles per second.

camera nadir — See **photo nadir.**

camouflage — (*) The use of natural or artificial material on personnel, objects, or tactical positions with the aim of confusing, misleading, or evading the enemy.

camouflage detection photography — (*) Photography utilizing a special type of film (usually infrared) designed for the detection of camouflage.

camouflet — (*) The resulting cavity in a deep underground burst when there is no rupture of the surface. See also **crater.**

campaign — A series of related military operations aimed at accomplishing a strategic or operational objective within a given time and space. See also **campaign plan.**

campaign plan — A plan for a series of related military operations aimed at accomplishing a strategic or operational objective within a given time and space. See also **campaign; campaign planning.** (JP 5-0)

campaign planning — The process whereby combatant commanders and subordinate joint force commanders translate national or theater strategic and operational concepts through the development of campaign plans. Campaign planning may begin during deliberate planning when the actual threat, national guidance, and available resources become evident, but is normally not completed until after the National Command Authorities select the course of action during crisis action planning. Campaign planning is conducted when contemplated military operations exceed the scope of a single major joint operation. See also **campaign; campaign plan.** (JP 5-0)

canalize — To restrict operations to a narrow zone by use of existing or reinforcing obstacles or by fire or bombing.

cannibalize — To remove serviceable parts from one item of equipment in order to install them on another item of equipment.

cannot observe — **(*)** A type of fire control which indicates that the observer or spotter will be unable to adjust fire, but believes a target exists at the given location and is of sufficient importance to justify firing upon it without adjustment or observation.

cantilever lifting frame — Used to move Navy lighterage causeway systems on to and off of lighter aboard ship (LASH) vessels. This device is suspended from the Morgan LASH barge crane and can lift one causeway section at a time. It is designed to allow the long sections to clear the rear of the ship as they are lowered into the water. Also called **CLF.** See also **causeway; lighterage.** (JP 4-01.6)

capability — The ability to execute a specified course of action. (A capability may or may not be accompanied by an intention.)

capacity load (Navy) — The maximum quantity of all supplies (ammunition; petroleum, oils, and lubricants; rations; general stores; maintenance stores; etc.) which each vessel can carry in proportions prescribed by proper authority. See also **wartime load.**

capstone publication — The top group of joint doctrine publications in the hierarchy of joint publications. Capstone publications link joint doctrine to national strategy and the contributions of other government agencies, alliances, and coalitions. See also **above-the-line publications; below-the-line publications; joint publication; keystone publications.** (JP 1-01)

capstone requirements document — A document that contains performance-based requirements to facilitate development of individual operational requirements documents by providing a common framework and operational concept to guide their development. Also called **CRD.**

capsule — **(*)** 1. A sealed, pressurized cabin for extremely high altitude or space flight which provides an acceptable environment for man, animal, or equipment. 2. An ejectable sealed cabin having automatic devices for safe return of the occupants to the surface.

captive firing — **(*)** A firing test of short duration, conducted with the missile propulsion system operating while secured to a test stand.

captured — See **missing.**

cardinal point effect — **(*)** The increased intensity of a line or group of returns on the radarscope occurring when the radar beam is perpendicular to the rectangular surface of a line or group of similarly aligned features in the ground pattern.

caretaker status — A nonoperating condition in which the installations, materiel, and facilities are in a care and limited preservation status. Only a minimum of personnel is required to safeguard against fire, theft, and damage from the elements.

cargo classification (combat loading) — The division of military cargo into categories for combat loading aboard ships.

cargo increment number — A seven-character alphanumeric field that uniquely describes a non-unit-cargo entry (line) in the Joint Operation Planning and Execution System time-phased force and deployment data.

cargo outturn message — A brief message report transmitted within 48 hours of completion of ship discharge to advise both the Military Sealift Command and the terminal of loading of the condition of the cargo, including any discrepancies in the form of overages, shortages, or damages between cargo as manifested and cargo as checked at time of discharge.

cargo outturn report — A detailed report prepared by a discharging terminal to record discrepancies in the form of over, short, and damaged cargo as manifested, and cargo checked at a time and place of discharge from ship.

cargo sling — **(*)** A strap, chain, or other material used to hold cargo items securely which are to be hoisted, lowered, or suspended.

cargo tie-down point — A point on military materiel designed for attachment of various means for securing the item for transport.

cargo transporter — A reusable metal shipping container designed for worldwide surface and air movement of suitable military supplies and equipment through the cargo transporter service.

carpet bombing — **(*)** The progressive distribution of a mass bomb load upon an area defined by designated boundaries, in such manner as to inflict damage to all portions thereof.

carrier air wing — Two or more aircraft squadrons formed under one commander for administrative and tactical control of operations from a carrier.

carrier battle group — A standing naval task group consisting of a carrier, surface combatants, and submarines as assigned in direct support, operating in mutual support with the task of destroying hostile submarine, surface, and air forces within the group's assigned operational area and striking at targets along hostile shore lines or projecting fire power inland. Also called **CVBG.** (JP 3-33)

carrier striking force — A naval task force composed of aircraft carriers and supporting combatant ships capable of conducting strike operations.

cartel — An association of independent businesses organized to control prices and production, eliminate competition, and reduce the cost of doing business. (JP 3-07.4)

CARVER — A special operations forces acronym used throughout the targeting and mission planning cycle to assess mission validity and requirements. The acronym stands for criticality, accessibility, recuperability, vulnerability, effect, and recognizability. (JP 3-05.5)

case — 1. An intelligence operation in its entirety. 2. Record of the development of an intelligence operation, including personnel, modus operandi, and objectives.

casual — See **transient.**

casualty — Any person who is lost to the organization by having been declared dead, duty status – whereabouts unknown, missing, ill, or injured. See also **casualty category; casualty status; casualty type; duty status – whereabouts unknown; hostile casualty; nonhostile casualty.**

casualty category — A term used to specifically classify a casualty for reporting purposes based upon the casualty type and the casualty status. Casualty categories include killed in action, died of wounds received in action, and wounded in action. See also **casualty; casualty status; casualty type; duty status - whereabouts unknown; missing.**

casualty evacuation — The movement of casualties. It includes movement both to and between medical treatment facilities. Any vehicle may be used to evacuate casualties. Also called **CASEVAC.** See also **casualty; evacuation; medical treatment facility.** (JP 4-02)

casualty receiving and treatment ship — In amphibious operations, a ship designated to receive, provide treatment for, and transfer casualties. (JP 3-02)

casualty status — A term used to classify a casualty for reporting purposes. There are seven casualty statuses: (1) deceased; (2) duty status - whereabouts unknown; (3) missing; (4) very seriously ill or injured; (5) seriously ill or injured; (6) incapacitating illness or injury; and (7) not seriously injured. See also **casualty; casualty category; casualty type; deceased; duty status - whereabouts unknown; incapacitating illness or injury; missing; not seriously injured; seriously ill or injured; very seriously ill or injured.**

casualty type — A term used to identify a casualty for reporting purposes as either a hostile casualty or a nonhostile casualty. See also **casualty; casualty category; casualty status; hostile casualty; nonhostile casualty.**

catalytic attack — An attack designed to bring about a war between major powers through the disguised machinations of a third power.

catalytic war — Not to be used. See **catalytic attack.**

catapult — **(*)** A structure which provides an auxiliary source of thrust to a missile or aircraft; must combine the functions of directing and accelerating the missile during its travel on the catapult; serves the same functions for a missile as does a gun tube for a shell.

categories of data — In the context of perception management and its constituent approaches, data obtained by adversary individuals, groups, intelligence systems, and officials. Such data fall in two categories: a. **information** — A compilation of data provided by protected or open sources that would provide a substantially complete picture of friendly intentions, capabilities, or activities. b. **indicators** — Data derived from open sources or from detectable actions that adversaries can piece together or interpret to reach personal conclusions or official estimates concerning friendly intentions, capabilities, or activities. (Note: In operations security, actions that convey indicators exploitable by adversaries, but that must be carried out regardless, to plan, prepare for, and execute activities, are called "observables.") See also **operations security.**

causeway — A craft similar in design to a barge, but longer and narrower, designed to assist in the discharge and transport of cargo from vessels. See also **barge; watercraft.** (JP 4-01.6)

causeway launching area — An area located near the line of departure but clear of the approach lanes, where ships can launch pontoon causeways. (JP 3-02)

caveat — A designator used with a classification to further limit the dissemination of restricted information. (JP 3-07.4)

C-day — See **times.**

CEASE BUZZER — An unclassified term to terminate electronic attack activities, including the use of electronic warfare expendables. See also **electronic attack; electronic warfare.** (JP 3-51)

cease fire — 1. A command given to any unit or individual firing any weapon to stop engaging the target. See also **call for fire; fire mission.** 2. A command given to air defense artillery units to refrain from firing on, but to continue to track, an airborne object. Missiles already in flight will be permitted to continue to intercept.

cease fire line — See **armistice demarcation line.** See also **armistice; cease fire.** (JP 3-07.3)

ceiling — The height above the Earth's surface of the lowest layer of clouds or obscuration phenomena that is reported as "broken," "overcast," or "obscured" and not classified as "thin" or "partial."

celestial guidance — The guidance of a missile or other vehicle by reference to celestial bodies.

celestial sphere — **(*)** An imaginary sphere of infinite radius concentric with the Earth, on which all celestial bodies except the Earth are imagined to be projected.

cell — Small group of individuals who work together for clandestine or subversive purposes.

cell system — See **net, chain, cell system.**

censorship — See **armed forces censorship; civil censorship; field press censorship; national censorship; primary censorship; prisoner of war censorship; secondary censorship.**

center of burst — See **mean point of impact.**

centers of gravity — Those characteristics, capabilities, or sources of power from which a military force derives its freedom of action, physical strength, or will to fight. Also called **COGs.** See also **capability; decisive point.** (JP 3-0)

centigray — **(*)** A unit of absorbed dose of radiation (one centigray equals one rad).

central control officer — The officer designated by the amphibious task force commander for the overall coordination of the waterborne ship-to-shore movement. The central control officer is embarked in the central control ship. Also called **CCO.** (JP 3-02.2)

centralized control — **(*)** In air defense, the control mode whereby a higher echelon makes direct target assignments to fire units. See also **decentralized control.**

centralized receiving and shipping point — Actual location where containers with cargo must be sorted before transshipment to the appropriate supply support activity or owning unit. Single consignee cargo and ammunition will not pass through the centralized receiving and shipping point. Cargo will be shipped directly to the owner with the movement organization maintaining visibility, and ammunition will go directly to the appropriate ammunition storage facility. Also called **CRSP.** (JP 4-01.7)

centrally managed item — An item of materiel subject to inventory control point (wholesale level) management.

central procurement — The procurement of materiel, supplies, or services by an officially designated command or agency with funds specifically provided for such procurement for the benefit and use of the entire component or, in the case of single managers, for the Military Departments as a whole.

chaff — Radar confusion reflectors, consisting of thin, narrow metallic strips of various lengths and frequency responses, which are used to reflect echoes for confusion purposes. Causes enemy radar guided missiles to lock on to it instead of the real aircraft, ship, or other platform. See also **deception; rope.**

chain — See **net, chain, cell system.**

chain of command — **(*)** The succession of commanding officers from a superior to a subordinate through which command is exercised. Also called **command channel.**

Chairman of the Joint Chiefs of Staff Instruction — A replacement document for all types of correspondence containing Chairman of the Joint Chiefs of Staff (CJCS) policy and guidance that does not involve the employment of forces. An instruction is of indefinite duration and is applicable to external agencies or both the Joint Staff and external agencies. It remains in effect until superseded, rescinded, or otherwise canceled. CJCS Instructions, unlike joint publications, will not contain joint doctrine and/or joint tactics, techniques, and procedures. Terminology used in these publications will be consistent with JP 1-02. Also called **CJCSI.** See also **Chairman of the Joint Chiefs of Staff Manual.** (JP 1-01)

Chairman of the Joint Chiefs of Staff Manual — A document containing detailed procedures for performing specific tasks that do not involve the employment of forces. A manual is of indefinite duration and is applicable to external agencies or both the Joint Staff and external agencies. It may supplement a Chairman of the Joint Chiefs of Staff Instruction or stand alone and remains in effect until superseded, rescinded, or otherwise canceled. Chairman of the Joint Chiefs of Staff Manuals, unlike joint publications, will not contain joint doctrine and/or joint tactics, techniques, and procedures. Terminology used in these publications will be consistent with JP 1-02. Also called **CJCSM.** See also **Chairman of the Joint Chiefs of Staff Instruction.** (JP 1-01)

Chairman of the Joint Chiefs of Staff memorandum of policy — A statement of policy approved by the Chairman of the Joint Chiefs of Staff and issued for the guidance of the Services, the combatant commands, and the Joint Staff.

Chairman's program assessment — Provides the Chairman of the Joint Chiefs of Staff's personal appraisal on alternative program recommendations and budget proposals to the Secretary of Defense for consideration in refining the defense program and budget in accordance with 10 United States Code. The Chairman's program assessment comments on the risk associated with the programmed allocation of Defense resources and evaluates the conformance of program objective memoranda to the priorities established in strategic plans and combatant commanders' priority requirements. Also called **CPA.**

Chairman's program recommendations — Provides the Chairman of the Joint Chiefs of Staff's personal recommendations to the Secretary of Defense for the programming and budgeting process before publishing the Defense Planning Guidance (DPG) in accordance with 10 United States Code. The Chairman's program recommendations articulates programs the Chairman deems critical for the Secretary of Defense to consider when identifying Department of Defense (DOD) priorities and performance goals in the DPG and emphasizes specific recommendations that will enhance joint readiness, promote joint doctrine and training, improve joint warfighting capabilities, and satisfy joint warfighting requirements within DOD resource constraints and within acceptable risk levels. Also called **CPR.**

chalk commander — **(*)** The commander of all troops embarked under one chalk number. See also **chalk number; chalk troops.**

chalk number — **(*)** The number given to a complete load and to the transporting carrier. See also **chalk commander; chalk troops.**

chalk troops — **(*)** A load of troops defined by a particular chalk number. See also **chalk commander; chalk number.**

challenge — **(*)** Any process carried out by one unit or person with the object of ascertaining the friendly or hostile character or identity of another. See also **countersign; password.**

chancery — The building upon a diplomatic or consular compound which houses the offices of the chief of mission or principal officer.

change of operational control — The date and time (Coordinated Universal Time) at which a force or unit is reassigned or attached from one commander to another where the gaining commander will exercise operational control over that force or unit. Also called **CHOP.** See also **operational control.** (JP 0-2)

channel airlift — Common-user airlift service provided on a scheduled basis between two points. There are two types of channel airlift. A requirements channel serves two or more points on a scheduled basis depending upon the volume of traffic; a frequency channel is time-based and serves two or more points at regular intervals.

characteristic actuation probability — In naval mine warfare, the average probability of a mine of a given type being actuated by one run of the sweep within the characteristic actuation width.

characteristic actuation width — In naval mine warfare, the width of path over which mines can be actuated by a single run of the sweep gear.

characteristic detection probability — In naval mine warfare, the ratio of the number of mines detected on a single run to the number of mines which could have been detected within the characteristic detection width.

characteristic detection width — In naval mine warfare, the width of path over which mines can be detected on a single run.

charged demolition target — **(*)** A demolition target on which all charges have been placed and which is in the states of readiness, either state 1--safe, or state

2--armed. See also **state of readiness--state 1--safe; state of readiness--state 2--armed.**

chart base — (*) A chart used as a primary source for compilation or as a framework on which new detail is printed. Also called **topographic base.**

chart index — See **map index.**

chart location of the battery — See **battery center.**

chart series — See **map; map series.**

chart sheet — See **map; map sheet.**

check firing — In artillery, mortar, and naval gunfire support, a command to cause a temporary halt in firing. See also **cease fire; fire mission.**

checkout — (*) A sequence of functional, operational, and calibrational tests to determine the condition and status of a weapon system or element thereof.

checkpoint — (*) 1. A predetermined point on the surface of the Earth used as a means of controlling movement, a registration target for fire adjustment, or reference for location. 2. Center of impact; a burst center. 3. Geographical location on land or water above which the position of an aircraft in flight may be determined by observation or by electrical means. 4. A place where military police check vehicular or pedestrian traffic in order to enforce circulation control measures and other laws, orders, and regulations.

check sweeping — (*) In naval mine warfare, sweeping to check that no moored mines are left after a previous clearing operation.

chemical agent — Any toxic chemical intended for use in military operations. See also **chemical ammunition; chemical defense; chemical dose; chemical environment; chemical warfare; riot control agent.** (JP 3-11)

chemical agent cumulative action — The building up, within the human body, of small ineffective doses of certain chemical agents to a point where eventual effect is similar to one large dose.

chemical ammunition — (*) A type of ammunition, the filler of which is primarily a chemical agent.

chemical ammunition cargo — Cargo such as white phosphorous munitions (shell and grenades).

chemical, biological, and radiological operation — (*) A collective term used only when referring to a combined chemical, biological, and radiological operation.

chemical contamination — See **contamination.** (JP 3-11)

chemical defense — (*) The methods, plans, and procedures involved in establishing and executing defensive measures against attack utilizing chemical agents. See also **nuclear, biological, and chemical defense.**

chemical dose — (*) The amount of chemical agent, expressed in milligrams, that is taken or absorbed by the body.

chemical environment — (*) Conditions found in an area resulting from direct or persisting effects of chemical weapons.

chemical horn — (*) In naval mine warfare, a mine horn containing an electric battery, the electrolyte for which is in a glass tube protected by a thin metal sheet. Also called **Hertz Horn.**

chemical monitoring — **(*)** The continued or periodic process of determining whether or not a chemical agent is present. See also **chemical survey.**

chemical operation — **(*)** Employment of chemical agents to kill, injure, or incapacitate for a significant period of time, man or animals, and deny or hinder the use of areas, facilities, or materiel; or defense against such employment.

chemical survey — **(*)** The directed effort to determine the nature and degree of chemical hazard in an area and to delineate the perimeter of the hazard area.

chemical warfare — All aspects of military operations involving the employment of lethal and incapacitating munitions/agents and the warning and protective measures associated with such offensive operations. Since riot control agents and herbicides are not considered to be chemical warfare agents, those two items will be referred to separately or under the broader term "chemical," which will be used to include all types of chemical munitions/agents collectively. Also called **CW.** See also **chemical agent; chemical defense; chemical dose; chemical environment; chemical weapon; riot control agent.** (JP 3-11)

chemical weapon — Together or separately, (a) a toxic chemical and its precursors, except when intended for a purpose not prohibited under the Chemical Weapons Convention; (b) a munition or device, specifically designed to cause death or other harm through toxic properties of those chemicals specified in (a), above, which would be released as a result of the employment of such munition or device; (c) any equipment specifically designed for use directly in connection with the employment of munitions or devices specified in (b), above. See also **chemical agent; chemical defense; chemical dose; chemical environment; chemical warfare; riot control agent.** (JP 3-11)

chief Army, Navy, Air Force, or Marine Corps censor — An officer appointed by the commander of the Army, Navy, Air Force, or Marine Corps component of a unified command to supervise all censorship activities of that Service.

chief of mission — A chief of mission (COM) (normally the ambassador) is the principal officer in charge of a diplomatic facility of the United States, including any individual assigned to be temporarily in charge of such a facility. The COM is the personal representative of the President to the country of accreditation. The COM is responsible for the direction, coordination, and supervision of all US Government executive branch employees in that country (except those under the command of a US area military commander). The security of the diplomatic post is the COM's direct responsibility. Also called **COM.**

chief of staff — The senior or principal member or head of a staff, or the principal assistant in a staff capacity to a person in a command capacity; the head or controlling member of a staff, for purposes of the coordination of its work; a position that in itself is without inherent power of command by reason of assignment, except that which is invested in such a position by delegation to exercise command in another's name.

chronic radiation dose — A dose of ionizing radiation received either continuously or intermittently over a prolonged period of time. A chronic radiation dose may be high enough to cause radiation sickness and death but, if received at a low dose rate, a significant portion of the acute cellular damage may be repaired. See also **acute radiation dose; radiation dose; radiation dose rate.**

chuffing — **(*)** The characteristic of some rockets to burn intermittently and with an irregular noise.

CINC's required date — The original date relative to C-day, specified by the combatant commander for arrival of forces or cargo at the destination; shown in the time-phased force and deployment data to assess the impact of later arrival. Also called **CRD.**

CINC's strategic concept — Final document produced in step 5 of the concept development phase of the deliberate planning process. The CINC's strategic concept is used as the vehicle to distribute the CINC's decision and planning guidance for accomplishing Joint Strategic Capabilities Plan or other Chairman of the Joint Chiefs of Staff (CJCS) taskings. CJCS approval of the strategic concept becomes the basis of the plan for development into an operation plan or operation plan in concept format. Also called **CSC**. (JP 5-0)

cipher — Any cryptographic system in which arbitrary symbols (or groups of symbols) represent units of plain text of regular length, usually single letters; units of plain text are rearranged; or both, in accordance with certain predetermined rules. See also **cryptosystem.**

circular error probable — An indicator of the delivery accuracy of a weapon system, used as a factor in determining probable damage to a target. It is the radius of a circle within which half of a missile's projectiles are expected to fall. Also called **CEP.** See also **delivery error; deviation; dispersion error; horizontal error.**

civic action — See **military civic action.**

civil administration — An administration established by a foreign government in (1) friendly territory, under an agreement with the government of the area concerned, to exercise certain authority normally the function of the local government; or (2) hostile territory, occupied by United States forces, where a foreign government exercises executive, legislative, and judicial authority until an indigenous civil government can be established. Also called **CA administration.** (JP 3-57)

civil affairs — Designated Active and Reserve component forces and units organized, trained, and equipped specifically to conduct civil affairs activities and to support civil-military operations. See also civil affairs activities; civil-military operations. Also called **CA.** (JP 3-57)

civil affairs activities — Activities performed or supported by civil affairs that (1) enhance the relationship between military forces and civil authorities in areas where military forces are present; and (2) involve application of civil affairs functional specialty skills, in areas normally the responsibility of civil government, to enhance conduct of civil-military operations. See also **civil affairs; civil-military operations.** (JP 3-57)

civil affairs agreement — An agreement that governs the relationship between allied armed forces located in a friendly country and the civil authorities and people of that country. See also **civil affairs.**

civil augmentation program — Standing, long-term contacts designed to augment Service logistic capabilities with contract support in both preplanned and short notice contingencies. Examples include US Army Logistics Civilian Augmentation Program, US Air Force Contract Augmentation Program, and US Navy Construction Capabilities Contract. See also **contingency.** (JP 4-07)

civil censorship — Censorship of civilian communications, such as messages, printed matter, and films entering, leaving, or circulating within areas or territories occupied or controlled by armed forces. See also **censorship.**

civil damage assessment — An appraisal of damage to a nation's population, industry, utilities, communications, transportation, food, water, and medical resources to support planning for national recovery. See also **damage assessment.**

civil defense — All those activities and measures designed or undertaken to: a. minimize the effects upon the civilian population caused or which would be caused by an enemy attack on the United States; b. deal with the immediate emergency conditions that would be created by any such attack; and c. effectuate emergency repairs to, or the emergency restoration of, vital utilities and facilities destroyed or damaged by any such attack.

civil defense emergency — See **domestic emergencies.**

civil defense intelligence — The product resulting from the collection and evaluation of information concerning all aspects of the situation in the United States and its territories that are potential or actual targets of any enemy attack including, in the preattack phase, the emergency measures taken and estimates of the civil populations' preparedness. In the event of an actual attack, the information will include a description of conditions in the affected area with emphasis on the extent of damage, fallout levels, and casualty and resource estimates. The product is required by civil and military authorities for use in the formulation of decisions, the conduct of operations, and the continuation of the planning processes.

civil disturbance — **(*)** Group acts of violence and disorder prejudicial to public law and order. See also **domestic emergencies.**

civil disturbance readiness conditions — Required conditions of preparedness to be attained by military forces in preparation for deployment to an objective area in response to an actual or threatened civil disturbance.

civil engineering — Those combat support and combat service support activities that identify, design, construct, lease, or provide facilities, and which operate, maintain, and perform war damage repair and other engineering functions in support of military operations. See also **civil engineering support plan; combat service support; combat support.** (JP 4-04)

civil engineering support plan — An appendix to the logistics annex or separate annex of an operation plan that identifies the minimum essential engineering services and construction requirements required to support the commitment of military forces. Also called **CESP.** See also **civil engineering; operation plan.** (JP 4-04)

civilian internee — 1. A civilian who is interned during armed conflict or occupation for security reasons or for protection or because he or she has committed an offense against the detaining power. 2. A term used to refer to persons interned and protected in accordance with the Geneva Convention Relative to the Protection of Civilian Persons in Time of War, 12 August 1949 (Geneva Convention). Also called **CI.** See also **prisoner of war.**

civilian internee camp — An installation established for the internment and administration of civilian internees.

civil-military operations — The activities of a commander that establish, maintain, influence, or exploit relations between military forces, governmental and nongovernmental civilian organizations and authorities, and the civilian populace in a friendly, neutral, or hostile operational area in order to facilitate military operations, to consolidate and achieve operational US objectives. Civil-military operations may include performance by military forces of activities and functions normally the responsibility of the local, regional, or national government. These activities may occur prior to, during, or subsequent to other military actions. They may also occur, if directed, in the absence of other military operations. Civil-military operations may be performed by designated civil affairs, by other military forces, or by a combination of civil affairs and other forces. Also called **CMO.** See also **civil affairs; operation.** (JP 3-57)

civil-military operations center — An ad hoc organization, normally established by the geographic combatant commander or subordinate joint force commander, to assist in the coordination of activities of engaged military forces, and other United States Government agencies, nongovernmental organizations, and regional and international organizations. There is no established structure, and its size and composition are situation dependent. Also called **CMOC.** See also **civil affairs activities; civil-military operations; operation.** (JP 3-57)

civil nuclear power — A nation that has the potential to employ nuclear technology for development of nuclear weapons but has deliberately decided against doing so.

civil requirements — The necessary production and distribution of all types of services, supplies, and equipment during periods of armed conflict or occupation to ensure the productive efficiency of the civilian economy and to provide to civilians the treatment and protection to which they are entitled under customary and conventional international law.

civil reserve air fleet — A program in which the Department of Defense uses aircraft owned by a US entity or citizen. The aircraft are allocated by the Department of Transportation to augment the military airlift capability of the Department of Defense. These aircraft are allocated, in accordance with DOD requirements, to segments, according to their capabilities, such as International Long Range and Short Range Cargo and Passenger sections, National (Domestic and Alaskan sections) and Aeromedical Evacuation and other segments as may be mutually agreed upon by the Department of Defense and the Department of Transportation. Also called **CRAF.** See also **reserve.**

civil transportation — The movement of persons, property, or mail by civil facilities, and the resources (including storage, except that for agricultural and petroleum products) necessary to accomplish the movement. (Excludes transportation operated or controlled by the military as well as petroleum and gas pipelines.)

clandestine operation — An operation sponsored or conducted by governmental departments or agencies in such a way as to assure secrecy or concealment. A clandestine operation differs from a covert operation in that emphasis is placed on concealment of the operation rather than on concealment of the identity of the sponsor. In special operations, an activity may be both covert and clandestine and may focus equally on operational considerations and intelligence-related activities. See also **covert operation; overt operation.** (JP 3-05.3)

classes of supply — There are ten categories into which supplies are grouped in order to facilitate supply management and planning. I. Rations and gratuitous issue of health, morale, and welfare items. II. Clothing, individual equipment, tentage, tool sets, and administrative and housekeeping supplies and equipment. III. Petroleum, oils, and lubricants. IV. Construction materiels. V. Ammunition. VI. Personal demand items. VII. Major end items, including tanks, helicopters, and radios. VIII. Medical. IX. Repair parts and components for equipment maintenance. X. Nonstandard items to support nonmilitary programs such as agriculture and economic development. See also **ammunition; petroleum, oils, and lubricants.** (JP 4-09)

classification — The determination that official information requires, in the interests of national security, a specific degree of protection against unauthorized disclosure, coupled with a designation signifying that such a determination has been made. See also **security classification.**

classification of bridges and vehicles — See **military load classification.**

classified contract — Any contract that requires or will require access to classified information by the contractor or the employees in the performance of the contract. (A contract may be classified even though the contract document itself is not classified.)

classified information — Official information that has been determined to require, in the interests of national security, protection against unauthorized disclosure and which has been so designated.

classified matter — (*) Official information or matter in any form or of any nature which requires protection in the interests of national security. See also **unclassified matter.**

clean aircraft — 1. An aircraft in flight configuration (versus landing configuration); i.e., landing gear and flaps retracted, etc. 2. An aircraft that does not have external stores.

cleansing station — See **decontamination station.**

clear — 1. To approve or authorize, or to obtain approval or authorization for: a. a person or persons with regard to their actions, movements, duties, etc.; b. an object or group of objects, as equipment or supplies, with regard to quality, quantity, purpose, movement, disposition, etc.; and c. a request, with regard to correctness of form, validity, etc. 2. To give one or more aircraft a clearance. 3. To give a person a security clearance. 4. To fly over an obstacle without touching it. 5. To pass a designated point, line, or object. The end of a column must pass the designated feature before the latter is cleared. 6. a. To operate a gun so as to unload it or make certain no ammunition remains; and b. to free a gun of stoppages. 7. To clear an engine; to open the throttle of an idling engine to free it from carbon. 8. To clear the air to gain either temporary or permanent air superiority or control in a given sector.

clearance capacity — An estimate expressed in terms of measurement or weight tons per day of the cargo that may be transported inland from a beach or port over the available means of inland communication, including roads, railroads, and inland waterways. The estimate is based on an evaluation of the physical characteristics of the transportation facilities in the area. See also **beach capacity; port capacity.**

clearance rate — **(*)** The area which would be cleared per unit time with a stated minimum percentage clearance, using specific minehunting and/or minesweeping procedures.

clearing operation — An operation designed to clear or neutralize all mines and obstacles from a route or area. (JP 3-15)

clock code position — The position of a target in relation to an aircraft or ship with dead-ahead position considered as 12 o'clock.

close air support — Air action by fixed- and rotary-wing aircraft against hostile targets that are in close proximity to friendly forces and that require detailed integration of each air mission with the fire and movement of those forces. Also called **CAS.** See also **air interdiction; air support; immediate mission request; preplanned mission request.** (JP 3-0)

close-controlled air interception — **(*)** An interception in which the interceptor is continuously controlled to a position from which the target is within visual range or radar contact. See also **air interception.**

closed area — **(*)** A designated area in or over which passage of any kind is prohibited. See also **prohibited area.**

close support — **(*)** That action of the supporting force against targets or objectives which are sufficiently near the supported force as to require detailed integration or coordination of the supporting action with the fire, movement, or other actions of the supported force. See also **direct support; general support; mutual support; support.**

close support area — Those parts of the ocean operating areas nearest to, but not necessarily in, the objective area. They are assigned to naval support carrier battle groups, surface action groups, surface action units, and certain logistic combat service support elements. (JP 3-02)

closure — In transportation, the process of a unit arriving at a specified location. It begins when the first element arrives at a designated location, e.g., port of entry and/or port of departure, intermediate stops, or final destination, and ends when the last element does likewise. For the purposes of studies and command post exercises, a unit is considered essentially closed after 95 percent of its movement requirements for personnel and equipment are completed.

closure minefield — **(*)** In naval mine warfare, a minefield which is planned to present such a threat that waterborne shipping is prevented from moving.

closure shortfall — The specified movement requirement or portion thereof that did not meet scheduling criteria and/or movement dates.

cloud amount — **(*)** The proportion of sky obscured by cloud, expressed as a fraction of sky covered.

cloud chamber effect — See **condensation cloud.**

cloud top height — The maximal altitude to which a nuclear mushroom cloud rises.

cluster bomb unit — **(*)** An aircraft store composed of a dispenser and submunitions. Also called **CBU.**

clutter — Permanent echoes, cloud, or other atmospheric echo on radar scope; as contact has entered scope clutter. See also **radar clutter.**

coalition — An ad hoc arrangement between two or more nations for common action. See also **alliance; multinational.** (JP 5-0)

coalition action — Multinational action outside the bounds of established alliances, usually for single occasions or longer cooperation in a narrow sector of common interest. See also **alliance; coalition; multinational operations.** (JP 5-0)

coalition coordination cell — An ad hoc unified or sub-unified staff organization composed of staff elements required to integrate coalition contributions (forces and capabilities) into a contingency operation. Also called **CCC.**

coarse mine — **(*)** In naval mine warfare, a relatively insensitive influence mine.

coassembly — With respect to exports, a cooperative arrangement (e.g., US Government or company with foreign government or company) by which finished parts, components, assemblies, or subassemblies are provided to an eligible foreign government, international organization, or commercial producer for the assembly of an end-item or system. This is normally accomplished under the provisions of a manufacturing license agreement per the US International Traffic in Arms Regulation and could involve the implementation of a government-to-government memorandum of understanding.

coastal convoy — **(*)** A convoy whose voyage lies in general on the continental shelf and in coastal waters.

coastal frontier — A geographic division of a coastal area, established for organization and command purposes in order to ensure the effective coordination of military forces employed in military operations within the coastal frontier area.

coastal refraction — **(*)** The change of the direction of travel of a radio ground wave as it passes from land to sea or from sea to land. Also called **land effect or shoreline effect.**

coastal sea control — The employment of forces to ensure the unimpeded use of an offshore coastal area by friendly forces and, as appropriate, to deny the use of the area to enemy forces. (JP 3-10)

code — 1. Any system of communication in which arbitrary groups of symbols represent units of plain text of varying length. Codes may be used for brevity or for security. 2. A cryptosystem in which the cryptographic equivalents (usually called "code groups"), typically consisting of letters or digits (or both) in otherwise meaningless combinations, are substituted for plain text elements which are primarily words, phrases, or sentences. See also **cryptosystem.**

code word — **(*)** 1. A word that has been assigned a classification and a classified meaning to safeguard intentions and information regarding a classified plan or operation. 2. A cryptonym used to identify sensitive intelligence data.

cold war — A state of international tension wherein political, economic, technological, sociological, psychological, paramilitary, and military measures short of overt armed conflict involving regular military forces are employed to achieve national objectives.

collaborative purchase — A method of purchase whereby, in buying similar commodities, buyers for two or more departments exchange information concerning planned purchases in order to minimize competition between them for commodities in the same market.

collapse depth — **(*)** The design depth, referenced to the axis of the pressure hull, beyond which the hull structure or hull penetrations are presumed to suffer catastrophic failure to the point of total collapse.

collate — 1. The grouping together of related items to provide a record of events and facilitate further processing. 2. To compare critically two or more items or documents concerning the same general subject; normally accomplished in the processing and exploitation phase in the intelligence cycle. See also **intelligence cycle.** (JP 2-0)

collateral damage — Unintentional or incidental injury or damage to persons or objects that would not be lawful military targets in the circumstances ruling at the time. Such damage is not unlawful so long as it is not excessive in light of the overall military advantage anticipated from the attack. (JP 3-60)

collateral mission — A mission other than those for which a force is primarily organized, trained, and equipped, that the force can accomplish by virtue of the inherent capabilities of that force. (JP 3-05.3)

collection — See **intelligence cycle, Subpart b.**

collection (acquisition) — The obtaining of information in any manner, including direct observation, liaison with official agencies, or solicitation from official, unofficial, or public sources.

collection agency — Any individual, organization, or unit that has access to sources of information and the capability of collecting information from them. See also **agency.**

collection asset — A collection system, platform, or capability that is supporting, assigned, or attached to a particular commander. See also **capability; collection.** (JP 2-01)

collection coordination facility line number — An arbitrary number assigned to contingency intelligence reconnaissance objectives by the Defense Intelligence Agency collection coordination facility to facilitate all-source collection.

collection management — In intelligence usage, the process of converting intelligence requirements into collection requirements, establishing priorities, tasking or coordinating with appropriate collection sources or agencies, monitoring results, and retasking, as required. See also **collection; collection requirement; collection requirements management; intelligence; intelligence cycle.** (JP 2-0)

collection management authority — Constitutes the authority to establish, prioritize, and validate theater collection requirements, establish sensor tasking guidance, and develop theater collection plans. Also called **CMA.** See also **collection manager; collection plan; collection requirement.** (JP 2-01)

collection manager — An individual with responsibility for the timely and efficient tasking of organic collection resources and the development of requirements for theater and national assets that could satisfy specific information needs in support of the mission. Also called **CM.** See also **collection; collection management authority.** (JP 2-01)

collection operations management — The authoritative direction, scheduling, and control of specific collection operations and associated processing, exploitation, and reporting resources. Also called **COM.** See also **collection management; collection requirements management.** (JP 2-0)

collection plan — **(*)** A plan for collecting information from all available sources to meet intelligence requirements and for transforming those requirements into orders and requests to appropriate agencies. See also **information; information requirements; intelligence cycle.**

collection planning — A continuous process that coordinates and integrates the efforts of all collection units and agencies. See also **collection.** (JP 2-0)

collection point — A point designated for the assembly of personnel casualties, stragglers, disabled materiel, salvage, etc., for further movement to collecting stations or rear installations.

collection requirement — An established intelligence need considered in the allocation of intelligence resources to fulfill the essential elements of information and other intelligence needs of a commander.

collection requirements management — The authoritative development and control of collection, processing, exploitation, and/or reporting requirements that normally result in either the direct tasking of assets over which the collection manager has authority, or the generation of tasking requests to collection management authorities at a higher, lower, or lateral echelon to accomplish the collection mission. Also called **CRM.** See also **collection; collection management; collection operations management.** (JP 2-0)

collection resource — A collection system, platform, or capability that is not assigned or attached to a specific unit or echelon which must be requested and coordinated through the chain of command. See also **collection management.** (JP 2-01)

collective call sign — **(*)** Any call sign which represents two or more facilities, commands, authorities, or units. The collective call sign for any of these includes the commander thereof and all subordinate commanders therein. See also **call sign.**

collective nuclear, biological, and chemical protection — **(*)** Protection provided to a group of individuals in a nuclear, biological, and chemical environment which permits relaxation of individual nuclear, biological, and chemical protection.

collective self-defense — Collective self-defense is the act of defending other designated non-US forces. Only the National Command Authorities may authorize US forces to exercise the right of collective self-defense.

collocation — **(*)** The physical placement of two or more detachments, units, organizations, or facilities at a specifically defined location.

colored beach — That portion of usable coastline sufficient for the assault landing of a regimental landing team or similar sized unit. In the event that the landing force consists of a single battalion landing team, a colored beach will be used and no further subdivision of the beach is required. See also **numbered beach.** (JP 3-02)

column formation — **(*)** A formation in which elements are placed one behind the other.

column gap — **(*)** The space between two consecutive elements proceeding on the same route. It can be calculated in units of length or in units of time measured from the rear of one element to the front of the following element.

column length — **(*)** The length of the roadway occupied by a column or a convoy in movement. See also **road space.**

combat air patrol — **(*)** An aircraft patrol provided over an objective area, the force protected, the critical area of a combat zone, or in an air defense area, for the purpose of intercepting and destroying hostile aircraft before they reach their targets. Also called **CAP.** See also **airborne alert; barrier combat air patrol; patrol; rescue combat air patrol.**

combat airspace control — See **airspace control in the combat zone.** (JP 3-52)

combat and operational stress — The expected and predictable emotional, intellectual, physical, and/or behavioral reactions of Service members who have been exposed to stressful events in war or military operations other than war. Combat stress reactions vary in quality and severity as a function of operational conditions, such as intensity, duration, rules of engagement, leadership, effective communication, unit morale, unit cohesion, and perceived importance of the mission. (JP 4-02)

combatant command — A unified or specified command with a broad continuing mission under a single commander established and so designated by the President, through the Secretary of Defense and with the advice and assistance of the Chairman of the Joint Chiefs of Staff. Combatant commands typically have geographic or functional responsibilities. See also **specified command; unified command.** (JP 5-0)

combatant command (command authority) — Nontransferable command authority established by title 10 ("Armed Forces"), United States Code, section 164, exercised only by commanders of unified or specified combatant commands unless otherwise directed by the President or the Secretary of Defense. Combatant command (command authority) cannot be delegated and is the authority of a combatant commander to perform those functions of command over assigned forces involving organizing and employing commands and forces, assigning tasks, designating objectives, and giving authoritative direction over all aspects of military operations, joint training, and logistics necessary to accomplish the missions assigned to the command. Combatant command (command authority) should be exercised through the commanders of subordinate organizations. Normally this authority is exercised through subordinate joint force commanders and Service and/or functional component commanders. Combatant command (command authority) provides full authority to organize and employ commands and forces as the combatant commander considers necessary to accomplish assigned missions. Operational control is inherent in combatant command (command authority). Also called **COCOM.** See also **combatant command; combatant commander; operational control; tactical control.** (JP 0-2)

combatant commander — A commander in chief of one of the unified or specified combatant commands established by the President. Also called **CINC.** See also **combatant command.** (JP 0-2)

combat area — A restricted area (air, land, or sea) that is established to prevent or minimize mutual interference between friendly forces engaged in combat operations. See also **combat zone.**

combat assessment — The determination of the overall effectiveness of force employment during military operations. Combat assessment is composed of three major components: (a) battle damage assessment; (b) munitions effectiveness assessment; and (c) reattack recommendation. Also called **CA.** See also **battle damage assessment; munitions**

effectiveness assessment; reattack recommendation. (JP 3-60)

combat camera — Visual information documentation covering air, sea, and ground actions of the Armed Forces of the United States in combat or combat support operations and in related peacetime training activities such as exercises, war games, and operations. Also called **COMCAM.** See also **visual information; visual information documentation.**

combat cargo officer — An embarkation officer assigned to major amphibious ships or naval staffs, functioning primarily as an adviser to and representative of the naval commander in matters pertaining to embarkation and debarkation of troops and their supplies and equipment. Also called **CCO.** See also **embarkation officer.**

combat chart — A special naval chart, at a scale of 1:50,000, designed for naval surface fire support and close air support during coastal or amphibious operations and showing detailed hydrography and topography in the coastal belt. See also **amphibious chart**.

combat control team — A small task organized team of Air Force parachute and combat diver qualified personnel trained and equipped to rapidly establish and control drop, landing, and extraction zone air traffic in austere or hostile conditions. They survey and establish terminal airheads as well as provide guidance to aircraft for airlift operations. They provide command and control, and conduct reconnaissance, surveillance, and survey assessments of potential objective airfields or assault zones. They also can perform limited weather observations and removal of obstacles or unexploded ordinance with demolitions. Also called **CCT.** (JP 3-17)

combat engineering — Those engineering tasks that assist the tactical and/or operational commander to "shape" the battlespace by enhancing mobility, creating the space and time necessary to generate mass and speed while protecting the force and denying mobility and key terrain to the enemy. These tasks include breaching, bridging, and emplacement of obstacles to deny mobility to the enemy. (JP 3-34)

combat forces — Those forces whose primary missions are to participate in combat. See also **operating forces.**

combat information — Unevaluated data, gathered by or provided directly to the tactical commander which, due to its highly perishable nature or the criticality of the situation, cannot be processed into tactical intelligence in time to satisfy the user's tactical intelligence requirements. See also **information.**

combat information center — (*) The agency in a ship or aircraft manned and equipped to collect, display, evaluate, and disseminate tactical information for the use of the embarked flag officer, commanding officer, and certain control agencies. Certain control, assistance, and coordination functions may be delegated by command to the combat information center. Also called **action information center; CIC.** See also **air defense control center.**

combat intelligence — That knowledge of the enemy, weather, and geographical features required by a commander in the planning and conduct of combat operations.

combat loading — (*) The arrangement of personnel and the stowage of equipment and supplies in a manner designed to conform to the anticipated tactical operation of the organization embarked. Each individual item is stowed so that it can be unloaded at the required time. See also **loading.**

combat power — **(*)** The total means of destructive and/or disruptive force which a military unit/formation can apply against the opponent at a given time.

combat readiness — Synonymous with operational readiness, with respect to missions or functions performed in combat.

combat search and rescue — A specific task performed by rescue forces to effect the recovery of distressed personnel during war or military operations other than war. Also called **CSAR.** See also **search and rescue.** (JP 3-50.2)

combat search and rescue mission coordinator — The designated person or organization selected to direct and coordinate support for a specific combat search and rescue mission. Also called **CSAR mission coordinator.** See also **combat search and rescue; component search and rescue controller; search and rescue; search and rescue mission coordinator.** (JP 3-50.2)

combat search and rescue task force — All forces committed to a specific combat search and rescue operation to search for, locate, identify, and recover isolated personnel during wartime or contingency operations. This includes those elements assigned to provide command and control and protect the recovery vehicle from enemy air or ground attack. Also called **CSARTF.** See also **combat search and rescue; search; search and rescue.** (JP 3-50.21)

combat service support — The essential capabilities, functions, activities, and tasks necessary to sustain all elements of operating forces in theater at all levels of war. Within the national and theater logistic systems, it includes but is not limited to that support rendered by service forces in ensuring the aspects of supply, maintenance, transportation, health services, and other services required by aviation and ground combat troops to permit those units to accomplish their missions in combat. Combat service support encompasses those activities at all levels of war that produce sustainment to all operating forces on the battlefield. Also called **CSS.** See also **combat support.** (JP 4-0)

combat service support area — An area ashore that is organized to contain the necessary supplies, equipment, installations, and elements to provide the landing force with combat service support throughout the operation. Also called **CSSA.** (JP 3-02)

combat service support element — The core element of a Marine air-ground task force (MAGTF) that is task-organized to provide the combat service support necessary to accomplish the MAGTF mission. The combat service support element varies in size from a small detachment to one or more force service support groups. It provides supply, maintenance, transportation, general engineering, health services, and a variety of other services to the MAGTF. The combat service support element itself is not a formal command. Also called **CSSE.** See also **aviation combat element; command element; ground combat element; Marine air-ground task force; Marine expeditionary force; Marine expeditionary force (forward); Marine expeditionary unit; special purpose Marine air-ground task force; task force.**

combat service support elements — Those elements whose primary missions are to provide service support to combat forces and which are a part, or prepared to become a part, of a theater, command, or task force formed for combat operations. See also **operating forces; service troops; troops.**

combat support — Fire support and operational assistance provided to combat elements. Also called **CS.** See also **combat service support.** (JP 4-0)

combat support elements — Those elements whose primary missions are to provide combat support to the combat forces and which are a part, or prepared to become a part, of a theater, command, or task force formed for combat operations. See also **operating forces.**

combat support troops — Those units or organizations whose primary mission is to furnish operational assistance for the combat elements. See also **troops.**

combat surveillance — A continuous, all-weather, day-and-night, systematic watch over the battle area in order to provide timely information for tactical combat operations.

combat surveillance radar — Radar with the normal function of maintaining continuous watch over a combat area.

combat survival — **(*)** Those measures to be taken by Service personnel when involuntarily separated from friendly forces in combat, including procedures relating to individual survival, evasion, escape, and conduct after capture.

combatting terrorism — Actions, including antiterrorism (defensive measures taken to reduce vulnerability to terrorist acts) and counterterrorism (offensive measures taken to prevent, deter, and respond to terrorism), taken to oppose terrorism throughout the entire threat spectrum. Also called **CBT.** See also **antiterrorism; counterterrorism.**

combat vehicle — A vehicle, with or without armor, designed for a specific fighting function. Armor protection or armament mounted as supplemental equipment on noncombat vehicles will not change the classification of such vehicles to combat vehicles.

combat visual information support center — A visual information support facility established at a base of operations during war or military operations other than war to provide limited visual information support to the base and its supported elements. Also called **CVISC.**

combat zone — 1. That area required by combat forces for the conduct of operations. 2. The territory forward of the Army rear area boundary. See also **combat area; communications zone.**

combination influence mine — **(*)** A mine designed to actuate only when two or more different influences are received either simultaneously or in a predetermined order. Also called **combined influence mine.** See also **mine.**

combination mission/level of effort-oriented items — Items for which requirement computations are based on the criteria used for both level of effort-oriented and mission-oriented items.

combined — Between two or more forces or agencies of two or more allies. (When all allies or services are not involved, the participating nations and services shall be identified, e.g., combined navies.) See also **joint.**

combined airspeed indicator — **(*)** An instrument which displays both indicated airspeed and mach number.

combined arms team — The full integration and application of two or more arms or elements of one Military Service into an operation. (JP 3-18)

combined force — A military force composed of elements of two or more allied nations. See also **force(s).**

combined influence mine — See **combination influence mine.**

combined operation — **(*)** An operation conducted by forces of two or more Allied nations acting together for the accomplishment of a single mission.

combustor — **(*)** A name generally assigned to the combination of flame holder or stabilizer, igniter, combustion chamber, and injection system of a ramjet or gas turbine.

command — 1. The authority that a commander in the Armed Forces lawfully exercises over subordinates by virtue of rank or assignment. Command includes the authority and responsibility for effectively using available resources and for planning the employment of, organizing, directing, coordinating, and controlling military forces for the accomplishment of assigned missions. It also includes responsibility for health, welfare, morale, and discipline of assigned personnel. 2. An order given by a commander; that is, the will of the commander expressed for the purpose of bringing about a particular action. 3. A unit or units, an organization, or an area under the command of one individual. Also called **CMD.** See also **area command; base command; combatant command; combatant command (command authority).** (JP 0-2)

command and control — The exercise of authority and direction by a properly designated commander over assigned and attached forces in the accomplishment of the mission. Command and control functions are performed through an arrangement of personnel, equipment, communications, facilities, and procedures employed by a commander in planning, directing, coordinating, and controlling forces and operations in the accomplishment of the mission. Also called **C2.** (JP 0-2)

command and control system — The facilities, equipment, communications, procedures, and personnel essential to a commander for planning, directing, and controlling operations of assigned forces pursuant to the missions assigned.

command and control warfare — The integrated use of operations security, military deception, psychological operations, electronic warfare, and physical destruction, mutually supported by intelligence, to deny information to, influence, degrade, or destroy adversary command and control capabilities, while protecting friendly command and control capabilities against such actions. Command and control warfare is an application of information operations in military operations. Also called **C2W.** C2W is both offensive and defensive: a. **C2-attack.** Prevent effective C2 of adversary forces by denying information to, influencing, degrading, or destroying the adversary C2 system. b. **C2-protect.** Maintain effective command and control of own forces by turning to friendly advantage or negating adversary efforts to deny information to, influence, degrade, or destroy the friendly C2 system. See also **command and control; electronic warfare; information operations; intelligence; military deception; operations security; psychological operations.** (JP 3-13.1)

command axis — (*) A line along which a headquarters will move.

command center — A facility from which a commander and his or her representatives direct operations and control forces. It is organized to gather, process, analyze, display, and disseminate planning and operational data and perform other related tasks. Also called **CC.**

command channel — See **chain of command.**

command chaplain — The senior chaplain assigned to or designated by a commander of a staff, command, or unit. See also **command chaplain of the combatant command; lay leader or lay reader; religious ministry support; religious ministry support plan; religious ministry support team; Service component command chaplain.** (JP 1-05)

command chaplain of the combatant command — The senior chaplain assigned to the staff of, or designated by, the combatant commander to provide advice on religion, ethics, and morale of assigned personnel and to coordinate religious ministries within the commander's area of responsibility. The command chaplain of the combatant command may be supported by a staff of chaplains and enlisted religious support personnel. See also **command chaplain; lay leader or lay reader; religious ministry support; religious ministry support plan; religious ministry support team; Service component command chaplain.** (JP 1-05)

command, control, communications, and computer systems — Integrated systems of doctrine, procedures, organizational structures, personnel, equipment, facilities, and communications designed to support a commander's exercise of command and control across the range of military operations. Also called **C4 systems**. See also **command and control; tactical command, control, communications, and computer system(s).** (JP 6-0)

command controlled stocks — (*) Stocks which are placed at the disposal of a designated NATO commander in order to provide him with a flexibility with which to influence the battle logistically. "Placed at the disposal of" implies responsibility for storage, maintenance, accounting, rotation or turnover, physical security, and subsequent transportation to a particular battle area.

command destruct signal — (*) A signal used to operate intentionally the destruction signal in a missile.

command detonated mine — (*) A mine detonated by remotely controlled means.

command ejection system — See **ejection systems.**

command element — The core element of a Marine air-ground task force (MAGTF) that is the headquarters. The command element is composed of the commander, general or executive and special staff sections, headquarters section, and requisite communications support, intelligence, and reconnaissance forces necessary to accomplish the MAGTF mission. The command element provides command and control, intelligence, and other support essential for effective planning and execution of operations by the other elements of the MAGTF. The command element varies in size and composition. Also called **CE.** See also **aviation combat element; combat service support element; ground combat element; Marine air-ground task force; Marine expeditionary force; Marine expeditionary force (forward); Marine**

expeditionary unit; special purpose Marine air-ground task force; task force.

commander, amphibious task force — The Navy officer designated in the order initiating the amphibious operation as the commander of the amphibious task force. Also called **CATF.** See also **amphibious operation; amphibious task force; commander, landing force.** (JP 3-02)

commander, landing force — The officer designated in the order initiating the amphibious operation as the commander of the landing force for an amphibious operation. Also called **CLF.** See also **amphibious operation; commander, amphibious task force; landing force.** (JP 3-02)

commander's concept — See **concept of operations.**

commander's critical information requirements — A comprehensive list of information requirements identified by the commander as being critical in facilitating timely information management and the decisionmaking process that affect successful mission accomplishment. The two key subcomponents are critical friendly force information and priority intelligence requirements. Also called **CCIR.** See also **critical information; information; information requirements; intelligence; priority intelligence requirements.** (JP 5-00.2)

commander's estimate of the situation — A logical process of reasoning by which a commander considers all the circumstances affecting the military situation and arrives at a decision as to a course of action to be taken in order to accomplish the mission. A commander's estimate that considers a military situation so far in the future as to require major assumptions is called a commander's long-range estimate of the situation.

commander's intent — A concise expression of the purpose of the operation and the desired end state that serves as the initial impetus for the planning process. It may also include the commander's assessment of the adversary commander's intent and an assessment of where and how much risk is acceptable during the operation. See also **assessment; end state.** (JP 5-00.1)

command guidance — **(*)** A guidance system wherein intelligence transmitted to the missile from an outside source causes the missile to traverse a directed flight path.

command information — Communication by a military organization with Service members, civilian employees, retirees, and family members of the organization that creates an awareness of the organization's goals, informs them of significant developments affecting them and the organization, increases their effectiveness as ambassadors of the organization, and keeps them informed about what is going on in the organization. Also called **internal information.** See also **command; information; public affairs.** (JP 3-61)

commanding officer of troops — On a ship that has embarked units, a designated officer (usually the senior embarking unit commander) who is responsible for the administration, discipline, and training of all embarked units. Also called **COT.** (JP 3-02.2)

command net — **(*)** A communications network which connects an echelon of command with some or all of its subordinate echelons for the purpose of command and control.

command post — **(*)** A unit's or subunit's headquarters where the commander and the staff perform their activities. In combat, a unit's or subunit's headquarters is often divided into echelons; the echelon in which the unit or subunit commander is located or from which such commander operates is called a command post. Also called **CP.**

command post exercise — An exercise in which the forces are simulated, involving the commander, the staff, and communications within and between headquarters. Also called **CPX.** See also **exercise; maneuver.**

command relationships — The interrelated responsibilities between commanders, as well as the operational authority exercised by commanders in the chain of command; defined further as combatant command (command authority), operational control, tactical control, or support. See also **chain of command; combatant command (command authority); command; operational control; support; tactical control.** (JP 0-2)

command select ejection system — See **ejection systems.**

command-sponsored dependent — A dependent entitled to travel to overseas commands at Government expense and endorsed by the appropriate military commander to be present in a dependent's status.

commercial items — Articles of supply readily available from established commercial distribution sources which the Department of Defense or inventory managers in the Military Services have designated to be obtained directly or indirectly from such sources.

commercial loading — See **administrative loading.**

commercial vehicle — A vehicle that has evolved in the commercial market to meet civilian requirements and which is selected from existing production lines for military use.

commission — 1. To put in or make ready for service or use, as to commission an aircraft or a ship. 2. A written order giving a person rank and authority as an officer in the armed forces. 3. The rank and the authority given by such an order. See also **constitute.**

commit — The process of committing one or more air interceptors or surface-to-air missiles for interception against a target track.

commodity loading — **(*)** A method of loading in which various types of cargoes are loaded together, such as ammunition, rations, or boxed vehicles, in order that each commodity can be discharged without disturbing the others. See also **combat loading; loading.**

commodity manager — An individual within the organization of an inventory control point or other such organization assigned management responsibility for homogeneous grouping of materiel items.

commonality — A quality that applies to materiel or systems: a. possessing like and interchangeable characteristics enabling each to be utilized, or operated and maintained, by personnel trained on the others without additional specialized training; b. having interchangeable repair parts and/or components; and c. applying to consumable items interchangeably equivalent without adjustment.

common control (artillery) — Horizontal and vertical map or chart location of points in the target area and position area, tied in with the horizontal and vertical control in use by two or more units. May be established by firing, survey, or combination of both, or by assumption. See also **control point; ground control.**

common infrastructure — (*) Infrastructure essential to the training of NATO forces or to the implementation of NATO operational plans which, owing to its degree of common use or interest and its compliance with criteria laid down from time to time by the North Atlantic Council, is commonly financed by NATO members. See also **infrastructure.**

common item — 1. Any item of materiel that is required for use by more than one activity. 2. Sometimes loosely used to denote any consumable item except repair parts or other technical items. 3. Any item of materiel that is procured for, owned by (Service stock), or used by any Military Department of the Department of Defense and is also required to be furnished to a recipient country under the grant-aid Military Assistance Program. 4. Readily available commercial items. 5. Items used by two or more Military Services of similar manufacture or fabrication that may vary between the Services as to color or shape (as vehicles or clothing). 6. Any part or component that is required in the assembly of two or more complete end-items.

common operating environment — Automation services that support the development of the common reusable software modules which enable interoperability across multiple combat support applications. This includes segmentation of common software modules from existing applications, integration of commercial products, development of a common architecture, and development of common tools for application developers. Also called **COE.** See also **global grid; node/command, control, communications, and computers node.** (JP 4-0)

common operational picture — A single identical display of relevant information shared by more than one command. A common operational picture facilitates collaborative planning and assists all echelons to achieve situational awareness. Also called **COP.** (JP 3-0)

common servicing — That function performed by one Military Service in support of another Military Service for which reimbursement is not required from the Service receiving support. See also **servicing.**

common supplies — Those supplies common to two or more Services.

common use — Services, materiel, or facilities provided by a Department of Defense agency or a Military Department on a common basis for two or more Department of Defense agencies, elements, or other organizations as directed.

common use alternatives — Systems, subsystems, devices, components, and materials, already developed or under development, that could be used to reduce the cost of new systems acquisition and support by reducing duplication of research and development effort and by limiting the addition of support base.

common-use container — Any Department of Defense-owned, -leased, or -controlled 20- or 40-foot International Organization for Standardization container managed by US Transportation Command as an element of the Department of Defense common-use container system. See also **component-owned container; Service-unique container.** (JP 4-01.7)

common-user airlift service — The airlift service provided on a common basis for all

Department of Defense agencies and, as authorized, for other agencies of the US Government.

common-user item — An item that is in common use by two or more nations or services of a nation.

common-user logistics — Materiel or service support shared with or provided by two or more Services, Department of Defense (DOD) agencies, or multinational partners to another Service, DOD agency, non-DOD agency, and/or multinational partner in an operation. Common-user logistics is usually restricted to a particular type of supply and/or service and may be further restricted to specific unit(s) or types of units, specific times, missions, and/or geographic areas. Also called **CUL.** See also **common use.** (JP 4-07)

common-user military land transportation — Point-to-point land transportation service operated by a single Service for common use by two or more Services.

common-user network — A system of circuits or channels allocated to furnish communication paths between switching centers to provide communication service on a common basis to all connected stations or subscribers. It is sometimes described as a general purpose network.

common-user ocean terminals — A military installation, part of a military installation, or a commercial facility operated under contract or arrangement by the Military Traffic Management Command that regularly provides for two or more Services terminal functions of receipt, transit storage or staging, processing, and loading and unloading of passengers or cargo aboard ships.

common-user sealift — The sealift services provided on a common basis for all Department of Defense agencies and, as authorized, for other agencies of the US Government. The Military Sealift Command, a transportation component command of the US Transportation Command, provides common-user sealift for which users reimburse the transportation accounts of the Transportation Working Capital Fund. See also **Military Sealift Command; transportation component command.** (JP 3-35)

common-user transportation — Transportation and transportation services provided on a common basis for two or more Department of Defense (DOD) agencies and, as authorized, non-DOD agencies. Common-user assets are under the combatant command (command authority) of the Commander in Chief, US Transportation Command, excluding Service-unique or theater-assigned transportation assets. See also **common use.**

communicate — To use any means or method to convey information of any kind from one person or place to another. (JP 6-02)

communication deception — Use of devices, operations, and techniques with the intent of confusing or misleading the user of a communications link or a navigation system.

communication operation instructions — See **signal operation instructions.**

communications center — **(*)** An agency charged with the responsibility for handling and controlling communications traffic. The center normally includes message center, transmitting, and receiving facilities. Also called **COMCEN.** See also **telecommunications center.**

communications intelligence — Technical information and intelligence derived from foreign communications by other than the intended recipients. Also called **COMINT.**

communications intelligence database — The aggregate of technical information and intelligence derived from the interception and analysis of foreign communications (excluding press, propaganda, and public broadcast) used in the direction and redirection of communications intelligence intercept, analysis, and reporting activities.

communications mark — An electronic indicator used for directing attention to a particular object or position of mutual interest within or between command and control systems.

communications net — **(*)** An organization of stations capable of direct communications on a common channel or frequency.

communications network — An organization of stations capable of intercommunications, but not necessarily on the same channel.

communications satellite — **(*)** An orbiting vehicle, which relays signals between communications stations. There are two types: a. **active communications satellite** — A satellite that receives, regenerates, and retransmits signals between stations; b. **passive communications satellite** — A satellite which reflects communications signals between stations. Also called **COMSAT.**

communications security — The protection resulting from all measures designed to deny unauthorized persons information of value that might be derived from the possession and study of telecommunications, or to mislead unauthorized persons in their interpretation of the results of such possession and study. Also called **COMSEC.** Communications security includes: cryptosecurity, transmission security, emission security, and physical security of communications security materials and information. a. **cryptosecurity** — The component of communications security that results from the provision of technically sound cryptosystems and their proper use. b. **transmission security** — The component of communications security that results from all measures designed to protect transmissions from interception and exploitation by means other than cryptanalysis. c. **emission security** — The component of communications security that results from all measures taken to deny unauthorized persons information of value that might be derived from intercept and analysis of compromising emanations from crypto-equipment and telecommunications systems. d. **physical security** — The component of communications security that results from all physical measures necessary to safeguard classified equipment, material, and documents from access thereto or observation thereof by unauthorized persons.

communications security equipment — Equipment designed to provide security to telecommunications by converting information to a form unintelligible to an unauthorized interceptor and by reconverting such information to its original form for authorized recipients, as well as equipment designed specifically to aid in (or as an essential element of) the conversion process. Communications security equipment is cryptoequipment, cryptoancillary equipment, cryptoproduction equipment, and authentication equipment.

communications security material — All documents, devices, equipment, or apparatus, including cryptomaterial, used in establishing or maintaining secure communications.

communications security monitoring — The act of listening to, copying, or recording transmissions of one's own circuits (or when specially agreed, e.g., in allied exercises, those of friendly forces) to provide material for communications security analysis in order to determine the degree of security being provided to those transmissions. In particular, the purposes include providing a basis for advising commanders on the security risks resulting from their transmissions, improving the security of communications, and planning and conducting manipulative communications deception operations.

communications terminal — Terminus of a communications circuit at which data can be either entered or received; located with the originator or ultimate addressee. Also called **CT.**

communications zone — Rear part of a theater of war or theater of operations (behind but contiguous to the combat zone) which contains the lines of communications, establishments for supply and evacuation, and other agencies required for the immediate support and maintenance of the field forces. Also called **COMMZ.** See also **combat zone; line of communications; rear area; theater of operations; theater of war.** (JP 4-0)

community relations — 1. The relationship between military and civilian communities. 2. Those public affairs programs that address issues of interest to the general public, business, academia, veterans, Service organizations, military-related associations, and other non-news media entities. These programs are usually associated with the interaction between US military installations and their surrounding or nearby civilian communities. Interaction with overseas non-news media civilians in a wartime or contingency theater will be handled by civil-military operations with public affairs support as required. See also **public affairs.**

community relations program — That command function that evaluates public attitudes, identifies the mission of a military organization with the public interest, and executes a program of action to earn public understanding and acceptance.

comparative cover — **(*)** Coverage of the same area or object taken at different times, to show any changes in details. See also **cover.**

compartmentation — 1. Establishment and management of an organization so that information about the personnel, internal organization, or activities of one component is made available to any other component only to the extent required for the performance of assigned duties. 2. Effects of relief and drainage upon avenues of approach so as to produce areas bounded on at least two sides by terrain features such as woods, ridges, or ravines that limit observation or observed fire into the area from points outside the area. (JP 3-05.3)

compass rose — **(*)** A graduated circle, usually marked in degrees, indicating directions and printed or inscribed on an appropriate medium.

complaint-type investigation — A counterintelligence investigation in which sabotage, espionage, treason, sedition, subversive activity, or disaffection is suspected.

complete round — A term applied to an assemblage of explosive and nonexplosive components designed to perform a specific function at the time and under the conditions desired. Examples of complete rounds of ammunition are: a. separate loading, consisting of a primer, propelling charge and, except for blank ammunition, a projectile and a fuze; b. fixed or semifixed, consisting of a primer, propelling charge, cartridge case, a projectile and, except when solid projectiles are used, a fuze; c. bomb, consisting of all component parts required to drop and function the bomb once; d. missile, consisting of a complete warhead section and a missile body with its associated components and propellants; and e. rocket, consisting of all components necessary to function.

complex contingency operations — Large-scale peace operations (or elements thereof) conducted by a combination of military forces and nonmilitary organizations that combine one or more of the elements of peace operations which include one or more elements of other types of operations such as foreign humanitarian assistance, nation assistance, support to insurgency, or support to counterinsurgency. See also **operation; peace operations.** (JP 3-57)

component — 1. One of the subordinate organizations that constitute a joint force. Normally a joint force is organized with a combination of Service and functional components. 2. In logistics, a part or combination of parts having a specific function, which can be installed or replaced only as an entity. Also called **COMP.** See also **functional component command; Service component command.** (JP 0-2)

component (materiel) — An assembly or any combination of parts, subassemblies, and assemblies mounted together in manufacture, assembly, maintenance, or rebuild.

component-owned container — A 20- or 40-foot International Organization for Standardization container procured and owned by a single Department of Defense component. May be either on an individual unit property book or contained within a component pool (e.g., Marine Corps maritime pre-positioning force containers). May be temporarily assigned to the Department of Defense common-use container system. Also called **Service-unique container.** See also **common-use container.** (JP 4-01.7)

component search and rescue controller — The designated search and rescue representative of a component commander of a joint force who is responsible for coordinating and controlling that component's search and rescue forces. See also **combat search and rescue; combat search and rescue mission coordinator; search and rescue; search and rescue mission coordinator.** (JP 3-50.2)

composite air photography — Air photographs made with a camera having one principal lens and two or more surrounding and oblique lenses. The several resulting photographs are corrected or transformed in printing to permit assembly as verticals with the same scale.

composite warfare commander — The officer in tactical command is normally the composite warfare commander. However the composite warfare commander concept allows an officer in tactical command to delegate tactical command to the composite warfare commander. The composite warfare commander wages combat operations to counter threats to the force and to maintain tactical sea control with assets assigned; while the officer in tactical command retains close control of power projection and strategic sea control operations. (JP 3-02)

composite wing — An Air Force wing that operates more than one type of weapon system. Some composite wings are built from the ground up and designed to put all resources required to meet a specific warfighting objective in a single wing under one commander at one location. Other composite wings need not be built from the ground up but combine different weapon systems operating at the same base into a single wing.

compression chamber — See **hyperbaric chamber.**

compromise — The known or suspected exposure of clandestine personnel, installations, or other assets or of classified information or material, to an unauthorized person.

compromised — **(*)** A term applied to classified matter, knowledge of which has, in whole or in part, passed to an unauthorized person or persons, or which has been subject to risk of such passing. See also **classified matter.**

computed air release point — **(*)** A computed air position where the first paratroop or cargo item is released to land on a specified impact point.

computer intrusion — An incident of unauthorized access to data or an automated information system.

computer intrusion detection — The process of identifying that a computer intrusion has been attempted, is occurring, or has occurred.

computer modeling — See **configuration management; independent review; validation; verification.**

computer network attack — Operations to disrupt, deny, degrade, or destroy information resident in computers and computer networks, or the computers and networks themselves. Electronic attack (EA) can be used against a computer, but it is not computer network attack (CNA). CNA relies on the data stream to execute the attack while EA relies on the electromagnetic spectrum. An example of the two operations is the following: sending a code or instruction to a central processing unit that causes the computer to short out the power supply is CNA. Using an electromagnetic pulse device to destroy a computer's electronics and causing the same result is EA. Also called **CNA.** See also **computer network defense; electronic attack; offensive information operations.** (JP 3-51)

computer network defense — Defensive measures to protect and defend information, computers, and networks from disruption, denial, degradation, or destruction. Also called **CND.** See also **computer network attack.** (JP 3-51)

computer security — The protection resulting from all measures to deny unauthorized access and exploitation of friendly computer systems. Also called **COMPUSEC.** See also **communications security.** (JP 6-02)

computer simulation — See **configuration management; independent review; validation; verification.**

concealment — **(*)** The protection from observation or surveillance. See also **camouflage; cover; screen.**

concentration area — **(*)** 1. An area, usually in the theater of operations, where troops are assembled before beginning active operations. 2. A limited area on which a volume of gunfire is placed within a limited time.

concept of intelligence operations — A verbal or graphic statement, in broad outline, of a Intelligence Directorate's assumptions or intent in regard to intelligence support of an operation or series of operations. The concept of intelligence operations, which complements the commander's concept of operations, is contained in the intelligence annex of operation plans. The concept of intelligence operations is designed to give an overall picture of intelligence support for joint operations. It is included primarily for additional clarity of purpose. See also **concept of operations.** (JP 2-0)

concept of logistic support — A verbal or graphic statement, in a broad outline, of how a commander intends to support and integrate with a concept of operations in an operation or campaign. (JP 4-0)

concept of operations — A verbal or graphic statement, in broad outline, of a commander's assumptions or intent in regard to an operation or series of operations. The concept of operations frequently is embodied in campaign plans and operation plans; in the latter case, particularly when the plans cover a series of connected operations to be carried out simultaneously or in succession. The concept is designed to give an overall picture of the operation. It is included primarily for additional clarity of purpose. Also called **commander's concept** or **CONOPS.**

concept plan — An operation plan in concept format. Also called **CONPLAN.** See also **operation plan.**

condensation cloud — A mist or fog of minute water droplets that temporarily surrounds the fireball following a nuclear (or atomic) detonation in a comparatively humid atmosphere. The expansion of the air in the negative phase of the blast wave from the explosion results in a lowering of the temperature, so that condensation of water vapor present in the air occurs and a cloud forms. The cloud is soon dispelled when the pressure returns to normal and the air warms up again. The phenomenon is similar to that used by physicists in the Wilson cloud chamber and is sometimes called the cloud chamber effect.

condensation trail — A visible cloud streak, usually brilliantly white in color, which trails behind a missile or other vehicle in flight under certain conditions. Also called **CONTRAIL.**

condition — Those variables of an operational environment or situation in which a unit, system, or individual is expected to operate and may affect performance. See also **joint mission-essential tasks; standard.**

conducting staff — See **exercise directing staff.**

configuration management — A discipline applying technical and administrative direction and surveillance to: (1) identify and document the functional and physical characteristics of a configuration item; (2) control changes to those characteristics; and (3) record and report changes to processing and implementation status.

confirmation of information (intelligence) — An information item is said to be confirmed when it is reported for the second time, preferably by another independent source whose reliability is considered when confirming information.

conflict — An armed struggle or clash between organized groups within a nation or between nations in order to achieve limited political or military objectives. Although regular forces are often involved, irregular forces frequently predominate. Conflict often is protracted, confined to a restricted geographic area, and constrained in weaponry and level of violence. Within this state, military power in response to threats may be exercised in an indirect manner while supportive of other instruments of national power. Limited objectives may be achieved by the short, focused, and direct application of force. (JP 3-0)

confusion agent — An individual who is dispatched by the sponsor for the primary purpose of confounding the intelligence or counterintelligence apparatus of another country rather than for the purpose of collecting and transmitting information.

confusion reflector — **(*)** A reflector of electromagnetic radiations used to create echoes for confusion purposes. Radar confusion reflectors include such devices as chaff, rope, and corner reflectors.

connecting route — **(*)** A route connecting axial and/or lateral routes. See also **route.**

connectivity — The ability to exchange information by electronic means. (JP 3-18)

consecutive voyage charter — A contract by which a commercial ship is chartered by the Military Sealift Command for a series of specified voyages. (JP 3-02.2)

consequence management — Those measures taken to protect public health and safety, restore essential government services, and provide emergency relief to governments, businesses, and individuals affected by the consequences of a chemical, biological, nuclear, and/or high-yield explosive situation. For domestic consequence management, the primary authority rests with the States to respond and the Federal Government to provide assistance as required. Also called **CM.** See also **nuclear, biological, and chemical defense.** (JP 3-0)

console — **(*)** A grouping of controls, indicators, and similar electronic or mechanical equipment, used to monitor readiness of, and/or control specific functions of, a system, such as missile checkout, countdown, or launch operations.

consolidated vehicle table — A summary of all vehicles loaded on a ship, listed by types and showing the units to which they belong.

consolidation — The combining or merging of elements to perform a common or related function.

consolidation of position — **(*)** Organizing and strengthening a newly captured position so that it can be used against the enemy.

consolidation psychological operations — Psychological operations conducted in foreign areas inhabited by an enemy or potentially hostile populace and occupied by United States forces, or in which United States forces are based, designed to produce behaviors by the foreign populace that support United States objectives in the area. See also **overt peacetime psychological operations programs; psychological operations.** (JP 3-53)

constitute — To provide the legal authority for the existence of a new unit of the Armed Services. The new unit is designated and listed, but it has no specific existence until it is activated. See also **commission.**

constructive presence — Doctrine of constructive presence allows a coastal State to exercise jurisdiction over a foreign flag vessel that remains seaward of coastal State waters but acts in concert with another vessel (contact vessel) or aircraft that violates coastal State laws in waters over which the coastal State may exercise jurisdiction. In order to exercise jurisdiction over a "mothership" located seaward of coastal State waters, the contact vessel must be physically present in coastal State waters or be subject to coastal State jurisdiction under the doctrine of hot pursuit. Once pursuit of the mothership has legitimately commenced, it may proceed until it ceases to be continuous or until the mothership enters foreign territorial waters. Cases potentially involving the doctrine of constructive presence can be complex and should be quickly referred to higher authority.

consumable supplies and materiel — See **expendable supplies and materiel.**

consumer — Person or agency that uses information or intelligence produced by either its own staff or other agencies.

consumer logistics — That part of logistics concerning reception of the initial product, storage, inspection, distribution, transport, maintenance (including repair and serviceability), and disposal of materiel as well as the provision of support and services. In consequence, consumer logistics includes materiel requirements determination, follow-on support, stock control, provision or construction of facilities (excluding any materiel element and those facilities needed to support production logistic activities), movement control, codification, reliability and defect reporting, storage, transport and handling safety standards, and related training.

consumption rate — **(*)** The average quantity of an item consumed or expended during a given time interval, expressed in quantities by the most appropriate unit of measurement per applicable stated basis.

contact — 1. In air intercept, a term meaning, "Unit has an unevaluated target." 2. In health services, an unevaluated individual who is known to have been sufficiently near an infected individual to have been exposed to the transfer of infectious material.

contact burst preclusion — A fuzing arrangement that prevents an unwanted surface burst in the event of failure of the air burst fuze.

contact mine — **(*)** A mine detonated by physical contact. See also **mine.**

contact point — **(*)** 1. In land warfare, a point on the terrain, easily identifiable, where two or more units are required to make contact. 2. In air operations, the position at which a mission leader makes radio contact with an air control agency. 3. **(DOD only)** In evasion and recovery operations, a location where an evader can establish contact with friendly forces. Also called **CP.** See also **checkpoint; control point; coordinating point.** (JP 3-50.3)

contact print — **(*)** A print made from a negative or a diapositive in direct contact with sensitized material.

contact procedure — Those predesignated actions taken by evaders and recovery forces that permit link-up between the two parties in hostile territory and facilitate the return of evaders to friendly control. See also **evader; hostile; recovery force.** (JP 3-50.3)

contact reconnaissance — Locating isolated units out of contact with the main force.

contact report — **(*)** A report indicating any detection of the enemy.

contain — To stop, hold, or surround the forces of the enemy or to cause the enemy to center activity on a given front and to prevent the withdrawal of any part of the enemy's forces for use elsewhere.

container — An article of transport equipment that meets American National

Standards Institute/International Organization for Standardization standards that is designed to be transported by various modes of transportation. These containers are also designed to facilitate and optimize the carriage of goods by one or more modes of transportation without intermediate handling of the contents and equipped with features permitting ready handling and transfer from one mode to another. Containers may be fully enclosed with one or more doors, open top, refrigerated, tank, open rack, gondola, flatrack, and other designs. See also **containerization.** (JP 4-01)

container anchorage terminal — (*) A sheltered anchorage (not a port) with the appropriate facilities for the transshipment of containerized cargo from containerships to other vessels.

container control officer — A designated official (E6 or above or civilian equivalent) within a command, installation, or activity who is responsible for control, reporting, use, and maintenance of all Department of Defense-owned and controlled intermodal containers and equipment. This officer has custodial responsibility for containers from time received until dispatched. (JP 4-01.7)

container-handling equipment — Items of materials-handling equipment required to specifically receive, maneuver, and dispatch International Organization for Standardization containers. Also called **CHE.** See also **materials handling equipment.** (JP 4-01.7)

containerization — The use of containers to unitize cargo for transportation, supply and storage. Containerization incorporates supply, transportation, packaging, storage and security together with visibility of container and its contents into a distribution system from source to user. (JP 4-01.7)

containership — A ship specially constructed and equipped to carry only containers without associated equipment, in all available cargo spaces, either below or above deck. Containerships are usually non-self-sustaining, do not have built-in capability to load or off-load containers, and require port crane service. A containership with shipboard-installed cranes capable of loading and off-loading containers without assistance of port crane service is considered self-sustaining. See also **non-self-sustaining containership; self-sustaining containership.** (JP 4-01.7)

container ship cargo stowage adapter — Serves as the bottom-most SEASHED and precludes the necessity of strengthening of tank tops or the installation of hard points on decks, thereby accelerating container ship readiness. See also **container ship; SEASHED; stowage.** (JP 4-01.6)

contaminate — See **contamination.** (JP 3-11)

contaminated remains — Remains of personnel which have absorbed or upon which have been deposited radioactive material, or biological or chemical agents. See also **mortuary affairs.** (JP 4-06)

contamination — (*) 1. The deposit, absorption, or adsorption of radioactive material, or of biological or chemical agents on or by structures, areas, personnel, or objects. See also **fallout; induced radiation; residual radiation.** 2. **(DOD only)** Food and/or water made unfit for consumption by humans or animals because of the presence of environmental chemicals, radioactive elements, bacteria or organisms, the byproduct of the growth of bacteria or organisms, the decomposing material (to include the food substance itself), or waste in the food or water.

contamination control — Procedures to avoid, reduce, remove, or render harmless (temporarily or permanently) nuclear, biological, and chemical contamination for the purpose of maintaining or enhancing the efficient conduct of military operations. See also **biological agent; biological ammunition; biological defense; biological environment; biological threat; chemical agent; chemical ammunition; chemical, biological, and radiological operation; chemical defense; chemical environment; contamination.** (JP 3-11)

contiguous zone — 1. A maritime zone adjacent to the territorial sea that may not extend beyond 24 nautical miles (nms) from the baselines from which the breadth of the territorial sea is measured. Within the contiguous zone the coastal state may exercise the control necessary to prevent and punish infringement of its customs, fiscal, immigration, or sanitary laws and regulations within its territory or territorial sea. In all other respects the contiguous zone is an area subject to high seas freedom of navigation, overflight, and related freedoms, such as the conduct of military exercises. 2. The zone of the ocean extending 3-12 nms from the US coastline.

continental United States — United States territory, including the adjacent territorial waters, located within North America between Canada and Mexico. Also called **CONUS.**

continental United States replacement center — The processing centers at selected Army installations through which individual personnel will be processed to ensure that soldier readiness processing actions have been completed prior to reporting to the aerial port of embarkation for deployment to a theater of operations. See also **continental United States; deployment.** (JP 1-0)

contingency — An emergency involving military forces caused by natural disasters, terrorists, subversives, or by required military operations. Due to the uncertainty of the situation, contingencies require plans, rapid response, and special procedures to ensure the safety and readiness of personnel, installations, and equipment. See also **contingency contracting.**

contingency contracting — Contracting performed in support of a peacetime contingency in an overseas location pursuant to the policies and procedures of the Federal Acquisition Regulatory System. See also **contingency.**

contingency engineering management organization — An organization that may be formed by the combatant commander, or subordinate joint force commander to augment the combatant command, or subordinate joint force staffs to provide additional Service engineering expertise to support both deliberate and crisis action planning and to provide construction management in contingency and wartime operations. The combatant commander may form a theater contingency engineering management cell, and similar organizations may be formed at subordinate levels of command (e.g., regional contingency engineering management cell and/or joint task force contingency engineering management cell). These organizations should be staffed with expertise in combat engineering, general engineering, and topographic engineering. See also **combat engineering; contingency; crisis action planning; topographic engineering**. (JP 3-34)

contingency operation — A military operation that is either designated by the Secretary of Defense as a contingency operation or becomes a contingency operation as a matter of law (10 United States code (USC) 101[a][13]). It is a military operation that: a. is designated by the Secretary of Defense as an operation in which members of the Armed Forces are or may become involved in military actions, operations, or hostilities against an enemy of the United States or against an opposing force; or b. is created by definition of law. Under 10 USC 101 (a)(13)(B), a contingency operation exists if a military operation results in the (1) callup to (or retention on) active duty of members of the uniformed Services under certain enumerated statutes (10 USC Sections 688, 12301(a), 12302, 12304, 12305, 12406, or 331-335); and (2) the callup to (or retention on) active duty of members of the uniformed Services under other (non-enumerated) statutes during war or national emergency declared by the President or Congress. See also **contingency; operation.** (JP 1-0)

contingency plan — A plan for major contingencies that can reasonably be anticipated in the principal geographic subareas of the command. See also **joint operation planning.**

contingency planning facilities list program — A joint Defense Intelligence Agency and unified and specified command program for the production and maintenance of current target documentation of all countries of contingency planning interest to US military planners.

Contingency Planning Guidance — The Contingency Planning Guidance (CPG) fulfills the statutory duty of the Secretary of Defense to furnish written policy guidance annually to the Chairman of the Joint Chiefs of Staff for contingency planning. The Secretary issues this guidance with the approval of the President after consultation with the Chairman of the Joint Chiefs of Staff. The CPG focuses the guidance given in the National Security Strategy and Defense Planning Guidance, and is the principal source document for the Joint Strategic Capabilities Plan. Also called **CPG.**

contingency response program — Fast reaction transportation procedures intended to provide for priority use of land transportation assets by Department of Defense when required. Also called **CORE.** (JP 4-01)

contingency retention stock — That portion of the quantity of an item excess to the approved force retention level for which there is no predictable demand or quantifiable requirement, and which normally would be allocated as potential DOD excess stock, except for a determination that the quantity will be retained for possible contingencies for United States forces. (Category C ships, aircraft, and other items being retained as contingency reserve are included in this stratum.)

contingency ZIP Code — A ZIP Code assigned by Military Postal Service Agency to a contingency post office for the tactical use of the Armed Forces on a temporary basis. The number consists of a five-digit base with a four-digit add-on to assist in routing and sorting. (JP 1-0)

contingent effects — The effects, both desirable and undesirable, that are in addition to the primary effects associated with a nuclear detonation.

continuity of command — The degree or state of being continuous in the exercise of the authority vested in an individual of the Armed Forces for the direction, coordination, and control of military forces.

continuity of operations — The degree or state of being continuous in the conduct of functions, tasks, or duties necessary to accomplish a military action or mission in carrying out the national military strategy. It includes the functions and duties of the commander, as well as the supporting functions and duties performed by the staff and others acting under the authority and direction of the commander. Also called **COOP.**

continuous fire — (*) 1. Fire conducted at a normal rate without interruption for application of adjustment corrections or for other causes. 2. In field artillery and naval gunfire support, loading and firing at a specified rate or as rapidly as possible consistent with accuracy within the prescribed rate of fire for the weapon. Firing will continue until terminated by the command "end of mission" or temporarily suspended by the command "cease loading" or "check firing."

continuous illumination fire — (*) A type of fire in which illuminating projectiles are fired at specified time intervals to provide uninterrupted lighting on the target or specified area.

continuous strip camera — (*) A camera in which the film moves continuously past a slit in the focal plane, producing a photograph in one unbroken length by virtue of the continuous forward motion of the aircraft.

continuous strip imagery — (*) Imagery of a strip of terrain in which the image remains unbroken throughout its length, along the line of flight.

contour flight — See **terrain flight.**

contour interval — (*) Difference in elevation between two adjacent contour lines.

contour line — (*) A line on a map or chart connecting points of equal elevation.

contracted logistic support — Support in which maintenance operations for a particular military system are performed exclusively by contract support personnel. Also called **CLS.** See also **logistic support; support.** (JP 4-07)

contracting officer — A US military officer or civilian employee who has a valid appointment as a contracting officer under the provisions of the Federal Acquisition Regulation. The individual has the authority to enter into and administer contracts and determinations as well as findings about such contracts. (JP 1-06)

contract maintenance — The maintenance of materiel performed under contract by commercial organizations (including prime contractors) on a one-time or continuing basis, without distinction as to the level of maintenance accomplished.

contract termination — Defense procurement: the cessation or cancellation, in whole or in part, of work under a prime contract or a subcontract thereunder for the convenience of, or at the option of, the government, or due to failure of the contractor to perform in accordance with the terms of the contract (default).

control — 1. Authority that may be less than full command exercised by a commander over part of the activities of subordinate or other organizations. 2. In mapping, charting, and photogrammetry, a collective term for a system of marks or objects on the Earth or on a map or a photograph, whose positions or elevations (or both) have been or will be determined. 3. Physical or psychological pressures exerted with the intent to assure that an agent or group will respond as directed. 4. An indicator governing the distribution and use of documents, information, or material. Such indicators are the subject of intelligence community agreement and are specifically defined in appropriate regulations. See also **administrative control; operational control; tactical control.**

control area — **(*)** A controlled airspace extending upwards from a specified limit above the Earth. See also **airway; controlled airspace; control zone; terminal control area.**

control group — Personnel, ships, and craft designated to control the waterborne ship-to-shore movement. (JP 3-02)

control (intelligence) — See **control, Parts 3 and 4.**

controllable mine — **(*)** A mine which after laying can be controlled by the user, to the extent of making the mine safe or live, or to fire the mine. See also **mine.**

controlled airspace — **(*)** An airspace of defined dimensions within which air traffic control service is provided to controlled flights.

controlled dangerous air cargo — **(*)** Cargo which is regarded as highly dangerous and which may only be carried by cargo aircraft operating within specific safety regulations.

controlled effects nuclear weapons — Nuclear weapons designed to achieve variation in the intensity of specific effects other than normal blast effect.

controlled exercise — **(*)** An exercise characterized by the imposition of constraints on some or all of the participating units by planning authorities with the principal intention of provoking types of interaction. See also **free play exercise.**

controlled firing area — An area in which ordnance firing is conducted under controlled conditions so as to eliminate hazard to aircraft in flight. See also **restricted area.**

controlled forces — Military or paramilitary forces under effective and sustained political and military direction.

controlled information — 1. Information conveyed to an adversary in a deception operation to evoke desired appreciations. 2. Information and indicators deliberately conveyed or denied to foreign targets to evoke invalid official estimates that result in foreign official actions advantageous to US interests and objectives.

controlled item — See **regulated item.**

controlled map — A map with precise horizontal and vertical ground control as a basis. Scale, azimuth, and elevation are accurate. See also **map.**

controlled mosaic — **(*)** A mosaic corrected for scale, rectified and laid to ground control to provide an accurate representation of distances and direction. See also **mosaic; rectification.**

controlled passing — **(*)** A traffic movement procedure whereby two lines of traffic travelling in opposite directions are enabled to traverse alternately a point or section of route which can take only one line of traffic at a time.

controlled port — **(*)** A harbor or anchorage at which entry and departure, assignment of berths, and traffic within the harbor or anchorage are controlled by military authorities.

controlled reprisal — Not to be used. See **controlled response.**

controlled response — The selection from a wide variety of feasible options one of which will provide the specific military response most advantageous in the circumstances.

controlled route — **(*)** A route, the use of which is subject to traffic or movement restrictions which may be supervised. See also **route.**

controlled shipping — Shipping that is controlled by the Military Sealift Command. Included in this category are Military Sealift Command ships (United States Naval Ships), government-owned ships operated under a general agency agreement, and commercial ships under charter to the Military Sealift Command. See also **Military Sealift Command; United States Naval Ship.** (JP 3-02.2)

controlled substance — A drug or other substance, or immediate precursor included in Schedule I, II, III, IV, or V of the Controlled Substances Act. (JP 3-07.4)

controlled war — Not to be used. See **limited war.**

control of electromagnetic radiation — A national operation plan to minimize the use of electromagnetic radiation in the United States and its possessions and the Panama Canal Zone in the event of attack or imminent threat thereof, as an aid to the navigation of hostile aircraft, guided missiles, or other devices. See also **emission control orders.**

control point — **(*)** 1. A position along a route of march at which men are stationed to give information and instructions for the regulation of supply or traffic. 2. A position marked by a buoy, boat, aircraft, electronic device, conspicuous terrain feature, or other identifiable object which is given a name or number and used as an aid to navigation or control of ships, boats, or aircraft. 3. In marking mosaics, a point located by ground survey with which a corresponding point on a photograph is matched as a check.

control zone — **(*)** A controlled airspace extending upwards from the surface of the Earth to a specified upper limit. See also **airway; control area; controlled airspace; terminal control area.**

conventional forces — Those forces capable of conducting operations using nonnuclear weapons.

conventional mines — Land mines, other than nuclear or chemical, that are not designed to self-destruct. They are designed to be emplaced by hand or mechanical means. Conventional mines can be buried or surface laid and are normally emplaced in a pattern to aid in recording. See also **mine.** (JP 3-15)

conventional recovery operation — Evader recovery operations conducted by conventional forces. See also **conventional forces; evader; recovery; recovery operations.** (JP 3-50.3)

conventional weapon — **(*)** A weapon which is neither nuclear, biological, nor chemical.

converge — A request or command used in a call for fire to indicate that the observer or spotter desires a sheaf in which the planes of fire intersect at a point.

converged sheaf — The lateral distribution of fire of two or more pieces so that the planes of fire intersect at a given point. See also **parallel sheaf.**

convergence — See **convergence factor; grid convergence; grid convergence factor; map convergence; true convergence.**

convergence factor — **(*)** The ratio of the angle between any two meridians on the chart to their actual change of longitude. See also **convergence.**

convergence zone — That region in the deep ocean where sound rays, refractured from the depths, return to the surface.

conversion angle — **(*)** The angle between a great circle (orthodromic) bearing and a rhumb line (loxodromic) bearing of a point, measured at a common origin.

conversion scale — **(*)** A scale indicating the relationship between two different units of measurement. See also **scale.**

convoy — 1. A number of merchant ships and/or naval auxiliaries usually escorted by warships and/or aircraft — or a single merchant ship or naval auxiliary under surface escort — assembled and organized for the purpose of passage together. 2. A group of vehicles organized for the purpose of control and orderly movement with or without escort protection that moves over the same route at the same time and under one commander. See also **coastal convoy; evacuation convoy; ocean convoy.**

convoy commodore — A naval officer, or master of one of the ships in a convoy, designated to command the convoy, subject to the orders of the officer in tactical command. If no surface escort is present, the convoy commodore takes entire command.

convoy dispersal point — **(*)** The position at sea where a convoy breaks up, each ship proceeding independently thereafter.

convoy escort — **(*)** 1. A naval ship(s) or aircraft in company with a convoy and responsible for its protection. 2. An escort to protect a convoy of vehicles from being scattered, destroyed, or captured. See also **escort.**

convoy joiner — See **joiner.** See also **joiner convoy; joiner section.**

convoy leaver — See **leaver.** See also **leaver convoy; leaver section.**

convoy loading — **(*)** The loading of troop units with their equipment and supplies in vessels of the same movement group, but not necessarily in the same vessel. See also **loading.**

convoy route — **(*)** The specific route assigned to each convoy by the appropriate routing authority.

convoy schedule — **(*)** Planned convoy sailings showing the shipping lanes, assembly and terminal areas, scheduled speed, and sailing interval.

convoy speed — **(*)** For ships, the speed which the convoy commodore orders the guide of the convoy to make good through the water.

convoy terminal area — **(*)** A geographical area, designated by the name of a port or anchorage on which it is centered, at which convoys or sections of convoys arrive and from which they will be dispersed to coastal convoy systems or as independents to their final destination.

convoy through escort — **(*)** Those ships of the close escort which normally remain with the convoy from its port of assembly to its port of arrival.

convoy title — **(*)** A combination of letters and numbers that gives the port of departure and arrival, speed, and serial number of each convoy.

cooperative logistics — The logistic support provided a foreign government or agency through its participation in the US

Department of Defense logistic system, with reimbursement to the United States for support provided.

cooperative logistic support arrangements — The combining term for procedural arrangements (cooperative logistic arrangements) and implementing procedures (supplementary procedures) that together support, define, or implement cooperative logistic understandings between the United States and a friendly foreign government under peacetime conditions.

coordinated draft plan — **(*)** A plan for which a draft plan has been coordinated with the nations involved. It may be used for future planning and exercises and may be implemented during an emergency. See also **draft plan; final plan; initial draft plan; operation plan.**

coordinated fire line — The coordinated fire line (CFL) is a line beyond which conventional, direct, and indirect surface fire support means may fire at any time within the boundaries of the establishing headquarters without additional coordination. The purpose of the CFL is to expedite the surface-to-surface attack of targets beyond the CFL without coordination with the ground commander in whose area the targets are located. Also called **CFL.** See also **fire support.** (JP 3-09)

coordinated procurement assignee — The agency or Military Service assigned purchase responsibility for all Department of Defense requirements of a particular Federal Supply Group/class, commodity, or item.

Coordinated Universal Time — An atomic time scale that is the basis for broadcast time signals. Coordinated Universal Time (UTC) differs from International Atomic Time by an integral number of seconds; it is maintained within 0.9 seconds of UT1 (see Universal Time) by introduction of Leap Seconds. The rotational orientation of the Earth, specified by UT1, may be obtained to an accuracy of a tenth of a second by applying the UTC to the increment DUT1 (where DUT1 = UT1 - UTC) that is broadcast in code with the time signals. Also called **UTC.** See also **International Atomic Time; Universal Time; ZULU Time.**

coordinates — **(*)** Linear or angular quantities which designate the position that a point occupies in a given reference frame or system. Also used as a general term to designate the particular kind of reference frame or system such as plane rectangular coordinates or spherical coordinates. See also **geographic coordinates; georef; grid coordinates.**

coordinating altitude — A procedural airspace control method to separate fixed- and rotary-wing aircraft by determining an altitude below which fixed-wing aircraft will normally not fly and above which rotary-wing aircraft normally will not fly. The coordinating altitude is normally specified in the airspace control plan and may include a buffer zone for small altitude deviations. (JP 3-52)

coordinating authority — A commander or individual assigned responsibility for coordinating specific functions or activities involving forces of two or more Military Departments, two or more joint force components, or two or more forces of the same Service. The commander or individual has the authority to require consultation between the agencies involved, but does not have the authority to compel agreement. In the event that essential agreement cannot be obtained, the matter shall be referred to the appointing authority. Coordinating authority is a consultation relationship, not an authority through which command may be exercised. Coordinating authority is more applicable to planning and similar activities than to operations. (JP 0-2)

coordinating point — (*) Designated point at which, in all types of combat, adjacent units/formations must make contact for purposes of control and coordination.

coordinating review authority — An agency appointed by a Service or combatant command to coordinate with and assist the primary review authority in joint doctrine development, and maintenance efforts. Each Service or combatant command must assign a coordinating review authority. When authorized by the appointing Service or combatant command, coordinating review authority comments provided to designated primary review authorities will represent the position of the appointing Service or combatant command with regard to the publication under development. Also called **CRA.** See also **joint doctrine; joint publication; joint tactics, techniques, and procedures; joint test publication; lead agent; primary review authority.** (JP 1-01)

coproduction — 1. With respect to exports, a cooperative manufacturing arrangement (e.g., US Government or company with foreign government or company) providing for the transfer of production information that enables an eligible foreign government, international organization, or commercial producer to manufacture, in whole or in part, an item of US defense equipment. Such an arrangement would include the functions of production engineering, controlling, quality assurance, and determination of resource requirements. This is normally accomplished under the provisions of a manufacturing license agreement per the US International Traffic in Arms Regulation and could involve the implementation of a government-to-government memorandum of understanding. 2. A cooperative manufacturing arrangement (US Government or company with foreign government or company) providing for the transfer of production information which enables the receiving government, international organization, or commercial producer to manufacture, in whole or in part, an item of defense equipment. The receiving party could be an eligible foreign government, international organization, or foreign producer; or the US Government or a US producer, depending on which direction the information is to flow. A typical coproduction arrangement would include the functions of production engineering, controlling, quality assurance, and determining of resource requirements. It may or may not include design engineering information and critical materials production and design information.

copy negative — (*) A negative produced from an original not necessarily at the same scale.

corner reflector — (*) 1. A device, normally consisting of three metallic surfaces or screens perpendicular to one another, designed to act as a radar target or marker. 2. In radar interpretation, an object which, by means of multiple reflections from smooth surfaces, produces a radar return of greater magnitude than might be expected from the physical size of the object.

corps support command — Provides corps logistic support and command and control of water supply battalions. (JP 4-01.6)

corps troops — (*) Troops assigned or attached to a corps, but not a part of one of the divisions that make up the corps.

correlation factor — (*) The ratio of a ground dose rate reading to a reading taken at approximately the same time at survey height over the same point on the ground.

cost contract — 1. A contract that provides for payment to the contractor of allowable costs, to the extent prescribed in the contract, incurred in performance of the contract. 2. A cost-reimbursement type contract under which the contractor receives no fee.

cost-plus a fixed-fee contract — A cost-reimbursement type contract that provides for the payment of a fixed fee to the contractor. The fixed fee, once negotiated, does not vary with actual cost but may be adjusted as a result of any subsequent changes in the scope of work or services to be performed under the contract.

cost sharing contract — A cost-reimbursement type contract under which the contractor receives no fee but is reimbursed only for an agreed portion of its allowable costs.

counterair — A mission that integrates offensive and defensive operations to attain and maintain a desired degree of air superiority. Counterair missions are designed to destroy or negate enemy aircraft and missiles, both before and after launch. See also **air superiority; mission; offensive counterair.** (JP 3-01)

counterattack — Attack by part or all of a defending force against an enemy attacking force, for such specific purposes as regaining ground lost or cutting off or destroying enemy advance units, and with the general objective of denying to the enemy the attainment of the enemy's purpose in attacking. In sustained defensive operations, it is undertaken to restore the battle position and is directed at limited objectives. See also **countermove; counteroffensive.**

counterbattery fire — **(*)** Fire delivered for the purpose of destroying or neutralizing indirect fire weapon systems.

counterdeception — Efforts to negate, neutralize, diminish the effects of, or gain advantage from a foreign deception operation. Counterdeception does not include the intelligence function of identifying foreign deception operations. See also **deception.**

counterdrug — Those active measures taken to detect, monitor, and counter the production, trafficking, and use of illegal drugs. Also called **CD.** (JP 3-05)

counterdrug nonoperational support — Support provided to law enforcement agencies or host nations that includes loan or lease of equipment without operators, use of facilities (such as buildings, training areas, and ranges), training conducted in formal schools, transfer of excess equipment, or other support provided by the Services from forces not assigned or made available to the combatant commanders. See also **counterdrug operational support; counterdrug operations.** (JP 3-07.4)

counterdrug operational support — Support to host nations and drug law enforcement agencies involving military personnel and their associated equipment, and provided by the geographic combatant commanders from forces assigned to them or made available to them by the Services for this purpose. Operational support does not include support in the form of equipment alone, nor the conduct of joint law enforcement investigations with cooperating civilian law enforcement agencies. See also **counterdrug nonoperational support; counterdrug operations.** (JP 3-07.4)

counterdrug operations — Civil or military actions taken to reduce or eliminate illicit drug trafficking. See also **counterdrug; counterdrug nonoperational support; counterdrug operational support.** (JP 3-07.4)

counterdrug support office — In counterdrug operations, offices under the office of the Department of Defense Coordinator for Drug Enforcement Policy and Support, responsible for processing, tracking, and coordinating all nonoperational support requests from drug law enforcement officials. Also called **CDSO.** See also **counterdrug; counterdrug operations.** (JP 3-07.4)

counterespionage — That aspect of counterintelligence designed to detect, destroy, neutralize, exploit, or prevent espionage activities through identification, penetration, manipulation, deception, and repression of individuals, groups, or organizations conducting or suspected of conducting espionage activities.

counterfire — **(*)** Fire intended to destroy or neutralize enemy weapons. **(DOD only)** Includes counterbattery, counterbombardment, and countermortar fire. See also **fire.**

counterforce — The employment of strategic air and missile forces in an effort to destroy, or render impotent, selected military capabilities of an enemy force under any of the circumstances by which hostilities may be initiated.

counterguerrilla warfare — **(*)** Operations and activities conducted by armed forces, paramilitary forces, or nonmilitary agencies against guerrillas.

counterinsurgency — Those military, paramilitary, political, economic, psychological, and civic actions taken by a government to defeat insurgency. Also called **COIN.**

counterintelligence — Information gathered and activities conducted to protect against espionage, other intelligence activities, sabotage, or assassinations conducted by or on behalf of foreign governments or elements thereof, foreign organizations, or foreign persons, or international terrorist activities. Also called **CI**. See also **counterespionage; countersabotage; countersubversion; security; security intelligence.** (JP 2-0)

counterintelligence activities — The four functions of counterintelligence: operations; investigations; collection and reporting; and analysis, production, and dissemination. See also **counterintelligence.** (JP 2-01.2)

counterintelligence collection — The systematic acquisition of information (through investigations, operations, or liaison) concerning espionage, sabotage, terrorism, other intelligence activities or assassinations conducted by or on behalf of foreign governments or elements thereof, foreign organizations, or foreign persons that are directed against or threaten Department of Defense interests. See also **counterintelligence.** (JP 2-01.2)

counterintelligence investigations — Counterintelligence investigations establish the elements of proof for prosecution or administrative action. Counterintelligence investigations can provide a basis for or be developed from conducting counterintelligence operations. Counterintelligence investigations are conducted against individuals or groups suspected of committing acts of espionage, sabotage, sedition, subversion, terrorism, and other major security violations as well as failure to follow Defense agency and Military Service directives governing reporting of contacts with foreign citizens and "out-of-channel" requests for defense information. Counterintelligence investigations provide military commanders and policymakers with information used to eliminate security vulnerabilities and otherwise to improve the security posture of threatened interests. See also **counterintelligence.** (JP 2-01.2)

counterintelligence production — The process of analyzing all-source information concerning espionage or other multidiscipline intelligence collection threats, sabotage, terrorism, and other related threats to US military commanders, the Department of Defense, and the US Intelligence Community and developing it into a final product that is disseminated. Counterintelligence production is used in formulating security policy, plans, and operations. See also **counterintelligence.** (JP 2-01.2)

counterintelligence support — Conducting counterintelligence activities to protect against espionage and other foreign intelligence activities, sabotage, international terrorist activities, or assassinations conducted for or on behalf of foreign powers, organizations, or persons. See also **counterintelligence.** (JP 2-01.2)

countermeasures — That form of military science that, by the employment of devices and/or techniques, has as its objective the impairment of the operational effectiveness of enemy activity. See also **electronic warfare.**

countermine — **(*)** To explode the main charge in a mine by the shock of a nearby explosion of another mine or independent explosive charge. The explosion of the main charge may be caused either by sympathetic detonation or through the explosive train and/or firing mechanism of the mine.

countermine operation — **(*)** In land mine warfare, an operation to reduce or eliminate the effects of mines or minefields. See also **countermine; countermining; mine warfare.**

countermining — 1. **Land mine warfare** — Tactics and techniques used to detect, avoid, breach, and/or neutralize enemy mines and the use of available resources to deny the enemy the opportunity to employ mines. 2. **Naval mine warfare** — The detonation of mines by nearby explosions, either accidental or deliberate.

countermobility operations — The construction of obstacles and emplacement of minefields to delay, disrupt, and destroy the enemy by reinforcement of the terrain. The primary purpose of countermobility operations is to slow or divert the enemy, to increase time for target acquisition, and to increase weapons effectiveness. See also **minefield; operation; target acquisition**. (JP 3-34)

countermove — **(*)** An operation undertaken in reaction to or in anticipation of a move by the enemy. See also **counterattack.**

counteroffensive — A large scale offensive undertaken by a defending force to seize the initiative from the attacking force. See also **counterattack.**

counterpreparation fire — **(*)** Intensive prearranged fire delivered when the imminence of the enemy attack is discovered. **(DOD only)** It is designed to: break up enemy formations; disorganize the enemy's systems of command, communications, and observation; decrease the effectiveness of artillery preparation;

and impair the enemy's offensive spirit. See also **fire.**

counterreconnaissance — All measures taken to prevent hostile observation of a force, area, or place.

countersabotage — That aspect of counterintelligence designed to detect, destroy, neutralize, or prevent sabotage activities through identification, penetration, manipulation, deception, and repression of individuals, groups, or organizations conducting or suspected of conducting sabotage activities.

countersign — (*) A secret challenge and its reply. See also **challenge; password.**

countersubversion — That aspect of counterintelligence designed to detect, destroy, neutralize, or prevent subversive activities through the identification, exploitation, penetration, manipulation, deception, and repression of individuals, groups, or organizations conducting or suspected of conducting subversive activities.

countersurveillance — All measures, active or passive, taken to counteract hostile surveillance. See also **surveillance.**

counterterrorism — Offensive measures taken to prevent, deter, and respond to terrorism. Also called **CT.** See also **antiterrorism; combatting terrorism; terrorism.**

country cover diagram — (*) A small scale index, by country, depicting the existence of air photography for planning purposes only.

country team — The senior, in-country, US coordinating and supervising body, headed by the chief of the US diplomatic mission, and composed of the senior member of each represented US department or agency, as desired by the chief of the US diplomatic mission. (JP 3-07.4)

coup de main — An offensive operation that capitalizes on surprise and simultaneous execution of supporting operations to achieve success in one swift stroke. (JP 3-0)

courier — A messenger (usually a commissioned or warrant officer) responsible for the secure physical transmission and delivery of documents and material. Generally referred to as a command or local courier. See also **armed forces courier.**

course — (*) The intended direction of movement in the horizontal plane.

course of action — 1. Any sequence of activities that an individual or unit may follow. 2. A possible plan open to an individual or commander that would accomplish, or is related to the accomplishment of the mission. 3. The scheme adopted to accomplish a job or mission. 4. A line of conduct in an engagement. 5. A product of the Joint Operation Planning and Execution System concept development phase. Also called **COA.**

course of action development — The phase of the Joint Operation Planning and Execution System within the crisis action planning process that provides for the development of military responses and includes, within the limits of the time allowed: establishing force and sustainment requirements with actual units; evaluating force, logistic, and transportation feasibility; identifying and resolving resource shortfalls; recommending resource allocations; and producing a course of action via a commander's estimate that contains a concept of operations, employment concept, risk assessments, prioritized courses of action, and supporting databases. See also **course of action; crisis action planning.**

cover — (*) 1. The action by land, air, or sea forces to protect by offense, defense, or threat of either or both. 2. Those measures necessary to give protection to a person, plan, operation, formation, or installation from the enemy intelligence effort and leakage of information. 3. The act of maintaining a continuous receiver watch with transmitter calibrated and available, but not necessarily available for immediate use. 4. Shelter or protection, either natural or artificial. 5. **(DOD only)** Photographs or other recorded images which show a particular area of ground. 6. **(DOD only)** A code meaning, "Keep fighters between force/base and contact designated at distance stated from force/base" (e.g., "cover bogey twenty-seven to thirty miles").

coverage — (*) 1. The ground area represented on imagery, photomaps, mosaics, maps, and other geographical presentation systems. 2. **(DOD only)** Cover or protection, as the coverage of troops by supporting fire. 3. **(DOD only)** The extent to which intelligence information is available in respect to any specified area of interest. 4. **(DOD only)** The summation of the geographical areas and volumes of aerospace under surveillance. See also **comparative cover.**

covering fire — (*) 1. Fire used to protect troops when they are within range of enemy small arms. 2. In amphibious usage, fire delivered prior to the landing to cover preparatory operations such as underwater demolition or minesweeping. See also **fire.**

covering force — (*) 1. A force operating apart from the main force for the purpose of intercepting, engaging, delaying, disorganizing, and deceiving the enemy before the enemy can attack the force covered. 2. Any body or detachment of troops which provides security for a larger force by observation, reconnaissance, attack, or defense, or by any combination of these methods. See also **force(s).**

covering force area — (*) The area forward of the forward edge of the battle area out to the forward positions initially assigned to the covering forces. It is here that the covering forces execute assigned tasks.

cover (military) — Actions to conceal actual friendly intentions, capabilities, operations, and other activities by providing a plausible yet erroneous explanation of the observable.

cover search — (*) In air photographic reconnaissance, the process of selection of the most suitable existing cover for a specific requirement.

covert operation — An operation that is so planned and executed as to conceal the identity of or permit plausible denial by the sponsor. A covert operation differs from a clandestine operation in that emphasis is placed on concealment of identity of sponsor rather than on concealment of the operation. See also **clandestine operation; overt operation.** (JP 3-05.3)

coxswain — A person in charge of a small craft (in the Army, a Class B or smaller craft) who often functions as the helmsman. For a causeway ferry, the pilot is in charge with the coxswain performing helmsman functions. See **causeway.** (JP 4-01.6)

crash locator beacon — (*) An automatic emergency radio locator beacon to help searching forces locate a crashed aircraft. See also **emergency locator beacon; personal locator beacon.**

crash position indicator — See **crash locator beacon.**

crash rescue and fire suppression — Extraction of aircrew members from crashed or burning aircraft and the control and extinguishing of aircraft and structural fires. (JP 4-04)

crater — The pit, depression, or cavity formed in the surface of the Earth by an explosion. It may range from saucer shaped to conical, depending largely on the depth of burst. In the case of a deep underground burst, no rupture of the surface may occur. The resulting cavity is termed a "camouflet."

crater depth — The maximum depth of the crater measured from the deepest point of the pit to the original ground level.

cratering charge — **(*)** A charge placed at an adequate depth to produce a crater.

crater radius — The average radius of the crater measured at the level corresponding to the original surface of the ground.

creeping barrage — **(*)** A barrage in which the fire of all units participating remains in the same relative position throughout and which advances in steps of one line at a time.

creeping mine — **(*)** In naval mine warfare, a buoyant mine held below the surface by a weight, usually in the form of a chain, which is free to creep along the seabed under the influence of stream or current.

crest — **(*)** A terrain feature of such altitude that it restricts fire or observation in an area beyond, resulting in dead space, or limiting the minimum elevation, or both.

crested — A report that indicates that engagement of a target or observation of an area is not possible because of an obstacle or intervening crest.

crisis — An incident or situation involving a threat to the United States, its territories, citizens, military forces, possessions, or vital interests that develops rapidly and creates a condition of such diplomatic, economic, political, or military importance that commitment of US military forces and resources is contemplated in order to achieve national objectives. (JP 5-0)

crisis action planning — 1. The Joint Operation Planning and Execution System process involving the time-sensitive development of joint operation plans and orders in response to an imminent crisis. Crisis action planning follows prescribed crisis action procedures to formulate and implement an effective response within the time frame permitted by the crisis. 2. The time-sensitive planning for the deployment, employment, and sustainment of assigned and allocated forces and resources that occurs in response to a situation that may result in actual military operations. Crisis action planners base their plan on the circumstances that exist at the time planning occurs. Also called **CAP.** See also **Joint Operation Planning and Execution System.** (JP 5-0)

crisis management — Measure to resolve a hostile situation and investigate and prepare a criminal case for prosecution under federal law. Crisis management will include a response to an incident involving a weapon of mass destruction, special improvised explosive device, or a hostage crisis that is beyond the capability of the lead federal agency. See also **crisis; hostage; hostile.** (JP 3-07.6)

critical information — Specific facts about friendly intentions, capabilities, and activities vitally needed by adversaries for them to plan and act effectively so as to guarantee failure or unacceptable consequences for friendly mission accomplishment.

critical intelligence — Intelligence that is crucial and requires the immediate attention of the commander. It is required to enable the commander to make decisions that will provide a timely and appropriate response to actions by the potential or actual enemy. It includes but is not limited to the following: a. strong indications of the imminent outbreak of hostilities of any type (warning of attack); b. aggression of any nature against a friendly country; c. indications or use of nuclear, biological, and chemical weapons (targets); and d. significant events within potential enemy countries that may lead to modification of nuclear strike plans.

critical item — An essential item which is in short supply or expected to be in short supply for an extended period. See also **critical supplies and materiel; regulated item.**

critical item list — Prioritized list, compiled from a subordinate commander's composite critical item lists, identifying supply items and weapon systems that assist Service and Defense Logistics Agency's selection of supply items and systems for production surge planning. Also may be used in operational situations by the combatant commander and/or subordinate joint force commander (within combatant commander directives) to cross-level critical supply items between Service components. Also called **CIL.** See also **critical item.** (JP 4-07)

critical joint duty assignment billet — A joint duty assignment position for which, considering the duties and responsibilities of the position, it is highly important that the assigned officer be particularly trained in, and oriented toward, joint matters. Critical billets are selected by heads of joint organizations, approved by the Secretary of Defense and documented in the Joint Duty Assignment List.

critical mass — The minimum amount of fissionable material capable of supporting a chain reaction under precisely specified conditions.

critical node — An element, position, or command and control entity whose disruption or destruction immediately degrades the ability of a force to command, control, or effectively conduct combat operations. Also called **target critical damage point.**

critical occupational specialty — A military occupational specialty selected from among the combat arms in the Army or equivalent military specialties in the Navy, Air Force, or Marine Corps. Equivalent military specialties are those engaged in operational art in order to attain strategic goals in an operational area through the design, organization, and conduct of campaigns and major operations. Critical occupational specialties are designated by the Secretary of Defense. Also called **COS.**

critical point — 1. A key geographical point or position important to the success of an operation. 2. In point of time, a crisis or a turning point in an operation. 3. A selected point along a line of march used for reference in giving instructions. 4. A point where there is a change of direction or change in slope in a ridge or stream. 5. Any point along a route of march where interference with a troop movement may occur.

critical safety item — A part, assembly, installation, or production system with one or more essential characteristics that, if not conforming to the design data or quality requirements, would result in an unsafe condition that could cause loss or serious damage to the end item or major components, loss of control, or serious injury to personnel. Also called **CSI.**

critical speed — **(*)** A speed or range of speeds which a ship cannot sustain due to vibration or other similar phenomena.

critical supplies and materiel — **(*)** Those supplies vital to the support of operations, which owing to various causes are in short supply or are expected to be in short supply. See also **critical item; regulated item.**

critical sustainability item — Any item described at National Stock Number level of detail, by federal supply class, as part of the logistic factors file, that significantly affect the commander's ability to execute an operation plan. Also called **CSI.**

critic report — See **critical intelligence.**

crossing area — **(*)** 1. A number of adjacent crossing sites under the control of one commander. 2. **(DOD only)** A controlled access area for a river crossing operation used to decrease traffic congestion at the river. It is normally a brigade-sized area defined by lateral boundaries and release lines 3 to 4 kilometers (based on mission, enemy, terrain and weather, troops and support available-time available) from each side of the river.

cross-leveling — The authority and ability to shift materiel inventory from one owner to meet the requirement of another. At the theater strategic level and operational level, it is the process of diverting en route or in-theater materiel from one military element to meet the higher priority of another within the combatant commander's directive authority for logistics. Cross-leveling plans must include specific reimbursement procedures. (JP 4-07)

cross-loading (personnel) — The distribution of leaders, key weapons, personnel, and key equipment among the aircraft, vessels, or vehicles of a formation to preclude the total loss of command and control or unit effectiveness if an aircraft, vessel, or vehicle is lost. It is also an important factor in aiding rapid assembly of units at the drop zone or landing zone. See also **loading.**

cross-servicing — A subset of common-user logistics in which a function is performed by one Military Service in support of another Military Service and for which reimbursement is required from the Service receiving support. See also **acquisition and cross-servicing agreement; common-user logistics; servicing.** (JP 4-07)

cross-targeting (nuclear) — The layering of weapons from different delivery platforms to increase the probability of target damage or destruction.

cross tell — **(*)** The transfer of information between facilities at the same operational level. See also **track telling.**

cruise missile — Guided missile, the major portion of whose flight path to its target is conducted at approximately constant velocity; depends on the dynamic reaction of air for lift and upon propulsion forces to balance drag.

cruising altitude — **(*)** A level determined by vertical measurement from mean sea level, maintained during a flight or portion thereof.

cruising level — **(*)** A level maintained during a significant portion of a flight. See also **altitude.**

crush depth — See **collapse depth.**

cryogenic liquid — Liquefied gas at very low temperature, such as liquid oxygen, nitrogen, or argon.

cryptanalysis — The steps and operations performed in converting encrypted messages into plain text without initial knowledge of the key employed in the encryption.

cryptochannel — 1. A complete system of crypto-communications between two or more holders. 2. The basic unit for naval cryptographic communication. It includes: a. the cryptographic aids prescribed; b. the holders thereof; c. the indicators or other means of identification; d. the area or areas in which effective; e. the special purpose, if any, for which provided; and f. pertinent notes as to distribution, usage, etc. A cryptochannel is analogous to a radio circuit.

cryptographic information — All information significantly descriptive of cryptographic techniques and processes or of cryptographic systems and equipment (or their functions and capabilities) and all cryptomaterial.

cryptologic — Of or pertaining to cryptology.

cryptology — The science that deals with hidden, disguised, or encrypted communications. It includes communications security and communications intelligence.

cryptomaterial — All material including documents, devices, equipment, and apparatus essential to the encryption, decryption, or authentication of telecommunications. When classified, it is designated CRYPTO and subject to special safeguards.

cryptopart — (*) A division of a message as prescribed for security reasons. The operating instructions for certain cryptosystems prescribe the number of groups which may be encrypted in the systems, using a single message indicator. Cryptoparts are identified in plain language. They are not to be confused with message parts.

cryptosecurity — See **communications security.**

cryptosystem — The associated items of cryptomaterial that are used as a unit and provide a single means of encryption and decryption. See also **cipher; code; decrypt; encipher.**

CSAR-capable assets — Those aircraft, vehicles, maritime craft, and assigned personnel possessing inherent capabilities to recover isolated personnel, but whose primary designed operational capability or mission is other than combat search and rescue (CSAR). CSAR-capable assets are mobile, responsive, and capable of physically recovering and/or returning isolated personnel to friendly forces.

CSAR-dedicated assets — Those aircraft, vehicles, maritime craft, and assigned personnel with a primary designed operational capability and Service- or component-designated mission of conducting or directly supporting combat search and rescue (CSAR) operations. Dedicated assets and assigned personnel are specifically trained, equipped, and employed to recover isolated personnel or provide direct support to assets conducting CSAR operations. "CSAR-capable" assets specifically tasked to conduct CSAR operations by a joint force commander or other appropriate authority are included in this category.

culminating point — The point at which a force no longer has the capability to continue its form of operations, offense or defense. a. In the offense, the point at which continuing the attack is no longer possible and the force must consider reverting to a defensive posture or attempting an operational pause. b. In the defense, the point at which counteroffensive action is no longer possible. (JP 3-0)

cultivation — A deliberate and calculated association with a person for the purpose of recruitment, obtaining information, or gaining control for these or other purposes.

culture — **(*)** A feature of the terrain that has been constructed by man. Included are such items as roads, buildings, and canals; boundary lines; and, in a broad sense, all names and legends on a map.

curb weight — Weight of a ground vehicle including fuel, lubricants, coolant, and on-vehicle materiel, excluding cargo and operating personnel.

current — A body of water moving in a certain direction and caused by wind and density differences in water. The effects of a current are modified by water depth, underwater topography, basin shape, land masses, and deflection from the earth's rotation. (JP 4-01.6)

current force — The force that exists today. The current force represents actual force structure and/or manning available to meet present contingencies. It is the basis for operations and contingency plans and orders. See also **force; Intermediate Force Planning Level; Programmed Forces.**

current intelligence — One of two categories of descriptive intelligence that is concerned with describing the existing situation.

current, offshore — Deep water movements caused by tides or seasonal changes in ocean water level. (JP 4-01.6)

current, rip — A water movement that flows from the beach through the surf zone in swiftly moving narrow channels. See also **surf zone.** (JP 4-01.6)

curve of pursuit — **(*)** The curved path described by a fighter plane making an attack on a moving target while holding the proper aiming allowance.

cusps — Ridges of beach material extending seaward from the beach face with intervening troughs. (JP 4-01.6)

custodian of postal effects — Members of the US Armed Forces or Department of Defense civilian employees accountable for administration of the postal effects entrusted to them by the United States Postal Service. Civilian custodians of postal effects are supervised by the members of the US Armed Forces. Also called **COPE.**

custody — 1. The responsibility for the control of, transfer and movement of, and access to, weapons and components. Custody also includes the maintenance of accountability for weapons and components. 2. Temporary restraint of a person.

customer ship — **(*)** The ship in a replenishment unit that receives the transferred personnel and/or supplies.

customer wait time — The total elapsed time between issuance of a customer order and satisfaction of that order. Also called **CWT.** (JP 4-09)

Customs Over-The-Horizon Enforcement Network — US Customs Service long-range voice communications system. Also called **COTHEN.** (JP 3-07.4)

cut-off — **(*)** The deliberate shutting off of a reaction engine.

cutoff attack — An attack that provides a direct vector from the interceptor's position to an intercept point with the target track.

cut-off velocity — **(*)** The velocity attained by a missile at the point of cut-off.

cutout — An intermediary or device used to obviate direct contact between members of a clandestine organization.

cutter — **(*)** 1. In naval mine warfare, a device fitted to a sweep wire to cut or part the moorings of mines or obstructors; it may also be fitted in the mooring of a mine or obstructor to part a sweep. 2. **(DOD only)** Coast Guard watercraft 65 feet long or larger. See also **mine warfare; watercraft.** (JP 3-33)

cutting charge — **(*)** A charge which produces a cutting effect in line with its plane of symmetry.

cyberspace — The notional environment in which digitized information is communicated over computer networks. (JP 2-01.3)

D

daily intelligence summary — A report prepared in message form at the joint force headquarters that provides higher, lateral, and subordinate headquarters with a summary of all significant intelligence produced during the previous 24-hour period. The "as of" time for information, content, and submission time for the report will be as specified by the joint force commander. Also called **DISUM.**

daily movement summary (shipping) — A tabulation of departures and arrivals of all merchant shipping (including neutrals) arriving or departing ports during a 24-hour period.

damage area — **(*)** In naval mine warfare, the plan area around a minesweeper inside which a mine explosion is likely to interrupt operations.

damage assessment — **(*)** 1. The determination of the effect of attacks on targets. 2. **(DOD only)** A determination of the effect of a compromise of classified information on national security. See **also civil damage assessment; military damage assessment.**

damage control — In naval usage, measures necessary aboard ship to preserve and reestablish watertight integrity, stability, maneuverability, and offensive power; to control list and trim; to effect rapid repairs of materiel; to limit the spread of and provide adequate protection from fire; to limit the spread of, remove the contamination by, and provide adequate protection from chemical, biological, and radiological agents; and to provide for care of wounded personnel. See also **area damage control; disaster control.**

damage criteria — The critical levels of various effects, such as blast pressure and thermal radiation, required to achieve specified levels of damage.

damage estimation — A preliminary appraisal of the potential effects of an attack. See also **attack assessment.**

damage expectancy (nuclear) — The probability that a weapon will arrive, detonate, and achieve at least a specified level of damage (severe or moderate) against a given target. Damage expectancy is a function of both probability of arrival and probability of damage of a weapon.

damage radius — **(*)** In naval mine warfare, the average distance from a ship within which a mine containing a given weight and type of explosive must detonate if it is to inflict a specified amount of damage.

damage threat — **(*)** The probability that a target ship passing once through a minefield will explode one or more mines and sustain a specified amount of damage.

danger area — **(*)** 1. In air traffic control, an airspace of defined dimensions within which activities dangerous to the flight of aircraft may exist at specified times. 2. **(DOD only)** A specified area above, below, or within which there may be potential danger. See also **closed area; prohibited area; restricted area.**

danger close — In close air support, artillery, mortar, and naval gunfire support fires, it is the term included in the method of engagement segment of a call for fire which indicates that friendly forces are within close proximity of the target. The close proximity distance is determined by the weapon and munition fired. See also **call for fire; final protective fire.**

dangerous cargo — **(*)** Cargo which, because of its dangerous properties, is subject to special regulations for its transport.

danger space — That space between the weapon and the target where the trajectory does not rise 1.8 meters (the average height of a standing human). This includes the area encompassed by the beaten zone. See also **beaten zone.**

data — Representation of facts, concepts, or instructions in a formalized manner suitable for communication, interpretation, or processing by humans or by automatic means. Any representations such as characters or analog quantities to which meaning is or might be assigned.

database — Information that is normally structured and indexed for user access and review. Databases may exist in the form of physical files (folders, documents, etc.) or formatted automated data processing system data files. (JP 2-0)

data block — Information presented on air imagery relevant to the geographical position, altitude, attitude, and heading of the aircraft and, in certain cases, administrative information and information on the sensors employed.

data code — A number, letter, character, or any combination thereof used to represent a data element or data item.

data element — 1. A basic unit of information built on standard structures having a unique meaning and distinct units or values. 2. In electronic recordkeeping, a combination of characters or bytes referring to one separate item of information, such as name, address, or age.

data item — A subunit of descriptive information or value classified under a data element. For example, the data element "military personnel grade" contains data items such as sergeant, captain, and colonel.

data link — **(*)** The means of connecting one location to another for the purpose of transmitting and receiving data. See also **tactical digital information link.**

data link coordination net — A voice coordination net of voice circuits used to coordinate technical operation of data terminal equipment. One voice circuit is required for each tactical digital information link (TADIL)-B pair, and one net is required for participants on each TADIL-A, TADIL-J, or interim Joint Tactical Information Distribution System message specification net. The net is normally secure or covered. Also called **DCN.**

data mile — A standard unit of distance

date line — See **international date line.**

date-time group — The date and time, expressed in digits and time zone suffix, at which the message was prepared for transmission. (Expressed as six digits followed by the time zone suffix; first pair of digits denotes the date, second pair the hours, third pair the minutes, followed by a three-letter month abbreviation and two-digit year abbreviation.) Also called **DTG.**

datum — **(*)** Any numerical or geometrical quantity or set of such quantities which may serve as reference or base for other quantities. Where the concept is geometric, the plural form is "datums" in contrast to the normal plural "data."

datum (antisubmarine warfare) — A datum is the last known position of a submarine, or suspected submarine, after contact has been lost.

datum error (antisubmarine warfare) — An estimate of the degree of accuracy in the reported position of datum.

datum (geodetic) — 1. A reference surface consisting of five quantities: the latitude and longitude of an initial point, the azimuth of a line from that point, and the parameters of the reference ellipsoid. 2. The mathematical model of the earth used to calculate the coordinates on any map. Different nations use different datums for printing coordinates on their maps. The datum is usually referenced in the marginal information of each map.

datum level — **(*)** A surface to which elevations, heights, or depths on a map or chart are related. See also **altitude.**

datum point — **(*)** Any reference point of known or assumed coordinates from which calculation or measurements may be taken. See also **pinpoint.**

datum time (antisubmarine warfare) — The time when contact with the submarine, or suspected submarine, was lost.

davit — A small crane on a vessel that is used to raise and lower small boats, such as lifeboats, an LCM-8 on landing ship, tanks, side loadable warping tugs, or causeway sections. (JP 4-01.6)

day of supply — See **one day's supply.**

dazzle — Temporary loss of vision or a temporary reduction in visual acuity; may also be applied to effects on optics. See also **directed-energy warfare; flash blindness.**

D-day — See **times.**

D-day consumption/production differential assets — As applied to the D-to-P concept, these assets are required to compensate for the inability of the production base to meet expenditure (consumption) requirements during the D-to-P period. See also **D-to-P concept.**

D-day materiel readiness gross capability — As applied to the D-to-P concept, this capability represents the sum of all assets on hand on D-day and the gross production capability (funded and unfunded) between D-day and P-day. When this capability equals the D-to-P materiel readiness gross requirement, requirements and capabilities are in balance. See also **D-to-P concept.**

D-day pipeline assets — As applied to the D-to-P concept, these assets represent the sum of continental United States and overseas operating and safety levels and intransit levels of supply. See also **D-to-P concept.**

deadline — To remove a vehicle or piece of equipment from operation or use for one of the following reasons: a. is inoperative due to damage, malfunctioning, or necessary repairs (the term does not include items temporarily removed from use by reason of routine maintenance and repairs that do not affect the combat capability of the item); b. is unsafe; and c. would be damaged by further use.

dead mine — **(*)** A mine which has been neutralized, sterilized, or rendered safe. See also **mine.**

dead space — **(*)** 1. An area within the maximum range of a weapon, radar, or observer, which cannot be covered by fire or observation from a particular position because of intervening obstacles, the nature of the ground, or the characteristics of the trajectory, or the limitations of the pointing capabilities of the weapon. 2. An area or zone which is within range of a radio transmitter, but in which a signal is not received. 3. The volume of space above and around a gun or guided missile system into which it cannot fire because of mechanical or electronic limitations.

de-arming — An operation in which a weapon is changed from a state of readiness for initiation to a safe condition. Also called **safing.** See also **arm or de-arm.** (JP 3-04.1)

debarkation — The unloading of troops, equipment, or supplies from a ship or aircraft.

debarkation net — A specially prepared type of cargo net employed for the debarkation of troops over the side of a ship.

debarkation schedule — **(*)** A schedule that provides for the timely and orderly debarkation of troops and equipment and emergency supplies for the waterborne ship-to-shore movement.

deceased — A casualty status applicable to a person who is either known to have died, determined to have died on the basis of conclusive evidence, or declared to be dead on the basis of a presumptive finding of death. The recovery of remains is not a prerequisite to determining or declaring a person deceased. See also **casualty status.**

decentralized control — **(*)** In air defense, the normal mode whereby a higher echelon monitors unit actions, making direct target assignments to units only when necessary to ensure proper fire distribution or to prevent engagement of friendly aircraft. See also **centralized control.**

decentralized execution — The delegation of execution authority to subordinate commanders. (JP 0-2)

decentralized items — Those items of supply for which appropriate authority has prescribed local management and procurement.

deception — Those measures designed to mislead the enemy by manipulation, distortion, or falsification of evidence to induce the enemy to react in a manner prejudicial to the enemy's interests. See also **counterdeception; military deception.**

deception action — A collection of related deception events that form a major component of a deception operation. (JP 3-58)

deception concept — The deception course of action forwarded to the Chairman of the Joint Chiefs of Staff for review as part of the CINC's strategic concept. (JP 3-58)

deception course of action — A deception scheme developed during the estimate process in sufficient detail to permit decisionmaking. At a minimum, a deception course of action will identify the deception objective, the deception target, the desired perception, the deception story, and tentative deception means. (JP 3-58)

deception event — A deception means executed at a specific time and location in support of a deception operation. (JP 3-58)

deception means — Methods, resources, and techniques that can be used to convey information to the deception target. There are three categories of deception means: a. **physical means** — Activities and resources used to convey or deny selected information to a foreign power. (Examples include military operations, including exercises, reconnaissance, training activities, and movement of forces; the use of dummy equipment and devices; tactics; bases, logistic actions, stockpiles, and repair activity; and test and evaluation activities.) b. **technical means** — Military materiel resources and their associated operating techniques used to convey or deny selected information to a foreign power through the deliberate radiation, re-radiation, alteration, absorption, or reflection of energy; the emission or suppression of chemical or biological odors; and the emission or suppression of nuclear particles. c. **administrative means** — Resources, methods, and techniques to convey or deny oral, pictorial, documentary, or other physical evidence to a foreign power. (JP 3-58)

deception objective — The desired result of a deception operation expressed in terms of what the adversary is to do or not to do at the critical time and/or location. (JP 3-58)

deception story — A scenario that outlines the friendly actions that will be portrayed to cause the deception target to adopt the desired perception. (JP 3-58)

deception target — The adversary decisionmaker with the authority to make the decision that will achieve the deception objective. (JP 3-58)

decision — In an estimate of the situation, a clear and concise statement of the line of action intended to be followed by the commander as the one most favorable to the successful accomplishment of the assigned mission.

decision altitude — **(*)** An altitude related to the highest elevation in the touchdown zone, specified for a glide slope approach, at which a missed-approach procedure must be initiated if the required visual reference has not been established. See also **decision height.**

decision height — **(*)** A height above the highest elevation in the touchdown zone, specified for a glide slope approach, at which a missed-approach procedure must be initiated if the required visual reference has not been established. See also **decision altitude.**

decision point — The point in space and time where the commander or staff anticipates making a decision concerning a specific friendly course of action. A decision point is usually associated with a specific target area of interest, and is located in time and space to permit the commander sufficient lead time to engage the adversary in the target area of interest. Decision points may also be associated with the friendly force and the status of ongoing operations. See also **course of actions; decision support template; target area of interest.** (JP 2-01.3)

decision support template — A graphic record of wargaming. The decision support template depicts decision points, timelines associated with movement of forces and the flow of the operation, and other key items of information required to execute a specific friendly course of action. See also **course of action; decision point.** (JP 2-01.3)

decisive engagement — In land and naval warfare, an engagement in which a unit is considered fully committed and cannot maneuver or extricate itself. In the absence of outside assistance, the action must be fought to a conclusion and either won or lost with the forces at hand.

decisive point — A geographic place, specific key event, critical system, or function that allows commanders to gain a marked advantage over an enemy and greatly influence the outcome of an attack. See also **centers of gravity.** (JP 3-0)

deck alert — See **ground alert.**

declared speed — The continuous speed which a master declares the ship can maintain on a forthcoming voyage under moderate weather conditions having due regard to the ship's present condition.

declassification — The determination that, in the interests of national security, classified information no longer requires any degree of protection against unauthorized disclosure, coupled with removal or cancellation of the classification designation.

declassify — **(*)** To cancel the security classification of an item of classified matter. Also called **DECL.** See also **downgrade.**

declination — **(*)** The angular distance to a body on the celestial sphere measured north or south through 90 degrees from the celestial equator along the hour circle of the body. Comparable to latitude on the terrestrial sphere. See also **magnetic declination; magnetic variation.**

decompression chamber — See **hyperbaric chamber.**

decompression sickness — A syndrome, including bends, chokes, neurological disturbances, and collapse, resulting from exposure to reduced ambient pressure and caused by gas bubbles in the tissues, fluids, and blood vessels.

decontamination — **(*)** The process of making any person, object, or area safe by absorbing, destroying, neutralizing, making harmless, or removing chemical or biological agents, or by removing radioactive material clinging to or around it.

decontamination station — **(*)** A building or location suitably equipped and organized where personnel and materiel are cleansed of chemical, biological, or radiological contaminants.

decoy — An imitation in any sense of a person, object, or phenomenon which is intended to deceive enemy surveillance devices or mislead enemy evaluation. Also called **dummy.**

decoy ship — **(*)** A ship camouflaged as a noncombatant ship with its armament and other fighting equipment hidden and with special provisions for unmasking its weapons quickly. Also called **Q-ship.**

decrypt — To convert encrypted text into its equivalent plain text by means of a cryptosystem. (This does not include solution by cryptanalysis.) (Note: The term "decrypt" covers the meanings of "decipher" and "decode.") See also **cryptosystem.**

deep fording capability — **(*)** The characteristic of a self-propelled gun or ground vehicle equipped with built-in waterproofing and/or a special waterproofing kit, to negotiate a water obstacle with its wheels or tracks in contact with the ground.

deep minefield — **(*)** An antisubmarine minefield which is safe for surface ships to cross. See also **minefield.**

de facto boundary — **(*)** An international or administrative boundary whose existence and legality is not recognized, but which is a practical division between separate national and provincial administering authorities.

defense area — **(*)** For any particular command, the area extending from the forward edge of the battle area to its rear boundary. It is here that the decisive defensive battle is fought.

Defense Business Operations Fund — A revolving industrial fund concept for a large number of Defense support functions, including transportation. Utilizes business-like cost accounting to determine total cost of a business activity. Defense Business Operations Fund-Transportation is comprised of those Defense Business

Operations Fund accounts assigned by the Office of the Secretary of Defense for Commander in Chief, United States Transportation Command control. Also called **DBOF.** (JP 4-01.7)

defense classification — See **security classification.**

Defense Communications System — Department of Defense long-haul voice, data, and record traffic system which includes the Defense Data Network, Defense Satellite Communications System, and Defense Switched Network. Also called **DCS.** See also **Defense Data Network; Defense Switched Network.** (JP 3-07.4)

Defense Data Network — Component of the Defense Communications System used for switching Department of Defense automated data processing systems. Also called **DDN.** See also **Defense Communications System; Defense Switched Network.** (JP 3-07.4)

defense emergency — An emergency condition that exists when: a. a major attack is made upon US forces overseas or on allied forces in any theater and is confirmed by either the commander of a command established by the Secretary of Defense or higher authority; or b. an overt attack of any type is made upon the United States and is confirmed either by the commander of a command established by the Secretary of Defense or higher authority.

defense in depth — The siting of mutually supporting defense positions designed to absorb and progressively weaken attack, prevent initial observations of the whole position by the enemy, and to allow the commander to maneuver the reserve.

defense information infrastructure — The shared or interconnected system of computers, communications, data applications, security, people, training, and other support structures serving Department of Defense (DOD) local, national, and worldwide information needs. The defense information infrastructure connects DOD mission support, command and control, and intelligence computers through voice, telecommunications, imagery, video, and multimedia services. It provides information processing and services to subscribers over the Defense Information Systems Network and includes command and control, tactical, intelligence, and commercial communications systems used to transmit DOD information. Also called **DII.** See also **global information infrastructure; information; infrastructure; national information infrastructure.** (JP 3-13)

Defense Information Systems Network — Integrated network, centrally managed and configured to provide long-haul information transfer services for all Department of Defense activities. It is an information transfer utility designed to provide dedicated point-to-point, switched voice and data, imagery, and video teleconferencing services. Also called **DISN.** (JP 2-01)

defense intelligence production — The integration, evaluation, analysis, and interpretation of information from single or multiple sources into finished intelligence for known or anticipated military and related national security consumer requirements. (JP 2-0)

defense message system — Consists of all hardware, software, procedures, standards, facilities, and personnel used to exchange messages electronically.

Defense Meteorological Satellite Program — Military weather satellite controlled by

National Oceanic and Atmospheric Administration. Also called **DMSP.**

Defense Planning Guidance — This document, issued by the Secretary of Defense, provides firm guidance in the form of goals, priorities, and objectives, including fiscal constraints, for the development of the Program Objective Memorandums by the Military Departments and Defense agencies. Also called **DPG.**

defense readiness condition — A uniform system of progressive alert postures for use between the Chairman of the Joint Chiefs of Staff and the commanders of unified and specified commands and for use by the Services. Defense readiness conditions are graduated to match situations of varying military severity (status of alert). Defense readiness conditions are identified by the short title DEFCON (5), (4), (3), (2), and (1), as appropriate. Also called **DEFCON.**

Defense Satellite Communications System — Geosynchronous military communications satellites that provide high data rate communications for military forces, diplomatic corps, and the White House. The Defense Satellite Communications System provides long-haul super-high frequency 7/8 gigahertz voice and high data rate communications for fixed and transportable terminals, and extends mobile service to a limited number of ships and aircraft. Also called **DSCS.**

Defense Support Program — Satellites that provide early warning of missile launches; the first line of defense against missile attack against North America. Also called **DSP.**

Defense Switched Network — Component of the Defense Communications System that handles Department of Defense voice, data, and video communications. Also called **DSN.** See also **Defense Communications System; Defense Data Network.** (JP 3-07.4)

Defense Transportation System — That portion of the Nation's transportation infrastructure that supports Department of Defense common-user transportation needs across the range of military operations. It consists of those common-user military and commercial assets, services, and systems organic to, contracted for, or controlled by the Department of Defense. Also called **DTS.** See also **common-user transportation; transportation system.**

defensive coastal area — **(*)** A part of a coastal area and of the air, land, and water area adjacent to the coastline within which defense operations may involve land, sea, and air forces.

defensive counterair — All defensive measures designed to detect, identify, intercept, and destroy or negate enemy forces attempting to attack or penetrate the friendly air environment. Also called **DCA.** See also **counterair; offensive counterair.** (JP 3-01)

defensive information operations — The integration and coordination of policies and procedures, operations, personnel, and technology to protect and defend information and information systems. Defensive information operations are conducted through information assurance, physical security, operations security, counter-deception, counter-psychological operations, counterintelligence, electronic warfare, and special information operations. Defensive information operations ensure timely, accurate, and relevant information access while denying adversaries the opportunity to exploit friendly information and information systems for their own purposes. See also **counterintelligence; electronic warfare; information**

assurance; information operations; information system; offensive information operations; operations security; physical security; special information operations. (JP 3-13)

defensive minefield — **(*)** 1. In naval mine warfare, a minefield laid in international waters or international straits with the declared intention of controlling shipping in defense of sea communications. 2. **(DOD only)** In land mine warfare, a minefield laid in accordance with an established plan to prevent a penetration between positions and to strengthen the defense of the positions themselves. See also **minefield.**

defensive sea area — A sea area, usually including the approaches to and the waters of important ports, harbors, bays, or sounds, for the control and protection of shipping; for the safeguarding of defense installations bordering on waters of the areas; and for provision of other security measures required within the specified areas. It does not extend seaward beyond the territorial waters. See also **maritime control area.**

defensive zone — A belt of terrain, generally parallel to the front, that includes two or more organized, or partially organized, battle positions.

defilade — **(*)** 1. Protection from hostile observation and fire provided by an obstacle such as a hill, ridge, or bank. 2. A vertical distance by which a position is concealed from enemy observation. 3. To shield from enemy fire or observation by using natural or artificial obstacles.

defoliant operation — **(*)** The employment of defoliating agents on vegetated areas in support of military operations.

defoliating agent — **(*)** A chemical which causes trees, shrubs, and other plants to shed their leaves prematurely.

degaussing — The process whereby a ship's magnetic field is reduced by the use of electromagnetic coils, permanent magnets, or other means.

degree of risk — As specified by the commander, the risk to which friendly forces may be subjected from the effects of the detonation of a nuclear weapon used in the attack of a close-in enemy target; acceptable degrees of risk under differing tactical conditions are emergency, moderate, and negligible. See also **emergency risk (nuclear); negligible risk (nuclear).**

de jure boundary — **(*)** An international or administrative boundary whose existence and legality is recognized.

delayed entry program — A program under which an individual may enlist in a Reserve Component of a military service and specify a future reporting date for entry on active duty in the Active Component that would coincide with availability of training spaces and with personal plans such as high school graduation. Also called **DEP.** See also **active duty; Reserve Components.** (JP 4-05)

delaying action — See **delaying operation.**

delaying operation — **(*)** An operation in which a force under pressure trades space for time by slowing down the enemy's momentum and inflicting maximum damage on the enemy without, in principle, becoming decisively engaged.

delay release sinker — **(*)** A sinker which holds a moored mine on the sea-bed for a predetermined time after laying.

delegation of authority — The action by which a commander assigns part of his or her authority commensurate with the assigned task to a subordinate commander. While ultimate responsibility cannot be relinquished, delegation of authority carries with it the imposition of a measure of responsibility. The extent of the authority delegated must be clearly stated.

deliberate attack — **(*)** A type of offensive action characterized by preplanned coordinated employment of firepower and maneuver to close with and destroy or capture the enemy.

deliberate breaching — **(*)** The creation of a lane through a minefield or a clear route through a barrier or fortification, which is systematically planned and carried out.

deliberate crossing — **(*)** The crossing of an inland water obstacle that requires extensive planning and detailed preparations. See also **hasty crossing.**

deliberate defense — **(*)** A defense normally organized when out of contact with the enemy or when contact with the enemy is not imminent and time for organization is available. It normally includes an extensive fortified zone incorporating pillboxes, forts, and communications systems. See also **hasty defense.**

deliberate planning — 1. The Joint Operation Planning and Execution System process involving the development of joint operation plans for contingencies identified in joint strategic planning documents. Deliberate planning is accomplished in prescribed cycles that complement other Department of Defense planning cycles in accordance with the formally established Joint Strategic Planning System. 2. A planning process for the deployment and employment of apportioned forces and resources that occurs in response to a hypothetical situation. Deliberate planners rely heavily on assumptions regarding the circumstances that will exist when the plan is executed. See also **Joint Operation Planning and Execution System; Joint Strategic Planning System.** (JP 5-00.1)

delivering ship — The ship in a replenishment unit that delivers the rig(s).

delivery error — **(*)** The inaccuracy associated with a given weapon system resulting in a dispersion of shots about the aiming point. See also **circular error probable; deviation; dispersion; dispersion error; horizontal error.**

delivery forecasts — 1. Periodic estimates of contract production deliveries used as a measure of the effectiveness of production and supply availability scheduling and as a guide to corrective actions to resolve procurement or production bottlenecks. 2. Estimates of deliveries under obligation against procurement from appropriated or other funds.

delivery requirements — The stipulation that requires that an item of materiel must be delivered in the total quantity required by the date required.

demilitarized zone — **(*)** A defined area in which the stationing or concentrating of military forces, or the retention or establishment of military installations of any description, is prohibited. (JP 3-07.3)

demobilization — The process of transitioning a conflict or wartime military establishment and defense-based civilian economy to a peacetime configuration while maintaining national security and economic vitality. See also **mobilization.** (JP 4-05)

demolition belt — A selected land area sown with explosive charges, mines, and other available obstacles to deny use of the land to enemy operations, and as a protection to friendly troops. There are two types of demolition belts: a. **primary**. A continuous series of obstacles across the whole front, selected by the division or higher commander. The preparation of such a belt is normally a priority engineer task. b. **subsidiary**. A supplement to the primary belt to give depth in front or behind or to protect the flanks.

demolition chamber — (*) Space intentionally provided in a structure for the emplacement of explosive charges.

demolition firing party — The party at the site that is technically responsible for the demolition and that actually initiates detonation or fires the demolitions. See also **demolition guard; state of readiness.**

demolition guard — A local force positioned to ensure that a target is not captured by an enemy before orders are given for its demolition and before the demolition has been successfully fired. The commander of the demolition guard is responsible for the tactical control of all troops at the demolition site, including the demolition firing party. The commander of the demolition guard is responsible for transmitting the order to fire to the demolition firing party.

demolition kit — (*) The demolition tool kit complete with explosives. See also **demolition tool kit.**

demolition target — (*) A target of known military interest identified for possible future demolition. See also **charged demolition target; preliminary demolition target; prewithdrawal demolition target; reserved demolition target; uncharged demolition target.**

demolition tool kit — (*) The tools, materials and accessories of a nonexplosive nature necessary for preparing demolition charges. See also **demolition kit.**

demonstration — (*) 1. An attack or show of force on a front where a decision is not sought, made with the aim of deceiving the enemy. See also **amphibious demonstration; diversion; diversionary attack.** 2. **(DOD only)** In military deception, a show of force in an area where a decision is not sought made to deceive an adversary. It is similar to a feint but no actual contact with the adversary is intended. (JP 3-58)

denial measure — (*) An action to hinder or deny the enemy the use of space, personnel, or facilities. It may include destruction, removal, contamination, or erection of obstructions.

denied area — An area under enemy or unfriendly control in which friendly forces cannot expect to operate successfully within existing operational constraints and force capabilities. (JP 3-05)

density altitude — (*) An atmospheric density expressed in terms of the altitude which corresponds with that density in the standard atmosphere.

departmental intelligence — Intelligence that any department or agency of the Federal Government requires to execute its own mission.

Department of Defense construction agent — The Corps of Engineers, Naval Facilities Engineering Command, or other such approved Department of Defense activity, that is assigned design or execution responsibilities associated with military construction programs, facilities support, or civil engineering support to the combatant commanders in contingency operations. See also **contingency operation.** (JP 3-34)

Department of Defense container system — All Department of Defense (DOD)-owned, leased, and controlled 20-

or 40-foot intermodal International Organization for Standardization containers and flatracks, supporting equipment such as generator sets and chassis, container handling equipment, information systems, and other infrastructure that supports DOD transportation and logistic operations, including commercially provided transportation services. This also includes 463L pallets, nets, and tie down equipment as integral components of the DOD Intermodal Container System. Size and configuration of the common-use portion of the DOD container system controlled by US Transportation Command (USTRANSCOM), will be determined by USTRANSCOM based on established requirements and availability of commercially owned containers and equipment. USTRANSCOM will lease or procure additional containers as required to augment the DOD container system. See also **container-handling equipment; containerization; International Organization for Standardization.** (JP 4-01.7)

Department of Defense Intelligence Information System — The combination of Department of Defense personnel, procedures, equipment, computer programs, and supporting communications that support the timely and comprehensive preparation and presentation of intelligence and information to military commanders and national-level decision makers. Also called **DODIIS.**

Department of Defense internal audit organizations — The Army Audit Agency; Naval Audit Service; Air Force Audit Agency; and the Office of the Assistant Inspector General for Auditing, Office of the Inspector General, Department of Defense.

Department of Defense single manager — A Military Department or Agency, designated by the Secretary of Defense as a Department of Defense (DOD) executive agent, that is responsible for management of specified logistic commodities or common service activities on a DOD-wide basis. See also **executive agent.** (JP 4-07)

Department of the Air Force — The executive part of the Department of the Air Force at the seat of government and all field headquarters, forces, Reserve Components, installations, activities, and functions under the control or supervision of the Secretary of the Air Force. Also called **DAF.** See also **Military Department.**

Department of the Army — The executive part of the Department of the Army at the seat of government and all field headquarters, forces, Reserve Components, installations, activities, and functions under the control or supervision of the Secretary of the Army. Also called **DA.** See also **Military Department.**

Department of the Navy — The executive part of the Department of the Navy at the seat of government; the headquarters, US Marine Corps; the entire operating forces of the United States Navy and of the US Marine Corps, including the Reserve Components of such forces; all field activities, headquarters, forces, bases, installations, activities, and functions under the control or supervision of the Secretary of the Navy; and the US Coast Guard when operating as a part of the Navy pursuant to law. Also called **DON.** See also **Military Department.**

departure airfield — An airfield on which troops and/or materiel are enplaned for flight. See also **airfield.**

departure area — The general area encompassing all base camps, bivouacs, and departure airfield facilities. (JP 3-17)

departure end — **(*)** That end of a runway nearest to the direction in which initial departure is made.

departure point — **(*)** 1. A navigational check point used by aircraft as a marker for setting course. 2. In amphibious operations, an air control point at the seaward end of the helicopter approach lane system from which helicopter waves are dispatched along the selected helicopter approach lane to the initial point.

deployable joint task force augmentation cell — A combatant commander (CINC) asset composed of personnel from the CINC's staff and component representatives. The members represent a multi-service, multi-disciplined group of planners and operators which operationally report to the CINC's Operations Directorate until deployed to a joint task force (JTF). It can be tailored to meet the needs of a commander, joint task force and deploy within 48 hours from notification. Members can also act as liaison officers between the CINC and the JTF. Also called **DJTFAC.**

deployed health surveillance — The identification of a population at risk, recognition and assessment of hazardous exposures, employment of specific countermeasures, and monitoring health outcomes.

deployed nuclear weapons — 1. When used in connection with the transfer of weapons between the Department of Energy and the Department of Defense, this term describes those weapons transferred to and in the custody of the Department of Defense. 2. Those nuclear weapons specifically authorized by the Joint Chiefs of Staff to be transferred to the custody of the storage facilities or carrying or delivery units of the Armed Forces.

deployment — 1. In naval usage, the change from a cruising approach or contact disposition to a disposition for battle. 2. The movement of forces within operational areas. 3. The positioning of forces into a formation for battle. 4. The relocation of forces and materiel to desired operational areas. Deployment encompasses all activities from origin or home station through destination, specifically including intra-continental United States, intertheater, and intratheater movement legs, staging, and holding areas. See also **deployment order; deployment planning; deployment preparation order.** (JP 4-0)

deployment database — The Joint Operation Planning and Execution System database containing the necessary information on forces, materiel, and filler and replacement personnel movement requirements to support execution. The database reflects information contained in the refined time-phased force and deployment data from the deliberate planning process or developed during the various phases of the crisis action planning process, and the movement schedules or tables developed by the transportation component commands to support the deployment of required forces, personnel, and materiel. See also **time-phased force and deployment data.**

deployment diagram — In the assault phase of an amphibious operation, a diagram showing the formation in which the boat group proceeds from the rendezvous area to the line of departure and the method of deployment into the landing formation.

deployment order — A planning directive from the Secretary of Defense, issued by the Chairman of the Joint Chiefs of Staff, that authorizes and directs the transfer of forces between combatant commands by reassignment or attachment. A deployment order normally specifies the authority that the gaining combatant commander will exercise over the transferred forces. See also **deployment; deployment planning; deployment preparation order.** (JP 5-0)

deployment planning — Operational planning directed toward the movement of forces and sustainment resources from their original locations to a specific operational area for conducting the joint operations contemplated in a given plan. Encompasses all activities from origin or home station through destination, specifically including intra-continental United States, intertheater, and intratheater movement legs, staging areas, and holding areas. See also **deployment; deployment order; deployment preparation order.** (JP 5-0)

deployment preparation order — An order issued by competent authority to move forces or prepare forces for movement (e.g., increase deployability posture of units). See also **deployment; deployment order; deployment planning.**

depot — 1. **supply** — An activity for the receipt, classification, storage, accounting, issue, maintenance, procurement, manufacture, assembly, research, salvage, or disposal of material. 2. **personnel** — An activity for the reception, processing, training, assignment, and forwarding of personnel replacements.

depot maintenance — That maintenance performed on materiel requiring major overhaul or a complete rebuild of parts, assemblies, subassemblies, and end-items, including the manufacture of parts, modifications, testing, and reclamation as required. Depot maintenance serves to support lower categories of maintenance by providing technical assistance and performing that maintenance beyond their responsibility. Depot maintenance provides stocks of serviceable equipment by using more extensive facilities for repair than are available in lower level maintenance activities.

depth — **(*)** In maritime/hydrographic use, the vertical distance from the plane of the hydrographic datum to the bed of the sea, lake, or river.

depth contour — **(*)** A line connecting points of equal depth below the hydrographic datum. Also called **bathymetric contour or depth curve.**

depth curve — See **depth contour.**

descriptive name — **(*)** Written indication on maps and charts, used to specify the nature of a feature (natural or artificial) shown by a general symbol.

desired appreciation — See **appreciations.**

desired effects — The damage or casualties to the enemy or materiel that a commander desires to achieve from a nuclear weapon detonation. Damage effects on materiel are classified as light, moderate, or severe. Casualty effects on personnel may be immediate, prompt, or delayed.

desired ground zero — **(*)** The point on the surface of the Earth at, or vertically below or above, the center of a planned nuclear detonation. Also called **DGZ.** See also **actual ground zero; ground zero.**

desired perception — In military deception, what the deception target must believe for it to make the decision that will achieve the deception objective. (JP 3-58)

destroyed — A condition of a target so damaged that it can neither function as intended nor be restored to a usable condition. In the case of a building, all vertical supports and spanning members are damaged to such an extent that nothing is salvageable. In the case of bridges, all spans must have dropped and all piers must require replacement.

destruction — A type of adjustment for destroying a given target.

destruction fire — Fire delivered for the sole purpose of destroying material objects. See also **fire.**

destruction fire mission — **(*)** In artillery, fire delivered for the purpose of destroying a point target. See also **fire.**

destruction radius — **(*)** In mine warfare, the maximum distance from an exploding charge of stated size and type at which a mine will be destroyed by sympathetic detonation of the main charge, with a stated probability of destruction, regardless of orientation.

detachment — **(*)** 1. A part of a unit separated from its main organization for duty elsewhere. 2. A temporary military or naval unit formed from other units or parts of units. Also called **DET.**

detailed photographic report — **(*)** A comprehensive, analytical, intelligence report written as a result of the interpretation of photography usually covering a single subject, a target, target complex, and of a detailed nature.

detained — See **missing.**

detainee — A term used to refer to any person captured or otherwise detained by an armed force.

detainee collecting point — A facility or other location where detainees are assembled for subsequent movement to a detainee processing station.

detainee processing station — A facility or other location where detainees are administratively processed and provided custodial care pending disposition and subsequent release, transfer, or movement to a prisoner-of-war or civilian internee camp.

detecting circuit — **(*)** The part of a mine firing circuit which responds to the influence of a target.

detection — 1. In tactical operations, the perception of an object of possible military interest but unconfirmed by recognition. 2. In surveillance, the determination and transmission by a surveillance system that an event has occurred. 3. In arms control, the first step in the process of ascertaining the occurrence of a violation of an arms control agreement. 4. In nuclear, biological, and chemical (NBC) environments, the act of locating NBC hazards by use of NBC detectors or monitoring and/or survey teams. See also **hazard; monitoring; nuclear, biological, and chemical environment.** (JP 3-11)

deterioration limit — **(*)** A limit placed on a particular product characteristic to define the minimum acceptable quality requirement for the product to retain its NATO code number.

deterrence — The prevention from action by fear of the consequences. Deterrence is a state of mind brought about by the existence of a credible threat of unacceptable counteraction.

deterrent options — A course of action, developed on the best economic, diplomatic, political, and military judgment, designed to dissuade an adversary from a current course of action or contemplated operations. (In constructing an operation plan, a range of options should be presented to effect deterrence. Each option requiring deployment of forces should be a separate force module.)

detonating cord — **(*)** A waterproof, flexible fabric tube containing a high explosive designed to transmit the detonation wave.

detonator — **(*)** A device containing a sensitive explosive intended to produce a detonation wave.

developmental assistance — US Agency for International Development function chartered under chapter one of the Foreign Assistance Act of 1961, primarily designed to promote economic growth and the equitable distribution of its benefits. (JP 3-08)

deviation — **(*)** 1. The distance by which a point of impact or burst misses the target. See also **circular error probable; delivery error; dispersion error; horizontal error.** 2. The angular difference between magnetic and compass headings.

diaphragm stop — See **relative aperture.**

diapositive — **(*)** A positive photograph on a transparent medium.

died of wounds received in action — A casualty category applicable to a hostile casualty, other than the victim of a terrorist activity, who dies of wounds or other injuries received in action after having reached a medical treatment facility. Also called **DWRIA.** See also **casualty category.**

differential ballistic wind — **(*)** In bombing, a hypothetical wind equal to the difference in velocity between the ballistic wind and the actual wind at a release altitude.

diffraction loading — **(*)** The total force which is exerted on the sides of a structure by the advancing shock front of a nuclear explosion.

dip — **(*)** In naval mine warfare, the amount by which a moored mine is carried beneath its set depth by a current or tidal stream acting on the mine casing and mooring.

diplomatic authorization — **(*)** Authority for overflight or landing obtained at government-to-government level through diplomatic channels.

diplomatic and/or consular facility — Any Foreign Service establishment maintained by the US Department of State abroad. It may be designated a "mission" or "consular office," or given a special designation for particular purposes, such as "United States Liaison Office." A "mission" is designated as an embassy and is maintained in order to conduct normal continuing diplomatic relations between the US Government and other governments. A "consular office" is any consulate general or consulate that may participate in most foreign affairs activities, and varies in size and scope.

dip needle circuit — **(*)** In naval mine warfare, a mechanism which responds to a change in the magnitude of the vertical component of the total magnetic field.

direct action — Short-duration strikes and other small-scale offensive actions by special operations forces or special operations-capable units to seize, destroy, capture, recover, or inflict damage on designated personnel or materiel. In the conduct of these operations, special operations forces or special operations-capable units may employ raid, ambush, or direct assault tactics; emplace mines and other munitions; conduct standoff attacks by fire from air, ground, or maritime platforms; provide terminal guidance for precision-guided munitions; conduct independent sabotage; and conduct anti-ship operations. Also called **DA.** See also

special operations; special operations forces. (JP 3-05)

direct action fuze — See **impact action fuze; proximity fuze; self-destroying fuse; time fuze.**

direct air support center — The principal air control agency of the US Marine air command and control system responsible for the direction and control of air operations directly supporting the ground combat element. It processes and coordinates requests for immediate air support and coordinates air missions requiring integration with ground forces and other supporting arms. It normally collocates with the senior fire support coordination center within the ground combat element and is subordinate to the tactical air command center. Also called **DASC.** See also **Marine air command and control system; tactical air operations center.** (JP 3-09.3)

direct air support center (airborne) — An airborne aircraft equipped with the necessary staff personnel, communications, and operations facilities to function as a direct air support center. Also called **DASC-A.** See also **direct air support center.**

directed energy — An umbrella term covering technologies that relate to the production of a beam of concentrated electromagnetic energy or atomic or subatomic particles. Also called **DE.** See also **directed-energy device; directed-energy weapon.**

directed-energy device — A system using directed energy primarily for a purpose other than as a weapon. Directed-energy devices may produce effects that could allow the device to be used as a weapon against certain threats; for example, laser rangefinders and designators used against sensors that are sensitive to light. See also **directed energy; directed-energy weapon.**

directed-energy protective measures — That division of directed-energy warfare involving actions taken to protect friendly equipment, facilities, and personnel to ensure friendly effective uses of the electromagnetic spectrum that are threatened by hostile directed-energy weapons and devices.

directed-energy warfare — Military action involving the use of directed-energy weapons, devices, and countermeasures to either cause direct damage or destruction of enemy equipment, facilities, and personnel, or to determine, exploit, reduce, or prevent hostile use of the electromagnetic spectrum through damage, destruction, and disruption. It also includes actions taken to protect friendly equipment, facilities, and personnel and retain friendly use of the electromagnetic spectrum. Also called **DEW.** See also **directed energy; directed-energy device; directed-energy weapon; electromagnetic spectrum; electronic warfare.**

directed-energy weapon — A system using directed energy primarily as a direct means to damage or destroy enemy equipment, facilities, and personnel. See also **directed energy; directed-energy device.**

direct exchange — A supply method of issuing serviceable materiel in exchange for unserviceable materiel on an item-for-item basis. Also called **DX.**

direct fire — Gunfire delivered on a target, using the target itself as a point of aim for either the gun or the director.

direct illumination — **(*)** Illumination provided by direct light from pyrotechnics or searchlights.

directing staff — See **exercise directing staff.**

direction — In artillery and naval gunfire support, a term used by a spotter and/or observer in a call for fire to indicate the bearing of the spotting line. See also **bearing; call for fire; naval gunfire support; spotter; spotting line.** (JP 2-0)

directional gyro indicator — An azimuth gyro with a direct display and means for setting the datum to a specified compass heading.

direction finding — A procedure for obtaining bearings of radio frequency emitters by using a highly directional antenna and a display unit on an intercept receiver or ancillary equipment.

direction of attack — A specific direction or route that the main attack or center of mass of the unit will follow. The unit is restricted, required to attack as indicated, and is not normally allowed to bypass the enemy. The direction of attack is used primarily in counterattacks or to ensure that supporting attacks make maximal contribution to the main attack.

directive — **(*)** 1. A military communication in which policy is established or a specific action is ordered. 2. A plan issued with a view to putting it into effect when so directed, or in the event that a stated contingency arises. 3. Broadly speaking, any communication which initiates or governs action, conduct, or procedure.

directive authority for logistics — Combatant commander authority to issue directives to subordinate commanders, including peacetime measures, necessary to ensure the effective execution of approved operation plans. Essential measures include the optimized use or reallocation of available resources and prevention or elimination of redundant facilities and/or overlapping functions among the Service component commands. See also **combatant command (command authority); logistics.** (JP 0-2)

direct laying — Laying in which the sights of weapons are aligned directly on the target. Normally used in conjunction with mortars and sometimes artillery. See also **lay.**

direct liaison authorized — That authority granted by a commander (any level) to a subordinate to directly consult or coordinate an action with a command or agency within or outside of the granting command. Direct liaison authorized is more applicable to planning than operations and always carries with it the requirement of keeping the commander granting direct liaison authorized informed. Direct liaison authorized is a coordination relationship, not an authority through which command may be exercised. Also called **DIRLAUTH.** (JP 0-2)

director of mobility forces — Normally a senior officer who is familiar with the area of responsibility or joint operations area and possesses an extensive background in airlift operations. When established, the director of mobility forces serves as the designated agent for all airlift issues in the area of responsibility or joint operations area, and for other duties as directed. The director of mobility forces exercises coordinating authority between the airlift coordination cell, the air mobility element, the Tanker Airlift Control Center, the joint movement center, and the air operations center in order to expedite the resolution of airlift problems. The director of mobility forces may be sourced from the theater's organizations, United States Transportation Command, or United States Atlantic Command. Also called **DIRMOBFOR.** See also **airlift coordination cell; air**

mobility element; air operations center; area of responsibility; coordinating authority; joint movement center; joint operations area; tanker airlift control center. (JP 4-01.1)

direct support — A mission requiring a force to support another specific force and authorizing it to answer directly to the supported force's request for assistance. Also called **DS.** See also **close support; general support; mission; mutual support; support.** (JP 3-09.1)

direct support artillery — **(*)** Artillery whose primary task is to provide fire requested by the supported unit.

direct supporting fire — **(*)** Fire delivered in support of part of a force, as opposed to general supporting fire which is delivered in support of the force as a whole. See also **supporting fire.**

direct vendor delivery — A materiel acquisition and distribution method that requires vendor delivery directly to the customer. Also called **DVD.** See also **distribution.** (JP 4-09)

disabling fire — The firing of ordnance by ships or aircraft at the steering or propulsion system of a vessel. The intent is to disable with minimum injury to personnel or damage to vessel.

disaffected person — A person who is alienated or estranged from those in authority or lacks loyalty to the government; a state of mind.

disarmament — The reduction of a military establishment to some level set by international agreement. See also **arms control; arms control agreement; arms control measure.**

disarmed mine — **(*)** A mine for which the arming procedure has been reversed, rendering the mine inoperative. It is safe to handle and transport and can be rearmed by simple action.

disaster assistance response team — United States Agency for International Development's (USAID) Office of United States Foreign Disaster Assistance provides this rapidly deployable team in response to international disasters. A disaster assistance response team provides specialists, trained in a variety of disaster relief skills, to assist US embassies and USAID missions with the management of US Government response to disasters. Also called **DART.** See also **foreign disaster; foreign disaster relief.** (JP 3-08)

disaster control — Measures taken before, during, or after hostile action or natural or manmade disasters to reduce the probability of damage, minimize its effects, and initiate recovery. See also **area damage control; damage control.**

discriminating circuit — **(*)** That part of the operating circuit of a sea mine which distinguishes between the response of the detecting circuit to the passage of a ship and the response to other disturbances (e.g., influence sweep, countermining, etc.)

disease and nonbattle injury casualty — A person who is not a battle casualty but who is lost to the organization by reason of disease or injury, including persons dying of disease or injury, by reason of being missing where the absence does not appear to be voluntary, or due to enemy action or being interned. Also called **DNBI casualty.** (JP 4-02)

disembarkation schedule — See **debarkation schedule.**

disengagement — In arms control, a general term for proposals that would result in the geographic separation of opposing nonindigenous forces without directly affecting indigenous military forces.

dislocated civilian — A broad term that includes a displaced person, a stateless person, an evacuee, an expellee, or a refugee. Also called **DC.** See also **displaced person; evacuee; expellee; refugee; stateless person.** (JP 3-07.6)

dispatch route — **(*)** In road traffic, a roadway over which full control, both as to priorities of use and the regulation of movement of traffic in time and space, is exercised. Movement authorization is required for its use, even by a single vehicle. See also **route.**

dispenser — **(*)** In air armament, a container or device which is used to carry and release submunitions. See also **cluster bomb unit.**

dispersal — Relocation of forces for the purpose of increasing survivability. See also **dispersion.**

dispersal airfield — An airfield, military or civil, to which aircraft might move before H-hour on either a temporary duty or permanent change of station basis and be able to conduct operations. See also **airfield.**

dispersed movement pattern — **(*)** A pattern for ship-to-shore movement which provides additional separation of landing craft both laterally and in depth. This pattern is used when nuclear weapon threat is a factor.

dispersed site — **(*)** A site selected to reduce concentration and vulnerability by its separation from other military targets or a recognized threat area.

dispersion — **(*)** 1. A scattered pattern of hits around the mean point of impact of bombs and projectiles dropped or fired under identical conditions. 2. In antiaircraft gunnery, the scattering of shots in range and deflection about the mean point of explosion. 3. The spreading or separating of troops, materiel, establishments, or activities which are usually concentrated in limited areas to reduce vulnerability. 4. In chemical and biological operations, the dissemination of agents in liquid or aerosol form. 5. In airdrop operations, the scatter of personnel and/or cargo on the drop zone. 6. In naval control of shipping, the reberthing of a ship in the periphery of the port area or in the vicinity of the port for its own protection in order to minimize the risk of damage from attack. See also **circular error probable; convoy dispersal point; delivery error; deviation; dispersion error; horizontal error.**

dispersion error — **(*)** The distance from the point of impact or burst of a round to the mean point of impact or burst.

dispersion pattern — **(*)** The distribution of a series of rounds fired from one weapon or a group of weapons under conditions as nearly identical as possible; the points of burst or impact being dispersed about a point called the mean point of impact.

displaced person — A civilian who is involuntarily outside the national boundaries of his or her country. See also **evacuee; refugee.**

display — In military deception, a static portrayal of an activity, force, or equipment intended to deceive the adversary's visual observation. (JP 3-58)

disposition — **(*)** 1. Distribution of the elements of a command within an area; usually the exact location of each unit headquarters and the deployment of the forces subordinate to it. 2. A prescribed arrangement of the stations to be occupied by the several formations and single ships of a fleet, or major subdivisions of a fleet, for any purpose, such as cruising, approach, maintaining contact, or battle. 3. A prescribed arrangement of all the tactical units composing a flight or group of aircraft. See also **deployment; dispersion.** 4. **(DOD only)** The removal of a patient from a medical treatment facility by reason of return to duty, transfer to another treatment facility, death, or other termination of medical case.

disruptive pattern — **(*)** In surveillance, an arrangement of suitably colored irregular shapes which, when applied to the surface of an object, is intended to enhance its camouflage.

dissemination — See **intelligence cycle.**

dissemination and integration — See **intelligence cycle.** See also **dissemination; integration.** (JP 2-0)

distance — 1. The space between adjacent individual ships or boats measured in any direction between foremasts. 2. The space between adjacent men, animals, vehicles, or units in a formation measured from front to rear. 3. The space between known reference points or a ground observer and a target, measured in meters (artillery), in yards (naval gunfire), or in units specified by the observer. See also **interval.**

distant retirement area — In amphibious operations, that sea area located to seaward of the landing area. This area is divided into a number of operating areas to which assault ships may retire and operate in the event of adverse weather or to prevent concentration of ships in the landing area. See also **amphibious operation; landing area; retirement.** (JP 3-02)

distant support area — In amphibious operations, the area located in the vicinity of the landing area but at considerable distance seaward of it. These areas are assigned to distant support forces, such as striking forces, surface action groups, surface action units, and their logistic groups. See also **amphibious operation; landing area.** (JP 3-02)

distributed fire — **(*)** Fire so dispersed as to engage most effectively an area target. See also **fire.**

distribution — 1. The arrangement of troops for any purpose, such as a battle, march, or maneuver. 2. A planned pattern of projectiles about a point. 3. A planned spread of fire to cover a desired frontage or depth. 4. An official delivery of anything, such as orders or supplies. 5. The operational process of synchronizing all elements of the logistic system to deliver the "right things" to the "right place" at the "right time" to support the geographic combatant commander. 6. The process of assigning military personnel to activities, units, or billets. (JP 4-0)

distribution manager — The executive agent for managing distribution with the combatant commander's area of responsibility. See also **area of responsibility; distribution.** (JP 4-01.4)

distribution pipeline — Continuum or channel through which the Department of Defense conducts distribution operations. The distribution pipeline represents the end-to-end flow of resources from supplier to consumer and, in some cases, back to the supplier in retrograde activities. See also **distribution; pipeline.** (JP 4-01.4)

distribution plan — A reporting system comprising reports, updates, and information systems feeds that articulate the requirements of the theater distribution system to the strategic and operational resources assigned responsibility for support to the theater. It portrays the interface of the physical, financial, information and communications networks for gaining visibility of the theater distribution system and communicates control activities necessary for optimizing capacity of the system. It depicts, and is continually updated to reflect changes in, infrastructure, support relationships, and customer locations to all elements of the distribution system (strategic operational, and tactical). See also **distribution; distribution system; theater distribution; theater distribution system.** (JP 4-01.4)

distribution point — (*) A point at which supplies and/or ammunition, obtained from supporting supply points by a division or other unit, are broken down for distribution to subordinate units. Distribution points usually carry no stocks; items drawn are issued completely as soon as possible.

distribution system — That complex of facilities, installations, methods, and procedures designed to receive, store, maintain, distribute, and control the flow of military materiel between the point of receipt into the military system and the point of issue to using activities and units.

ditching — Controlled landing of a distressed aircraft on water.

diversion — 1. The act of drawing the attention and forces of an enemy from the point of the principal operation; an attack, alarm, or feint that diverts attention. 2. A change made in a prescribed route for operational or tactical reasons. A diversion order will not constitute a change of destination. 3. A rerouting of cargo or passengers to a new transshipment point or destination or on a different mode of transportation prior to arrival at ultimate destination. 4. In naval mine warfare, a route or channel bypassing a dangerous area. A diversion may connect one channel to another or it may branch from a channel and rejoin it on the other side of the danger. See also **demonstration.**

diversion airfield — (*) An airfield with at least minimum essential facilities, which may be used as an emergency airfield or when the main or redeployment airfield is not usable or as required to facilitate tactical operations. Also called **divert field.** See also **airfield; departure airfield; main airfield; redeployment airfield.**

diversionary attack — (*) An attack wherein a force attacks, or threatens to attack, a target other than the main target for the purpose of drawing enemy defenses away from the main effort. See also **demonstration.**

diversionary landing — An operation in which troops are actually landed for the purpose of diverting enemy reaction away from the main landing.

divert field — See **diversion airfield.**

diving chamber — See **hyperbaric chamber.**

division — (*) 1. A tactical unit/formation as follows: a. A major administrative and tactical unit/formation which combines in itself the necessary arms and services required for sustained combat, larger than a regiment/brigade and smaller than a corps. b. A number of naval vessels of similar type grouped together for operational and administrative command, or a tactical unit of a naval aircraft squadron, consisting of two or more sections. c. An air division is an air combat organization normally consisting of two or more wings with appropriate service units. The combat wings of an air division will normally contain similar type units. 2. An organizational part of a headquarters that handles military matters of a particular nature, such as personnel, intelligence, plans, and training, or supply and evacuation. 3. **(DOD only)** A number of personnel of a ship's complement grouped together for tactical and administrative control.

division artillery — Artillery that is permanently an integral part of a division. For tactical purposes, all artillery placed under the command of a division commander is considered division artillery.

doctrinal template — A model based on known or postulated adversary doctrine. Doctrinal templates illustrate the disposition and activity of adversary forces and assets conducting a particular operation unconstrained by the effects of the battlespace. They represent the application of adversary doctrine under ideal conditions. Ideally, doctrinal templates depict the threat's normal organization for combat, frontages, depths, boundaries and other control measures, assets available from other commands, objective depths, engagement areas, battle positions, and so forth. Doctrinal templates are usually scaled to allow ready use with geospatial products. See also **doctrine**. (JP 2-01.3)

doctrine — Fundamental principles by which the military forces or elements thereof guide their actions in support of national objectives. It is authoritative but requires judgment in application. See also **multinational doctrine; joint doctrine; multi-Service doctrine.**

DOD civilian — A Federal civilian employee of the Department of Defense directly hired and paid from appropriated or nonappropriated funds, under permanent or temporary appointment. Specifically excluded are contractors and foreign host nationals as well as third country civilians. (JP 1-03.17)

DOD support to counterdrug operations — Support provided by the Department of Defense to law enforcement agencies to detect, monitor, and counter the production, trafficking, and use of illegal drugs. See also **counterdrug operations.** (JP 3-07)

dolly — Airborne data link equipment.

dome — See **spray dome.**

domestic air traffic — Air traffic within the continental United States.

domestic emergencies — Emergencies affecting the public welfare and occurring within the 50 states, District of Columbia, Commonwealth of Puerto Rico, US possessions and territories, or any political subdivision thereof, as a result of enemy attack, insurrection, civil disturbance, earthquake, fire, flood, or other public disasters or equivalent emergencies that endanger life and property or disrupt the usual process of government. The term domestic emergency includes any or all of the emergency conditions defined below: a. **civil defense emergency** — A domestic emergency disaster situation resulting from devastation created by an enemy attack and requiring emergency operations during and following that attack. It may be proclaimed by appropriate authority in anticipation of an attack. b. **civil disturbances** — Riots, acts of violence, insurrections, unlawful obstructions or assemblages, or other disorders prejudicial to public law and order. The term civil disturbance includes all domestic conditions requiring or likely to require the use of Federal Armed Forces pursuant to the provisions of Chapter 15 of Title 10, United States Code. c. **major disaster** — Any flood, fire, hurricane, tornado, earthquake, or other catastrophe which, in the determination of the President, is or threatens to be of sufficient severity and magnitude to warrant disaster assistance by the Federal Government under Public Law 606, 91st Congress (42 United States Code 58) to supplement the efforts and available resources of State and local governments in alleviating the damage, hardship, or suffering caused thereby. d. **natural disaster** — All domestic emergencies except those created as a result of enemy attack or civil disturbance. See also **civil defense emergency; civil disturbance; major disaster; natural disaster.**

domestic intelligence — Intelligence relating to activities or conditions within the United States that threaten internal security and that might require the employment of troops; and intelligence relating to activities of individuals or agencies potentially or actually dangerous to the security of the Department of Defense.

domestic support operations — Those activities and measures taken by the Department of Defense to foster mutual assistance and support between the Department of Defense and any civil government agency in planning or preparedness for, or in the application of resources for response to, the consequences of civil emergencies or attacks, including national security emergencies. Also called **DSOs.** (JP 3-57)

dominant user — The Service or multinational partner who is the principal consumer of a particular common-user logistic supply or service within a joint or multinational operation. The dominant user will normally act as the lead Service to provide this particular common-user logistic supply or service to other Service components, multinational partners, other governmental agencies, or nongovernmental agencies as directed by the combatant commander. See also **common-user logistics; lead Service or agency for common-user logistics.** (JP 4-07)

dominant user concept — The concept that the Service that is the principal consumer will have the responsibility for performance of a support workload for all using Services.

doppler effect — **(*)** The phenomenon evidenced by the change in the observed frequency of a sound or radio wave caused by a time rate of change in the effective length of the path of travel between the source and the point of observation.

doppler radar — A radar system that differentiates between fixed and moving targets by detecting the apparent change in frequency of the reflected wave due to motion of target or the observer.

dormant — In mine warfare, the state of a mine during which a time delay feature in a mine prevents it from being actuated.

dose rate contour line — **(*)** A line on a map, diagram, or overlay joining all points at which the radiation dose rate at a given time is the same.

dosimetry — **(*)** The measurement of radiation doses. It applies to both the devices used (dosimeters) and to the techniques.

double agent — Agent in contact with two opposing intelligence services, only one of which is aware of the double contact or quasi-intelligence services.

double flow route — **(*)** A route of at least two lanes allowing two columns of vehicles to proceed simultaneously, either in the same direction or in opposite directions. See also **single flow route.**

downgrade — To determine that classified information requires, in the interests of national security, a lower degree of protection against unauthorized disclosure than currently provided, coupled with a changing of the classification designation to reflect such a lower degree.

downloading — An operation that removes airborne weapons or stores from an aircraft. (JP 3-04.1)

down lock — (*) A device for locking retractable landing gear in the down or extended position.

draft — 1. The conscription of qualified citizens in military service. 2. The depth of water that a vessel requires to float freely; the depth of a vessel from the water line to the keel. See also **active duty; Military Service; watercraft.** (JP 4-01.6)

draft plan — (*) A plan for which a draft plan has been coordinated and agreed with the other military headquarters and is ready for coordination with the nations involved, that is those nations who would be required to take national actions to support the plan. It may be used for future planning and exercises and may form the basis for an operation order to be implemented in time of emergency. See also **coordinated draft plan; final plan; initial draft plan; operation plan.**

drag — Force of aerodynamic resistance caused by the violent currents behind the shock front.

drag loading — The force on an object or structure due to transient winds accompanying the passage of a blast wave. The drag pressure is the product of the dynamic pressure and the drag coefficient which is dependent upon the shape (or geometry) of the structure or object.

drift — (*) In ballistics, a shift in projectile direction due to gyroscopic action which results from gravitational and atmospherically induced torques on the spinning projectile.

drift angle — (*) The angle measured in degrees between the heading of an aircraft or ship and the track made good.

drill mine — (*) An inert filled mine or mine-like body, used in loading, laying, or discharge practice and trials. See also **mine.**

drone — A land, sea, or air vehicle that is remotely or automatically controlled. See also **remotely piloted vehicle; unmanned aerial vehicle.** (JP 4-01.5)

droop stop — (*) A device to limit downward vertical motion of helicopter rotor blades upon rotor shutdown.

drop altitude — (*) The altitude above mean sea level at which airdrop is executed. See also **altitude; drop height.**

drop height — (*) The vertical distance between the drop zone and the aircraft. See also **altitude; drop altitude.**

dropmaster — 1. An individual qualified to prepare, perform acceptance inspection, load, lash, and eject material for airdrop. 2. An aircrew member who, during parachute operations, will relay any required information between pilot and jumpmaster.

drop message — (*) A message dropped from an aircraft to a ground or surface unit.

drop zone — (*) A specific area upon which airborne troops, equipment, or supplies are airdropped. Also called **DZ.**

drug interdiction — The interception of illegal drugs being smuggled by air, sea, or land. See also **counterdrug operations.** (JP 3-07.4)

dry deck shelter — A shelter module that attaches to the hull of a specially configured submarine to provide the submarine with the capability to launch and recover special operations personnel, vehicles, and equipment while submerged. The dry deck shelter provides a working environment at one atmosphere for the special operations element during transit and has structural integrity to the collapse depth of the host submarine. Also called **DDS.** (JP 3-05.3)

D-to-P assets required on D-day — As applied to the D-to-P concept, this asset requirement represents those stocks that must be physically available on D-day to meet initial allowance requirements, to fill the wartime pipeline between the producers and users (even if P-day and D-day occur simultaneously), and to provide any required D-to-P consumption or production differential stockage. The D-to-P assets required on D-day are also represented as the difference between the D-to-P materiel readiness gross requirements and the cumulative sum of all production deliveries during the D-to-P period. See also **D-to-P concept.**

D-to-P concept — A logistic planning concept by which the gross materiel readiness requirement in support of approved forces at planned wartime rates for conflicts of indefinite duration will be satisfied by a balanced mix of assets on hand on D-day and assets to be gained from production through P-day when the planned rate of production deliveries to the users equals the planned wartime rate of expenditure (consumption). See also **D-day consumption/production differential assets; D-day pipeline assets; D-to-P assets required on D-day; D-to-P materiel readiness gross requirement.**

D-to-P materiel readiness gross requirement — As applied to the D-to-P concept, the gross requirement for all supplies and materiel needed to meet all initial pipeline and anticipated expenditure (consumption) requirements between D-day and P-day. Includes initial allowances, continental United States and overseas operating and safety levels, intransit levels of supply, and the cumulative sum of all items expended (consumed) during the D-to-P period. See also **D-to-P concept.**

dual agent — One who is simultaneously and independently employed by two or more intelligence agencies, covering targets for both.

dual-capable aircraft — Allied and US fighter aircraft tasked and configured to perform either conventional or theater nuclear missions. Also called **DCA.**

dual-capable forces — Forces capable of employing dual-capable weapons.

dual capable unit — **(*)** A nuclear certified delivery unit capable of executing both conventional and nuclear missions.

dual-firing circuit — **(*)** An assembly comprising two independent firing systems, both electric or both non-electric, so that the firing of either system will detonate all charges.

dual (multi)-capable weapons — 1. Weapons, weapon systems, or vehicles capable of selective equipage with different types or mixes of armament or firepower. 2. Sometimes restricted to weapons capable of handling either nuclear or non-nuclear munitions.

dual (multi)-purpose weapons — Weapons which possess the capability for effective application in two or more basically different military functions and/or levels of conflict.

dual-purpose weapon — A weapon designed for delivering effective fire against air or surface targets.

dud — **(*)** Explosive munition which has not been armed as intended or which has failed to explode after being armed. See also **absolute dud; dwarf dud; flare dud; nuclear dud.**

dud probability — The expected percentage of failures in a given number of firings.

due in — Quantities of materiel scheduled to be received from vendors, repair facilities, assembly operation, interdepot transfers, and other sources.

dummy — See **decoy.**

dummy message — **(*)** A message sent for some purpose other than its content, which may consist of dummy groups or may have a meaningless text.

dummy minefield — **(*)** In naval mine warfare, a minefield containing no live mines and presenting only a psychological threat.

dummy run — Any simulated firing practice, particularly a dive bombing approach made without release of a bomb. Also called **dry run.**

dump — **(*)** A temporary storage area, usually in the open, for bombs, ammunition, equipment, or supplies.

duplicate negative — **(*)** A negative reproduced from a negative or diapositive.

durable materiel — See **nonexpendable supplies and materiel.**

duty status - whereabouts unknown — A transitory casualty status, applicable only to military personnel, that is used when the responsible commander suspects the member may be a casualty whose absence is involuntary, but does not feel sufficient evidence currently exists to make a definite determination of missing or deceased. Also called **DUSTWUN.** See also **casualty status.**

dwarf dud — A nuclear weapon that, when launched at or emplaced on a target, fails to provide a yield within a reasonable range of that which could be anticipated with normal operation of the weapon. This constitutes a dud only in a relative sense.

dwell time — The time cargo remains in a terminal's in-transit storage area while awaiting shipment by clearance transportation. See also **storage.** (JP 4-01.6)

Intentionally Blank

E

earliest anticipated launch time — The earliest time expected for a special operations tactical element and its supporting platform to depart the staging or marshalling area together en route to the operations area. Also called **EALT.** (JP 3-05.5)

earliest arrival date — A day, relative to C-day, that is specified by a planner as the earliest date when a unit, a resupply shipment, or replacement personnel can be accepted at a port of debarkation during a deployment. Used with the latest arrival data, it defines a delivery window for transportation planning. Also called **EAD.** See also **latest arrival date.**

Early Spring — An antireconnaissance satellite weapon system.

early time — See **span of detonation (atomic demolition munition employment).**

early warning — **(*)** Early notification of the launch or approach of unknown weapons or weapons carriers. Also called **EW.** See also **attack assessment; tactical warning.**

earmarking of stocks — **(*)** The arrangement whereby nations agree, normally in peacetime, to identify a proportion of selected items of their war reserve stocks to be called for by specified NATO commanders.

earthing — **(*)** The process of making a satisfactory electrical connection between the structure, including the metal skin, of an object or vehicle, and the mass of the Earth, to ensure a common potential with the Earth. See also **bonding.**

echelon — **(*)** 1. A subdivision of a headquarters, i.e., forward echelon, rear echelon. 2. Separate level of command. As compared to a regiment, a division is a higher echelon, a battalion is a lower echelon. 3. A fraction of a command in the direction of depth to which a principal combat mission is assigned; i.e., attack echelon, support echelon, reserve echelon. 4. A formation in which its subdivisions are placed one behind another, with a lateral and even spacing to the same side.

echeloned displacement — **(*)** Movement of a unit from one position to another without discontinuing performance of its primary function. **(DOD only)** Normally, the unit divides into two functional elements (base and advance); and, while the base continues to operate, the advance element displaces to a new site where, after it becomes operational, it is joined by the base element.

economic action — The planned use of economic measures designed to influence the policies or actions of another state, e.g., to impair the war-making potential of a hostile power or to generate economic stability within a friendly power.

economic mobilization — **(*)** The process of preparing for and carrying out such changes in the organization and functioning of the national economy as are necessary to provide for the most effective use of resources in a national emergency.

economic order quantity — That quantity derived from a mathematical technique used to determine the optimum (lowest) total variable costs required to order and hold inventory.

economic potential — **(*)** The total capacity of a nation to produce goods and services.

economic potential for war — That share of the total economic capacity of a nation that can be used for the purposes of war.

economic retention stock — That portion of the quantity of an item excess of the approved force retention level that has been determined will be more economical to retain for future peacetime issue in lieu of replacement of future issues by procurement. To warrant economic retention, items must have a reasonably predictable demand rate.

economic warfare — Aggressive use of economic means to achieve national objectives.

effective damage — That damage necessary to render a target element inoperative, unserviceable, nonproductive, or uninhabitable.

effective US controlled ships — US-owned foreign flagships that can be tasked by the Maritime Administration to support Department of Defense requirements when necessary. Also called **EUSCS.**

ejection — **(*)** 1. Escape from an aircraft by means of an independently propelled seat or capsule. 2. In air armament, the process of forcefully separating an aircraft store from an aircraft to achieve satisfactory separation.

ejection systems — **(*)** a. **command ejection system** — A system in which the pilot of an aircraft or the occupant of the other ejection seat(s) initiates ejection resulting in the automatic ejection of all occupants. b. **command select ejection system** — A system permitting the optional transfer from one crew station to another of the control of a command ejection system for automatic ejection of all occupants. c. **independent ejection system** — An ejection system which operates independently of other ejection systems installed in one aircraft. d. **sequenced ejection system** — A system which ejects the aircraft crew in sequence to ensure a safe minimum total time of escape without collision.

electrode sweep — In naval mine warfare, a magnetic cable sweep in which the water forms part of the electric circuit.

electro-explosive device — **(*)** An explosive or pyrotechnic component that initiates an explosive, burning, electrical, or mechanical train and is activated by the application of electrical energy. Also called **EED.**

electromagnetic compatibility — The ability of systems, equipment, and devices that utilize the electromagnetic spectrum to operate in their intended operational environments without suffering unacceptable degradation or causing unintentional degradation because of electromagnetic radiation or response. It involves the application of sound electromagnetic spectrum management; system, equipment, and device design configuration that ensures interference-free operation; and clear concepts and doctrines that maximize operational effectiveness. Also called **EMC.** See also **electromagnetic spectrum; electronic warfare; spectrum management.**

electromagnetic deception — The deliberate radiation, re-radiation, alteration, suppression, absorption, denial, enhancement, or reflection of electromagnetic energy in a manner intended to convey misleading information to an enemy or to enemy electromagnetic-dependent weapons, thereby degrading or neutralizing the enemy's combat capability. Among the types of electromagnetic deception are: a. **manipulative electromagnetic deception**

— Actions to eliminate revealing, or convey misleading, electromagnetic telltale indicators that may be used by hostile forces; b. **simulative electromagnetic deception** — Actions to simulate friendly, notional, or actual capabilities to mislead hostile forces; and c. **imitative electromagnetic deception** — The introduction of electromagnetic energy into enemy systems that imitates enemy emissions. See also **electronic warfare.**

electromagnetic environment — The resulting product of the power and time distribution, in various frequency ranges, of the radiated or conducted electromagnetic emission levels that may be encountered by a military force, system, or platform when performing its assigned mission in its intended operational environment. It is the sum of electromagnetic interference; electromagnetic pulse; hazards of electromagnetic radiation to personnel, ordnance, and volatile materials; and natural phenomena effects of lightning and precipitation static. Also called **EME.**

electromagnetic environmental effects — The impact of the electromagnetic environment upon the operational capability of military forces, equipment, systems, and platforms. It encompasses all electromagnetic disciplines, including electromagnetic compatibility and electromagnetic interference; electromagnetic vulnerability; electromagnetic pulse; electronic protection, hazards of electromagnetic radiation to personnel, ordnance, and volatile materials; and natural phenomena effects of lightning and precipitation static. Also called **E3.**

electromagnetic hardening — Action taken to protect personnel, facilities, and/or equipment by filtering, attenuating, grounding, bonding, and/or shielding against undesirable effects of electromagnetic energy. See also **electronic warfare.**

electromagnetic interference — Any electromagnetic disturbance that interrupts, obstructs, or otherwise degrades or limits the effective performance of electronics and electrical equipment. It can be induced intentionally, as in some forms of electronic warfare, or unintentionally, as a result of spurious emissions and responses, intermodulation products, and the like. Also called **EMI.**

electromagnetic intrusion — The intentional insertion of electromagnetic energy into transmission paths in any manner, with the objective of deceiving operators or of causing confusion. See also **electronic warfare.**

electromagnetic jamming — The deliberate radiation, reradiation, or reflection of electromagnetic energy for the purpose of preventing or reducing an enemy's effective use of the electromagnetic spectrum, and with the intent of degrading or neutralizing the enemy's combat capability. See also **electromagnetic spectrum; electronic warfare; spectrum management.**

electromagnetic pulse — The electromagnetic radiation from a strong electronic pulse, most commonly caused by a nuclear explosion that may couple with electrical or electronic systems to produce damaging current and voltage surges. Also called **EMP.** See also **electromagnetic radiation.** (JP 3-51)

electromagnetic radiation — Radiation made up of oscillating electric and magnetic fields and propagated with the speed of light. Includes gamma radiation, X-rays, ultraviolet, visible, and infrared radiation, and radar and radio waves.

electromagnetic radiation hazards — Hazards caused by transmitter or antenna installation that generates electromagnetic radiation in the vicinity of ordnance, personnel, or fueling operations in excess of established safe levels or increases the existing levels to a hazardous level; or a personnel, fueling, or ordnance installation located in an area that is illuminated by electromagnetic radiation at a level that is hazardous to the planned operations or occupancy. Also called **EMR hazards or RADHAZ.**

electromagnetic spectrum — The range of frequencies of electromagnetic radiation from zero to infinity. It is divided into 26 alphabetically designated bands. See also **electronic warfare.**

electromagnetic vulnerability — The characteristics of a system that cause it to suffer a definite degradation (incapability to perform the designated mission) as a result of having been subjected to a certain level of electromagnetic environmental effects. Also called **EMV.**

electronic attack — See **electronic warfare.**

electronic imagery dissemination — The transmission of imagery or imagery products by any electronic means. This includes the following four categories. a. **primary imagery dissemination system** — The equipment and procedures used in the electronic transmission and receipt of un-exploited original or near-original quality imagery in near real time. b. **primary imagery dissemination** — The electronic transmission and receipt of unexploited original or near-original quality imagery in near real time through a primary imagery dissemination system. c. **secondary imagery dissemination system** — The equipment and procedures used in the electronic transmission and receipt of exploited non-original quality imagery and imagery products in other than real or near real time. d. **secondary imagery dissemination** — The electronic transmission and receipt of exploited non-original quality imagery and imagery products in other than real or near real time through a secondary imagery dissemination system.

electronic intelligence — Technical and geolocation intelligence derived from foreign non-communications electromagnetic radiations emanating from other than nuclear detonations or radioactive sources. Also called **ELINT.** See also **electronic warfare; foreign instrumentation signals intelligence; intelligence; signals intelligence.** (JP 2-01)

electronic line of sight — The path traversed by electromagnetic waves that is not subject to reflection or refraction by the atmosphere.

electronic masking — **(*)** The controlled radiation of electromagnetic energy on friendly frequencies in a manner to protect the emissions of friendly communications and electronic systems against enemy electronic warfare support measures/signals intelligence without significantly degrading the operation of friendly systems.

electronic probing — Intentional radiation designed to be introduced into the devices or systems of potential enemies for the purpose of learning the functions and operational capabilities of the devices or systems.

electronic protection — See **electronic warfare.**

electronic reconnaissance — The detection, location, identification, and evaluation of foreign electromagnetic radiations. See also **electromagnetic radiation; reconnaissance.** (JP 3-51)

electronics security — The protection resulting from all measures designed to deny unauthorized persons information of value that might be derived from their interception and study of noncommunications electromagnetic radiations, e.g., radar.

electronic warfare — Any military action involving the use of electromagnetic and directed energy to control the electromagnetic spectrum or to attack the enemy. Also called **EW.** The three major subdivisions within electronic warfare are: electronic attack, electronic protection, and electronic warfare support. a. **electronic attack.** That division of electronic warfare involving the use of electromagnetic energy, directed energy, or antiradiation weapons to attack personnel, facilities, or equipment with the intent of degrading, neutralizing, or destroying enemy combat capability and is considered a form of fires. Also called **EA.** EA includes: 1) actions taken to prevent or reduce an enemy's effective use of the electromagnetic spectrum, such as jamming and electromagnetic deception, and 2) employment of weapons that use either electromagnetic or directed energy as their primary destructive mechanism (lasers, radio frequency weapons, particle beams). b. **electronic protection.** That division of electronic warfare involving passive and active means taken to protect personnel, facilities, and equipment from any effects of friendly or enemy employment of electronic warfare that degrade, neutralize, or destroy friendly combat capability. Also called **EP.** c. **electronic warfare support.** That division of electronic warfare involving actions tasked by, or under direct control of, an operational commander to search for, intercept, identify, and locate or localize sources of intentional and unintentional radiated electromagnetic energy for the purpose of immediate threat recognition, targeting, planning and conduct of future operations. Thus, electronic warfare support provides information required for decisions involving electronic warfare operations and other tactical actions such as threat avoidance, targeting, and homing. Also called **ES.** Electronic warfare support data can be used to produce signals intelligence, provide targeting for electronic or destructive attack, and produce measurement and signature intelligence. See also **directed energy; electromagnetic spectrum.** (JP 3-51)

electronic warfare frequency deconfliction — Actions taken to integrate those frequencies used by electronic warfare systems into the overall frequency deconfliction process. See also **electronic warfare.** (JP 3-51)

electronic warfare reprogramming — The deliberate alteration or modification of electronic warfare (EW) or target sensing systems (TSS), or the tactics and procedures that employ them, in response to validated changes in equipment, tactics, or the electromagnetic environment. These changes may be the result of deliberate actions on the part of friendly, adversary or third parties; or may be brought about by electromagnetic interference or other inadvertent phenomena. The purpose of EW reprogramming is to maintain or enhance the effectiveness of EW and TSS equipment. EW reprogramming includes changes to self-defense systems, offensive weapons systems, and intelligence collection systems. See also **electronic warfare.** (JP 3-51)

electronic warfare support — See **electronic warfare.**

electro-optical intelligence — Intelligence other than signals intelligence derived from the optical monitoring of the electromagnetic spectrum from ultraviolet (0.01 micrometers) through far infrared

(1,000 micrometers). Also called **ELECTRO-OPTINT.** See also **intelligence; laser intelligence.** (JP 2-0)

electro-optics — (*) The technology associated with those components, devices and systems which are designed to interact between the electromagnetic (optical) and the electric (electronic) state.

element set — Three lines of data which define the location of a satellite in space. Also called **ELSET.**

elements of national power — All the means that are available for employment in the pursuit of national objectives.

elevated causeway system — An elevated causeway pier that provides a means of delivering containers, certain vehicles, and bulk cargo ashore without the lighterage contending with the surf zone. See also **causeway.** (JP 4-01.6)

elevation — (*) The vertical distance of a point or level on or affixed to the surface of the Earth measured from mean sea level. See also **altitude.**

elevation tint — See **hypsometric tinting.**

elicitation (intelligence) — Acquisition of information from a person or group in a manner that does not disclose the intent of the interview or conversation. A technique of human source intelligence collection, generally overt, unless the collector is other than he or she purports to be.

eligible traffic — Traffic for which movement requirements are submitted and space is assigned or allocated. Such traffic must meet eligibility requirements specified in Joint Travel Regulations for the Uniformed Services and publications of the Department of Defense and Military Departments governing eligibility for land, sea, and air transportation, and be in accordance with the guidance of the Joint Chiefs of Staff.

embarkation — (*) The process of putting personnel and/or vehicles and their associated stores and equipment into ships and/or aircraft. See also **loading.**

embarkation and tonnage table — A consolidated table showing personnel and cargo, by troop or naval units, loaded aboard a combat-loaded ship.

embarkation area — (*) An area ashore, including a group of embarkation points, in which final preparations for embarkation are completed and through which assigned personnel and loads for craft and ships are called forward to embark. See also **mounting area.**

embarkation element (unit) (group) — A temporary administrative formation of personnel with supplies and equipment embarking or to be embarked (combat loaded) aboard the ships of one transport element (unit) (group). It is dissolved upon completion of the embarkation. An embarkation element normally consists of two or more embarkation teams: a unit, of two or more elements; and a group, of two or more units. See also **embarkation organization; embarkation team.**

embarkation officer — An officer on the staff of units of the landing force who advises the commander thereof on matters pertaining to embarkation planning and loading ships. See also **combat cargo officer.**

embarkation order — (*) An order specifying dates, times, routes, loading diagrams, and methods of movement to shipside or aircraft for troops and their equipment. See also **movement table.**

embarkation organization — A temporary administrative formation of personnel with supplies and equipment embarking or to be embarked (combat loaded) aboard amphibious shipping. See also **embarkation element (unit) (group); embarkation team.**

embarkation phase — In amphibious operations, the phase that encompasses the orderly assembly of personnel and materiel and their subsequent loading aboard ships and/or aircraft in a sequence designed to meet the requirements of the landing force concept of operations ashore. (JP 3-02.2)

embarkation plans — The plans prepared by the landing force and appropriate subordinate commanders containing instructions and information concerning the organization for embarkation, assignment to shipping, supplies and equipment to be embarked, location and assignment of embarkation areas, control and communication arrangements, movement schedules and embarkation sequence, and additional pertinent instructions relating to the embarkation of the landing force. (JP 3-02)

embarkation team — A temporary administrative formation of all personnel with supplies and equipment embarking or to be embarked (combat loaded) aboard one ship. See also **embarkation element (unit) (group); embarkation organization.**

emergency anchorage — **(*)** An anchorage, which may have a limited defense organization, for naval vessels, mobile support units, auxiliaries, or merchant ships. See also **assembly anchorage; holding anchorage; working anchorage.**

emergency barrier — See **aircraft arresting barrier.**

emergency-essential employee — A Department of Defense civilian employee whose assigned duties and responsibilities must be accomplished following the evacuation of non-essential personnel (including dependents) during a declared emergency or outbreak of war. The position occupied cannot be converted to a military billet because it requires uninterrupted performance so as to provide immediate and continuing support for combat operations and/or combat systems support functions. See also **evacuation.** (JP 1-0)

emergency interment — An interment, usually on the battlefield, when conditions do not permit either evacuation for interment in an interment site or interment according to national or international legal regulations. See also **group interment; mortuary affairs; temporary interment; trench interment.** (JP 4-06)

emergency locator beacon — **(*)** A generic term for all radio beacons used for emergency locating purposes. See also **crash locator beacon; personal locator beacon.**

emergency priority — A category of immediate mission request that takes precedence over all other priorities, e.g., an enemy breakthrough. See also **immediate mission request; priority of immediate mission requests.**

emergency relocation site — A site located where practicable outside a prime target area to which all or portions of a civilian or military headquarters may be moved. As a minimum, it is manned to provide for the maintenance of the facility, communications, and database. It should be capable of rapid activation, of supporting the initial requirements of the relocated headquarters for a predetermined period, and of expansion to meet wartime requirements of the relocated headquarters.

emergency repair — The least amount of immediate repair to damaged facilities necessary for the facilities to support the mission. These repairs will be made using expedient materials and methods (such as AM-2 aluminum matting, cold-mix asphalt, plywood scabs, temporary utility lines, and emergency generators). Modular or kit-type facility substitutes would be appropriate if repairs cannot be made in time to meet mission requirements. See also **facility substitutes.** (JP 4-04)

emergency resupply — A resupply mission that occurs based on a predetermined set of circumstances and time interval should radio contact not be established or, once established, is lost between a special operations tactical element and its base. See also **automatic resupply; on-call resupply.** (JP 3-05.3)

emergency risk (nuclear) — A degree of risk where anticipated effects may cause some temporary shock, casualties and may significantly reduce the unit's combat efficiency. See also **degree of risk; negligible risk (nuclear).**

emergency substitute — **(*)** A product which may be used, in an emergency only, in place of another product, but only on the advice of technically qualified personnel of the nation using the product, who will specify the limitations.

emission control — The selective and controlled use of electromagnetic, acoustic, or other emitters to optimize command and control capabilities while minimizing, for operations security: a. detection by enemy sensors; b. mutual interference among friendly systems; and/or c. enemy interference with the ability to execute a military deception plan. Also called **EMCON.** See also **electronic warfare.**

emission control orders — Orders used to authorize, control, or prohibit the use of electronic emission equipment. Also called **EMCON orders.** See also **control of electromagnetic radiation.**

emplacement — **(*)** 1. A prepared position for one or more weapons or pieces of equipment, for protection against hostile fire or bombardment, and from which they can execute their tasks. 2. The act of fixing a gun in a prepared position from which it may be fired.

employment — The strategic, operational, or tactical use of forces. See also **employment planning.** (JP 5-0)

employment planning — Planning that prescribes how to apply force and/or forces to attain specified military objectives. Employment planning concepts are developed by combatant commanders through their component commanders. See also **employment.** (JP 5-0)

enabling mine countermeasures — Countermeasures designed to counter mines once they have been laid. This includes both passive and active mine countermeasures. See also **mine countermeasures.** (JP 3-15)

encipher — To convert plain text into unintelligible form by means of a cipher system.

end evening civil twilight — The time period when the sun has dropped 6 degrees beneath the western horizon; it is the instant at which there is no longer sufficient light to see objects with the unaided eye. Light intensification devices are recommended from this time until begin morning civil twilight. Also called **EECT.**

end item — A final combination of end products, component parts, and/or materials that is ready for its intended use, e.g., ship, tank, mobile machine shop, or aircraft.

end of evening nautical twilight — Occurs when the sun has dropped 12 degrees below the western horizon, and is the instant of last available daylight for the visual control of limited ground operations. At end of evening nautical twilight there is no further sunlight available. See also **horizon.** (JP 2-01.3)

end of mission — In artillery, mortar, and naval gunfire support, an order given to terminate firing on a specific target. See also **cease loading; call for fire; fire mission.**

end state — The set of required conditions that defines achievement of the commander's objectives. (JP 3-18)

endurance — **(*)** The time an aircraft can continue flying, or a ground vehicle or ship can continue operating, under specified conditions, e.g., without refueling. See also **endurance distance.**

endurance distance — **(*)** Total distance that a ground vehicle or ship can be self-propelled at any specified endurance speed.

endurance loading — The stocking aboard ship for a period of time, normally covering the number of months between overhauls, of items with all of the following characteristics: a. low price; b. low weight and cube; c. a predictable usage rate; and d. nondeteriorative. See also **loading.**

enemy capabilities — Those courses of action of which the enemy is physically capable and that, if adopted, will affect accomplishment of the friendly mission. The term "capabilities" includes not only the general courses of action open to the enemy, such as attack, defense, reinforcement, or withdrawal, but also all the particular courses of action possible under each general course of action. "Enemy capabilities" are considered in the light of all known factors affecting military operations, including time, space, weather, terrain, and the strength and disposition of enemy forces. In strategic thinking, the capabilities of a nation represent the courses of action within the power of the nation for accomplishing its national objectives throughout the range of military operations. See also **capability; course of action; mission.** (JP 2-01.3)

engage — **(*)** 1. In air defense, a fire control order used to direct or authorize units and/or weapon systems to fire on a designated target. See also **cease engagement; hold fire.** 2. **(DOD only)** To bring the enemy under fire.

engagement — 1. In air defense, an attack with guns or air-to-air missiles by an interceptor aircraft, or the launch of an air defense missile by air defense artillery and the missile's subsequent travel to intercept. 2. A tactical conflict, usually between opposing lower echelons maneuver forces. See also **battle; campaign.**

enlisted terminal attack controller — Tactical air party member who assists in mission planning and provides final control of close air support aircraft in support of ground forces. Also called **ETAC.** See also **close air support; mission; terminal.** (JP 3-09.1)

en route care — The care required to maintain the phase treatment initiated prior to evacuation and the sustainment of the patient's medical condition during evacuation. See also **evacuation; patient.** (JP 4-02)

envelopment — **(*)** An offensive maneuver in which the main attacking force passes around or over the enemy's principal defensive positions to secure objectives to the enemy's rear. See also **turning movement.**

environmental cleanup — The process of removing solid, liquid, and hazardous wastes, except for unexploded ordnance, resulting from the joint operation of US forces to a condition that approaches the one existing prior to operation as determined by the environmental baseline survey, if one was conducted. The extent of this process will depend upon the operational situation at the time that cleanup is accomplished.

environmental considerations — The spectrum of environmental media, resources, or programs that may impact on, or are affected by, the planning and execution of military operations. Factors may include, but are not limited to, environmental compliance, pollution prevention, conservation, protection of historical and cultural sites, and protection of flora and fauna. (JP 3-34)

environmental services — The various combinations of scientific, technical, and advisory activities (including modification processes, i.e., the influence of manmade and natural factors) required to acquire, produce, and supply information on the past, present, and future states of space, atmospheric, oceanographic, and terrestrial surroundings for use in military planning and decisionmaking processes, or to modify those surroundings to enhance military operations.

environmental stewardship — The integration and application of environmental values into the military mission in order to sustain readiness, improve quality of life, strengthen civil relations, and preserve valuable natural resources.

equipment — In logistics, all nonexpendable items needed to outfit or equip an individual or organization. See also **assembly; component; subassembly; supplies.**

equipment operationally ready — The status of an item of equipment in the possession of an operating unit that indicates it is capable of fulfilling its intended mission and in a system configuration that offers a high assurance of an effective, reliable, and safe performance.

escalation — A deliberate or unpremeditated increase in scope or violence of a conflict.

escapee — Any person who has been physically captured by the enemy and succeeds in getting free. See also **evasion and escape.**

escape line — A planned route to allow personnel engaged in clandestine activity to depart from a site or area when possibility of compromise or apprehension exists.

escape route — See **evasion and escape route.**

escort — **(*)** 1. A combatant unit(s) assigned to accompany and protect another force or convoy. 2. Aircraft assigned to protect other aircraft during a mission. 3. An armed guard that accompanies a convoy, a train, prisoners, etc. 4. An armed guard accompanying persons as a mark of honor. 5. **(DOD only)** To convoy. 6. **(DOD only)** A member of the Armed Forces assigned to accompany, assist, or guide an individual or group, e.g., an escort officer.

escort forces — Combat forces of various types provided to protect other forces against enemy attack.

espionage — The act of obtaining, delivering, transmitting, communicating, or receiving information about the national defense with an intent, or reason to believe, that the information may be used to the injury of the United States or to the advantage of any foreign nation. See also **counterintelligence.** (JP 2-01.2)

espionage against the United States — Overt, covert, or clandestine activity designed to obtain information relating to the national defense with intent or reason to believe that it will be used to the injury of the United States or to the advantage of a foreign nation. For espionage crimes see Chapter 37 of Title 18, United States Code.

essential care — That care received within a theater that is dependent upon the mission, enemy, terrain, troops, time available, and other civilian considerations. It includes first responder care, forward resuscitative surgery, and en route care as well as treatment and hospitalization to return the patient to duty or to stabilize for movement to a higher level of care. See also **en route care; first responder phase; forward resuscitative surgery; patient; theater.** (JP 4-02)

essential chemicals — In counterdrug operations, compounds that are required in the synthetic or extraction processes of drug production, but in most cases do not become part of the drug molecule. Essential chemicals are used in the production of cocaine or heroin. (JP 3-07.4)

essential communications traffic — Transmissions (record or voice) of any precedence that must be sent electrically in order for the command or activity concerned to avoid a serious impact on mission accomplishment or safety or life.

essential elements of friendly information — Key questions likely to be asked by adversary officials and intelligence systems about specific friendly intentions, capabilities, and activities, so they can obtain answers critical to their operational effectiveness. Also called **EEFI.**

essential industry — Any industry necessary to the needs of a civilian or war economy. The term includes the basic industries as well as the necessary portions of those other industries that transform the crude basic raw materials into useful intermediate or end products, e.g., the iron and steel industry, the food industry, and the chemical industry.

essential secrecy — The condition achieved from the denial of critical information to adversaries.

establishment — **(*)** An installation, together with its personnel and equipment, organized as an operating entity. See also **activity; base; equipment.**

estimate — 1. An analysis of a foreign situation, development, or trend that identifies its major elements, interprets the significance, and appraises the future possibilities and the prospective results of the various actions that might be taken. 2. An appraisal of the capabilities, vulnerabilities, and potential courses of action of a foreign nation or combination of nations in consequence of a specific national plan, policy, decision, or contemplated course of action. 3. An analysis of an actual or contemplated clandestine operation in relation to the situation in which it is or would be conducted in order to identify and appraise such factors as available as well as needed assets and potential obstacles, accomplishments, and consequences. See also **intelligence estimate.**

evacuation — 1. The process of moving any person who is wounded, injured, or ill to and/or between medical treatment facilities.

2. The clearance of personnel, animals, or materiel from a given locality. 3. The controlled process of collecting, classifying, and shipping unserviceable or abandoned materiel, US or foreign, to appropriate reclamation, maintenance, technical intelligence, or disposal facilities. 4. The ordered or authorized departure of noncombatants from a specific area by Department of State, Department of Defense, or appropriate military commander. This refers to the movement from one area to another in the same or different countries. The evacuation is caused by unusual or emergency circumstances and applies equally to command or non-command sponsored family members. See also **evacuee; noncombatant evacuation operations.**

evacuation control ship — (*) In an amphibious operation, a ship designated as a control point for landing craft, amphibious vehicles, and helicopters evacuating casualties from the beaches. Medical personnel embarked in the evacuation control ship effect distribution of casualties throughout the attack force in accordance with ship's casualty capacities and specialized medical facilities available, and also perform emergency surgery.

evacuation convoy — (*) A convoy which is used for evacuation of dangerously exposed waters. See also **evacuation of dangerously exposed waters.**

evacuation of dangerously exposed waters — (*) The movement of merchant ships under naval control from severely threatened coastlines and dangerously exposed waters to safer localities. See also **dangerously exposed waters.**

evacuation of port equipment — (*) The transfer of mobile/movable equipment from a threatened port to another port or to a working anchorage.

evacuation policy — 1. Command decision establishing the maximum number of days that patients may be held within the command for treatment. Patients who, in the opinion of responsible medical officers, cannot be returned to a duty status within the period prescribed are evacuated by the first available means, provided the travel involved will not aggravate their disabilities. 2. A command decision concerning the movement of civilians from the proximity of military operations for security and safety reasons and involving the need to arrange for movement, reception, care, and control of such individuals. 3. Command policy concerning the evacuation of unserviceable or abandoned materiel and including designation of channels and destinations for evacuated materiel, the establishment of controls and procedures, and the dissemination of condition standards and disposition instructions. See also **evacuation; patient.** (JP 4-02)

evacuee — A civilian removed from a place of residence by military direction for reasons of personal security or the requirements of the military situation. See also **displaced person; expellee; refugee.**

evader — Any person isolated in hostile or unfriendly territory who eludes capture.

evaluation — In intelligence usage, appraisal of an item of information in terms of credibility, reliability, pertinence, and accuracy.

evaluation agent — That command or agency designated in the program directive to be responsible for the planning, coordination, and conduct of the required evaluation. The evaluation agent, normally the Joint Doctrine Center , J-7, identifies evaluation criteria and the media to be used, develops a proposed evaluation directive, coordinates exercise-related evaluation requirements with the sponsoring commands, and provides required evaluation reports to the Director, J-7. See also **joint doctrine; joint tactics, techniques, and procedures; joint test publication.**

evaluation and feedback — See **intelligence cycle.** See also **evaluation.** (JP 2-0)

evasion — The process whereby individuals who are isolated in hostile or unfriendly territory avoid capture with the goal of successfully returning to areas under friendly control. See also **evasion and recovery.** (JP 3-50.3)

evasion aid — In evasion and recovery operations, any piece of information or equipment designed to assist an individual in evading capture. Evasion aids include, but are not limited to, blood chits, pointee-talkees, evasion charts, barter items, and equipment designed to complement issued survival equipment. See also **blood chit; evasion; evasion and recovery; evasion chart; pointee-talkee; recovery; recovery operations.** (JP 3-50.3)

evasion and escape — **(*)** The procedures and operations whereby military personnel and other selected individuals are enabled to emerge from an enemy-held or hostile area to areas under friendly control. Also called **E&E.**

evasion and escape intelligence — Processed information prepared to assist personnel to escape if captured by the enemy or to evade capture if lost in enemy-dominated territory.

evasion and escape net — The organization within enemy-held or hostile areas that operates to receive, move, and exfiltrate military personnel or selected individuals to friendly control. See also **unconventional warfare.**

evasion and escape route — A course of travel, preplanned or not, that an escapee or evader uses in an attempt to depart enemy territory in order to return to friendly lines.

evasion and recovery — The full spectrum of coordinated actions carried out by evaders, recovery forces, and operational recovery planners to effect the successful return of personnel isolated in hostile territory to friendly control. See also **evader; evasion; hostile; recovery force.** (JP 3-50.3)

evasion chart — Special map or chart designed as an evasion aid. See also **evasion; evasion aid.** (JP 3-50.3)

evasion plan of action — A course of action, developed before executing a combat mission, that is intended to improve a potential evader's chances of successful evasion and recovery by providing recovery forces with an additional source of information that can increase the predictability of the evader's actions and movement. Also called **EPA.** See also **course of action; evader; evasion; evasion and recovery; recovery force.** (JP 3-50.3)

event matrix — A description of the indicators and activity expected to occur in each named area of interest. It normally cross-references each named area of interest and indicator with the times they are expected to occur and the courses of action they will confirm or deny. There is no prescribed format. See also **activity; area of interest; indicator.** (JP 2-01.3)

event template — A guide for collection planning. The event template depicts the named areas of interest where activity, or its lack of activity, will indicate which course of action the adversary has adopted. See also **activity; area of interest; collection planning; course of action.** (JP 2-01.3)

exaggerated stereoscopy — See **hyperstereoscopy.**

exceptional transport — **(*)** In railway terminology, transport of a load whose size, weight, or preparation entails special difficulties vis-a-vis the facilities or equipment of even one of the railway systems to be used. See also **ordinary transport.**

excess property — The quantity of property in possession of any component of the Department of Defense that exceeds the quantity required or authorized for retention by that component.

exclusive economic zone — A maritime zone adjacent to the territorial sea that may not extend beyond 200 nautical miles from the baselines from which the breadth of the territorial sea is measured. Within the exclusive economic zone (EEZ), the coastal state has sovereign rights for the purpose of exploring, exploiting, conserving, and managing natural resources, both living and nonliving, of the seabed, subsoil, and the subjacent waters and, with regard to other activities, for the economic exploitation and exploration of the zone (e.g., the production of energy from the water, currents, and winds). Within the EEZ, the coastal state has jurisdiction with regard to establishing and using artificial islands, installations, and structures having economic purposes as well as for marine scientific research and the protection and preservation of the marine environment. Other states may, however, exercise traditional high seas freedoms of navigation, overflight, and related freedoms, such as conducting military exercises in the EEZ. Also called **EEZ.**

exclusion zone — A zone established by a sanctioning body to prohibit specific activities in a specific geographic area. The purpose may be to persuade nations or groups to modify their behavior to meet the desires of the sanctioning body or face continued imposition of sanctions, or use or threat of force. (JP 3-07)

execute order — 1. An order issued by the Chairman of the Joint Chiefs of Staff, by the authority and at the direction of the Secretary of Defense, to implement a National Command Authorities decision to initiate military operations. 2. An order to initiate military operations as directed. Also called **EXORD.** (JP 5-0)

executing commander (nuclear weapons) — A commander to whom nuclear weapons are released for delivery against specific targets or in accordance with approved plans. See also **commander(s); releasing commander (nuclear weapons).**

execution planning — The phase of the Joint Operation Planning and Execution System crisis action planning process that provides for the translation of an approved course of action into an executable plan of action through the preparation of a complete operation plan or operation order. Execution planning is detailed planning for the commitment of specified forces and resources. During crisis action planning, an approved operation plan or other National Command Authorities-approved course of action is adjusted, refined, and translated into an operation order. Execution planning can proceed on the basis of prior deliberate planning, or it can take place in the absence of prior planning. Also called **EP.** See **also Joint Operation Planning and Execution System.** (JP 5-0)

executive agent — A term used to indicate a delegation of authority by the Secretary of Defense to a subordinate to act on the Secretary's behalf. An agreement between equals does not create an executive agent. For example, a Service cannot become a Department of Defense executive agent for a particular matter with simply the agreement of the other Services; such authority must be delegated by the Secretary of Defense. Designation as executive agent, in and of itself, confers no authority. The exact nature and scope of the authority delegated must be stated in the document designating the executive agent. An executive agent may be limited to providing only administration and support or coordinating common functions, or it may be delegated authority, direction, and control over specified resources for specified purposes. Also called **EA.** See also **agent.** (JP 0-2)

exercise — A military maneuver or simulated wartime operation involving planning, preparation, and execution. It is carried out for the purpose of training and evaluation. It may be a multinational, joint, or single-Service exercise, depending on participating organizations. See also **command post exercise; field exercise; maneuver.**

exercise directing staff — (*) A group of officers who by virtue of experience, qualifications, and a thorough knowledge of the exercise instructions, are selected to direct or control an exercise.

exercise filled mine — (*) In naval mine warfare, a mine containing an inert filling and an indicating device. See also **explosive filled mine; fitted mine; mine.**

exercise incident — (*) An occurrence injected by directing staffs into the exercise which will have an effect on the forces being exercised, or their facilities, and which will require action by the appropriate commander and/or staff being exercised.

exercise mine — (*) In naval mine warfare, a mine suitable for use in mine warfare exercises, fitted with visible or audible indicating devices to show where and when it would normally fire. See also **drill mine; mine; practice mine.**

exercise specifications — (*) The fundamental requirements for an exercise, providing in advance an outline of the concept, form, scope, setting, aim, objectives, force requirements, political implications, analysis arrangements, and costs.

exercise sponsor — (*) The commander who conceives a particular exercise and orders that it be planned and executed either by the commander's staff or by a subordinate headquarters.

exercise study — (*) An activity which may take the form of a map exercise, a war game, a series of lectures, a discussion group, or an operational analysis.

exercise term — A combination of two words, normally unclassified, used exclusively to designate a test, drill, or exercise. An exercise term is employed to preclude the possibility of confusing exercise directives with actual operations directives.

exfiltration — The removal of personnel or units from areas under enemy control by stealth, deception, surprise, or clandestine means. See also **special operations; unconventional warfare.**

existence load — Consists of items other than those in the fighting load that are required to sustain or protect the combat soldier. These items may be necessary for increased personal and environmental protection and are not normally carried by the individual. See also **fighting load.**

exoatmosphere — See **nuclear exoatmospheric burst.**

expedition — A military operation conducted by an armed force to accomplish a specific objective in a foreign country. (JP 3-0)

expeditionary force — An armed force organized to accomplish a specific objective in a foreign country.

expellee — A civilian outside the boundaries of the country of his or her nationality or ethnic origin who is being forcibly repatriated to that country or to a third country for political or other purposes. See also **displaced person; evacuee; refugee.**

expendable property — Property that may be consumed in use or loses its identity in use and may be dropped from stock record accounts when it is issued or used.

expendable supplies and materiel — Supplies that are consumed in use, such as ammunition, paint, fuel, cleaning and preserving materials, surgical dressings, drugs, medicines, etc., or that lose their identity, such as spare parts, etc. Also called **consumable supplies and materiel.**

exploder — **(*)** A device designed to generate an electric current in a firing circuit after deliberate action by the user in order to initiate an explosive charge or charges.

exploitation — **(*)** 1. **(DOD only)** Taking full advantage of success in military operations, following up initial gains, and making permanent the temporary effects already achieved. 2. Taking full advantage of any information that has come to hand for tactical, operational, or strategic purposes. 3. An offensive operation that usually follows a successful attack and is designed to disorganize the enemy in depth. See also **attack; pursuit.**

exploratory hunting — **(*)** In naval mine warfare, a parallel operation to search sweeping, in which a sample of the route or area is subjected to minehunting procedures to determine the presence or absence of mines.

explosive filled mine — **(*)** In mine warfare, a mine containing an explosive charge but not necessarily the firing train needed to detonate it. See also **exercise filled mine; fitted mine.**

explosive ordnance — **(*)** All munitions containing explosives, nuclear fission or fusion materials, and biological and chemical agents. This includes bombs and warheads; guided and ballistic missiles; artillery, mortar, rocket, and small arms ammunition; all mines, torpedoes, and depth charges; demolition charges; pyrotechnics; clusters and dispensers; cartridge and propellant actuated devices; electro-explosive devices; clandestine and improvised explosive devices; and all similar or related items or components explosive in nature.

explosive ordnance disposal — **(*)** The detection, identification, on-site evaluation, rendering safe, recovery, and final disposal of unexploded explosive ordnance. It may also include explosive ordnance which has become hazardous by damage or deterioration. Also called **EOD.**

explosive ordnance disposal incident — **(*)** The suspected or detected presence of unexploded or damaged explosive ordnance which constitutes a hazard to operations, installations, personnel, or material. Not included in this definition are the accidental arming or other conditions that develop during the manufacture of high explosive material, technical service assembly operations or the laying of mines and demolition charges.

explosive ordnance disposal procedures — **(*)** Those particular courses or modes of action taken by explosive ordnance disposal personnel for access to, diagnosis, rendering safe, recovery, and final disposal of explosive ordnance or any hazardous material associated with an explosive ordnance disposal incident. a. **access procedures** — Those actions taken to locate exactly and gain access to unexploded explosive ordnance. b. **diagnostic procedures** — Those actions taken to identify and evaluate unexploded explosive ordnance. c. **render safe procedures** — The portion of the explosive ordnance disposal procedures involving the application of special explosive ordnance disposal methods and tools to provide for the interruption of functions or separation of essential components of unexploded explosive ordnance to prevent an unacceptable detonation. d. **recovery procedures** — Those actions taken to recover unexploded explosive ordnance. e. **final disposal procedures** — The final disposal of explosive ordnance which may include demolition or burning in place, removal to a disposal area, or other appropriate means.

explosive ordnance disposal unit — Personnel with special training and equipment who render explosive ordnance safe (such as bombs, mines, projectiles, and booby traps), make intelligence reports on such ordnance, and supervise the safe removal thereof.

explosive train — **(*)** A succession of initiating and igniting elements arranged to cause a charge to function.

exposure dose — **(*)** The exposure dose at a given point is a measurement of radiation in relation to its ability to produce ionization. The unit of measurement of the exposure dose is the roentgen.

exposure station — See **air station.**

extended communications search — In search and rescue operations, consists of contacting all possible sources of information on the missing craft, including physically checking possible locations such as harbors, marinas, and airport ramps. An extended communications search is normally conducted after a preliminary communications search has yielded no results and when the mission is upgraded to the alert phase. Also called **EXCOM.** See also **preliminary communications search; search and rescue incident classification, Subpart b.**

extent of a military exercise — **(*)** The scope of an exercise in relation to the involvement of NATO and/or national commands. See also **intra-command exercise.**

extent of damage — The visible plan area of damage to a target element, usually expressed in units of 1,000 square feet, in detailed damage analysis and in approximate percentages in immediate-type damage assessment reports; e.g., 50 percent structural damage.

external audience — All people who are not part of the internal audience of US military members and civilian employees and their immediate families. Part of the concept of "publics." Includes many varied subsets that may be referred to as "audiences" or "publics." See also **internal audience; public.**

external reinforcing force — **(*)** A reinforcing force which is principally stationed in peacetime outside its intended Major NATO Command area of operations.

external support contractors — US national or third party contract personnel hired from outside the operational area. See also **systems support contractors; theater support contractors.** (JP 4-07)

extraction parachute — An auxiliary parachute designed to release and extract and deploy cargo from aircraft in flight and deploy cargo parachutes. See also **gravity extraction.**

extraction zone — **(*)** A specified drop zone used for the delivery of supplies and/or equipment by means of an extraction technique from an aircraft flying very close to the ground.

F

fabricator — Individuals or groups who, without genuine resources, invent information or inflate or embroider over news for personal gain or for political purposes.

facility — A real property entity consisting of one or more of the following: a building, a structure, a utility system, pavement, and underlying land. See also **air facility.**

facility substitutes — Items such as tents and prepackaged structures requisitioned through the supply system that may be used to substitute for constructed facilities. (JP 4-04)

fairway — A channel either from offshore, in a river, or in a harbor that has enough depth to accommodate the draft of large vessels. See also **draft; watercraft.** (JP 4-01.6)

fallout — The precipitation to Earth of radioactive particulate matter from a nuclear cloud; also applied to the particulate matter itself.

fallout contours — **(*)** Lines joining points which have the same radiation intensity that define a fallout pattern, represented in terms of roentgens per hour.

fallout pattern — **(*)** The distribution of fallout as portrayed by fallout contours.

fallout prediction — An estimate, made before and immediately after a nuclear detonation, of the location and intensity of militarily significant quantities of radioactive fallout.

fallout safe height of burst — The height of burst at or above which no militarily significant fallout will be reproduced as a result of a nuclear weapon detonation. See also **types of burst.**

fallout wind vector plot — **(*)** A wind vector diagram based on the wind structure from the surface of the Earth to the highest altitude of interest.

false origin — **(*)** A fixed point to the south and west of a grid zone from which grid distances are measured eastward and northward.

fan camera photography — **(*)** Photography taken simultaneously by an assembly of three or more cameras systematically installed at fixed angles relative to each other so as to provide wide lateral coverage with overlapping images. See also **tri-camera photography.**

fan cameras — **(*)** An assembly of three or more cameras systematically disposed at fixed angles relative to each other so as to provide wide lateral coverage with overlapping images. See also **split cameras.**

fan marker beacon — **(*)** A type of radio beacon, the emissions of which radiate in a vertical, fan-shaped pattern. The signal can be keyed for identification purposes. See also **radio beacon.**

farm gate type operations — Operational assistance and specialized tactical training provided to a friendly foreign air force by the Armed Forces of the United States to include, under certain specified conditions, the flying of operational missions in combat by combined United States and foreign aircrews as a part of the training being given when such missions are beyond the capability of the foreign air force.

feasibility — Operation plan review criterion. The determination as to whether the assigned tasks could be accomplished by using available resources. See also **acceptability; adequacy.**

feasibility assessment — A basic target analysis that provides an initial determination of the viability of a proposed target for special operations forces employment. Also called **FA.** (JP 3-05.5)

feasibility test — An operation plan review criteria to determine whether or not a plan is within the capacity of the resources that can be made available. See also **logistic implications test.**

federal coordinating officer — Appointed by the Director of the Federal Emergency Management Agency, on behalf of the President, to coordinate federal assistance to a state affected by a disaster or emergency. The source and level of the federal coordinating officer will likely depend on the nature of the federal response. Also called **FCO.** (JP 3-08)

federal modal agencies — See **transportation operating agencies.**

federal service — A term applied to National Guard members and units when called to active duty to serve the Federal Government under Article I, Section 8 and Article II, Section 2 of the Constitution and the US Code, title 10 (Department of Defense), sections 12401 to 12408. See also **active duty; Reserve Components.** (JP 4-05)

federal supply class management — Those functions of materiel management that can best be accomplished by federal supply classification, such as cataloging, characteristic screening, standardization, interchangeability and substitution grouping, multi-item specification management, and engineering support of the foregoing.

federal transport agencies — See **transportation operating agencies.**

feint — In military deception, an offensive action involving contact with the adversary conducted for the purpose of deceiving the adversary as to the location and/or time of the actual main offensive action. (JP 3-58)

fender — An object, usually made of rope or rubber, hung over the side of a vessel to protect the sides from damage caused by impact with wharves or other craft. (JP 4-01.6)

ferret — An aircraft, ship, or vehicle especially equipped for the detection, location, recording, and analyzing of electromagnetic radiation.

F-hour — See **times.**

field army — Administrative and tactical organization composed of a headquarters, certain organic Army troops, service support troops, a variable number of corps, and a variable number of divisions. See also **Army corps.**

field artillery — Equipment, supplies, ammunition, and personnel involved in the use of cannon, rocket, or surface-to-surface missile launchers. Field artillery cannons are classified according to caliber as follows.

Light — 120mm and less.
Medium — 121-160mm.
Heavy — 161-210mm.
Very heavy — greater than 210mm.

Also called **FA.** See also **direct support artillery; general support artillery.**

field artillery observer — A person who watches the effects of artillery fire, adjusts the center of impact of that fire onto a target, and reports the results to the firing agency. See also **naval gunfire spotting team; spotter.**

field exercise — (*) An exercise conducted in the field under simulated war conditions in which troops and armament of one side are actually present, while those of the other side may be imaginary or in outline. See also **command post exercise.**

field fortifications — (*) An emplacement or shelter of a temporary nature which can be constructed with reasonable facility by units requiring no more than minor engineer supervisory and equipment participation.

field headquarters — See **command post.**

field of fire — (*) The area which a weapon or a group of weapons may cover effectively with fire from a given position.

field of view — (*) 1. In photography, the angle between two rays passing through the perspective center (rear nodal point) of a camera lens to the two opposite sides of the format. Not to be confused with "angle of view." 2. The total solid angle available to the gunner when looking through the gunsight. Also called **FOV.**

field of vision — (*) The total solid angle available to the gunner from his or her normal position. See also **field of view.**

field press censorship — The security review of news material subject to the jurisdiction of the Armed Forces of the United States, including all information or material intended for dissemination to the public. Also called **FPC.** See also **censorship**.

field training exercise — An exercise in which actual forces are used to train commanders, staffs, and individual units in basic, intermediate, and advanced-level warfare skills. Also called **FTX.** See also **exercise; maneuver.**

fighter cover — (*) The maintenance of a number of fighter aircraft over a specified area or force for the purpose of repelling hostile air activities. See also **airborne alert; cover.**

fighter engagement zone — See **weapon engagement zone.**

fighter sweep — (*) An offensive mission by fighter aircraft to seek out and destroy enemy aircraft or targets of opportunity in an allotted area of operations.

fighting load — Consists of items of individual clothing, equipment, weapons, and ammunition that are carried by and are essential to the effectiveness of the combat soldier and the accomplishment of the immediate mission of the unit when the soldier is on foot. See also **existence load.**

filler — A substance carried in an ammunition container such as a projectile, mine, bomb, or grenade. A filler may be an explosive, chemical, or inert substance.

filler personnel — Individuals of suitable grade and skill initially required to bring a unit or organization to its authorized strength.

film badge — (*) A photographic film packet to be carried by personnel, in the form of a badge, for measuring and permanently recording (usually) gamma-ray dosage.

filter — (*) In electronics, a device which transmits only part of the incident energy and may thereby change the spectral distribution of energy: a. High pass filters transmit energy above a certain frequency; b. Low pass filters transmit energy below a certain frequency; c. Band pass filters transmit energy of a certain bandwidth; d. Band stop filters transmit energy outside a specific frequency band.

final approach — **(*)** That part of an instrument approach procedure in which alignment and descent for landing are accomplished. a. In a non-precision approach it normally begins at the final approach fix or point and ends at the missed approach point or fix. b. In a precision approach the final approach commences at the glide path intercept point and ends at the decision height/altitude.

final bearing — The magnetic bearing assigned by an air operations center, helicopter direction center, or carrier air traffic control center for final approach; an extension of the landing area centerline. See also **air operations center; final approach; helicopter direction center.** (JP 3-04.1)

final destination — **(*)** In naval control of shipping, the final destination of a convoy or of an individual ship (whether in convoy or independent) irrespective of whether or not routing instructions have been issued.

final disposal procedures — See **explosive ordnance disposal procedures.**

final governing standards — A comprehensive set of country-specific substantive environmental provisions, typically technical limitations on effluent, discharges, etc., or a specific management practice. (JP 3-34)

final plan — **(*)** A plan for which drafts have been coordinated and approved and which has been signed by or on behalf of a competent authority. See also **operation plan.**

final protective fire — **(*)** An immediately available prearranged barrier of fire designed to impede enemy movement across defensive lines or areas.

finance operations — The execution of the joint finance mission to provide financial advice and guidance, support of the procurement process, providing pay support, and providing disbursing support. See also **financial management.** (JP 1-06)

financial management — Financial management encompasses the two core processes of resource management and finance operations. Also called **FM.** See also **finance operations; resource management operations.** (JP 1-06)

financial property accounting — The establishment and maintenance of property accounts in monetary terms; the rendition of property reports in monetary terms.

fire — **(*)** 1. The command given to discharge a weapon(s). 2. To detonate the main explosive charge by means of a firing system. See also **barrage fire; call fire; counterfire; counterpreparation fire; covering fire; destruction fire; direct fire; direct supporting fire; distributed fire; grazing fire; harassing fire; indirect fire; neutralization fire; observed fire; preparation fire; radar fire; registration fire; scheduled fire; searching fire; supporting fire; suppressive fire.**

fireball — **(*)** The luminous sphere of hot gases which forms a few millionths of a second after detonation of a nuclear weapon and immediately starts expanding and cooling.

fire barrage (specify) — An order to deliver a prearranged barrier of fire. Specification of the particular barrage may be by code name, numbering system, unit assignment, or other designated means.

fire capabilities chart — **(*)** A chart, usually in the form of an overlay, showing the areas which can be reached by the fire of the bulk of the weapons of a unit.

fire control — **(*)** The control of all operations in connection with the application of fire on a target.

fire control radar — **(*)** Radar used to provide target information inputs to a weapon fire control system.

fire control system — **(*)** A group of interrelated fire control equipments and/or instruments designed for use with a weapon or group of weapons.

fire coordination — See **fire support coordination.**

fire direction center — That element of a command post, consisting of gunnery and communications personnel and equipment, by means of which the commander exercises fire direction and/or fire control. The fire direction center receives target intelligence and requests for fire, and translates them into appropriate fire direction. The fire direction center provides timely and effective tactical and technical fire control in support of current operations. Also called **FDC.**

fire for effect — That volume of fires delivered on a target to achieve the desired effect. Also called **FFE.** See also **final protective fire; fire mission; neutralize; suppression.**

fire message — See **call for fire.**

fire mission — **(*)** 1. Specific assignment given to a fire unit as part of a definite plan. 2. Order used to alert the weapon/battery area and indicate that the message following is a call for fire.

fire plan — **(*)** A tactical plan for using the weapons of a unit or formation so that their fire will be coordinated.

firepower — **(*)** 1. The amount of fire which may be delivered by a position, unit, or weapon system. 2. Ability to deliver fire.

fires — The effects of lethal or nonlethal weapons. (JP 3-09)

fire storm — **(*)** Stationary mass fire, generally in built-up urban areas, generating strong, inrushing winds from all sides; the winds keep the fires from spreading while adding fresh oxygen to increase their intensity.

fire support — Fires that directly support land, maritime, amphibious, and special operation forces to engage enemy forces, combat formations, and facilities in pursuit of tactical and operational objectives. See also **fires.** (JP 3-09)

fire support area — An appropriate maneuver area assigned to fire support ships by the naval force commander from which they can deliver gunfire support to an amphibious operation. Also called **FSA.** See also **amphibious operation; fire support; naval support area.** (JP 3-09)

fire support coordinating measure — A measure employed by land or amphibious commanders to facilitate the rapid engagement of targets and simultaneously provide safeguards for friendly forces. See also **fire support coordination.** (JP 3-0)

fire support coordination — **(*)** The planning and executing of fire so that targets are adequately covered by a suitable weapon or group of weapons.

fire support coordination center — A single location in which are centralized communications facilities and personnel incident to the coordination of all forms of fire support. Also called **FSCC.** See also **fire; fire support; fire support coordination; support; supporting arms coordination center.** (JP 3-09.1)

fire support coordination line — A fire support coordinating measure that is established and adjusted by appropriate land or amphibious force commanders within their boundaries in consultation with superior, subordinate, supporting, and affected commanders. Fire support coordination lines (FSCLs) facilitate the expeditious attack of surface targets of opportunity beyond the coordinating measure. An FSCL does not divide an area of operations by defining a boundary between close and deep operations or a zone for close air support. The FSCL applies to all fires of air, land, and sea-based weapons systems using any type of ammunition. Forces attacking targets beyond an FSCL must inform all affected commanders in sufficient time to allow necessary reaction to avoid fratricide. Supporting elements attacking targets beyond the FSCL must ensure that the attack will not produce adverse attacks on, or to the rear of, the line. Short of an FSCL, all air-to-ground and surface-to-surface attack operations are controlled by the appropriate land or amphibious force commander. The FSCL should follow well-defined terrain features. Coordination of attacks beyond the FSCL is especially critical to commanders of air, land, and special operations forces. In exceptional circumstances, the inability to conduct this coordination will not preclude the attack of targets beyond the FSCL. However, failure to do so may increase the risk of fratricide and could waste limited resources. Also called **FSCL.** See also **fires; fire support.** (JP 3-0)

fire support element — That portion of the force tactical operations center at every echelon above company or troop (to corps) that is responsible for targeting coordination and for integrating fires delivered on surface targets by fire-support means under the control, or in support, of the force. Also called **FSE.** See also **fire; fire support; force; support.** (JP 3-09.1)

fire support group — **(*)** A temporary grouping of ships under a single commander charged with supporting troop operations ashore by naval gunfire. A fire support group may be further subdivided into fire support units and fire support elements.

fire support officer — Senior field artillery officer assigned to Army maneuver battalions and brigades. Advises commander on fire-support matters. Also called **FSO.** See also **field artillery; fire; fire support; support.** (JP 3-09.1)

fire support station — An exact location at sea within a fire support area from which a fire support ship delivers fire.

fire support team — An Army team provided by the field artillery component to each maneuver company and troop to plan and coordinate all indirect fire means available to the unit, including mortars, field artillery, close air support, and naval gunfire. Also called **FIST.** See also **close air support; field artillery; fire; fire support; support.** (JP 3-09.1)

firing area — **(*)** In a sweeper-sweep combination it is the horizontal area at the depth of a particular mine in which the mine will detonate. The firing area has exactly the same dimensions as the interception area but will lie astern of it unless the mine detonates immediately when actuated.

firing chart — Map, photo map, or grid sheet showing the relative horizontal and vertical positions of batteries, base points, base point lines, check points, targets, and other details needed in preparing firing data.

firing circuit — **(*)** 1. In land operations, an electrical circuit and/or pyrotechnic loop designed to detonate connected charges from a firing point. 2. In naval mine warfare, that part of a mine circuit which either completes the detonator circuit or operates a ship counter.

firing mechanism — See **firing circuit.**

firing point — **(*)** That point in the firing circuit where the device employed to initiate the detonation of the charges is located. Also called **FP.**

firing system — In demolition, a system composed of elements designed to fire the main charge or charges.

first light — The beginning of morning nautical twilight; i.e., when the center of the morning sun is 12 degrees below the horizon.

first responder phase — A phase of medical care in which health care providers' focus is to save life and limb and stabilize the patient sufficiently to withstand evacuation to the next level of care. This first response may include first aid (self-aid and buddy aid, combat lifesavers) or medical assistance by combat medics, hospital corpsmen, physician assistants, or physicians. See also **essential care; evacuation; patient.** (JP 4-02)

first strike — The first offensive move of a war. (Generally associated with nuclear operations.)

fission products — **(*)** A general term for the complex mixture of substances produced as a result of nuclear fission.

fission to yield ratio — **(*)** The ratio of the yield derived from nuclear fission to the total yield; it is frequently expressed in percent.

fitted mine — **(*)** In naval mine warfare, a mine containing an explosive charge, a primer, detonator, and firing system. See also **exercise filled mine; explosive filled mine.**

fix — **(*)** A position determined from terrestrial, electronic, or astronomical data.

fixed ammunition — **(*)** Ammunition in which the cartridge case is permanently attached to the projectile. See also **munition.**

fixed capital property — 1. Assets of a permanent character having continuing value. 2. As used in military establishments, includes real estate and equipment installed or in use, either in productive plants or in field operations. Synonymous with fixed assets.

fixed medical treatment facility — **(*)** A medical treatment facility which is designed to operate for an extended period of time at a specific site.

fixed port — Water terminals with an improved network of cargo-handling facilities designed for the transfer of oceangoing freight. See also **water terminal.** (JP 4-01.5)

fixed price incentive contract — A fixed price type of contract with provision for the adjustment of profit and price by a formula based on the relationship that final negotiated total cost bears to negotiated target cost as adjusted by approved changes.

fixed price type contract — A type of contract that generally provides for a firm price or, under appropriate circumstances, may provide for an adjustable price for the supplies or services being procured. Fixed price contracts are of several types so designed as to facilitate proper pricing under varying circumstances.

fixed station patrol — **(*)** One in which each scout maintains station relative to an assigned point on a barrier line while searching the surrounding area. Scouts are not stationary but remain underway and patrol near the center of their assigned stations. A scout is a surface ship, submarine, or aircraft.

fixer system — See **fixer network.**

flag days (red or green) — Red flag days are those during which movement requirements cannot be met; green flag days are those during which the requisite amount or a surplus of transportation capability exists.

flag officer — A term applied to an officer holding the rank of general, lieutenant general, major general, or brigadier general in the US Army, Air Force or Marine Corps or admiral, vice admiral, or rear admiral in the US Navy or Coast Guard.

flame field expedients — Simple, handmade devices used to produce flame or illumination. Also called **FFE.** (JP 3-15)

flame thrower — **(*)** A weapon that projects incendiary fuel and has provision for ignition of this fuel.

flammable cargo — See **inflammable cargo.**

flank guard — **(*)** A security element operating to the flank of a moving or stationary force to protect it from enemy ground observation, direct fire, and surprise attack.

flanking attack — **(*)** An offensive maneuver directed at the flank of an enemy. See also **frontal attack.**

flare — **(*)** The change in the flight path of an aircraft so as to reduce the rate of descent for touchdown.

flare dud — A nuclear weapon that, when launched at a target, detonates with anticipated yield but at an altitude appreciably greater than intended. This is not a dud insofar as yield is concerned, but it is a dud with respect to the effects on the target and the normal operation of the weapon.

flash blindness — **(*)** Impairment of vision resulting from an intense flash of light. It includes temporary or permanent loss of visual functions and may be associated with retinal burns. See also **dazzle.**

flash burn — **(*)** A burn caused by excessive exposure (of bare skin) to thermal radiation.

flash message — A category of precedence reserved for initial enemy contact messages or operational combat messages of extreme urgency. Brevity is mandatory. See also **precedence.**

flash ranging — Finding the position of the burst of a projectile or of an enemy gun by observing its flash.

flash report — Not to be used. See **inflight report.**

flash suppressor — **(*)** Device attached to the muzzle of the weapon which reduces the amount of visible light or flash created by burning propellant gases.

flash-to-bang time — **(*)** The time from light being first observed until the sound of the nuclear detonation is heard.

flatrack — Portable, open-topped, open-sided units that fit into existing below-deck container cell guides and provide a capability for container ships to carry oversized cargo and wheeled and tracked vehicles. (JP 4-01.7)

flatted cargo — Cargo placed in the bottom of the holds, covered with planks and dunnage, and held for future use. Flatted cargo usually has room left above it for the loading of vehicles that may be moved without interfering with the flatted cargo. Frequently, flatted cargo serves in lieu of ballast. Sometimes called understowed cargo.

fleet — An organization of ships, aircraft, Marine forces, and shore-based fleet activities all under the command of a commander or commander in chief who may exercise operational as well as administrative control. See also **major fleet; numbered fleet.**

fleet ballistic missile submarine — A nuclear-powered submarine designed to deliver ballistic missile attacks against assigned targets from either a submerged or surfaced condition. Designated as **SSBN.**

fleet in being — A fleet (force) that avoids decisive action, but, because of its strength and location, causes or necessitates counter-concentrations and so reduces the number of opposing units available for operations elsewhere.

Fleet Marine Force — A balanced force of combined arms comprising land, air, and service elements of the US Marine Corps. A Fleet Marine Force is an integral part of a US fleet and has the status of a type command. Also called **FMF.**

flexible deterrent option — A planning construct intended to facilitate early decision by laying out a wide range of interrelated response paths that begin with deterrent-oriented options carefully tailored to send the right signal. The flexible deterrent option is the means by which the various deterrent options available to a commander (such as economic, diplomatic, political, and military measures) are implemented into the planning process. Also called **FDO.** See also **deterrent options.** (JP 4-05.1)

flexible response — The capability of military forces for effective reaction to any enemy threat or attack with actions appropriate and adaptable to the circumstances existing.

flight — 1. In Navy and Marine Corps usage, a specified group of aircraft usually engaged in a common mission. 2. The basic tactical unit in the Air Force, consisting of four or more aircraft in two or more elements. 3. A single aircraft airborne on a nonoperational mission.

flight advisory — A message dispatched to aircraft in flight or to interested stations to advise of any deviation or irregularity.

flight deck — 1. In certain airplanes, an elevated compartment occupied by the crew for operating the airplane in flight. 2. The upper deck of an aircraft carrier that serves as a runway.

flight following — **(*)** The task of maintaining contact with specified aircraft for the purpose of determining en route progress and/or flight termination.

flight information center — **(*)** A unit established to provide flight information service and alerting service.

flight information region — **(*)** An airspace of defined dimensions within which flight information service and alerting service are provided. Also called **FIR.** See also **air traffic control center; area control center**.

flight information service — **(*)** A service provided for the purpose of giving advice and information useful for the safe and efficient conduct of flights. Also called **FIS.**

flight levels — **(*)** Surfaces of constant atmospheric pressure which are related to a specific pressure datum, 1013.2 mb (29.92 in), and are separated by specific pressure intervals. (Flight levels are expressed in three digits that represent hundreds of feet; e.g., flight level 250 represents a barometric altimeter indication of 25,000 feet and flight level 255 is an indication of 25,500 feet.)

flight path — **(*)** The line connecting the successive positions occupied, or to be occupied, by an aircraft, missile, or space vehicle as it moves through air or space.

flight plan — **(*)** Specified information provided to air traffic services units relative to an intended flight or portion of a flight of an aircraft.

flight plan correlation — A means of identifying aircraft by association with known flight plans.

flight profile — Trajectory, or its graphic representation, followed by its altitude, speed, distance flown, and maneuver.

flight quarters — A ship configuration that assigns and stations personnel at critical positions to conduct safe flight operations. (JP 3-04.1)

flight readiness firing — A missile system test of short duration conducted with the propulsion system operating while the missile is secured to the launcher. Such a test is performed to determine the readiness of the missile system and launch facilities prior to flight test.

flight surgeon — **(*)** A physician specially trained in aviator medical practice whose primary duty is the medical examination and medical care of aircrew.

flight test — **(*)** Test of an aircraft, rocket, missile, or other vehicle by actual flight or launching. Flight tests are planned to achieve specific test objectives and gain operational information.

flight visibility — The average forward horizontal distance from the cockpit of an aircraft in flight at which prominent unlighted objects may be seen and identified by day and prominent lighted objects may be seen and identified by night.

floating base support — **(*)** A form of logistic support in which supplies, repairs, maintenance, and other services are provided in harbor or at an anchorage for operating forces from ships.

floating craft company — A company-sized unit made up of various watercraft teams such as tugs, barges, and barge cranes. See also **watercraft.** (JP 4-01.6)

floating dump — Emergency supplies preloaded in landing craft, amphibious vehicles, or in landing ships. Floating dumps are located in the vicinity of the appropriate control officer, who directs their landing as requested by the troop commander concerned. (JP 3-02)

floating mine — **(*)** In naval mine warfare, a mine visible on the surface. See also **free mine; mine; watching mine.**

floating reserve — **(*)** In an amphibious operation, reserve troops which remain embarked until needed. See also **general reserve.**

flooder — **(*)** In naval mine warfare, a device fitted to a buoyant mine which, on operation after a preset time, floods the mine case and causes it to sink to the bottom.

flotation — **(*)** The capability of a vehicle to float in water.

fly-in echelon — Includes the balance of the initial assault force, not included in the assault echelon, and some aviation support equipment. Also called **FIE.** (JP 4-01.2)

foam path — A path of fire extinguisher foam laid on a runway to assist aircraft in an emergency landing.

follow-up — In amphibious operations, the reinforcements and stores carried on transport ships and aircraft (not originally part of the amphibious force) that are offloaded after the assault and assault follow-on echelons have been landed. See also **amphibious operation; assault; assault follow-on echelon.** (JP 3-02)

follow-up echelon — **(*)** In air transport operations, elements moved into the objective area after the assault echelon.

follow-up shipping — Ships not originally a part of the amphibious task force but which deliver troops and supplies to the objective area after the assault phase has begun. (JP 3-02.2)

follow-up supplies — Supplies delivered after the initial landings or airdrop to resupply units until routine supply procedures can be instituted. These supplies may be delivered either automatically or on an on-call basis and are prepared for delivery by supporting supply units. See also **resupply; routine supplies; supplies.** (JP 3-17)

footprint — 1. The area on the surface of the earth within a satellite's transmitter or sensor field of view. 2. The amount of personnel, spares, resources, and capabilities physically present and occupying space at a deployed location.

force — 1. An aggregation of military personnel, weapon systems, equipment, and necessary support, or combination thereof. 2. A major subdivision of a fleet. (JP 0-2)

force activity designators — Numbers used in conjunction with urgency of need designators to establish a matrix of priorities used for supply requisitions. Defines the relative importance of the unit to accomplish the objectives of the Department of Defense. Also called **FADs.** See also **force.** (JP 4-09)

force beddown — The provision of expedient facilities for troop support to provide a platform for the projection of force. These facilities may include modular or kit-type facility substitutes. See also **facility substitutes.** (JP 4-04)

force closure — The point in time when a supported joint force commander determines that sufficient personnel and equipment resources are in the assigned operational area to carry out assigned tasks. See also **closure; force.** (JP 3-35)

force health protection — All services performed, provided, or arranged by the Services to promote, improve, conserve, or restore the mental or physical well-being of personnel. These services include, but are not limited to, the management of health services resources, such as manpower, monies, and facilities; preventive and curative health measures; evacuation of the wounded, injured, or sick; selection of the medically fit and disposition of the medically unfit; blood management; medical supply, equipment, and maintenance thereof; combat stress control; and medical, dental, veterinary, laboratory, optometry, medical food, and medical intelligence services. See also **force; protection.** (JP 4-02)

force list — A total list of forces required by an operation plan, including assigned forces, augmentation forces, and other forces to be employed in support of the plan.

force module — A grouping of combat, combat support, and combat service support forces, with their accompanying supplies and the required nonunit resupply and personnel necessary to sustain forces for a minimum of 30 days. The elements of force modules are linked together or are uniquely identified so that they may be extracted from or adjusted as an entity in the Joint Operation Planning and Execution System databases to enhance flexibility and usefulness of the operation plan during a crisis. Also called **FM.** See also **force module package.**

force module package — A force module with a specific functional orientation (e.g. air superiority, close air support, reconnaissance, ground defense) that include combat, associated combat support, and combat service support forces. Additionally, force module packages will contain sustainment in accordance with logistic policy contained in Joint Strategic Capabilities Plan Annex B. Also called **FMP.** See also **force module.**

force movement control center — A temporary organization activated by the Marine air-ground task force to control and coordinate all deployment support activities. Also called **FMCC.** See also **Marine air-ground task force**. (JP 4-01.8)

force multiplier — A capability that, when added to and employed by a combat force, significantly increases the combat potential of that force and thus enhances the probability of successful mission accomplishment. (JP 3-05.3)

force planning — Planning associated with the creation and maintenance of military capabilities. It is primarily the responsibility of the Military Departments and Services and is conducted under the administrative control that runs from the Secretary of Defense to the Military Departments and Services. (JP 5-0)

force projection — The ability to project the military element of national power from the continental United States (CONUS) or another theater, in response to requirements for military operations. Force projection operations extend from mobilization and deployment of forces to redeployment to CONUS or home theater. See also **force.** (JP 3-35)

force protection — Actions taken to prevent or mitigate hostile actions against Department of Defense personnel (to include family members), resources, facilities, and critical information. These actions conserve the force's fighting potential so it can be applied at the decisive time and place and incorporate the coordinated and synchronized offensive and defensive measures to enable the effective employment of the joint force while degrading opportunities for the enemy. Force protection does not include actions to defeat the enemy or protect against accidents, weather, or disease. Also called **FP.** See also **force; protection; terrorist threat condition.** (JP 3-0)

force rendezvous — **(*)** A checkpoint at which formations of aircraft or ships join and become part of the main force. Also called **group rendezvous.**

force requirement number — An alphanumeric code used to uniquely identify force entries in a given operation plan time-phased force and deployment data. Also called **FRN.**

force(s) — See **airborne force; armed forces; covering force; garrison force; multinational force; Navy cargo handling force; task force; underway replenishment force.**

force shortfall — A deficiency in the number of types of units available for planning within the time required for the performance of an assigned task. (JP 4-05)

forces in being — **(*)** Forces classified as being in state of readiness "A" or "B" as prescribed in the appropriate Military Committee document.

force sourcing — The identification of the actual units, their origins, ports of embarkation, and movement characteristics to satisfy the time-phased force requirements of a supported commander.

force structure — See **military capability.**

force tabs — With reference to war plans, the statement of time-phased deployments of major combat units by major commands and geographical areas.

force tracking — The identification of units and their specific modes of transport during movement to an objective area. (JP 4-01.3)

forcible entry — Seizing and holding of a military lodgment in the face of armed opposition. See also **lodgment.** (JP 3-18)

fordability — See **shallow fording.**

foreign armed force — An armed force belonging to a government or organizational entity other than the United States.

foreign assistance — Assistance to foreign nations ranging from the sale of military equipment to donations of food and medical supplies to aid survivors of natural and manmade disasters. US assistance takes three forms — development assistance, humanitarian assistance, and security assistance. See also **domestic emergencies; foreign disaster; foreign humanitarian assistance; security assistance.** (JP 3-08)

foreign disaster — An act of nature (such as a flood, drought, fire, hurricane, earthquake, volcanic eruption, or epidemic), or an act of man (such as a riot, violence, civil strife, explosion, fire, or epidemic), which is or threatens to be of sufficient severity and magnitude to warrant United States foreign disaster relief to a foreign country, foreign persons, or to an international organization. See also **foreign disaster relief.** (JP 3-08)

foreign disaster relief — Prompt aid that can be used to alleviate the suffering of foreign disaster victims. Normally it includes humanitarian services and transportation; the provision of food, clothing, medicine, beds, and bedding; temporary shelter and housing; the furnishing of medical materiel and medical and technical personnel; and making repairs to essential services. See also **foreign disaster.** (JP 3-07.6)

foreign humanitarian assistance — Programs conducted to relieve or reduce the results of natural or manmade disasters or other endemic conditions such as human pain, disease, hunger, or privation that might present a serious threat to life or that can result in great damage to or loss of property. Foreign humanitarian assistance (FHA) provided by US forces is limited in scope and duration. The foreign assistance provided is designed to supplement or complement the efforts of the host nation civil authorities or agencies that may have the primary responsibility for providing FHA. FHA operations are those conducted outside the United States, its territories, and possessions. Also called **FHA.** See also **foreign assistance.** (JP 3-07.6)

foreign instrumentation signals intelligence — Technical information and intelligence derived from the intercept of foreign electromagnetic emissions associated with the testing and operational deployment of non-US aerospace, surface, and subsurface systems. Foreign instrumentation signals intelligence is a subcategory of signals intelligence. Foreign instrumentation signals include but are not limited to telemetry, beaconry, electronic interrogators, and video data links. Also called **FISINT**. See also **signals intelligence.** (JP 2-01)

foreign intelligence — Intelligence relating to capabilities, intentions, and activities of foreign powers, organizations, or persons (not including counterintelligence), except for information on international terrorist activities. See also **intelligence**. (JP 2-0)

foreign internal defense — Participation by civilian and military agencies of a government in any of the action programs taken by another government to free and protect its society from subversion, lawlessness, and insurgency. Also called **FID**.

foreign military sales — That portion of United States security assistance authorized by the Foreign Assistance Act of 1961, as amended, and the Arms Export Control Act of 1976, as amended. This assistance differs from the Military Assistance Program and the International Military Education and Training Program in that the recipient provides reimbursement for defense articles and services transferred. Also called **FMS.**

foreign military sales trainees — Foreign nationals receiving training conducted by the Department of Defense on a reimbursable basis, at the country's request.

foreign national — Any person other than a US citizen, US permanent or temporary legal resident alien, or person in US custody.

foreign object damage — Rags, pieces of paper, line, articles of clothing, nuts, bolts, or tools that, when misplaced or caught by air currents normally found around aircraft operations (jet blast, rotor or prop wash, engine intake), cause damage to aircraft systems or weapons or injury to personnel. Also called **FOD.** (JP 3-04.1)

foreshore — That portion of a beach extending from the low water (datum) shoreline to the limit of normal high water wave wash. (JP 4-01.6)

format — **(*)** 1. In photography, the size and/or shape of a negative or of the print therefrom. 2. In cartography, the shape and size of a map or chart.

formation — **(*)** 1. An ordered arrangement of troops and/or vehicles for a specific purpose. 2. An ordered arrangement of two or more ships, units, or aircraft proceeding together under a commander.

formatted message text — **(*)** A message text composed of several sets ordered in a specified sequence, each set characterized by an identifier and containing information of a specified type, coded and arranged in an ordered sequence of character fields in accordance with the NATO message text formatting rules. It is designed to permit both manual and automated handling and processing. See also **free form message text; structured message text.**

formerly restricted data — Information removed from the restricted data category upon a joint determination by the Department of Energy (or antecedent agencies) and Department of Defense that such information relates primarily to the military utilization of atomic weapons and that such information can be adequately safeguarded as classified defense information. (Section 142d, Atomic Energy Act of 1954, as amended.) See also **restricted data.**

form lines — (*) Lines resembling contours, but representing no actual elevations, which have been sketched from visual observation or from inadequate or unreliable map sources, to show collectively the configuration of the terrain.

forward aeromedical evacuation — (*) That phase of evacuation which provides airlift for patients between points within the battlefield, from the battlefield to the initial point of treatment, and to subsequent points of treatment within the combat zone.

forward air controller — An officer (aviator/pilot) member of the tactical air control party who, from a forward ground or airborne position, controls aircraft in close air support of ground troops. Also called **FAC.** See also **close air support.** (JP 3-09.1)

forward air controller (airborne) — A specifically trained and qualified aviation officer who exercises control from the air of aircraft engaged in close air support of ground troops. The forward air controller (airborne) is normally an airborne extension of the tactical air control party. Also called **FAC(A).** (JP 3-09.3)

forward area — An area in proximity to combat.

forward arming and refueling point — A temporary facility — organized, equipped, and deployed by an aviation commander, and normally located in the main battle area closer to the area where operations are being conducted than the aviation unit's combat service area — to provide fuel and ammunition necessary for the employment of aviation maneuver units in combat. The forward arming and refueling point permits combat aircraft to rapidly refuel and rearm simultaneously. Also called **FARP**.

forward aviation combat engineering — A mobility operation in which engineers perform tasks in support of forward aviation ground facilities. Tasks include reconnaissance; construction of low altitude parachute extraction zones, landing strips, and airstrips; and providing berms, revetments, and trenches for forward arming and refueling points. See also **combat engineering; reconnaissance**. (JP 3-34)

forward edge of the battle area — (*) The foremost limits of a series of areas in which ground combat units are deployed, excluding the areas in which the covering or screening forces are operating, designated to coordinate fire support, the positioning of forces, or the maneuver of units. Also called **FEBA.**

forward line of own troops — A line that indicates the most forward positions of friendly forces in any kind of military operation at a specific time. The forward line of own troops (FLOT) normally identifies the forward location of covering and screening forces. The FLOT may be at, beyond, or short of the forward edge of the battle area. An enemy FLOT indicates the forward-most position of hostile forces. Also called **FLOT.**

forward-looking infrared — An airborne, electro-optical thermal imaging device that detects far-infrared energy, converts the energy into an electronic signal, and provides a visible image for day or night viewing. Also called **FLIR.** (JP 3-09.3)

forward oblique air photograph — Oblique photography of the terrain ahead of the aircraft.

forward observer — An observer operating with front line troops and trained to adjust ground or naval gunfire and pass back battlefield information. In the absence of a forward air controller, the observer may control close air support strikes. Also called **FO.** See also **forward air controller; spotter.** (JP 3-09.1)

forward operating base — An airfield used to support tactical operations without establishing full support facilities. The base may be used for an extended time period. Support by a main operating base will be required to provide backup support for a forward operating base. Also called **FOB.** (JP 3-09.3)

forward operating location — Primarily used for counterdrug operations. Similar to a forward operating base (FOB) but without the in-place infrastructure associated with a FOB. Also called **FOL.**

forward operations base — In special operations, a base usually located in friendly territory or afloat that is established to extend command and control or communications or to provide support for training and tactical operations. Facilities may be established for temporary or longer duration operations and may include an airfield or an unimproved airstrip, an anchorage, or a pier. A forward operations base may be the location of special operations component headquarters or a smaller unit that is controlled and/or supported by a main operations base. Also called **FOB.** See also **advanced operations base; main operations base.** (JP 3-05.3)

forward recovery mission profile — A mission profile that involves the recovery of an aircraft at a neutral or friendly forward area airfield or landing site.

forward resuscitative surgery — The urgent initial surgery required to render patients transportable for further evacuation to medical treatment facilities staffed and equipped to provide for their care. Forward resuscitative surgery is performed on patients with signs and symptoms of initial airway compromise, difficult breathing, and circulatory shock and who do not respond to initial emergency medical treatment and advanced trauma management procedures. See also **essential care; evacuation; medical treatment facility; patient.** (JP 4-02)

forward slope — **(*)** Any slope which descends towards the enemy.

forward tell — **(*)** The transfer of information to a higher level of command. See also **track telling.**

four-round illumination diamond — **(*)** A method of distributing the fire of illumination shells which, by a combination of lateral spread and range spread, provides illumination of a large area.

463L system — Aircraft pallets, nets, tie down, and coupling devices, facilities, handling equipment, procedures, and other components designed to interface with military and civilian aircraft cargo restraint systems. Though designed for airlift, system components may have to move intermodally via surface to support geographic combatant commander objectives. (JP 4-01.7)

fragmentary order — An abbreviated form of an operation order (verbal, written or digital) usually issued on a day-to-day basis that eliminates the need for restating information contained in a basic operation order. It may be issued in sections. It is issued after an operation order to change or modify that order or to execute a branch or sequel to that order. Also called **FRAG order.**

frame — (*) In photography, any single exposure contained within a continuous sequence of photographs.

free air anomaly — The difference between observed gravity and theoretical gravity that has been computed for latitude and corrected for elevation of the station above or below the geoid, by application of the normal rate of change of gravity for change of elevation, as in free air.

free air overpressure — (*) The unreflected pressure, in excess of the ambient atmospheric pressure, created in the air by the blast wave from an explosion. See also **overpressure.**

freedom of navigation operations — Operations conducted to demonstrate US or international rights to navigate air or sea routes. (JP 3-07)

free drop — (*) The dropping of equipment or supplies from an aircraft without the use of parachutes. See also **airdrop; air movement; free fall; high velocity drop; low velocity drop.**

free fall — A parachute maneuver in which the parachute is manually activated at the discretion of the jumper or automatically at a preset altitude. See also **airdrop; air movement; free drop; high velocity drop; low velocity drop.**

free field overpressure — See **free air overpressure.**

free-fire area — A specific area into which any weapon system may fire without additional coordination with the establishing headquarters. Also called **FFA.** See also **fire.** (JP 3-09)

free form message text — (*) A message text without prescribed format arrangements. It is intended for fast drafting as well as manual handling and processing. See also **formatted message text; structured message text.**

free issue — Materiel provided for use or consumption without charge to the fund or fund subdivision that finances the activity to which it is issued.

free mail — Correspondence of a personal nature that weighs less than 11 ounces, to include audio and video recording tapes, from a member of the Armed Forces or designated civilian, mailed postage free from a Secretary of Defense approved free mail zone. (JP 1-0)

free mine — (*) In naval mine warfare, a moored mine whose mooring has parted or been cut.

free play exercise — (*) An exercise to test the capabilities of forces under simulated contingency and/or wartime conditions, limited only by those artificialities or restrictions required by peacetime safety regulations. See also **controlled exercise.**

free rocket — (*) A rocket not subject to guidance or control in flight.

freight consolidating activity — A transportation activity that receives less than car- or truckload shipments of materiel for the purpose of assembling them into car- or truckload lots for onward movement to the ultimate consignee or to a freight distributing activity or other break bulk point. See also **freight distributing activity.**

freight distributing activity — A transportation activity that receives and unloads consolidated car- or truckloads of less than car- or truckload shipments of material and forwards the individual shipments to the ultimate consignee. See also **freight consolidating activity.**

frequency deconfliction — A systematic management procedure to coordinate the use of the electromagnetic spectrum for operations, communications, and intelligence functions. Frequency deconfliction is one element of electromagnetic spectrum management. See also **electromagnetic spectrum; electronic warfare; spectrum management.** (JP 3-51)

friendly — A contact positively identified as friendly. See also **hostile.**

friendly fire — In casualty reporting, a casualty circumstance applicable to persons killed in action or wounded in action mistakenly or accidentally by friendly forces actively engaged with the enemy, who are directing fire at a hostile force or what is thought to be a hostile force. See also **casualty.**

front — (*) 1. The lateral space occupied by an element measured from the extremity of one flank to the extremity of the other flank. 2. The direction of the enemy. 3. The line of contact of two opposing forces. 4. When a combat situation does not exist or is not assumed, the direction toward which the command is faced.

frontal attack — (*) 1. An offensive maneuver in which the main action is directed against the front of the enemy forces. 2. **(DOD only)** In air intercept, an attack by an interceptor aircraft that terminates with a heading crossing angle greater than 135 degrees.

frustrated cargo — Any shipment of supplies and/or equipment which, while en route to destination, is stopped prior to receipt and for which further disposition instructions must be obtained.

full charge — The larger of the two propelling charges available for naval guns.

full mission-capable — Material condition of any piece of military equipment, aircraft, or training device indicating that it can perform all of its missions. Also called **FMC.** See also **deadline; mission-capable; partial mission-capable; partial mission-capable, maintenance; partial mission-capable, supply.**

full mobilization — See **mobilization.**

functional component command — A command normally, but not necessarily, composed of forces of two or more Military Departments which may be established across the range of military operations to perform particular operational missions that may be of short duration or may extend over a period of time. See also **component; Service component command.** (JP 0-2)

functional damage assessment — The estimate of the effect of military force to degrade or destroy the functional or operational capability of the target to perform its intended mission and on the level of success in achieving operational objectives established against the target. This assessment is based upon all-source information, and includes an estimation of the time required for recuperation or replacement of the target function. See also **damage assessment; target.** (JP 3-60)

functional kill — To render a targeted installation, facility, or target system unable to fulfill its primary function.

functional plans — Plans involving the conduct of military operations in a peacetime or permissive environment developed by combatant commanders to address requirements such as disaster relief, nation assistance, logistics, communications, surveillance, protection of US citizens, nuclear weapon recovery and evacuation, and continuity of operations or similar discrete tasks. They may be developed in response to the requirements of the Joint Strategic Capabilities Plan, at the initiative of the combatant commander (CINC), or as tasked by the supported combatant commander, Joint Staff, Service, or Defense agency. Chairman of the Joint Chiefs of Staff review of CINC-initiated plans is not normally required. (JP 5-0)

functions — The appropriate or assigned duties, responsibilities, missions, or tasks of an individual, office, or organization. As defined in the National Security Act of 1947, as amended, the term "function" includes functions, powers, and duties (5 United States Code 171n (a)).

fusion — In intelligence usage, the process of examining all sources of intelligence and information to derive a complete assessment of activity. (JP 2-0)

fusion center — In intelligence usage, a physical location to accomplish fusion. It normally has sufficient intelligence automated data processing capability to assist in the process. (JP 2-0)

fuze cavity — **(*)** A recess in a charge for receiving a fuze.

Intentionally Blank

G

gadget — Radar equipment (type of equipment may be indicated by a letter as listed in operation orders). May be followed by a color to indicate state of jamming. Colors will be used as follows: a. **green** — Clear of jamming. b. **amber** — Sector partially jammed. c. **red** — Sector completely jammed. d. **blue** — Completely jammed.

gap — An area within a minefield or obstacle belt, free of live mines or obstacles, whose width and direction will allow a friendly force to pass through in tactical formation. See also **phoney minefield.**

gap filler radar — **(*)** A radar used to supplement the coverage of the principal radar in areas where coverage is inadequate.

gap (imagery) — Any space where imagery fails to meet minimum coverage requirements. This might be a space not covered by imagery or a space where the minimum specified overlap was not obtained.

gap marker — **(*)** In landmine warfare, markers used to indicate a minefield gap. Gap markers at the entrance to, and exit from, the gap will be referenced to a landmark or intermediate marker. See also **marker.**

garble — An error in transmission, reception, encryption, or decryption that changes the text of a message or any portion thereof in such a manner that it is incorrect or undecryptable.

garnishing — **(*)** In surveillance, natural or artificial material applied to an object to achieve or assist camouflage.

garrison force — **(*)** All units assigned to a base or area for defense, development, operation, and maintenance of facilities. See also **force(s).**

gear — A general term for a collection of spars, ropes, blocks, and equipment used for lifting and stowing cargo and ships stores. (JP 4-01.6)

general agency agreement — A contract between the Maritime Administration and a steamship company which, as general agent, exercises administrative control over a government-owned ship for employment by the Military Sealift Command. Also called **GAA**. See also **Military Sealift Command.** (JP 3-02.2)

general air cargo — **(*)** Cargo without hazardous or dangerous properties and not requiring extra precautions for air transport.

general and complete disarmament — Reductions of armed forces and armaments by all states to levels required for internal security and for an international peace force. Connotation is "total disarmament" by all states.

general cargo — Cargo that is susceptible for loading in general, nonspecialized stowage areas or standard shipping containers; e.g., boxes, barrels, bales, crates, packages, bundles, and pallets.

general engineering — Encompasses the construction and repair of lines of communications, main supply routes, airfields, and logistic facilities to support joint military operations and may be performed in direct support of combat operations, such as battle damage repair. These operations include both horizontal and vertical construction, and may include use of both expedient repair methods and more deliberate construction methods characterized by the application of design criteria, advanced planning, and preparation, depending on the mission requirements. Also called **GE.** (JP 3-34)

general map — A map of small scale used for general planning purposes. See also **map.**

general military intelligence — Intelligence concerning the (1) military capabilities of foreign countries or organizations or (2) topics affecting potential US or multinational military operations, relating to the following subjects: armed forces capabilities, including order of battle, organization, training, tactics, doctrine, strategy, and other factors bearing on military strength and effectiveness; area and terrain intelligence, including urban areas, coasts and landing beaches, and meteorological, oceanographic, and geological intelligence; transportation in all modes; military materiel production and support industries; military and civilian command, control, communications, computers, and intelligence systems; military economics, including foreign military assistance; insurgency and terrorism; military-political-sociological intelligence; location, identification, and description of military-related installations; government control; escape and evasion; and threats and forecasts. (Excludes scientific and technical intelligence.) Also called **GMI.** See also **intelligence; military intelligence.** (JP 2-0)

general orders — 1. Permanent instructions, issued in order form, that apply to all members of a command, as compared with special orders, which affect only individuals or small groups. General orders are usually concerned with matters of policy or administration. 2. A series of permanent guard orders that govern the duties of a sentry on post.

general purchasing agents — Agents who have been appointed in the principal overseas areas to supervise, control, coordinate, negotiate, and develop the local procurement of supplies, services, and facilities by Armed Forces of the United States, in order that the most effective utilization may be made of local resources and production.

general quarters — A condition of readiness when naval action is imminent. All battle stations are fully manned and alert; ammunition is ready for instant loading; guns and guided missile launchers may be loaded.

general staff — A group of officers in the headquarters of Army or Marine divisions, Marine brigades, and aircraft wings, or similar or larger units that assist their commanders in planning, coordinating, and supervising operations. A general staff may consist of four or more principal functional sections: personnel (G-1), military intelligence (G-2), operations and training (G-3), logistics (G-4), and (in Army organizations) civil affairs and military government (G-5). (A particular section may be added or eliminated by the commander, dependent upon the need that has been demonstrated.) The comparable Air Force staff is found in the wing and larger units, with sections designated personnel, operations, etc. G-2 Air and G-3 Air are Army officers assigned to G-2 or G-3 at division, corps, and Army headquarters level who assist in planning and coordinating joint operations of ground and air units. Naval staffs ordinarily are not organized on these lines, but when they are, they are designated N-1, N-2, etc. Similarly, a joint staff may be designated J-1, J-2, etc. In Army brigades and smaller units and in Marine Corps units smaller than a brigade or aircraft wing, staff sections are designated S-1, S-2, etc., with corresponding duties; referred to as a unit staff in the Army and as an executive staff in the Marine Corps. See also **staff.**

general stopping power — **(*)** The percentage of a group of vehicles in battle formation likely to be stopped by mines when attempting to cross a minefield.

general support — **(*)** 1. That support which is given to the supported force as a whole and not to any particular subdivision thereof. See also **close support; direct support; mutual support; support.** 2. **(DOD only)** A tactical artillery mission. Also called **GS.** See also **direct support; general support-reinforcing; reinforcing.**

general support artillery — **(*)** Artillery which executes the fire directed by the commander of the unit to which it organically belongs or is attached. It fires in support of the operation as a whole rather than in support of a specific subordinate unit. Also called **GSA.** See also **direct support artillery; general support-reinforcing; reinforcing.**

general support-reinforcing — General support-reinforcing artillery has the mission of supporting the force as a whole and of providing reinforcing fires for other artillery units. Also called **GSR.** See also **direct support artillery; reinforcing.**

general unloading period — **(*)** In amphibious operations, that part of the ship-to-shore movement in which unloading is primarily logistic in character, and emphasizes speed and volume of unloading operations. It encompasses the unloading of units and cargo from the ships as rapidly as facilities on the beach permit. It proceeds without regard to class, type, or priority of cargo, as permitted by cargo handling facilities ashore. See also **initial unloading period.**

general war — Armed conflict between major powers in which the total resources of the belligerents are employed, and the national survival of a major belligerent is in jeopardy.

generation (photography) — The preparation of successive positive and/or negative reproductions from an original negative and/or positive (first-generation). For example, the first positive produced from an original negative is a second-generation product; the negative made from this positive is a third-generation product; and the next positive or print from that negative is a fourth-generation product.

geographic coordinates — **(*)** The quantities of latitude and longitude which define the position of a point on the surface of the Earth with respect to the reference spheroid. See also **coordinates.**

geographic reference points — A means of indicating position, usually expressed either as double letters or as code words that are established in operation orders or by other means.

georef — **(*)** A worldwide position reference system that may be applied to any map or chart graduated in latitude and longitude regardless of projection. It is a method of expressing latitude and longitude in a form suitable for rapid reporting and plotting. (This term is derived from the words "The World Geographic Reference System.")

geospatial information and services — The concept for collection, information extraction, storage, dissemination, and exploitation of geodetic, geomagnetic, imagery (both commercial and national source), gravimetric, aeronautical, topographic, hydrographic, littoral, cultural, and toponymic data accurately referenced to a precise location on the earth's surface. These data are used for military planning, training, and operations including navigation, mission planning, mission rehearsal, modeling, simulation and precise targeting. Geospatial information provides the basic framework for battlespace visualization. It is information produced by multiple sources to common interoperable data standards. It may be presented in the form of printed maps, charts, and publications; in digital simulation and modeling databases; in photographic form; or in the form of digitized maps and charts or attributed centerline data. Geospatial services include tools that enable users to access and manipulate data, and also includes instruction, training, laboratory support, and guidance for the use of geospatial data. Also called **GI&S.** (JP 2-03)

geospatial information and services priorities — The priorities defined by the Joint Chiefs of Staff for indicating the relative importance of geospatial information and services geographical area as well as weapons systems support requirements. The priorities are used as one of the factors in allocating National Imagery and Mapping Agency production resources. Priority definitions are contained in the joint strategic planning document.

glide bomb — A bomb fitted with airfoils to provide lift and which is carried and released in the direction of a target by an airplane.

glide mode — In a flight control system, a control mode in which an aircraft is automatically positioned to the center of the glide slope course.

Global Combat Support System — A strategy that provides information interoperability across combat support functions and between combat support and command and control functions through the Global Command and Control System. Also called **GCSS.** See also **combat forces; combat support.** (JP 4-0)

Global Command and Control System — Highly mobile, deployable command and control system supporting forces for joint and multinational operations across the range of military operations, any time and anywhere in the world with compatible, interoperable, and integrated command, control, communications, computers, and intelligence systems. Also called **GCCS.** See also **command and control; command and control system.** (JP 2-01)

global distribution — The process that synchronizes and integrates fulfillment of joint force requirements with employment of the joint force. It provides national resources (personnel and materiel) to support execution of joint operations. The ultimate objective of this process is the effective and efficient accomplishment of the joint force mission. See also **distribution.** (JP 4-09)

global distribution of materiel — The process of providing materiel from the source of supply to its point of consumption or use on a worldwide basis. See also **global distribution.** (JP 4-09)

global grid — An open systems architecture that provides global connectivity instantaneously on warrior demand. The global grid can support both vertical and horizontal information flow to joint and multinational forces. See also **common operating environment; node/command, control, communications, and computers node.** (JP 6-02)

Global Information Grid — The globally interconnected, end-to-end set of information capabilities, associated processes and personnel for collecting, processing, storing, disseminating and managing information on demand to warfighters, policy makers, and support personnel. The Global Information Grid (GIG) includes all owned and leased communications and computing systems and services, software (including applications), data, security services and other associated services necessary to achieve information superiority. It also includes National Security Systems as defined in section 5142 of the Clinger-Cohen Act of 1996. The GIG supports all Department of Defense (DOD), National Security, and related intelligence community missions and functions (strategic, operational, tactical and business), in war and in peace. The GIG provides capabilities from all operating locations (bases, posts, camps, stations, facilities, mobile platforms and deployed sites). The GIG provides interfaces to coalition, allied, and non-DOD users and systems. Also called **GIG.** See also **grid; information.** (JP 3-05.1)

global information infrastructure — The worldwide interconnection of communications networks, computers, databases, and consumer electronics that make vast amounts of information available to users. The global information infrastructure encompasses a wide range of equipment, including cameras, scanners, keyboards, facsimile machines, computers, switches, compact disks, video and audio tape, cable, wire, satellites, fiber-optic transmission lines, networks of all types, televisions, monitors, printers, and much more. The friendly and adversary personnel who make decisions and handle the transmitted information constitute a critical component of the global information infrastructure. Also called **GII.** See also **defense information infrastructure; information; information system; national information infrastructure.** (JP 3-13)

Global Patient Movement Requirements Center — A joint activity reporting directly to the Commander in Chief, US Transportation Command, the Department of Defense single manager for the regulation of movement of uniformed services patients. The Global Patient Movement Requirements Center authorizes transfers to medical treatment facilities of the Military Departments or the Department of Veterans Affairs and coordinates intertheater and inside continental United States patient movement requirements with the appropriate transportation component commands of US Transportation Command. Also called **GPMRC.** See also **medical treatment facility.** (JP 4-02)

global positioning system — A satellite constellation that provides highly accurate position, velocity, and time navigation information to users. Also called **GPS.**

Global Satellite Communications Support Center — United States Space Command operational element responsible for: providing global satellite communications system status; maintaining global situational awareness to include each combatant commander's planned and current operations as well as deliberate plans; supporting radio frequency interference resolution management; supporting satellite anomaly resolution and management; facilitating satellite communications interface to the defense information infrastructure; and managing the Regional Satellite Communications Support Centers. Also called **GSSC.**

global transportation management — The integrated process of satisfying transportation requirements using the Defense Transportation System to meet national security objectives. The process begins with planning, programming, and budgeting for transportation assets, services, and associated systems and continues through delivery of the users' transportation movement requirements. Also called **GTM.** See also **Defense Transportation System; Global Transportation Network.** (JP 4-01)

Global Transportation Network — The designated Department of Defense (DOD) in-transit visibility system, providing customers with the ability to track the identity, status, and location of DOD units and non-unit cargo, passengers, patients, forces, and military and commercial airlift, sealift, and surface assets from origin to destination across the range of military operations. The Global Transportation Network (GTN) collects, integrates, and distributes transportation information to combatant commanders, Services, and other DOD customers. GTN provides the US Transportation Command with the ability to perform command and control operations, planning and analysis, and business operations in tailoring customer requirements throughout the requirements process. Also called **GTN.** See also **in-transit visibility.** (JP 4-0)

go no-go — The condition or state of operability of a component or system: "go," functioning properly; or "no-go," not functioning properly. Alternatively, a critical point at which a decision to proceed or not must be made.

government-owned, contract-operated ships — Those ships to which the US Government holds title and which the Military Sealift Command operates under a contract (i.e., nongovernment-manned). These ships are designated United States Naval Ships and use the prefix "USNS" with the ship name and the letter "T" as a prefix to the ship classification (e.g., T-AKR). See also **Military Sealift Command; United States Naval Ship.** (JP 3-02.2)

government-owned, Military Sealift Command-operated ships — Those ships to which the US Government holds title and which the Military Sealift Command operates with US Government (civil service) employees. These ships are designated United States Naval Ships and use the prefix "USNS" with the ship name and the letter "T" as a prefix to the ship classification (e.g., T-AKR). See also **Military Sealift Command; United States Naval Ship.** (JP 3-02.2)

gradient — The rate of inclination to horizontal expressed as a ratio, such as 1:25, indicating a one unit rise to 25 units of horizontal distance. (JP 4-01.6)

gradient circuit — **(*)** In mine warfare, a circuit which is actuated when the rate of change, with time, of the magnitude of the influence is within predetermined limits.

grand strategy — See **national security strategy; national strategy.**

graphic — **(*)** Any and all products of the cartographic and photogrammetric art. A graphic may be a map, chart, or mosaic or even a film strip that was produced using cartographic techniques.

graphic scale — **(*)** A graduated line by means of which distances on the map, chart, or photograph may be measured in terms of ground distance. See also **scale.**

grapnel — **(*)** In naval mine warfare, a device fitted to a mine mooring designed to grapple the sweep wire when the mooring is cut.

graticule — **(*)** 1. In cartography, a network of lines representing the Earth's parallels of latitude and meridians of longitude. 2. In imagery interpretation, see **reticle.**

graticule ticks — **(*)** In cartography, short lines indicating where selected meridians and parallels intersect.

graves registration program — A program that provides for search, recovery, tentative identification, and evacuation or temporary interment. Temporary interment is only authorized by the geographic combatant commander. Disposition of personal effects is included in this program. See also **personal effects.** (JP 4-06)

gravity extraction — **(*)** The extraction of cargoes from the aircraft by influence of their own weight. See also **extraction parachute.**

grazing fire — **(*)** Fire approximately parallel to the ground where the center of the cone of fire does not rise above one meter from the ground. See also **fire.**

Greenwich Mean Time — See **Universal Time.** Also called **GMT.**

grey propaganda — Propaganda that does not specifically identify any source. See also **propaganda.**

grid — 1. Two sets of parallel lines intersecting at right angles and forming squares; the grid is superimposed on maps, charts, and other similar representations of the Earth's surface in an accurate and consistent manner in order to permit identification of ground locations with respect to other locations and the computation of direction and distance to other points. 2. A term used in giving the location of a geographic point by grid coordinates. See also **military grid; military grid reference system.**

grid bearing — Bearing measured from grid north.

grid convergence — The horizontal angle at a place between true north and grid north. It is proportional to the longitude difference between the place and the central meridian. See also **convergence.**

grid convergence factor — **(*)** The ratio of the grid convergence angle to the longitude difference. In the Lambert Conical Orthomorphic projection, this ratio is constant for all charts based on the same two standard parallels. See also **convergence; grid convergence.**

grid coordinates — **(*)** Coordinates of a grid coordinate system to which numbers and letters are assigned for use in designating a point on a gridded map, photograph, or chart. See also **coordinates.**

grid coordinate system — **(*)** A plane-rectangular coordinate system usually based on, and mathematically adjusted to, a map projection in order that geographic positions (latitudes and longitudes) may be readily transformed into plane coordinates and the computations relating to them may be made by the ordinary method of plane surveying. See also **coordinates.**

grid interval — **(*)** The distance represented between the lines of a grid.

grid magnetic angle — **(*)** Angular difference in direction between grid north and magnetic north. It is measured east or west from grid north. Also called **grid variation; grivation.**

grid navigation — **(*)** A method of navigation using a grid overlay for direction reference. See also **navigational grid.**

grid north — **(*)** The northerly or zero direction indicated by the grid datum of directional reference.

grid ticks — **(*)** Small marks on the neatline of a map or chart indicating additional grid reference systems included on that sheet. Grid ticks are sometimes shown on the interior grid lines of some maps for ease of referencing.

grid variation — See **grid magnetic angle.**

grivation — See **grid magnetic angle.**

grossly transportation feasible — A determination made by the supported commander that a draft operation plan can be supported with the apportioned transportation assets. This determination is made by using a transportation feasibility estimator to simulate movement of personnel and cargo from port of embarkation to port of debarkation within a specified time frame. (JP 5-03.1)

gross weight — **(*)** 1. Weight of a vehicle, fully equipped and serviced for operation, including the weight of the fuel, lubricants, coolant, vehicle tools and spares, crew, personal equipment, and load. 2. Weight of a container or pallet including freight and binding. Also called **WT.** See also **net weight.**

ground alert — **(*)** That status in which aircraft on the ground/deck are fully serviced and armed, with combat crews in readiness to take off within a specified short period of time (usually 15 minutes) after receipt of a mission order. See also **airborne alert; alert.**

ground combat element — The core element of a Marine air-ground task force (MAGTF) that is task-organized to conduct ground operations. It is usually constructed around an infantry organization but can vary in size from a small ground unit of any type, to one or more Marine divisions that can be independently maneuvered under the direction of the MAGTF commander. The ground combat element itself is not a formal command. Also called **GCE.** See also **aviation combat element; combat service support element; command element; Marine air-ground task force; Marine expeditionary force; Marine expeditionary force (forward); Marine expeditionary unit; special purpose Marine air-ground task force; task force.**

ground control — **(*)** A system of accurate measurements used to determine the distances and directions or differences in elevation between points on the Earth. See also **common control (artillery); control point; traverse.**

ground-controlled approach procedure — **(*)** The technique for talking down, through the use of both surveillance and precision approach radar, an aircraft during its approach so as to place it in a position for landing. See also **automatic approach and landing.**

ground-controlled interception — **(*)** A technique which permits control of friendly aircraft or guided missiles for the purpose of effecting interception. See also **air interception.**

ground fire — Small arms ground-to-air fire directed against aircraft.

ground liaison officer — An officer trained in offensive air support activities. Ground liaison officers are normally organized into parties under the control of the appropriate Army commander to provide liaison to Air Force and naval units engaged in training and combat operations. Also called **GLO.**

ground mine — See **bottom mine.**

ground observer center — A center to which ground observer teams report and which in turn will pass information to the appropriate control and/or reporting agency.

ground return — **(*)** The radar reflection from the terrain as displayed and/or recorded as an image.

ground speed — **(*)** The horizontal component of the speed of an aircraft relative to the Earth's surface. Also called **GS.**

ground visibility — Prevailing horizontal visibility near the Earth's surface as reported by an accredited observer.

ground zero — **(*)** The point on the surface of the Earth at, or vertically below or above, the center of a planned or actual nuclear detonation. See also **actual ground zero; desired ground zero.**

group — 1. A flexible administrative and tactical unit composed of either two or more battalions or two or more squadrons. The term also applies to combat support and combat service support units. 2. A number of ships and/or aircraft, normally a subdivision of a force, assigned for a specific purpose. Also called **GP.**

group interment — An interment in a common grave of two or more individually unidentified remains. See also **emergency interment; mortuary affairs; temporary interment; trench interment.** (JP 4-06)

group of targets — **(*)** Two or more targets on which fire is desired simultaneously. A group of targets is designated by a letter/number combination or a nickname.

group rendezvous — A check point at which formations of the same type will join before proceeding. See also **force rendezvous.**

guard — 1. A form of security operation whose primary task is to protect the main force by fighting to gain time while also observing and reporting information, and to prevent enemy ground observation of and direct fire against the main body by reconnoitering, attacking, defending, and delaying. A guard force normally operates within the range of the main body's indirect fire weapons. 2. A radio frequency that is normally used for emergency transmissions and is continuously monitored. UHF band: 243.0 MHZ; VHF band: 121.5 MHZ. See also **cover; flank guard; screen.** 3. A military or civilian individual assigned to protect personnel, equipment, or installations, or to oversee a prisoner.

guarded frequencies — Enemy frequencies that are currently being exploited for combat information and intelligence. A guarded frequency is time-oriented in that the guarded frequency list changes as the enemy assumes different combat postures. These frequencies may be jammed after the commander has weighed the potential operational gain against the loss of the technical information. See also **electronic warfare.** (JP 3-51)

guerrilla — A combat participant in guerrilla warfare. See also **unconventional warfare.**

guerrilla force — A group of irregular, predominantly indigenous personnel organized along military lines to conduct military and paramilitary operations in enemy-held, hostile, or denied territory. (JP 3-05)

guerrilla warfare — **(*)** Military and paramilitary operations conducted in enemy-held or hostile territory by irregular, predominantly indigenous forces. Also called **GW.** See also **unconventional warfare**.

guidance station equipment — **(*)** The ground-based portion of a missile guidance system necessary to provide guidance during missile flight.

guided missile — An unmanned vehicle moving above the surface of the Earth whose trajectory or flight path is capable of being altered by an external or internal mechanism. See also **aerodynamic missile; ballistic missile.**

guide specification — **(*)** Minimum requirements to be used as a basis for the evaluation of a national specification covering a fuel, lubricant or associated product proposed for standardization action.

guinea-pig — **(*)** In naval mine warfare, a ship used to determine whether an area can be considered safe from influence mines under certain conditions or, specifically, to detonate pressure mines.

gull — **(*)** In electronic warfare, a floating radar reflector used to simulate a surface target at sea for deceptive purposes.

gun — 1. A cannon with relatively long barrel, operating with relatively low angle of fire, and having a high muzzle velocity. 2. A cannon with tube length 30 calibers or more. See also **howitzer; mortar.**

gun carriage — **(*)** A mobile or fixed support for a gun. It sometimes includes the elevating and traversing mechanisms. Also called **carriage.**

gun-target line — **(*)** An imaginary straight line from gun to target. Also called **GTL.**

gun-type weapon — **(*)** A device in which two or more pieces of fissionable material, each less than a critical mass, are brought together very rapidly so as to form a supercritical mass that can explode as the result of a rapidly expanding fission chain.

gyromagnetic compass — **(*)** A directional gyroscope whose azimuth scale is maintained in alignment with the magnetic meridian by a magnetic detector unit.

H

half-life — **(*)** The time required for the activity of a given radioactive species to decrease to half of its initial value due to radioactive decay. The half-life is a characteristic property of each radioactive species and is independent of its amount or condition. The effective half-life of a given isotope is the time in which the quantity in the body will decrease to half as a result of both radioactive decay and biological elimination.

half-residence time — **(*)** As applied to delayed fallout, it is the time required for the amount of weapon debris deposited in a particular part of the atmosphere to decrease to half of its initial value.

half thickness — **(*)** Thickness of absorbing material necessary to reduce by one-half the intensity of radiation which passes through it.

handling (ordnance) — Applies to those individuals who engage in the breakout, lifting, or repositioning of ordnance or explosive devices in order to facilitate storage or stowage, assembly or disassembly, loading or downloading, or transporting. See also **assembly; downloading; loading; ordnance.** (JP 3-04.1)

handover — The passing of control authority of an aircraft from one control agency to another control agency. Handover action may be accomplished between control agencies of separate Services when conducting joint operations or between control agencies within a single command and control system. Handover action is complete when the receiving controller acknowledges assumption of control authority. Also called **hand-off.**

handover/crossover — In evasion and recovery operations, the transfer of evaders between two recovery forces. See also **evader; evasion; evasion and recovery; recovery; recovery operations.** (JP 3-50.3)

handover line — **(*)** A control feature, preferably following easily defined terrain features, at which responsibility for the conduct of combat operations is passed from one force to another.

hang fire — A malfunction that causes an undesired delay in the functioning of a firing system.

harassing fire — **(*)** Fire designed to disturb the rest of the enemy troops, to curtail movement, and, by threat of losses, to lower morale. See also **fire.**

harassment — An incident in which the primary objective is to disrupt the activities of a unit, installation, or ship, rather than to inflict serious casualties or damage.

harbor — A restricted body of water, an anchorage, or other limited coastal water area and its mineable water approaches, from which shipping operations are projected or supported. Generally, a harbor is part of a base, in which case the harbor defense force forms a component element of the base defense force established for the local defense of the base and its included harbor.

harbor defense — The defense of a harbor or anchorage and its water approaches against external threats such as: a. submarine, submarine-borne, or small surface craft attack; b. enemy minelaying operations; and c. sabotage. The defense of a harbor from guided missiles while such missiles are airborne is considered to be a part of air defense. See also **port security.**

hard beach — A portion of a beach especially prepared with a hard surface extending into the water, employed for the purpose of loading or unloading directly into or from landing ships or landing craft.

hardened site — **(*)** A site, normally constructed under rock or concrete cover, designed to provide protection against the effects of conventional weapons. It may also be equipped to provide protection against the side effects of a nuclear attack and against a chemical or a biological attack.

hard missile base — **(*)** A launching base that is protected against a nuclear explosion.

hardstand — **(*)** 1. A paved or stabilized area where vehicles are parked. 2. Open ground area having a prepared surface and used for the storage of materiel.

hardware — 1. The generic term dealing with physical items as distinguished from its capability or function such as equipment, tools, implements, instruments, devices, sets, fittings, trimmings, assemblies, subassemblies, components, and parts. The term is often used in regard to the stage of development, as in the passage of a device or component from the design stage into the hardware stage as the finished object. 2. In data automation, the physical equipment or devices forming a computer and peripheral components. See also **software.**

harmonization — The process and/or results of adjusting differences or inconsistencies to bring significant features into agreement.

hasty attack — **(*)** In land operations, an attack in which preparation time is traded for speed in order to exploit an opportunity. See also **deliberate attack.**

hasty breaching — **(*)** The rapid creation of a route through a minefield, barrier, or fortification by any expedient method.

hasty breaching (land mine warfare) — The creation of lanes through enemy minefields by expedient methods such as blasting with demolitions, pushing rollers or disabled vehicles through the minefields when the time factor does not permit detailed reconnaissance, deliberate breaching, or bypassing the obstacle.

hasty crossing — **(*)** The crossing of an inland water obstacle using the crossing means at hand or those readily available, and made without pausing for elaborate preparations. See also **deliberate crossing.**

hasty defense — **(*)** A defense normally organized while in contact with the enemy or when contact is imminent and time available for the organization is limited. It is characterized by improvement of the natural defensive strength of the terrain by utilization of foxholes, emplacements, and obstacles. See also **deliberate defense.**

hatch — An opening in a ship's deck giving access to cargo holds. (JP 4-01.6)

hatch list — A list showing, for each hold section of a cargo ship, a description of the items stowed, their volume and weight, the consignee of each, and the total volume and weight of materiel in the hold.

havens (moving) — See **moving havens.**

hazard — A condition with the potential to cause injury, illness, or death of personnel; damage to or loss of equipment or property; or mission degradation. See also **injury; risk.** (JP 5-00.2)

hazards of electromagnetic radiation to ordnance — The danger of accidental actuation of electro-explosive devices or otherwise electrically activating ordnance because of radio frequency electromagnetic fields. This unintended actuation could have safety (premature firing) or reliability (dudding) consequences. Also called **HERO.** See also **electromagnetic radiation; HERO SAFE ordnance; HERO UNSAFE ordnance; ordnance.** (JP 3-04.1)

heading hold mode — In a flight control system, a control mode that automatically maintains an aircraft heading that exists at the instant of completion of a maneuver.

heading indicator — **(*)** An instrument which displays heading transmitted electrically from a remote compass system.

heading select feature — A flight control system feature that permits selection or preselection of desired automatically controlled heading or headings of an aircraft.

head-up display — **(*)** A display of flight, navigation, attack, or other information superimposed upon the pilot's forward field of view. Also called **HUD.** See also **flight; horizontal situation display.** (JP 3-09.1)

health service logistic support — A functional area of logistic support that supports the joint force surgeon's health service support mission. It includes supplying Class VIII medical supplies (medical materiel to include medical peculiar repair parts used to sustain the health service support system), optical fabrication, medical equipment maintenance, blood storage and distribution, and medical gases. Also called **HSLS.** See also **health service support; joint force surgeon.** (JP 4-02.1)

health service support — All services performed, provided, or arranged by the Services to promote, improve, conserve, or restore the mental or physical well-being of personnel. These services include but are not limited to the management of health services resources, such as manpower, monies, and facilities; preventive and curative health measures; evacuation of the wounded, injured, or sick; selection of the medically fit and disposition of the medically unfit; blood management; medical supply, equipment, and maintenance thereof; combat stress control; and medical, dental, veterinary, laboratory, optometric, medical food, and medical intelligence services. Also called **HSS.** (JP 4-02)

health threat — A composite of ongoing or potential enemy actions; environmental, occupational, and geographic and meteorological conditions; endemic diseases; and employment of nuclear, biological, and chemical weapons (to include weapons of mass destruction) that can reduce the effectiveness of joint forces through wounds, injuries, illness, and psychological stressors. (JP 4-02)

heavy antitank weapon — A weapon capable of operating from ground or vehicle, used to defeat armor and other material targets.

heavy artillery — See **field artillery.**

heavy drop — A system of delivery of heavy supplies and equipment by parachute.

heavy-lift cargo — 1. Any single cargo lift, weighing over 5 long tons, and to be handled aboard ship. 2. In Marine Corps usage, individual units of cargo that exceed 800 pounds in weight or 100 cubic feet in volume.

heavy-lift ship — **(*)** A ship specially designed and capable of loading and unloading heavy and bulky items. It has booms of sufficient capacity to accommodate a single lift of 100 tons.

height datum — See **altitude datum.**

height delay — See **altitude delay.**

height hole — See **altitude hole.**

height of burst — **(*)** The vertical distance from the Earth's surface or target to the point of burst. Also called **HOB.** See also **optimum height of burst; safe burst height; types of burst.**

helicopter assault force — **(*)** A task organization combining helicopters, supporting units, and helicopter-borne troop units for use in helicopter-borne assault operations.

helicopter control station — A shipboard aircraft control tower or, on ships not equipped with a control tower, the communications installation that serves as such. On all Coast Guard cutters, the helicopter control station is located in the pilot house. Also called **HCS.** See also **station.** (JP 3-04.1)

helicopter direction center — **(*)** In amphibious operations, the primary direct control agency for the helicopter group/unit commander operating under the overall control of the tactical air control center.

helicopter drop point — A designated point within a landing zone where helicopters are unable to land because of the terrain, but in which they can discharge cargo or troops while hovering.

helicopter landing site — A designated subdivision of a helicopter landing zone in which a single flight or wave of assault helicopters land to embark or disembark troops and/or cargo.

helicopter landing zone — A specified ground area for landing assault helicopters to embark or disembark troops and/or cargo. A landing zone may contain one or more landing sites. Also called **HLZ.**

helicopter lane — **(*)** A safety air corridor in which helicopters fly to or from their destination during helicopter operations. See also **helicopter retirement route.**

helicopter retirement route — **(*)** The track or series of tracks along which helicopters move from a specific landing site or landing zone. See also **helicopter lane.**

helicopter support team — **(*)** A task organization formed and equipped for employment in a landing zone to facilitate the landing and movement of helicopter-borne troops, equipment, and supplies, and to evacuate selected casualties and enemy prisoners of war. Also called **HST.**

helicopter transport area — Areas to the seaward and on the flanks of the outer transport and landing ship areas, but preferably inside the area screen, used for launching and/or recovering helicopters. (JP 3-02)

helicopter wave — See **wave.**

helipad — **(*)** A prepared area designated and used for takeoff and landing of helicopters. (Includes touchdown or hover point.)

heliport — **(*)** A facility designated for operating, basing, servicing, and maintaining helicopters.

herbicide — A chemical compound that will kill or damage plants.

HERO SAFE ordnance — Any ordnance item that is percussion initiated, sufficiently shielded or otherwise so protected that all electro-explosive devices contained by the item are immune to adverse effects (safety or reliability) when the item is employed in its expected radio frequency environments, provided that the general hazards of electromagnetic radiation to ordnance requirements defined in the hazards from electromagnetic radiation manual are observed. See also **electromagnetic radiation; hazards of electromagnetic radiation to ordnance; HERO SUSCEPTIBLE ordnance; HERO UNSAFE ordnance; ordnance.** (JP 3-04.1)

HERO SUSCEPTIBLE ordnance — Any ordnance item containing electro-explosive devices proven by test or analysis to be adversely affected by radio frequency energy to the point that the safety and/or reliability of the system is in jeopardy when the system is employed in its expected radio frequency environment. See also **electromagnetic radiation; hazards of electromagnetic radiation to ordnance; HERO SAFE ordnance; HERO UNSAFE ordnance; ordnance.** (JP 3-04.1)

HERO UNSAFE ordnance — Any ordnance item containing electro-explosive devices that has not been classified as HERO SAFE or HERO SUSCEPTIBLE ordnance as a result of a hazards of electromagnetic radiation to ordnance (HERO) analysis or test is considered HERO UNSAFE ordnance. Additionally, any ordnance item containing electro-explosive devices (including those previously classified as HERO SAFE or HERO SUSCEPTIBLE ordnance) that has its internal wiring exposed; when tests are being conducted on that item that result in additional electrical connections to the item; when electro-explosive devices having exposed wire leads are present and handled or loaded in any but the tested condition; when the item is being assembled or disassembled; or when such ordnance items are damaged causing exposure of internal wiring or components or destroying engineered HERO protective devices. See also **electromagnetic radiation; hazards of electromagnetic radiation to ordnance; HERO SAFE ordnance; HERO SUSCEPTIBLE ordnance; ordnance.** (JP 3-04.1)

Hertz-Horn — See **chemical horn**.

H-hour — See **times.**

high airburst — The fallout safe height of burst for a nuclear weapon that increases damage to or casualties on soft targets, or reduces induced radiation contamination at actual ground zero. See also **types of burst.**

high altitude bombing — Horizontal bombing with the height of release over 15,000 feet.

high altitude burst — **(*)** The explosion of a nuclear weapon which takes place at a height in excess of 100,000 feet (30,000 meters). Also called **HAB.** See also **types of burst.**

high-altitude low-opening parachute technique — A method of delivering personnel, equipment, or supplies from airlift aircraft that must fly at altitudes above the threat umbrella. Also called **HALO.** (JP 3-17)

high-altitude missile engagement zone — See **weapon engagement zone.** (JP 3-52)

high angle — **(*)** In artillery and naval gunfire support, an order or request to obtain high angle fire.

high angle fire — **(*)** Fire delivered at angles of elevation greater than the elevation that corresponds to the maximum range of the gun and ammunition concerned; fire, the range of which decreases as the angle of elevation is increased.

high-density airspace control zone — Airspace designated in an airspace control plan or airspace control order, in which there is a concentrated employment of numerous and varied weapons and airspace users. A high-density airspace control zone has defined dimensions which usually coincide with geographical features or navigational aids. Access to a high-density airspace control zone is normally controlled by the maneuver commander. The maneuver commander can also direct a more restrictive weapons status within the high-density airspace control zone. Also called **HIDACZ.** (JP 3-52)

high explosive cargo — Cargo such as artillery ammunition, bombs, depth charges, demolition material, rockets, and missiles.

high oblique — See **oblique air photograph.**

high-payoff target — A target whose loss to the enemy will significantly contribute to the success of the friendly course of action. High-payoff targets are those high-value targets that must be acquired and successfully attacked for the success of the friendly commander's mission. Also called **HPT.** See also **high-value target; target.** (JP 3-60)

high-payoff target list — A prioritized list of high-payoff targets by phase of the joint operation. Also called **HPTL.** See also **high-payoff target; target.** (JP 3-60)

high-risk-of-capture personnel — US personnel whose position or assignment makes them particularly vulnerable to capture by hostile forces in combat, by terrorists, or by unfriendly governments. See also **hostile; terrorist.** (JP 3-50.3)

high-risk personnel — Personnel who, by their grade, assignment, symbolic value, or relative isolation, are likely to be attractive or accessible terrorist targets. See also **antiterrorism.** (JP 3-07.2)

high value airborne asset protection — A defensive counterair mission that defends airborne national assets which are so important that the loss of even one could seriously impact US warfighting capabilities or provide the enemy with significant propaganda value. Examples of high value airborne assets are Airborne Warning and Control System, Rivet Joint, Joint Surveillance and Target Attack Radar System, and Compass Call. Also called **HVAA protection.** See also **defensive counterair.** (JP 3-01)

high value asset control items — Items of supply identified for intensive management control under approved inventory management techniques designed to maintain an optimum inventory level of high investment items. Also called **hi-value asset control items.**

high-value target — A target the enemy commander requires for the successful completion of the mission. The loss of high-value targets would be expected to seriously degrade important enemy functions throughout the friendly commander's area of interest. Also called **HVT.** See also **high-payoff target; target.** (JP 3-09)

high velocity drop — **(*)** A drop procedure in which the drop velocity is greater than 30 feet per second (low velocity drop) and lower than free drop velocity. See also **airdrop.**

high-water mark — Properly, a mark left on a beach by wave wash at the preceding high water. It does not necessarily correspond to the high-water line. Because it can be determined by simple observation, it is frequently used in place of the high-water line, which can be determined only by a survey. When so used, it is called the high-water line. (JP 3-10)

hill shading — **(*)** A method of representing relief on a map by depicting the shadows that would be cast by high ground if light were shining from a certain direction.

hinterland, far — That region surrounding a beach or terminal operation to the extent that it has characteristics that affect the operation — normally within 100 miles. (JP 4-01.6)

hinterland, near — The area of land within an operational area of a specific beach or terminal operation — usually within 5 miles. (JP 4-01.6)

hi-value asset control item — See **high value asset control items.**

hoist — **(*)** In helicopters, the mechanism by which external loads may be raised or lowered vertically.

hold — **(*)** 1. A cargo stowage compartment aboard ship. 2. To maintain or retain possession of by force, as a position or an area. 3. In an attack, to exert sufficient pressure to prevent movement or redisposition of enemy forces. 4. As applied to air traffic, to keep an aircraft within a specified space or location which is identified by visual or other means in accordance with Air Traffic Control instructions. See also **fix**; **retain.**

holding anchorage — **(*)** An anchorage where ships may lie: a. if the assembly or working anchorage, or port, to which they have been assigned is full; b. when delayed by enemy threats or other factors from proceeding immediately on their next voyage; c. when dispersed from a port to avoid the effects of a nuclear attack. See also **assembly anchorage; emergency anchorage; working anchorage.**

holding attack — An attack designed to hold the enemy in position, to deceive the enemy as to where the main attack is being made, to prevent the enemy from reinforcing the elements opposing the main attack, and/or to cause the enemy to commit the reserves prematurely at an indecisive location.

holding point — **(*)** A geographically or electronically defined location used in stationing aircraft in flight in a predetermined pattern in accordance with air traffic control clearance. See also **orbit point.**

holding position — **(*)** A specified location on the airfield, close to the active runway and identified by visual means, at which the position of a taxiing aircraft is maintained in accordance with air traffic control instructions.

hollow charge — **(*)** A shaped charge producing a deep cylindrical hole of relatively small diameter in the direction of its axis of rotation.

home recovery mission profile — A mission profile that involves the recovery of an aircraft at its permanent or temporarily assigned operating base.

home station — The permanent location of active duty units and Reserve Component units (e.g., location of armory or reserve center). See also **active duty; Reserve Components.** (JP 4-05)

homing — **(*)** The technique whereby a mobile station directs itself, or is directed, towards a source of primary or reflected energy, or to a specified point.

homing guidance — A system by which a missile or torpedo steers itself towards a target by means of a self-contained mechanism which is activated by some distinguishing characteristics of the target. See also **active homing guidance; passive homing guidance; semi-active homing guidance.**

homing mine — **(*)** In naval mine warfare, a mine fitted with propulsion equipment which homes on to a target. See also **mine.**

horizon — In general, the apparent or visible junction of the Earth and sky, as seen from any specific position. Also called **the apparent, visible, or local horizon.** A horizontal plane passing through a point of vision or perspective center. The apparent or visible horizon approximates the true horizon only when the point of vision is very close to sea level.

horizontal action mine — **(*)** In land mine warfare, a mine designed to produce a destructive effect in a plane approximately parallel to the ground.

horizontal error — **(*)** The error in range, deflection, or in radius, which a weapon may be expected to exceed as often as not. Horizontal error of weapons making a nearly vertical approach to the target is described in terms of circular error probable. Horizontal error of weapons producing elliptical dispersion pattern is expressed in terms of probable error. See also **circular error probable; delivery error; deviation; dispersion error.**

horizontal loading — **(*)** Loading of items of like character in horizontal layers throughout the holds of a ship. See also **loading.**

horizontal situation display — **(*)** An electronically generated display on which navigation information and stored mission and procedural data can be presented. Radar information and television picture can also be displayed either as a map overlay or as a separate image. See also **head-up display.**

horizontal situation indicator — **(*)** An instrument which may display bearing and distance to a navigation aid, magnetic heading, track/course and track/course deviation.

horizontal stowage — The lateral distribution of unit equipment or categories of supplies so that they can be unloaded simultaneously from two or more holds. (JP 3-02.2)

horn — **(*)** In naval mine warfare, a projection from the mine shell of some contact mines which, when broken or bent by contact, causes the mine to fire.

hospital — A medical treatment facility capable of providing inpatient care. It is appropriately staffed and equipped to provide diagnostic and therapeutic services, as well as the necessary supporting services required to perform its assigned mission and functions. A hospital may, in addition, discharge the functions of a clinic.

hostage — A person held as a pledge that certain terms or agreements will be kept. (The taking of hostages is forbidden under the Geneva Conventions, 1949.)

host country — A nation in which representatives or organizations of another state are present because of government invitation and/or international agreement.

hostile — In combat and combat support operations, an identity applied to a track declared to belong to any opposing nation, party, group, or entity, which by virtue of its behavior or information collected on it such as characteristics, origin, or nationality contributes to the threat to friendly forces. See also **neutral; suspect; unknown.**

hostile act — 1. A hostile act is an attack or other use of force by any civilian, paramilitary, or military force or terrorist(s) (with or without national designation) against the United States, US forces and, in certain circumstances, US nationals, their property, US commercial assets, or other designated non-US forces, foreign nationals, and their property. 2. Force used directly to preclude or impede the mission and/or duties of US forces, including the recovery of US personnel and vital US Government property. When a hostile act is in progress the right exists to use proportional force, including armed force, in self-defense by all necessary means available to deter or neutralize the potential attacker or, if necessary, to destroy the threat.

hostile casualty — A person who is the victim of a terrorist activity or who becomes a casualty "in action." "In action" characterizes the casualty as having been the direct result of hostile action, sustained in combat or relating thereto, or sustained going to or returning from a combat mission provided that the occurrence was directly related to hostile action. Included are persons killed or wounded mistakenly or accidentally by friendly fire directed at a hostile force or what is thought to be a hostile force. However, not to be considered as sustained in action and not to be interpreted as hostile casualties are injuries or death due to the elements, self-inflicted wounds, combat fatigue, and except in unusual cases, wounds or death inflicted by a friendly force while the individual is in an absent-without-leave, deserter, or dropped-from-rolls status or is voluntarily absent from a place of duty. See also **casualty; casualty type; nonhostile casualty.**

hostile environment — See **operational environment.** (JP 3-05.3)

hostile force — Any civilian, paramilitary, or military force or terrorist(s), with or without national designation, that have committed a hostile act, exhibited hostile intent, or have been declared hostile by appropriate US authority.

hostile intent — The threat of imminent use of force by a foreign force, terrorist(s), or organization against the United States and US national interests, US forces and, in certain circumstances, US nationals, their property, US commercial assets, and other designated non-US forces, foreign nationals, and their property. When hostile intent is present, the right exists to use proportional force, including armed force, in self-defense by all necessary means available to deter or neutralize the potential attacker or, if necessary, to destroy the threat. A determination that hostile intent exists and requires the use of proportional force in self-defense must be based on evidence that an attack is imminent. Evidence necessary to determine hostile intent will vary depending on the state of international and regional political tension, military preparations, intelligence, and indications and warning information.

hostile track — See **hostile.**

host nation — A nation that receives the forces and/or supplies of allied nations, coalition partners, and/or NATO organizations to be located on, to operate in, or to transit through its territory. Also called **HN.**

host-nation support — Civil and/or military assistance rendered by a nation to foreign forces within its territory during peacetime, crises or emergencies, or war based on agreements mutually concluded between nations. Also called **HNS.** See also **host nation.** (JP 4-0)

host-nation support agreement — Basic agreement normally concluded at government-to-government or government-to-combatant commander level. These agreements may include general agreements, umbrella agreements, and memoranda of understanding. See also **host nation; host-nation support**. (JP 4-01.8)

hot photo interpretation report — A preliminary unformatted report of significant information from tactical reconnaissance imagery dispatched prior to compilation of the initial photo interpretation report. It should pertain to a single objective, event, or activity of significant interest to justify immediate reporting. Also called **HOTPHOTOREP.**

hot pursuit — Pursuit commenced within the territory, internal waters, the archipelagic waters, the territorial sea, or territorial airspace of the pursuing state and continued without interruption beyond the territory, territorial sea, or airspace. Hot pursuit also exists if pursuit commences within the contiguous or exclusive economic zones or on the continental shelf of the pursuing state, continues without interruption, and is undertaken based on a violation of the rights for the protection of which the zone was established. The right of hot pursuit ceases as soon as the ship or hostile force pursued enters the territory or territorial sea of its own state or of a third state. This definition does not imply that force may or may not be used in connection with hot pursuit. NOTE: This term applies only to law enforcement activities.

hot spot — **(*)** Region in a contaminated area in which the level of radioactive contamination is considerably greater than in neighboring regions in the area.

hovering — **(*)** A self-sustaining maneuver whereby a fixed, or nearly fixed, position is maintained relative to a spot on the surface of the Earth or underwater.

hovering ceiling — **(*)** The highest altitude at which the helicopter is capable of hovering in standard atmosphere. It is usually stated in two figures: hovering in ground effect and hovering out of ground effect.

howitzer — 1. A cannon that combines certain characteristics of guns and mortars. The howitzer delivers projectiles with medium velocities, either by low or high trajectories. 2. Normally a cannon with a tube length of 20 to 30 calibers; however, the tube length can exceed 30 calibers and still be considered a howitzer when the high angle fire zoning solution permits range overlap between charges. See also **gun; mortar.**

hub — An organization that sorts and distributes inbound cargo from wholesale supply sources (airlifted, sealifted, and ground transportable) and/or from within the theater. See also **hub and spoke distribution; spoke.** (JP 4-01.4)

hub and spoke distribution — A physical distribution system developed and modeled on industry standards to provide cargo management for a theater. It is based on a "hub" moving cargo to and between several "spokes". It is designed to increase transportation efficiencies and in-transit visibility and reduce order ship time. See also **distribution; distribution system; hub; in-transit visibility; spoke.** (JP 4-01.4)

human intelligence — **(*)** A category of intelligence derived from information collected and provided by human sources. Also called **HUMINT.** See also **human resources intelligence.**

humanitarian and civic assistance — Assistance to the local populace provided by predominantly US forces in conjunction with military operations and exercises. This assistance is specifically authorized by title 10, United States Code, section 401, and funded under separate authorities. Assistance provided under these provisions is limited to (1) medical, dental, and veterinary care provided in rural areas of a country; (2) construction of rudimentary surface transportation systems; (3) well drilling and construction of basic sanitation facilities; and (4) rudimentary construction and repair of public facilities. Assistance must fulfill unit training requirements that incidentally create humanitarian benefit to the local populace. Also called **HCA.** See also **foreign humanitarian assistance.** (JP 3-05.3)

humanitarian assistance coordination center — A temporary center established by a geographic combatant commander to assist with interagency coordination and planning. A humanitarian assistance coordination center operates during the early planning and coordination stages of foreign humanitarian assistance operations by providing the link between the geographic combatant commander and other United States Government agencies, nongovernmental organizations, and international and regional organizations at the strategic level. Also called **HACC.** See also **foreign humanitarian assistance; interagency coordination.** (JP 3-57)

humanitarian demining — Department of Defense and Department of State program to promote the foreign policy interests of the United States by assisting other nations in protecting their populations from landmines and clearing land of the threat posed by landmines remaining after conflict has ended. The humanitarian demining program includes training of host nation deminers, establishment of national demining organizations, provision of demining equipment, mine awareness training, and research development. (JP 3-07.6)

humanitarian operations center — An interagency policymaking body that coordinates the overall relief strategy and unity of effort among all participants in a large foreign humanitarian assistance operation. It normally is established under the direction of the government of the affected country or the United Nations, or a United States Government agency during a United States unilateral operation. The humanitarian operations center should consist of representatives from the affected country, the United States Embassy or Consulate, the joint force, the United Nations, nongovernmental and international organizations, and other major players in the operation. Also called **HOC.** See also **operation.** (JP 3-57)

human resources intelligence — The intelligence derived from the intelligence collection discipline that uses human beings as both sources and collectors, and where the human being is the primary collection instrument. Also called **HUMINT.**

hung weapons — Those weapons or stores on an aircraft that the pilot has attempted to drop or fire but could not because of a malfunction of the weapon, rack or launcher, or aircraft release and control system. (JP 3-04.1)

hunter track — **(*)** In naval mine warfare, the track to be followed by the hunter (or sweeper) to ensure that the hunting (or sweeping) gear passes over the lap track.

hydrogen bomb — See **thermonuclear weapon.**

hydrographic chart — **(*)** A nautical chart showing depths of water, nature of bottom, contours of bottom and coastline, and tides and currents in a given sea or sea and land area.

hydrographic reconnaissance — Reconnaissance of an area of water to determine depths, beach gradients, the nature of the bottom, and the location of coral reefs, rocks, shoals, and manmade obstacles.

hydrography — **(*)** The science which deals with the measurements and description of the physical features of the oceans, seas, lakes, rivers, and their adjoining coastal areas, with particular reference to their use for navigational purposes.

hyperbaric chamber — **(*)** A chamber used to induce an increase in ambient pressure as would occur in descending below sea level, in a water or air environment. It is the only type of chamber suitable for use in the treatment of decompression sickness in flying or diving. Also called **compression chamber; diving chamber; recompression chamber.**

hyperbolic navigation system — **(*)** A radio navigation system which enables the position of an aircraft equipped with a suitable receiver to be fixed by two or more intersecting hyperbolic position lines. The system employs either a time difference measurement of pulse transmissions or a phase difference measurement of phase-locked continuous wave transmissions. See also **loran.**

hypergolic fuel — **(*)** Fuel which will spontaneously ignite with an oxidizer, such as aniline with fuming nitric acid. It is used as the propulsion agent in certain missile systems.

hypersonic — **(*)** Of or pertaining to speeds equal to, or in excess of, five times the speed of sound. See also **speed of sound.**

hyperspectral imagery — Term used to describe the imagery derived from subdividing the electromagnetic spectrum into very narrow bandwidths. These narrow bandwidths may be combined with or subtracted from each other in various ways to form images useful in precise terrain or target analysis. Also called **HSI.**

hyperstereoscopy — **(*)** Stereoscopic viewing in which the relief effect is noticeably exaggerated, caused by the extension of the camera base. Also called **exaggerated stereoscopy.**

hypobaric chamber — **(*)** A chamber used to induce a decrease in ambient pressure as would occur in ascending to altitude. This type of chamber is primarily used for training and experimental purposes. Also called **altitude chamber; decompression chamber.**

hypsometric tinting — **(*)** A method of showing relief on maps and charts by coloring in different shades those parts which lie between selected levels. Also called **altitude tint; elevation tint; layer tint.**

I

identification — 1. The process of determining the friendly or hostile character of an unknown detected contact. 2. In arms control, the process of determining which nation is responsible for the detected violations of any arms control measure. 3. In ground combat operations, discrimination between recognizable objects as being friendly or enemy, or the name that belongs to the object as a member of a class. Also called **ID.**

identification, friend or foe — A device that emits a signal positively identifying it as a friendly. Also called **IFF.** See also **air defense.**

identification friend or foe personal identifier — The discrete identification friend or foe code assigned to a particular aircraft, ship, or other vehicle for identification by electronic means.

identification, friend or foe/selective identification feature procedures — The directives that govern the use of identification, friend or foe selective identification feature equipment. See also **identification, friend or foe.**

identification maneuver — A maneuver performed for identification purposes.

igloo space — Area in an earth-covered structure of concrete and/or steel designed for the storage of ammunition and explosives. See also **storage.**

ignition system — See **firing system.**

illuminate — Directing radar energy at an aircraft or surface vessel sufficient to obtain radar targeting information (fire control solution).

image format — Actual size of negative, scope, or other medium on which image is produced.

image motion compensation — **(*)** Movement intentionally imparted to film at such a rate as to compensate for the forward motion of an air or space vehicle when photographing ground objects.

imagery — **(*)** Collectively, the representations of objects reproduced electronically or by optical means on film, electronic display devices, or other media.

imagery collateral — **(*)** The reference materials which support the imagery interpretation function.

imagery correlation — **(*)** The mutual relationship between the different signatures on imagery from different types of sensors in terms of position and the physical characteristics signified.

imagery data recording — **(*)** The transposing of information relating to the airborne vehicle and sensor, such as speed, height, tilt, position, and time, to the matrix block on the sensor record at the moment of image acquisition.

imagery exploitation — **(*)** The cycle of processing and printing imagery to the positive or negative state, assembly into imagery packs, identification, interpretation, mensuration, information extraction, the preparation of reports, and the dissemination of information.

imagery intelligence — Intelligence derived from the exploitation of collection by visual photography, infrared sensors, lasers, electro-optics, and radar sensors such as synthetic aperture radar wherein images of objects are reproduced optically or electronically on film, electronic display devices, or other media. Also called **IMINT.** See also **intelligence; photographic intelligence.** (JP 2-0)

imagery interpretation — (*) 1. The process of location, recognition, identification, and description of objects, activities, and terrain represented on imagery. 2. The extraction of information from photographs or other recorded images. Also called **photographic interpretation.**

imagery interpretation key — (*) Any diagram, chart, table, list, or set of examples, etc., which is used to aid imagery interpreters in the rapid identification of objects visible on imagery.

imagery pack — (*) An assembly of the records from different imagery sensors covering a common target area.

imitative communications deception — That division of deception involving the introduction of false or misleading but plausible communications into target systems that mimics or imitates the targeted communications. See also **deception; target system.** (JP 3-51)

imitative electromagnetic deception — See **electromagnetic deception.**

immediate airlift requests — Requests generated that, due to their time-critical nature, cannot be filled by a planned mission. (JP 3-17)

immediate air support — (*) Air support to meet specific requests which arise during the course of a battle and which by their nature cannot be planned in advance. See also **air support.**

immediate decontamination — Decontamination carried out by individuals immediately upon becoming contaminated. It is performed in an effort to minimize casualties, save lives, and limit the spread of contamination. Also called **emergency decontamination.** See also **contamination; decontamination.** (JP 3-11)

immediate destination — (*) The next destination of a ship or convoy, irrespective of whether or not onward routing instructions have been issued to it.

immediately vital cargo — (*) A cargo already loaded which the consignee country regards as immediately vital for the prosecution of the war or for national survival, notwithstanding the risk to the ship. If the cargo is carried in a ship of another nation, then that nation must agree to the delivery of the cargo. The use of this term is limited to the period of implementation of the shipping movement policy.

immediate message — A category of precedence reserved for messages relating to situations that gravely affect the security of national and multinational forces or populace and that require immediate delivery to the addressee(s). See also **precedence.**

immediate mission request — A request for an air strike on a target that, by its nature, could not be identified sufficiently in advance to permit detailed mission coordination and planning. See also **preplanned mission request.**

immediate nuclear support — Nuclear support to meet specific requests that arise during the course of a battle, and that by their nature, cannot be planned in advance. See also **nuclear support; preplanned nuclear support.**

immediate operational readiness — Those operations directly related to the assumption of an alert or quick-reaction posture. Typical operations include strip alert, airborne alert and/or indoctrination, no-notice launch of an alert force, and the maintenance of missiles in an alert configuration. See also **nuclear weapon exercise; nuclear weapon maneuver.**

immediate targets — Targets that have been identified too late, or not selected for action in time to be included in the normal targeting process, and therefore have not been scheduled. Immediate targets have two subcategories: unplanned and unanticipated. See also **target.** (JP 3-60)

impact action fuze — **(*)** A fuze that is set in action by the striking of a projectile or bomb against an object, e.g., percussion fuze, contact fuze. Also called **direct action fuze.**

impact area — An area having designated boundaries within the limits of which all ordnance will detonate or impact.

impact pressure — **(*)** The difference between pitot pressure and static pressure.

implementation — Procedures governing the mobilization of the force and the deployment, employment, and sustainment of military operations in response to execution orders issued by the National Command Authorities. Also called **IMP.**

implementation planning — Operational planning associated with the conduct of a continuing operation, campaign, or war to attain defined objectives. At the national level, it includes the development of strategy and the assignment of strategic tasks to the combatant commanders. At the theater level, it includes the development of campaign plans to attain assigned objectives and the preparation of operation plans and operation orders to prosecute the campaign. At lower levels, implementation planning prepares for the execution of assigned tasks or logistic missions. See also **joint operation planning.**

implosion weapon — A weapon in which a quantity of fissionable material, less than a critical mass at ordinary pressure, has its volume suddenly reduced by compression (a step accomplished by using chemical explosives) so that it becomes supercritical, producing a nuclear explosion.

imprest fund — A cash fund of a fixed amount established through an advance of funds, without appropriation change, to an authorized imprest fund cashier to effect immediate cash payments of relatively small amounts for authorized purchases of supplies and nonpersonal services.

imprint — **(*)** Brief note in the margin of a map giving all or some of the following: date of publication, printing, name of publisher, printer, place of publication, number of copies printed, and related information.

improved conventional munitions — Munitions characterized by the delivery of two or more antipersonnel or antimateriel and/or antiarmor submunitions by a warhead or projectile.

improvised early resupply — **(*)** The onward movement of commodities which are available on land and which can be readily loaded into ships.

improvised explosive device — **(*)** A device placed or fabricated in an improvised manner incorporating destructive, lethal, noxious, pyrotechnic, or incendiary chemicals and designed to destroy, incapacitate, harass, or distract. It may incorporate military stores, but is normally devised from nonmilitary components. Also called **IED.** (JP 3-07.2)

improvised mine — A mine fabricated from available materials at or near its point of use.

improvised nuclear device — A device incorporating radioactive materials designed to result in the dispersal of radioactive material or in the formation of nuclear-yield reaction. Such devices may be fabricated in a completely improvised manner or may be an improvised modification to a US or foreign nuclear weapon. Also called **IND.**

inactive aircraft inventory — Aircraft in storage or bailment and/or government-furnished equipment on loan or lease outside of the Defense establishment or otherwise not available to the Military Services.

inactive duty training — Authorized training performed by a member of a Reserve Component not on active duty or active duty for training and consisting of regularly scheduled unit training assemblies, additional training assemblies, periods of appropriate duty or equivalent training, and any special additional duties authorized for Reserve Component personnel by the Secretary concerned, and performed by them in connection with the prescribed activities of the organization in which they are assigned with or without pay. Does not include work or study associated with correspondence courses. Also called **IDT.** See also **active duty for training.**

Inactive National Guard — Army National Guard personnel in an inactive status not in the Selected Reserve who are attached to a specific National Guard unit but do not participate in training activities. Upon mobilization, they will mobilize with their units. In order for these personnel to remain members of the Inactive National Guard, they must muster once a year with their assigned unit. Like the Individual Ready Reserve, all members of the Inactive National Guard have legal, contractual obligations. Members of the Inactive National Guard may not train for retirement credit or pay and are not eligible for promotion. Also called **ING.** See also **Individual Ready Reserve; Selected Reserve.** (JP 4-05)

inactive status — Status of reserve members on an inactive status list of a Reserve Component or assigned to the Inactive Army National Guard. Those in an inactive status may not train for points or pay, and may not be considered for promotion.

inbound traffic — Traffic originating in an area outside the continental United States destined for or moving in the general direction of the continental United States.

incapacitating agent — An agent that produces temporary physiological or mental effects, or both, which will render individuals incapable of concerted effort in the performance of their assigned duties.

incapacitating illness or injury — The casualty status of a person (a) whose illness or injury requires hospitalization but medical authority does not classify as very seriously ill or injured; or (b) seriously ill or injured and the illness or injury makes the person physically or mentally unable to communicate with the next of kin. Also called **III.** See also **casualty status.**

incentive type contract — A contract that may be of either a fixed price or cost reimbursement nature, with a special provision for adjustment of the fixed price or fee. It provides for a tentative target price and a maximum price or maximum fee, with price or fee adjustment after completion of the contract for the purpose of establishing a final price or fee based on the contractor's actual costs plus a sliding scale of profit or fee that varies inversely with the cost but which in no event shall permit the final price or fee to exceed the maximum price or fee stated in the contract. See also **cost contract; fixed price type contract.**

incident — In information operations, an assessed event of attempted entry, unauthorized entry, or an information attack on an automated information system. It includes unauthorized probing and browsing; disruption or denial of service; altered or destroyed input, processing, storage, or output of information; or changes to information system hardware, firmware, or software characteristics with or without the users' knowledge, instruction, or intent. See also **information operations.** (JP 3-13)

incident classification — See **search and rescue incident classification.**

incident control point — A designated point close to an incident where crisis management forces will rendezvous and establish control capability before initiating a tactical reaction. Also called **ICP.** See also **antiterrorism**. (JP 3-07.2)

incidents — Brief clashes or other military disturbances generally of a transitory nature and not involving protracted hostilities.

in-company — Two or more units proceeding together under the command of a designated senior.

inclination angle — See **pitch angle.**

incremental costs — Costs which are additional costs to the Service appropriations that would not have been incurred absent support of the contingency operation. See also **financial management.** (JP 1-06)

indefinite call sign — **(*)** A call sign which does not represent a specific facility, command, authority, activity, or unit, but which may represent any one or any group of these. See also **call sign.**

indefinite delivery type contract — A type of contract used for procurements where the exact time of delivery is not known at time of contracting.

independent — **(*)** A merchant ship under naval control sailed singly and unescorted by a warship. See also **military independent.**

independent ejection system — See **ejection systems.**

independent mine — **(*)** A mine which is not controlled by the user after laying. See also **mine.**

independent review — In computer modeling and simulation, a review performed by competent, objective reviewers who are independent of the model developer. Independent review includes either (a) a detailed verification and/or validation of the model or simulation; or (b) an examination of the verification and/or validation performed by the model or simulation developer. See also **configuration management; validation; verification.**

indicated airspeed — See **airspeed.**

indications and warning — Those intelligence activities intended to detect and report time-sensitive intelligence information on foreign developments that could involve a threat to the United States or allied and/or coalition military, political, or economic interests or to US citizens abroad. It includes forewarning of enemy actions or intentions; the imminence of hostilities; insurgency; nuclear or nonnuclear attack on the United States, its overseas forces, or allied and/or coalition nations; hostile reactions to US reconnaissance activities; terrorists' attacks; and other similar events. Also called **I&W.** See also **information; intelligence.** (JP 3-13)

indications (intelligence) — Information in various degrees of evaluation, all of which bear on the intention of a potential enemy to adopt or reject a course of action.

indicator — (*) In intelligence usage, an item of information which reflects the intention or capability of a potential enemy to adopt or reject a course of action.

indirect fire — Fire delivered on a target that is not itself used as a point of aim for the weapons or the director.

indirect laying — (*) Aiming a gun either by sighting at a fixed object, called the aiming point, instead of the target or by using a means of pointing other than a sight, such as a gun director, when the target cannot be seen from the gun position.

individual equipment — Referring to method of use: signifies personal clothing and equipment, for the personal use of the individual. See also **equipment.**

individual mobilization augmentee — An individual reservist attending drills who receives training and is preassigned to an Active Component organization, a Selective Service System, or a Federal Emergency Management Agency billet that must be filled on, or shortly after, mobilization. Individual mobilization augmentees train on a part-time basis with these organizations to prepare for mobilization. Inactive duty training for individual mobilization augmentees is decided by component policy and can vary from 0 to 48 drills a year. Also called **IMA.**

individual mobilization augmentee detachment — An administrative unit organized to train and manage individual mobilization augmentees.

individual protection — Actions taken by individuals to survive and continue the mission under nuclear, biological, and chemical conditions. See also **protection**. (JP 3-11)

individual protective equipment — (*) In nuclear, biological, and chemical warfare, the personal clothing and equipment required to protect an individual from biological and chemical hazards and some nuclear effects.

Individual Ready Reserve — A manpower pool consisting of individuals who have had some training, who have served previously in the Active Component or in the Selected Reserve, and have some period of their military service obligation remaining. Members may voluntarily participate in training for retirement points and promotion with or without pay. Also called **IRR.** See also **Selected Reserve.** (JP 4-05)

individual reserves — The supplies carried on a soldier, animal, or vehicle for individual use in an emergency. See also **reserve supplies.**

individual self-defense — The individual's inherent right of self-defense is an element of unit self-defense. It is critical that individuals are aware of and train to the principle that they have the authority to use all available means and to take all appropriate action to defend themselves and other US personnel in their vicinity. In the implementation of these standing and other rules of engagement (ROE), commanders have the obligation to ensure that the individuals within that commander's unit understand when and how they may use force in self-defense. While individuals assigned to a unit respond to a hostile act or hostile intent in the exercise of self-defense, their use of force must remain consistent with lawful orders of their superiors, the rules contained in joint doctrine, and other applicable ROE promulgated for the mission or area of responsibility.

individual sponsored dependent — A dependent not entitled to travel to the overseas command at Government expense or who enters the command without endorsement of the appropriate overseas commander.

induced environment — Any manmade or equipment-made environment that directly or indirectly affects the performance of man or materiel.

induced radiation — (*) Radiation produced as a result of exposure to radioactive materials, particularly the capture of neutrons. See also **contamination; initial radiation; residual radiation; residual radioactivity.**

induction circuit — (*) In naval mine warfare, a circuit actuated by the rate of change in a magnetic field due to the movement of the ship or the changing current in the sweep.

industrial chemicals — Chemicals developed or manufactured for use in industrial operations or research by industry, government, or academia. These chemicals are not primarily manufactured for the specific purpose of producing human casualties or rendering equipment, facilities, or areas dangerous for human use. Hydrogen cyanide, cyanogen chloride, phosgene, and chloropicrin are industrial chemicals that also can be military chemical agents. See also **chemical warfare.** (JP 3-11)

industrial mobilization — The transformation of industry from its peacetime activity to the industrial program necessary to support the national military objectives. It includes the mobilization of materials, labor, capital, production facilities, and contributory items and services essential to the industrial program. See also **mobilization.**

industrial preparedness — The state of preparedness of industry to produce essential materiel to support the national military objectives.

industrial preparedness program — Plans, actions, or measures for the transformation of the industrial base, both government-owned and civilian-owned, from its peacetime activity to the emergency program necessary to support the national military objectives. It includes industrial preparedness measures such as modernization, expansion, and preservation of the production facilities and contributory items and services for planning with industry. Also called **IPP.**

industrial property — As distinguished from military property, any contractor-acquired or government-furnished property, including materials, special tooling, and industrial facilities, furnished or acquired in the performance of a contract or subcontract.

industrial readiness — See **industrial preparedness.**

inert filling — (*) A prepared non-explosive filling of the same weight as the explosive filling.

inertial guidance — A guidance system designed to project a missile over a predetermined path, wherein the path of the missile is adjusted after launching by devices wholly within the missile and independent of outside information. The system measures and converts accelerations experienced to distance traveled in a certain direction.

inertial navigation system — **(*)** A self-contained navigation system using inertial detectors, which automatically provides vehicle position, heading, and velocity. Also called **INS.**

inert mine — **(*)** A mine or replica of a mine incapable of producing an explosion.

in extremis — A situation of such exceptional urgency that immediate action must be taken to minimize imminent loss of life or catastrophic degradation of the political or military situation. (JP 3-05)

infiltration — 1. The movement through or into an area or territory occupied by either friendly or enemy troops or organizations. The movement is made, either by small groups or by individuals, at extended or irregular intervals. When used in connection with the enemy, it infers that contact is avoided. 2. In intelligence usage, placing an agent or other person in a target area in hostile territory. Usually involves crossing a frontier or other guarded line. Methods of infiltration are: black (clandestine); grey (through legal crossing point but under false documentation); and white (legal).

inflammable cargo — Cargo such as drummed gasoline and oils.

inflight phase — The flight of a missile or space vehicle from launch to detonation or impact.

inflight report — The transmission from the airborne system of information obtained both at the target and en route.

influence field — **(*)** The distribution in space of the influence of a ship or minesweeping equipment.

influence mine — **(*)** A mine actuated by the effect of a target on some physical condition in the vicinity of the mine or on radiations emanating from the mine. See also **mine.**

influence release sinker — A sinker which holds a moored or rising mine at the sea-bed and releases it when actuated by a suitable ship influence.

influence sweep — A sweep designed to produce an influence similar to that produced by a ship and thus actuate mines.

information — 1. Facts, data, or instructions in any medium or form. 2. The meaning that a human assigns to data by means of the known conventions used in their representation. (JP 3-13.1)

information assurance — Information operations that protect and defend information and information systems by ensuring their availability, integrity, authentication, confidentiality, and nonrepudiation. This includes providing for restoration of information systems by incorporating protection, detection, and reaction capabilities. Also called **IA.** See also **information; information operations; information system.** (JP 3-13)

information-based processes — Processes that collect, analyze, and disseminate information using any medium or form. These processes may be stand-alone processes or sub-processes that, taken together, comprise a larger system or systems of processes. See also **information system.** (JP 3-13)

information box — **(*)** A space on an annotated overlay, mosaic, map, etc., which is used for identification, reference, and scale information.

information environment — The aggregate of individuals, organizations, or systems that collect, process, or disseminate information; also included is the information itself. See also **information system.** (JP 3-13)

information operations — Actions taken to affect adversary information and information systems while defending one's own information and information systems. Also called **IO.** See also **defensive information operations; information; information system; offensive information operations; operation.** (JP 3-13)

information report — Report used to forward raw information collected to fulfill intelligence requirements.

information requirements — **(*)** Those items of information regarding the enemy and his environment which need to be collected and processed in order to meet the intelligence requirements of a commander. See also **priority intelligence requirements.**

information resources — Information and related resources, such as personnel, equipment, and information technology. See also **information.** (JP 4-01.8)

information security — The protection of information and information systems against unauthorized access or modification of information, whether in storage, processing, or transit, and against denial of service to authorized users. Information security includes those measures necessary to detect, document, and counter such threats. Information security is composed of computer security and communications security. Also called **INFOSEC.** See also **communications security; computer security; information security; information system.** (JP 3-13)

information superiority — That degree of dominance in the information domain which permits the conduct of operations without effective opposition. See also **information operations.** (JP 2-01.3)

information system — The entire infrastructure, organization, personnel, and components that collect, process, store, transmit, display, disseminate, and act on information. See also **information; information warfare.** (JP 3-13)

information warfare — Information operations conducted during time of crisis or conflict to achieve or promote specific objectives over a specific adversary or adversaries. Also called **IW.** See also **crisis; information; information operations; operation.** (JP 3-13)

infrared film — Film carrying an emulsion especially sensitive to "near-infrared." Used to photograph through haze because of the penetrating power of infrared light and in camouflage detection to distinguish between living vegetation and dead vegetation or artificial green pigment.

infrared imagery — That imagery produced as a result of sensing electromagnetic radiations emitted or reflected from a given target surface in the infrared position of the electromagnetic spectrum (approximately 0.72 to 1,000 microns).

infrared linescan system — **(*)** A passive airborne infrared recording system which scans across the ground beneath the flight path, adding successive lines to the record as the vehicle advances along the flight path.

infrared photography — Photography employing an optical system and direct image recording on film sensitive to near-infrared wavelength (infrared film). (Note: Not to be confused with "infrared imagery.")

infrared pointer — A low power laser device operating in the near infrared light spectrum that is visible with light amplifying night vision devices. Also called **IR pointer.** (JP 3-09.3)

infrared radiation — Radiation emitted or reflected in the infrared portion of the electromagnetic spectrum.

infrastructure — All building and permanent installations necessary for the support, redeployment, and military forces operations (e.g., barracks, headquarters, airfields, communications, facilities, stores, port installations, and maintenance stations). See also **bilateral infrastructure; common infrastructure; national infrastructure.** (JP 4-01.8)

initial active duty for training — Basic military training and technical skill training required for all accessions. For nonprior service male enlistees between the ages of 18 1/2 and 26, initial active duty for training shall be not less than 12 weeks and start insofar as practical within 270 days after enlistment. Initial active duty for training for all other enlistees and inductees shall be prescribed by the Secretary concerned and start insofar as practical within 360 days of entry into the Service, except in time of war or national emergency declared by Congress or the President when basic training shall be not less than 12 weeks or its equivalent. Reservists may not be assigned to active duty on land outside the United States or its territories and possessions until basic training has been completed.

initial approach — (*) a. That part of an instrument approach procedure in which the aircraft has departed an initial approach fix or point and is maneuvering to enter the intermediate or final approach. It ends at the intermediate fix or point or, where no intermediate segment is established, at the final approach fix or point. b. That part of a visual approach of an aircraft immediately prior to arrival over the airfield of destination, or over the reporting point from which the final approach to the airfield is commenced.

initial approach area — (*) An area of defined width lying between the last preceding navigational fix or dead reckoning position and either the facility to be used for making an instrument approach or a point associated with such a facility that is used for demarcating the termination of initial approach.

initial assessment — An assessment that provides a basic determination of the viability of the infiltration and exfiltration portion of a proposed special operations forces mission. Also called **IA.** (JP 3-05.5)

initial contact report — See **contact report.**

initial draft plan — (*) A plan which has been drafted and coordinated by the originating headquarters, and is ready for external coordination with other military headquarters. It cannot be directly implemented by the issuing commander, but it may form the basis for an operation order issued by the commander in the event of an emergency. See also **coordinated draft plan; draft plan; final plan; operation plan.**

initial early resupply — The onward movement of ships which are already loaded with cargoes which will serve the requirements after D-day. This includes such shipping deployed from major ports/major water terminals and subsequently dispersed to secondary ports/alternate water terminals and anchorages.

initial entry into Military Service — Entry for the first time into military status (active duty or reserve) by induction, enlistment, or appointment in any Service of the Armed Forces of the United States. Appointment may be as a commissioned or warrant officer; as a cadet or midshipman at the Service academy of one of the armed forces; or as a midshipman, US Naval Reserve, for US Naval Reserve Officers' Training Corps training at a civilian institution.

initial issues — The issue of materiel not previously furnished to an individual or organization, including new inductees and newly activated organizations, and the issue of newly authorized items of materiel.

initial operational capability — The first attainment of the capability to employ effectively a weapon, item of equipment, or system of approved specific characteristics that is manned or operated by an adequately trained, equipped, and supported military unit or force. Also called **IOC.**

initial path sweeping — (*) In naval mine warfare, initial sweeping to clear a path through a mined area dangerous to the following mine sweepers. See also **precursor sweeping.**

initial photo interpretation report — A first-phase interpretation report, subsequent to the Joint Tactical Air Reconnaissance/Surveillance Mission Report, presenting the results of the initial readout of new imagery to answer the specific requirements for which the mission was requested.

initial point — 1. The first point at which a moving target is located on a plotting board. 2. A well-defined point, easily distinguishable visually and/or electronically, used as a starting point for the bomb run to the target. 3. **airborne** — A point close to the landing area where serials (troop carrier air formations) make final alterations in course to pass over individual drop or landing zones. 4. **helicopter** — An air control point in the vicinity of the landing zone from which individual flights of helicopters are directed to their prescribed landing sites. 5. Any designated place at which a column or element thereof is formed by the successive arrival of its various subdivisions, and comes under the control of the commander ordering the move. Also called **IP.** See also **target approach point.** (JP 3-09.1)

initial programmed interpretation report — (*) A standardized imagery interpretation report providing information on programmed mission objectives or other vital intelligence information which can be readily identified near these objectives, and which has not been reported elsewhere. Also called **IPIR.**

initial provisioning — The process of determining the range and quantity of items (i.e., spares and repair parts, special tools, test equipment, and support equipment) required to support and maintain an item for an initial period of service. Its phases include the identification of items of supply, the establishment of data for catalog, technical manual, and allowance list preparation, and the preparation of instructions to assure delivery of necessary support items with related end articles.

initial radiation — (*) The radiation, essentially neutrons and gamma rays, resulting from a nuclear burst and emitted from the fireball within one minute after burst. See also **induced radiation; residual radiation.**

initial reserves — In amphibious operations, those supplies that normally are unloaded immediately following the assault waves; usually the supplies for the use of the beach organization, battalion landing teams, and other elements of regimental combat teams for the purpose of initiating and sustaining combat until higher supply installations are established. See also **reserve supplies.**

initial response force — The first unit, usually military police, on the scene of a terrorist incident. See also **antiterrorism.** (JP 3-07.2)

initial unloading period — **(*)** In amphibious operations, that part of the ship-to-shore movement in which unloading is primarily tactical in character and must be instantly responsive to landing force requirements. All elements intended to land during this period are serialized. See also **general unloading period.**

initiating directive — An order to a subordinate commander to conduct military operations as directed. It is issued by the unified commander, subunified commander, Service component commander, or joint force commander delegated overall responsibility for the operation. (JP 3-18)

initiation of procurement action — That point in time when the approved document requesting procurement and citing funds is forwarded to the procuring activity. See also **procurement lead time.**

injury — A term comprising such conditions as fractures, wounds, sprains, strains, dislocations, concussions, and compressions. In addition, it includes conditions resulting from extremes of temperature or prolonged exposure. Acute poisonings (except those due to contaminated food) resulting from exposure to a toxic or poisonous substance are also classed as injuries. See also **casualty; wounded.**

inland petroleum distribution system — A multi-product petroleum pipeline designed to move bulk fuel forward in a theater of operation. (JP 4-01.6)

inland search and rescue region — The inland areas of the continental United States, except waters under the jurisdiction of the United States. See also **search and rescue region.**

inner transport area — In amphibious operations, an area as close to the landing beach as depth of water, navigational hazards, boat traffic, and enemy action permit, to which transports may move to expedite unloading. See also **outer transport area; transport area.**

innocent passage — The right of all ships to engage in continuous and expeditious surface passage through the territorial sea and archipelagic waters of foreign coastal states in a manner not prejudicial to its peace, good order, or security. Passage includes stopping and anchoring, but only if incidental to ordinary navigation or necessary by force majeure or distress, or for the purpose of rendering assistance to persons, ships, or aircraft in danger or distress.

in-place force — 1. A North Atlantic Treaty Organization (NATO)-assigned force that, in peacetime, is principally stationed in the designated combat zone of the NATO command to which it is committed. 2. Force within a combatant commander's area of responsibility and under the combatant commander's combatant command (command authority).

inshore patrol — **(*)** A naval defense patrol operating generally within a naval defense coastal area and comprising all elements of harbor defenses, the coastal lookout system, patrol craft supporting bases, aircraft, and Coast Guard stations.

inspection — In arms control, physical process of determining compliance with arms control measures.

installation — A grouping of facilities, located in the same vicinity, which support particular functions. Installations may be elements of a base. See also **base; base complex.**

installation commander — The individual responsible for all operations performed by an installation. See also **antiterrorism; base commander; installation.** (JP 3-07.2)

installation complex — In the Air Force, a combination of land and facilities comprised of a main installation and its noncontiguous properties (auxiliary air fields, annexes, and missile fields) that provide direct support to or are supported by that installation. Installation complexes may comprise two or more properties, e.g., a major installation, a minor installation, or a support site, each with its associated annex(es) or support property(ies). See also **minor installation.**

instructional mine — **(*)** An inert mine used for instruction and normally sectionalized for this purpose. See also **inert mine.**

instrument approach procedure — **(*)** A series of predetermined maneuvers for the orderly transfer of an aircraft under instrument flight conditions from the beginning of the initial approach to a landing or to a point from which a landing may be made visually or the missed approach procedure is initiated.

instrument flight — **(*)** Flight in which the path and attitude of the aircraft are controlled solely by reference to instruments.

instrument landing system — **(*)** A system of radio navigation intended to assist aircraft in landing which provides lateral and vertical guidance, which may include indications of distance from the optimum point of landing. Also called **ILS.**

instrument meteorological conditions — Meteorological conditions expressed in terms of visibility, distance from cloud, and ceiling; less than minimums specified for visual meteorological conditions. Also called **IMC.** See also **visual meteorological conditions.** (JP 3-04.1)

in support — **(*)** An expression used to denote the task of providing artillery supporting fire to a formation or unit. Liaison and observation are not normally provided. See also **at priority call; direct support.**

in support of — Assisting or protecting another formation, unit, or organization while remaining under original control.

insurgency — **(*)** An organized movement aimed at the overthrow of a constituted government through use of subversion and armed conflict.

insurgent — Member of a political party who rebels against established leadership. See also **antiterrorism; counterinsurgency; insurgency.** (JP 3-07.2)

integrated fire control system — A system that performs the functions of target acquisition, tracking, data computation, and engagement control, primarily using electronic means and assisted by electromechanical devices.

integrated logistic support — A composite of all the support considerations necessary to assure the effective and economical support of a system for its life cycle. It is an integral part of all other aspects of system acquisition and operation. Also called **ILS.**

integrated materiel management — The exercise of total Department of Defense-level management responsibility for a federal supply group or class, commodity, or item for a single agency. It normally includes computation of requirements, funding, budgeting, storing, issuing, cataloging, standardizing, and procuring functions. Also called **IMM.** See also **materiel; materiel management.** (JP 4-07)

integrated planning — In amphibious operations, the planning accomplished by commanders and staffs of corresponding echelons from parallel chains of command within the amphibious task force. See also **amphibious operation; amphibious task force.** (JP 3-02)

integrated priority list — A list of a combatant commander's highest priority requirements, prioritized across Service and functional lines, defining shortfalls in key programs that, in the judgment of the combatant commander, adversely affect the capability of the combatant commander's forces to accomplish their assigned mission. The integrated priority list provides the combatant commander's recommendations for programming funds in the planning, programming, and budgeting system process. Also called **IPL.**

integrated staff — **(*)** A staff in which one officer only is appointed to each post on the establishment of the headquarters, irrespective of nationality and Service. See also **multinational staff; joint staff; parallel staff; staff.**

integrated tactical warning — See **tactical warning.**

integrated warfare — The conduct of military operations in any combat environment wherein opposing forces employ non-conventional weapons in combination with conventional weapons.

integration — 1. In force protection, the synchronized transfer of units into an operational commander's force prior to mission execution. 2. The arrangement of military forces and their actions to create a force that operates by engaging as a whole. 3. In photography, a process by which the average radar picture seen on several scans of the time base may be obtained on a print, or the process by which several photographic images are combined into a single image. See also **force protection.** (JP 0-2)

intelligence — 1. The product resulting from the collection, processing, integration, analysis, evaluation, and interpretation of available information concerning foreign countries or areas. 2. Information and knowledge about an adversary obtained through observation, investigation, analysis, or understanding. See also **acoustic intelligence; all-source intelligence; basic intelligence; civil defense intelligence; combat intelligence; communications intelligence; critical intelligence; current intelligence; departmental intelligence; domestic intelligence; electronic intelligence; electro-optical intelligence; foreign intelligence; foreign instrumentation signals intelligence; general military intelligence; human resources intelligence; imagery intelligence; joint intelligence; laser intelligence; measurement and signature intelligence; medical intelligence; merchant**

intelligence; military intelligence; national intelligence; nuclear intelligence; open-source intelligence; operational intelligence; photographic intelligence; political intelligence; radar intelligence; radiation intelligence; scientific and technical intelligence; security intelligence; strategic intelligence; tactical intelligence; target intelligence; technical intelligence; technical operational intelligence; terrain intelligence; unintentional radiation intelligence. (JP 2-0)

intelligence annex — A supporting document of an operation plan or order that provides detailed information on the enemy situation, assignment of intelligence tasks, and intelligence administrative procedures.

intelligence collection plan — A plan for gathering information from all available sources to meet an intelligence requirement. Specifically, a logical plan for transforming the essential elements of information into orders or requests to sources within a required time limit. See also **intelligence cycle.**

intelligence contingency funds — Appropriated funds to be used for intelligence activities when the use of other funds is not applicable or would either jeopardize or impede the mission of the intelligence unit.

intelligence cycle — The process by which information is converted into intelligence and made available to users. There are six phases in the cycle: a. **planning and direction** — Determination of intelligence requirements, development of appropriate intelligence architecture, preparation of a collection plan, and issuance of orders and requests to information collection agencies. b. **collection** — Acquisition of information and the provision of this information to processing elements. c. **processing and exploitation** — Conversion of collected information into forms suitable to the production of intelligence. d. **analysis and production** — Conversion of processed information into intelligence through the integration, analysis, evaluation, and interpretation of all source data and the preparation of intelligence products in support of known or anticipated user requirements. e. **dissemination and integration** — Delivery of intelligence to users in a suitable form and the application of the intelligence to appropriate missions, tasks, and functions. f. **evaluation and feedback** — Continuous assessment of intelligence operations during each phase of the intelligence cycle to ensure that the commander's intelligence requirements are being met. See also **intelligence.** (JP 2-0)

intelligence database — The sum of holdings of intelligence data and finished intelligence products at a given organization.

intelligence data handling systems — Information systems that process and manipulate raw information and intelligence data as required. They are characterized by the application of general purpose computers, peripheral equipment, and automated storage and retrieval equipment for documents and photographs. While automation is a distinguishing characteristic of intelligence data handling systems, individual system components may be either automated or manually operated. Also called **IDHS.**

intelligence discipline — A well defined area of intelligence collection, processing, exploitation, and reporting using a specific category of technical or human resources. There are seven major disciplines: human intelligence, imagery intelligence, measurement and signature intelligence, signals intelligence (communications intelligence, electronic intelligence, and foreign instrumentation signals intelligence), open-source intelligence, technical intelligence, and counterintelligence. See also **counterintelligence; human intelligence; imagery intelligence; intelligence; measurement and signature intelligence; open-source intelligence; signals intelligence; technical intelligence.** (JP 2-0)

intelligence estimate — (*) The appraisal, expressed in writing or orally, of available intelligence relating to a specific situation or condition with a view to determining the courses of action open to the enemy or potential enemy and the order of probability of their adoption.

intelligence gathering — Collection of intelligence on other units or forces by own units or forces.

intelligence journal — A chronological log of intelligence activities covering a stated period, usually 24 hours. It is an index of reports and messages that have been received and transmitted, important events that have occurred, and actions taken. The journal is a permanent and official record.

intelligence operations — The variety of intelligence tasks that are carried out by various intelligence organizations and activities. Predominantly, it refers to either intelligence collection or intelligence production activities. When used in the context of intelligence collection activities, intelligence operations refer to collection, processing, exploitation, and reporting of information. When used in the context of intelligence production activities, it refers to collation, integration, interpretation, and analysis, leading to the dissemination of a finished product. (JP 2-0)

intelligence preparation of the battlespace — An analytical methodology employed to reduce uncertainties concerning the enemy, environment, and terrain for all types of operations. Intelligence preparation of the battlespace builds an extensive database for each potential area in which a unit may be required to operate. The database is then analyzed in detail to determine the impact of the enemy, environment, and terrain on operations and presents it in graphic form. Intelligence preparation of the battlespace is a continuing process. Also called **IPB.** (JP 2-0)

intelligence-related activities — Those activities outside the consolidated defense intelligence program that: respond to operational commanders' tasking for time-sensitive information on foreign entities; respond to national intelligence community tasking of systems whose primary mission is support to operating forces; train personnel for intelligence duties; provide an intelligence reserve; or are devoted to research and development of intelligence or related capabilities. (Specifically excluded are programs that are so closely integrated with a weapon system that their primary function is to provide immediate-use targeting data.)

intelligence report — A specific report of information, usually on a single item, made at any level of command in tactical operations and disseminated as rapidly as possible in keeping with the timeliness of the information. Also called **INTREP.**

intelligence reporting — The preparation and conveyance of information by any means. More commonly, the term is restricted to reports as they are prepared by the collector and as they are transmitted by the collector to the latter's headquarters and by this component of the intelligence structure to one or more intelligence-producing components. Thus, even in this limited sense, reporting embraces both collection and dissemination. The term is applied to normal and specialist intelligence reports. See also **normal intelligence reports; specialist intelligence report.**

intelligence requirement — 1. Any subject, general or specific, upon which there is a need for the collection of information, or the production of intelligence. 2. A requirement for intelligence to fill a gap in the command's knowledge or understanding of the battlespace or threat forces. See also **battlespace; intelligence; priority intelligence requirements.** (JP 2-0)

intelligence source — The means or system that can be used to observe and record information relating to the condition, situation, or activities of a targeted location, organization, or individual. An intelligence source can be people, documents, equipment, or technical sensors. See also **intelligence; source.** (JP 2-0)

intelligence subject code — A system of subject and area references to index the information contained in intelligence reports as required by a general intelligence document reference service.

intelligence summary — A specific report providing a summary of items of intelligence at frequent intervals. Also called **INTSUM.** See also **intelligence.**

intelligence system — Any formal or informal system to manage data gathering, to obtain and process the data, to interpret the data, and to provide reasoned judgments to decision makers as a basis for action. The term is not limited to intelligence organizations or services but includes any system, in all its parts, that accomplishes the listed tasks.

intensity factor — **(*)** A multiplying factor used in planning activities to evaluate the foreseeable intensity or the specific nature of an operation in a given area for a given period of time. It is applied to the standard day of supply in order to calculate the combat day of supply.

intensity mine circuit — **(*)** A circuit whose actuation is dependent on the field strength reaching a level differing by some pre-set minimum from that experienced by the mine when no ships are in the vicinity.

intensive management — The continuous process by which the supported and supporting commanders, the Services, transportation component commands, and appropriate Defense agencies ensure that movement data in the Joint Operation Planning and Execution System time-phased force and deployment data for the initial days of deployment and/or mobilization are current to support immediate execution.

intention — An aim or design (as distinct from capability) to execute a specified course of action.

interagency coordination — Within the context of Department of Defense involvement, the coordination that occurs between elements of Department of Defense, and engaged US Government agencies, nongovernmental organizations, and regional and international organizations for the purpose of accomplishing an objective. See also **international organization; nongovernmental organizations.** (JP 3-57)

interceptor — **(*)** A manned aircraft utilized for identification and/or engagement of airborne objects.

intercept point — **(*)** The point to which an airborne vehicle is vectored or guided to complete an interception.

intercept receiver — **(*)** A receiver designed to detect and provide visual and/or aural indication of electromagnetic emissions occurring within the particular portion of the electromagnetic spectrum to which it is tuned.

inter-chart relationship diagram — **(*)** A diagram on a map or chart showing names and/or numbers of adjacent sheets in the same (or related) series. Also called **index to adjoining sheets.** See also **map index.**

interconnection — The linking together of interoperable systems.

intercount dormant period — **(*)** In naval mine warfare, the period after the actuation of a ship counter before it is ready to receive another actuation.

interdepartmental or agency support — Provision of logistic and/or administrative support in services or materiel by one or more Military Services to one or more departments or agencies of the United States Government (other than military) with or without reimbursement. See also **international logistic support; inter-Service support; support.**

interdepartmental intelligence — Integrated departmental intelligence that is required by departments and agencies of the United States Government for the execution of their missions but which transcends the exclusive competence of a single department or agency to produce.

interdiction — An action to divert, disrupt, delay, or destroy the enemy's surface military potential before it can be used effectively against friendly forces. See also **air interdiction.**

interface — A boundary or point common to two or more similar or dissimilar command and control systems, sub-systems, or other entities against which or at which necessary information flow takes place.

interim financing — Advance payments, partial payments, loans, discounts, advances, and commitments in connection therewith; guarantees of loans, discounts, advances, and commitments in connection therewith; and any other type of financing necessary for both performance and termination of contracts.

interim overhaul — An availability for the accomplishment of necessary repairs and urgent alterations at a naval shipyard or other shore-based repair activity, normally scheduled halfway through the established regular overhaul cycle.

inter-look dormant period — **(*)** In mine warfare, the time interval after each look in a multi-look mine, during which the firing mechanism will not register.

intermediate approach — **(*)** That part of an instrument approach procedure in which aircraft configuration, speed, and positioning adjustments are made. It blends the initial approach segment into the final approach segment. It begins at the intermediate fix or point and ends at the final approach fix or point.

Intermediate Force Planning Level — The force level established during Planning Force development to depict the buildup from the Current Force to the Planning Force. The Intermediate Force Planning Level is insufficient to carry out strategy with a reasonable assurance of success and consequently cannot be referred to as the Planning Force. See also **current force; force; Programmed Forces.**

intermediate maintenance (field) — Maintenance that is the responsibility of and performed by designated maintenance activities for direct support of using organizations. Its phases normally consist of: a. calibration, repair, or replacement of damaged or unserviceable parts, components, or assemblies; b. the emergency manufacture of nonavailable parts; and c. providing technical assistance to using organizations.

intermediate marker (land mine warfare) — **(*)** A marker, natural, artificial or specially installed, which is used as a point of reference between the landmark and the minefield.

intermediate objective — **(*)** In land warfare, an area or feature between the line of departure and an objective which must be seized and/or held.

intermediate-range bomber aircraft — A bomber designed for a tactical operating radius of between 1,000 to 2,500 nautical miles at design gross weight and design bomb load.

intermediate staging base — A temporary location used to stage forces prior to inserting the forces into the host nation. Also called **ISB.** See also **base; staging base.** (JP 3-07.5)

intermittent arming device — **(*)** A device included in a mine so that it will be armed only at set times.

intermittent illumination — **(*)** A type of fire in which illuminating projectiles are fired at irregular intervals.

intermodal — Type of international freight system that permits transshipping among sea, highway, rail, and air modes of transportation through use of American National Standards Institute and International Organization for Standardization containers, line-haul assets, and handling equipment. See also **American National Standards Institute; International Organization for Standardization.** (JP 4-01.7)

intermodal support equipment — Fixed and deployable assets required to assist container operations throughout the intermodal container system. Included are straddle cranes, chassis, rough terrain container handlers, container cranes and spreader bars. See also **intermodal.** (JP 4-01.7)

intermodal systems — Specialized transportation facilities, assets, and handling procedures designed to create a seamless transportation system by combining multimodal operations and facilities during the shipment of cargo. See also **intermodal; transportation system.** (JP 4-01)

internal audience — US military members and civilian employees and their immediate families. One of the audiences comprising the concept of "publics." See also **external audience.**

internal defense and development — The full range of measures taken by a nation to promote its growth and to protect itself from subversion, lawlessness, and insurgency. It focuses on building viable institutions (political, economic, social, and military) that respond to the needs of society. Also called **IDAD.** See also **foreign internal defense.** (JP 3-07.1)

internal information — See **command information.**

internally displaced person — Any person who has left their residence by reason of real or imagined danger but has not left the territory of their own country. (JP 3-07.6)

internal radiation — **(*)** Nuclear radiation (alpha and beta particles and gamma radiation) resulting from radioactive substances in the body.

internal security — The state of law and order prevailing within a nation.

internal waters — All waters, other than lawfully claimed archipelagic waters, landward of the baseline from which the territorial sea is measured. Archipelagic states may also delimit internal waters consistent with the 1982 convention on the law of the sea. All states have complete sovereignty over their internal waters.

international arms control organization — An appropriately constituted organization established to supervise and verify the implementation of arms control measures.

International Atomic Time — The time reference scale established by the *Bureau International des Poids et Mesures* on the basis of atomic clock readings from various laboratories around the world. Also called **TAI.**

international call sign — **(*)** A call sign assigned in accordance with the provisions of the International Telecommunications Union to identify a radio station. The nationality of the radio station is identified by the first or the first two characters. (When used in visual signaling, international call signs are referred to as "signal letters.") See also **call sign.**

International Convention for Safe Containers — A convention held in Geneva, Switzerland, on 2 Dec 1972, which resulted in setting standard safety requirements for containers moving in international transport. These requirements were ratified by the United States on 3 January 1978. Also called **CSC.** (JP 4-01.7)

international cooperative logistics — **(*)** Cooperation and mutual support in the field of logistics through the coordination of policies, plans, procedures, development activities, and the common supply and exchange of goods and services arranged on the basis of bilateral and multilateral agreements with appropriate cost reimbursement provisions.

international date line — **(*)** The line coinciding approximately with the anti-meridian of Greenwich, modified to avoid certain habitable land. In crossing this line there is a date change of one day. Also called **date line.**

international identification code — **(*)** In railway terminology, a code which identifies a military train from point of origin to final destination. The code consists of a series of figures, letters, or symbols indicating the priority, country of origin, day of departure, national identification code number, and country of destination of the train.

international loading gauge (GIC) — **(*)** The loading gauge upon which international railway agreements are based. A load whose dimensions fall within the limits of this gauge may move without restriction on most of the railways of Continental Western Europe. GIC is an abbreviation for "*gabarit international de chargement*," formerly called PPI.

international logistics — The negotiating, planning, and implementation of supporting logistic arrangements between nations, their forces, and agencies. It includes furnishing logistic support (major end items, materiel, and/or services) to, or receiving logistic support from, one or more friendly foreign governments, international organizations, or military forces, with or without reimbursement. It also includes planning and actions related to the intermeshing of a significant element, activity, or component of the military logistic systems or procedures of the United States with those of one or more foreign governments, international organizations, or military forces on a temporary or permanent basis. It includes planning and actions related to the utilization of United States logistic policies, systems, and/or procedures to meet requirements of one or more foreign governments, international organizations, or forces.

international logistic support — The provision of military logistic support by one participating nation to one or more participating nations, either with or without reimbursement. See also **interdepartmental or agency support; inter-Service support; support.**

international military education and training — Formal or informal instruction provided to foreign military students, units, and forces on a nonreimbursable (grant) basis by offices or employees of the United States, contract technicians, and contractors. Instruction may include correspondence courses; technical, educational, or informational publications; and media of all kinds. Also called **IMET.** See also **United States Military Service funded foreign training.**

international narcotics activities — Those activities outside the United States that produce, transfer, or sell narcotics or other substances controlled in accordance with Title 21, "Food and Drugs" — United States Code, sections 811 and 812. (JP 3-07.4)

international organization — Organizations with global mandates, generally funded by contributions from national governments. Examples include the International Committee of the Red Cross, the International Organization for Migration, and United Nation agencies. Also called **IO.** See also **nongovernmental organizations.** (JP 3-07.6)

International Organization for Standardization — A worldwide federation of national standards bodies from some 100 countries, one from each country. The International Organization for Standardization (ISO) is a non-governmental organization, established to promote the development of standardization and related activities in the world with a view to facilitating the international exchange of goods and services, and to developing cooperation in the spheres of intellectual, scientific, technological, and economic activity. ISO's work results in international agreements which are published as international standards. Also called **ISO.**

interned — See **missing.**

interocular distance — The distance between the centers of rotation of the eyeballs of an individual or between the oculars of optical instruments.

interoperability — **(*)** 1. The ability of systems, units, or forces to provide services to and accept services from other systems, units, or forces and to use the services so exchanged to enable them to operate effectively together. 2. **(DOD only)** The condition achieved among communications-electronics systems or items of communications-electronics equipment when information or services can be exchanged directly and satisfactorily between them and/or their users. The degree of interoperability should be defined when referring to specific cases.

interoperation — The use of interoperable systems, units, or forces.

interpretability — **(*)** Suitability of imagery for interpretation with respect to answering adequately requirements on a given type of target in terms of quality and scale. a. **poor** — Imagery is unsuitable for interpretation to answer adequately requirements on a given type of target. b. **fair** — Imagery is suitable for interpretation to answer requirements on a given type of target but with only average detail. c. **good** — Imagery is suitable for interpretation to answer requirements on a given type of target in considerable detail. d. **excellent** — Imagery is suitable for interpretation to answer requirements on a given type of target in complete detail.

interpretation — A part of the analysis and production phase in the intelligence cycle in which the significance of information is judged in relation to the current body of knowledge. See also **intelligence cycle.** (JP 2-0)

interrogation (intelligence) — Systematic effort to procure information by direct questioning of a person under the control of the questioner.

inter-Service education — Military education provided by one Service to members of another Service. See also **military education; military training.**

inter-Service, intragovernmental agreements — Formal long-term or operational specific support agreements between Services, Department of Defense (DOD), and/or non-DOD agencies governed by DOD Instruction 4000.19, *Interservice and Intragovernmental Support.* These agreements, normally developed at the Service Secretariat and governmental agency director level, document funding and reimbursement procedures as well as standards of support between the supplying and receiving Service or agencies. Inter-Service, intragovernmental agreements, while binding Service level agreements, do not connote DOD-level executive agent responsibilities. See also **inter-Service support.** (JP 4-07)

inter-Service support — Action by one Military Service or element thereof to provide logistic and/or administrative support to another Military Service or element thereof. Such action can be recurring or nonrecurring in character on an installation, area, or worldwide basis. See also **interdepartmental or agency support; international logistic support; support.**

inter-Service training — Military training provided by one Service to members of another Service. See also **military education; military training.**

intertheater — Between theaters or between the continental United States and theaters. See also **intertheater traffic.**

intertheater airlift — See **strategic airlift.** (JP 4-01.1)

intertheater evacuation — Evacuation of stabilized patients between the originating theater and points outside the theater, to include the continental United States and other theaters. En route care is provided by medical attendants qualified for the specific mode of transportation. See also **en route care; evacuation; intratheater evacuation; patient.** (JP 4-02)

intertheater traffic — Traffic between theaters exclusive of that between the continental United States and theaters.

interval — **(*)** 1. The space between adjacent groups of ships or boats measured in any direction between the corresponding ships or boats in each group. 2. The space between adjacent individuals, ground vehicles, or units in a formation that are placed side by side, measured abreast. 3. The space between adjacent aircraft measured from front to rear in units of time or distance. 4. The time lapse between photographic exposures. 5. At battery right or left, an interval ordered in seconds is the time between one gun firing and the next gun firing. Five seconds is the standard interval. 6. At rounds of fire for effect the interval is the time in seconds between successive rounds from each gun.

intervention — Action taken to divert a unit or force from its track, flight path, or mission.

interview (intelligence) — To gather information from a person who is aware that information is being given although there is ignorance of the true connection and purposes of the interviewer. Generally overt unless the collector is other than purported to be.

intracoastal sealift — Shipping used primarily for the carriage of personnel and/ or cargo along a coast or into river ports to support operations within a given area.

intransit aeromedical evacuation facility — A medical facility, on or in the vicinity of an air base, that provides limited medical care for intransit patients awaiting air transportation. This type of medical facility is provided to obtain effective utilization of transport airlift within operating schedules. It includes "remain overnight" facilities, intransit facilities at aerial ports of embarkation and debarkation, and casualty staging facilities in an overseas combat area. See also **aeromedical evacuation unit.**

intransit inventory — That materiel in the military distribution system that is in the process of movement from point of receipt from procurement and production (either contractor's plant or first destination, depending upon point of delivery) and between points of storage and distribution.

intransit stock — See **intransit inventory.**

in-transit visibility — The ability to track the identity, status, and location of Department of Defense units, and non-unit cargo (excluding bulk petroleum, oils, and lubricants) and passengers; medical patients; and personal property from origin to consignee or destination across the range of military operations. Also called **ITV.** See also **Global Transportation Network; total asset visibility**. (JP 4-01.8)

intratheater — Within a theater. See also **intratheater traffic.**

intratheater airlift — See **theater airlift.** (JP 4-01.1)

intratheater evacuation — Evacuation of stabilized patients between points within the theater. En route care is provided by medical attendants qualified for the specific mode of transportation. See also **en route care; evacuation; intertheater evacuation; patient.** (JP 4-02)

intratheater traffic — Traffic within a theater.

intruder — An individual, unit, or weapon system, in or near an operational or exercise area, which presents the threat of intelligence gathering or disruptive activity.

intrusion — Movement of a unit or force within another nation's specified operational area outside of territorial seas and territorial airspace for surveillance or intelligence gathering in time of peace or tension.

invasion currency — See **military currency.**

inventory control — **(*)** That phase of military logistics which includes managing, cataloging, requirements determinations, procurement, distribution, overhaul, and disposal of materiel. Also called **inventory**

management; materiel control; materiel management; supply management.

inventory control point — An organizational unit or activity within a Department of Defense supply system that is assigned the primary responsibility for the materiel management of a group of items either for a particular Service or for the Defense Department as a whole. Materiel inventory management includes cataloging direction, requirements computation, procurement direction, distribution management, disposal direction and, generally, rebuild direction. Also called **ICP.**

inventory management — See **inventory control.**

inventory managers — See **inventory control point.**

investment costs — Those program costs required beyond the development phase to introduce into operational use a new capability; to procure initial, additional, or replacement equipment for operational forces; or to provide for major modifications of an existing capability. They exclude research, development, test and evaluation, military personnel, and operation and maintenance appropriation costs.

ionosphere — That part of the atmosphere, extending from about 70 to 500 kilometers, in which ions and free electrons exist in sufficient quantities to reflect electromagnetic waves.

IR pointer — See **infrared pointer.** (JP 3-09.3)

irregular forces — Armed individuals or groups who are not members of the regular armed forces, police, or other internal security forces.

irregular outer edge — **(*)** In land mine warfare, short mine rows or strips laid in an irregular manner in front of a minefield facing the enemy to deceive the enemy as to the type or extent of the minefield. Generally, the irregular outer edge will only be used in minefields with buried mines.

isodose rate line — See **dose rate contour line.**

isolated personnel — Military or civilian personnel separated from their unit or organization in an environment requiring them to survive, evade, or escape while awaiting rescue or recovery. See also **combat search and rescue; search and rescue.** (JP 3-50.2)

isolated personnel report — A Department of Defense Form (DD 1833) containing information designed to facilitate the identification and authentication of an evader by a recovery force. Also called **ISOPREP.** See also **authentication; evader; recovery force.** (JP 3-50.3)

issue control group — A detachment that operates the staging area, consisting of holding areas and loading areas, in an operation. See also **staging area.** (JP 4-01.6)

issue priority designator — See **priority designator.**

item manager — An individual within the organization of an inventory control point or other such organization assigned management responsibility for one or more specific items of materiel.

J

J-2X — Umbrella organization consisting of the human intelligence operations cell and the task force counterintelligence coordinating authority. The J-2X is responsible for coordination and deconfliction of all human source-related activity. See also **counterintelligence; human intelligence.** (JP 2-01)

jamming — See **barrage jamming; electronic attack; electromagnetic jamming; selective jamming; spot jamming.**

j-axis — A vertical axis in a system of rectangular coordinates; that line on which distances above or below (north or south) the reference line are marked, especially on a map, chart, or graph.

jet advisory service — The service provided certain civil aircraft while operating within radar and nonradar jet advisory areas. Within radar jet advisory areas, civil aircraft receiving this service are provided radar flight following, radar traffic information, and vectors around observed traffic. In nonradar jet advisory areas, civil aircraft receiving this service are afforded standard instrument flight rules separation from all other aircraft known to air traffic control to be operating within these areas.

jet propulsion — Reaction propulsion in which the propulsion unit obtains oxygen from the air, as distinguished from rocket propulsion, in which the unit carries its own oxygen-producing material. In connection with aircraft propulsion, the term refers to a gasoline or other fuel turbine jet unit that discharges hot gas through a tail pipe and a nozzle which provides a thrust that propels the aircraft. See also **rocket propulsion.**

jet stream — A narrow band of high velocity wind in the upper troposphere or in the stratosphere.

jettison — The selective release of stores from an aircraft other than normal attack.

jettisoned mines — **(*)** Mines which are laid as quickly as possible in order to empty the minelayer of mines, without regard to their condition or relative positions.

joiner — **(*)** An independent merchant ship sailed to join a convoy. See also **joiner convoy; joiner section.**

joiner convoy — **(*)** A convoy sailed to join the main convoy. See also **joiner; joiner section.**

joiner section — **(*)** A joiner or joiner convoy, after rendezvous, and while maneuvering to integrate with the main convoy.

joint — Connotes activities, operations, organizations, etc., in which elements of two or more Military Departments participate. (JP 0-2)

joint after action report — A report consisting of summary joint universal lessons learned. It describes a real world operation or training exercise and identifies significant lessons learned. Also called **JAAR.**

joint airborne advance party — An advance ground party that provides terminal guidance, air traffic control, ground control measures, intelligence gathering, and surface weather observation in the objective area of an airlift operation. It may consist of US Air Force combat control team members and a US Army long-range surveillance team or similar forces. Also called **JAAP.** (JP 3-17)

joint airborne training — Training operations or exercises involving airborne and appropriate troop carrier units. This training includes: a. air delivery of personnel and equipment; b. assault operations by airborne troops and/or air transportable units; c. loading exercises and local orientation fights of short duration; and d. maneuvers and/or exercises as agreed upon by Services concerned and/or as authorized by the Joint Chiefs of Staff.

joint air operations — Air operations performed with air capabilities/forces made available by components in support of the joint force commander's operation or campaign objectives, or in support of other components of the joint force. (JP 3-56.1)

joint air operations center — A jointly staffed facility established for planning, directing, and executing joint air operations in support of the joint force commander's operation or campaign objectives. Also called **JAOC.** See also **joint air operations.** (JP 3-56.1)

joint air operations plan — A plan for a connected series of joint air operations to achieve the joint force commander's objectives within a given time and theater of operations. See also **joint air operations.** (JP 3-56.1)

joint amphibious operation — **(*)** An amphibious operation conducted by significant elements of two or more Services.

joint amphibious task force — A temporary grouping of units of two or more Services under a single commander, organized for the purpose of engaging in an amphibious landing for assault on hostile shores. Also called **JATF.**

joint base — For purposes of base defense operations, a joint base is a locality from which operations of two or more of the Military Departments are projected or supported and which is manned by significant elements of two or more Military Departments or in which significant elements of two or more Military Departments are located. See also **base.** (JP 3-10)

joint captured materiel exploitation center — Physical location for deriving intelligence information from captured enemy materiel. It is normally subordinate to the joint force/J-2. Also called **JCMEC.** (JP 2-0)

joint civil-military operations task force — A joint task force composed of civil-military operations units from more than one Service. It provides support to the joint force commander in humanitarian or nation assistance operations, theater campaigns, or a civil-military operations concurrent with or subsequent to regional conflict. It can organize military interaction among many governmental and nongovernmental humanitarian agencies within the theater. Also called **JCMOTF.** See also **civil-military operations; joint task force; task force.** (JP 3-57)

joint combat search and rescue operation — A combat search and rescue operation in support of a component's military operations that has exceeded the combat search and rescue capabilities of that component and requires the efforts of two or more components of the joint force. Normally, the operation is conducted by the joint force commander or a component commander that has been designated by joint force commander tasking. See also

combat search and rescue; search and rescue. (JP 3-50.2)

joint communications control center — An element of the J-6 established to support a joint force commander. The joint communications control center (JCCC) serves as the single control agency for the management and direction of the joint force command, control, communications, and computer systems. The JCCC may include plans and operations, administration, system control, and frequency management sections. Also called **JCCC.** (JP 6-02)

joint communications network — The aggregation of all the joint communications systems in a theater. The joint communications network includes the joint multi-channel trunking and switching system and the joint command and control communications system(s). Also called **JCN.**

joint decision support tools — A compilation of processes and systems developed from the application of maturing leading edge information systems technologies that provide the warfighter and the logistician with the means to rapidly plan, execute, monitor, and replan logistic operations in a collaborative environment that is responsive to operational requirements. Also called **JDST.** (JP 4-0)

joint deployable intelligence support system — A transportable workstation and communications suite that electronically extends a joint intelligence center to a joint task force or other tactical user. Also called **JDISS.** (JP 2-0)

joint doctrine — Fundamental principles that guide the employment of forces of two or more Military Departments in coordinated action toward a common objective. It is authoritative; as such, joint doctrine will be followed except when, in the judgment of the commander, exceptional circumstances dictate otherwise. It will be promulgated by or for the Chairman of the Joint Chiefs of Staff, in coordination with the combatant commands and Services. See also **Chairman of the Joint Chiefs of Staff Instruction; Chairman of the Joint Chiefs of Staff Manual; doctrine; joint publication; joint tactics, techniques, and procedures; joint test publication; multinational doctrine; multi-Service doctrine.** (JP 1-01)

Joint Doctrine Publication System — The system of lead agents, Joint Staff doctrine sponsors, primary review authorities, coordinating review authorities, technical review authorities, Joint Doctrine Working Party, procedures, and hierarchical framework designed to organize, develop, maintain, print, and distribute joint publications. See also **coordinating review authority; Joint Doctrine Working Party; joint publication; Joint Staff doctrine sponsor; lead agent; primary review authority**. (JP 1-01)

Joint Doctrine Working Party — A forum to include representatives of the Services, combatant commands, and the Joint Staff (represented by the Operational Plans and Joint Force Development Directorate, J-7) which meets semiannually to address and vote on project proposals; discuss key joint doctrinal or operational issues; keep up to date on the status of the joint publication projects and emerging publications; and keep abreast of other initiatives of interest to the members. The Joint Doctrine Working Party meets under the sponsorship of the Director, J-7, Joint Staff. Also called **JDWP.** See also **joint doctrine; joint publication; joint tactics, techniques, and procedures; joint test publication.** (JP 1-01)

joint document exploitation center — Physical location for deriving intelligence information from captured enemy documents. It is normally subordinate to the joint force/J-2. Also called **JDEC.** See also **intelligence.** (JP 2-01)

joint duty assignment — An assignment to a designated position in a multi-Service, joint or multinational command or activity that is involved in the integrated employment or support of the land, sea, and air forces of at least two of the three Military Departments. Such involvement includes, but is not limited to, matters relating to national military strategy, joint doctrine and policy, strategic planning, contingency planning, and command and control of combat operations under a unified or specified command. Also called **JDA.**

Joint Duty Assignment List — Positions designated as joint duty assignments are reflected in a list approved by the Secretary of Defense and maintained by the Joint Staff. The Joint Duty Assignment List is reflected in the Joint Duty Assignment Management Information System. Also called **JDAL.**

joint engagement zone — See **weapon engagement zone.** (JP 3-52)

Joint Facilities Utilization Board — A joint board that evaluates and reconciles component requests for real estate, use of existing facilities, inter-Service support, and construction to ensure compliance with Joint Civil-Military Engineering Board priorities. (JP 4-04)

joint fires — Fires produced during the employment of forces from two or more components in coordinated action toward a common objective. See also **fires.** (JP 3-09)

joint fires element — An optional staff element that provides recommendations to the operations directorate to accomplish fires planning and synchronization. Also called **JFE.** See also **fire support; joint fires.** (JP 3-60)

joint fire support — Joint fires that assist air, land, maritime, amphibious, and special operations forces to move, maneuver, and control territory, populations, airspace, and key waters. See also **fire support; joint fires.** (JP 3-0)

joint flow and analysis system for transportation — System that determines the transportation feasibility of a course of action or operation plan; provides daily lift assets needed to move forces and resupply; advises logistic planners of channel and port inefficiencies; and interprets shortfalls from various flow possibilities. Also called **JFAST.** See also **course of action; operation plan; system**. (JP 4-01.8)

joint force — A general term applied to a force composed of significant elements, assigned or attached, of two or more Military Departments operating under a single joint force commander. See also **joint force commander.** (JP 3-0)

joint force air component commander — The commander within a unified command, subordinate unified command, or joint task force responsible to the establishing commander for making recommendations on the proper employment of assigned, attached, and/or made available for tasking air forces; planning and coordinating air operations; or accomplishing such operational missions as may be assigned. The joint force air component commander is given the authority necessary to accomplish missions and tasks assigned by the establishing commander. Also called **JFACC.** See also **joint force commander.** (JP 3-0)

joint force commander — A general term applied to a combatant commander, subunified commander, or joint task force commander authorized to exercise combatant command (command authority) or operational control over a joint force. Also called **JFC.** See also **joint force.** (JP 0-2)

joint force land component commander — The commander within a unified command, subordinate unified command, or joint task force responsible to the establishing commander for making recommendations on the proper employment of assigned, attached, and/or made available for tasking land forces; planning and coordinating land operations; or accomplishing such operational missions as may be assigned. The joint force land component commander is given the authority necessary to accomplish missions and tasks assigned by the establishing commander. Also called **JFLCC.** See also **joint force commander.** (JP 3-0)

joint force maritime component commander — The commander within a unified command, subordinate unified command, or joint task force responsible to the establishing commander for making recommendations on the proper employment of assigned, attached, and/or made available for tasking maritime forces and assets; planning and coordinating maritime operations; or accomplishing such operational missions as may be assigned. The joint force maritime component commander is given the authority necessary to accomplish missions and tasks assigned by the establishing commander. Also called **JFMCC.** See also **joint force commander.** (JP 3-0)

joint force meteorological and oceanographic officer — Officer designated to provide direct meteorological and oceanographic support to a joint force commander. Also called **JMO.** See also **meteorological and oceanographic.** (JP 3-59)

joint force special operations component commander — The commander within a unified command, subordinate unified command, or joint task force responsible to the establishing commander for making recommendations on the proper employment of assigned, attached, and/or made available for tasking special operations forces and assets; planning and coordinating special operations; or accomplishing such operational missions as may be assigned. The joint force special operations component commander is given the authority necessary to accomplish missions and tasks assigned by the establishing commander. Also called **JFSOCC.** See also **joint force commander.** (JP 3-0)

joint force surgeon — A general term applied to a medical officer appointed by the joint force commander to serve as the joint force special staff officer responsible for establishing, monitoring, or evaluating joint force health service support. Also called **JFS.** See also **health service support; joint force.** (JP 4-02)

joint guidance, apportionment, and targeting team — A group that makes recommendations for air apportionment to engage targets, and provides other targeting support requiring component input at the joint force air component commander level. See also **air apportionment; apportionment; joint force air component commander; targeting.** (JP 3-60)

joint information bureau — Facilities established by the joint force commander to serve as the focal point for the interface between the military and the media during the conduct of joint operations. When operated in support of multinational operations, a joint information bureau is called a "combined information bureau" or an "allied press information center." Also called **JIB.** See also **public affairs.** (JP 3-61)

joint integrated prioritized target list — A prioritized list of targets and associated data approved by the joint force commander or designated representative and maintained by a joint force. Targets and priorities are derived from the recommendations of components in conjunction with their proposed operations supporting the joint force commander's objectives and guidance. Also called **JIPTL.** See also **target; target list.** (JP 3-60)

joint intelligence — Intelligence produced by elements of more than one Service of the same nation.

joint intelligence architecture — A dynamic, flexible structure that consists of the National Military Joint Intelligence Center, the theater joint intelligence centers, and subordinate joint force joint intelligence support elements. This architecture encompasses automated data processing equipment capabilities, communications and information requirements, and responsibilities to provide national, theater, and tactical commanders with the full range of intelligence required for planning and conducting operations. See also **architecture; intelligence.** (JP 2-0)

joint intelligence center — The intelligence center of the combatant command headquarters. The joint intelligence center is responsible for providing and producing the intelligence required to support the combatant commander and staff, components, subordinate joint forces and elements, and the national intelligence community. Also called **JIC.** See also **intelligence; joint intelligence; joint intelligence architecture.** (JP 2-0)

joint intelligence liaison element — A liaison element provided by the Central Intelligence Agency in support of a unified command or joint task force.

joint intelligence preparation of the battlespace — The analytical process used by joint intelligence organizations to produce intelligence assessments, estimates and other intelligence products in support of the joint force commander's decisionmaking process. It is a continuous process that includes defining the total battlespace environment; describing the battlespace's effects; evaluating the adversary; and determining and describing adversary potential courses of action. The process is used to analyze the air, land, sea, space, electromagnetic, cyberspace, and human dimensions of the environment and to determine an opponent's capabilities to operate in each. Joint intelligence preparation of the battlespace products are used by the joint force and component command staffs in preparing their estimates and are also applied during the analysis and selection of friendly courses of action. Also called **JIPB.** See also **battlespace; intelligence; joint intelligence.** (JP 2-0)

joint intelligence support element — A subordinate joint force forms a joint intelligence support element as the focus for intelligence support for joint operations, providing the joint force commander, joint staff, and components with the complete air, space, ground, and maritime adversary situation. Also called **JISE.** See also **intelligence; joint force; joint operations.** (JP 2-01)

joint interrogation and debriefing center — Physical location for the exploitation of intelligence information from enemy prisoners of war and other non-prisoner sources. It is normally subordinate to the joint force/J-2. Also called **JIDC.** See also **information; intelligence.** (JP 2-01)

joint logistics — The art and science of planning and carrying out, by a joint force commander and staff, logistic operations to support the protection, movement, maneuver, firepower, and sustainment of operating forces of two or more Military Departments of the same nation. See also **logistics.** (JP 3-10)

joint logistics over-the-shore commander — The joint logistics over-the-shore (JLOTS) commander is selected by the joint force commander (JFC) and is usually from either the Army or Navy components that are part of the JFC's task organization. This individual then builds a joint headquarters from personnel and equipment in theater to organize the efforts of all elements participating in accomplishing the JLOTS mission having either wet or dry cargo or both. JLOTS commanders will usually integrate members from each participating organization to balance the overall knowledge base in their headquarters. See also **joint logistics over-the-shore operations.** (JP 4-01.6)

joint logistics over-the-shore operations — Operations in which Navy and Army logistics over-the-shore (LOTS) forces conduct LOTS operations together under a joint force commander. Also called **JLOTS operations.** See also **joint logistics; logistics over-the-shore operations.** (JP 4-01.6)

joint manpower program — The document that reflects an activity's mission, functions, organization, current and projected manpower needs and, when applicable, its required mobilization augmentation. A recommended joint manpower program also identifies and justifies any changes proposed by the commander or director of a joint activity for the next five fiscal years. Also called **JMP.**

Joint Materiel Priorities and Allocation Board — The agency charged with performing duties for the Chairman of the Joint Chiefs of Staff in matters that establish materiel priorities or allocate resources. Also called **JMPAB.** See also **materiel.** (JP 4-09)

joint meteorological and oceanographic forecast unit — An organization consisting of a jointly supported collective of meteorological and oceanographic personnel and equipment formed to provide meteorological and oceanographic support to the joint force commander. Also called **JMFU.** See also **meteorological and oceanographic.** (JP 3-59)

joint mission-essential task — A mission task selected by a joint force commander deemed essential to mission accomplishment and defined using the common language of the universal joint task list in terms of task, condition, and standard. Also called **JMET.** See also **condition, universal joint task list.**

Joint Mobility Control Group — The Joint Mobility Control Group is the focal point for coordinating and optimizing transportation operations. This group is comprised of seven essential elements. The primary elements are United States Transportation Command's (USTRANSCOM's) Mobility Control Center, Joint Operational Support Airlift Center, Global Patient Movement Requirements Center, Tanker Airlift Control Center, Military Sealift Command Command Center, Military Traffic Management Command Command Operations, and the Joint Intelligence Center-USTRANSCOM. Also called **JMCG.** See also **Global Patient Movement Requirements Center; United States Transportation Command.**

joint mortuary affairs office — Plans and executes all mortuary affairs programs within a theater. Provides guidance to facilitate the conduct of all mortuary programs and to maintain data (as required) pertaining to recovery, identification, and disposition of all US dead and missing in the assigned theater. Serves as the central clearing point for all mortuary affairs and monitors the deceased and missing personal effects program. Also called **JMAO.** See also **mortuary affairs; personal effects.** (JP 4-06)

joint movement center — The center established to coordinate the employment of all means of transportation (including that provided by allies or host nations) to support the concept of operations. This coordination is accomplished through establishment of transportation policies within the assigned operational area, consistent with relative urgency of need, port and terminal capabilities, transportation asset availability, and priorities set by a joint force commander. Also called **JMC.** See also **concept of operations.** (JP 4-0)

joint multi-channel trunking and switching system — That composite multi-channel trunking and switching system formed from assets of the Services, the Defense Information Systems Agency, other available systems, and/or assets controlled by the Joint Chiefs of Staff to provide an operationally responsive, survivable communication system, preferably in a mobile, transportable, and/or recoverable configuration, for the joint force commander in an operational area.

Joint Munitions Effectiveness Manual-Special Operations — A publication providing a single, comprehensive source of information covering weapon effectiveness, selection, and requirements for special operations munitions. In addition, the closely related fields of weapon characteristics and effects, target characteristics, and target vulnerability are treated in limited detail required by the mission planner. Although emphasis is placed on weapons that are currently in the inventory, information is also included for some weapons not immediately available but projected for the near future. Also called **JMEM-SO.** (JP 3-05.5)

joint nuclear accident coordinating center — A combined Defense Special Weapons Agency and Department of Energy centralized agency for exchanging and maintaining information concerned with radiological assistance capabilities and coordinating that assistance in response to an accident or incident involving radioactive materials. Also called **JNACC.**

joint operation planning — Planning for contingencies that can reasonably be anticipated in an area of responsibility or joint operations area of the command. Planning activities exclusively associated with the preparation of operation plans, operation plans in concept format, campaign plans, and operation orders (other than the Single Integrated Operational Plan) for the conduct of military operations by the combatant commanders in response to requirements established by the Chairman of the Joint Chiefs of Staff. Joint operation planning is coordinated at the national level to support Secretary of Defense Contingency Planning Guidance, strategic requirements in the National Military Strategy, and emerging crises. As such, joint operation planning includes mobilization planning, deployment planning, employment planning, sustainment planning, and redeployment planning procedures. Joint operation planning is performed in accordance with formally established planning and execution procedures. See also **contingency plan; execution planning; implementation planning; Joint Operation Planning and Execution**

System; joint operation planning process. (JP 5-0)

Joint Operation Planning and Execution System — A system that provides the foundation for conventional command and control by national- and combatant command-level commanders and their staffs. It is designed to satisfy their information needs in the conduct of joint planning and operations. Joint Operation Planning and Execution System (JOPES) includes joint operation planning policies, procedures, and reporting structures supported by communications and automated data processing systems. JOPES is used to monitor, plan, and execute mobilization, deployment, employment, sustainment, and redeployment activities associated with joint operations. Also called **JOPES.** See also **joint operation planning; joint operations.** (JP 5-00.1)

joint operation planning process — A coordinated Joint Staff procedure used by a commander to determine the best method of accomplishing assigned tasks and to direct the action necessary to accomplish the mission. See also **joint operation planning; Joint Operation Planning and Execution System.** (JP 5-0)

joint operations — A general term to describe military actions conducted by joint forces or by Service forces in relationships (e.g., support, coordinating authority) which, of themselves, do not create joint forces. (JP 0-2)

joint operations area — An area of land, sea, and airspace, defined by a geographic combatant commander or subordinate unified commander, in which a joint force commander (normally a joint task force commander) conducts military operations to accomplish a specific mission. Joint operations areas are particularly useful when operations are limited in scope and geographic area or when operations are to be conducted on the boundaries between theaters. Also called **JOA.** See also **area of responsibility; joint special operations area.** (JP 0-2)

joint operations center — A jointly manned facility of a joint force commander's headquarters established for planning, monitoring, and guiding the execution of the commander's decisions. Also called **JOC.**

joint patient movement requirements center — A joint force health service support center under the control of the subordinate joint force surgeon, established to coordinate and control, in terms of identifying bed space requirements, the movement of patients within and out of the joint operations area. The joint patient movement requirements center also generates subordinate joint force commander (JFC) plans and schedules to evacuate the subordinate JFC's patients to medical treatment facilities in accordance with the supported combatant commander's theater patient movement requirements center theater plans and schedules for movement of the patient to the medical treatment facility. Also called **JPMRC.** See also **health service support; joint force surgeon; joint operations area; medical treatment facility; patient.** (JP 5-00.2)

joint personnel training and tracking activity — The continental US center established (upon request of the supported combatant commander) to facilitate the reception, accountability, processing, training, and onward movement of both military and civilian individual augmentees preparing for overseas movement to support a joint military operation. Also called **JPTTA.** (JP 1-0)

joint planning and execution community — Those headquarters, commands, and agencies involved in the training, preparation, movement, reception, employment, support, and sustainment of military forces assigned or committed to a theater of operations or objective area. It usually consists of the Joint Staff, Services, Service major commands (including the Service wholesale logistic commands), unified commands (and their certain Service component commands), subunified commands, transportation component commands, joint task forces (as applicable), Defense Logistics Agency, and other Defense agencies (e.g., Defense Intelligence Agency) as may be appropriate to a given scenario. Also called **JPEC.** (JP 5-0)

joint planning group — A joint force planning organization consisting of designated representatives of the joint force headquarters principal and special staff sections, joint force components (Service and/or functional), and other supporting organizations or agencies as deemed necessary by the joint force commander (JFC). Joint planning group membership should be a long-term assignment and members should be designated spokespersons for their respective sections or organizations. Responsibilities and authority of the joint planning group are assigned by the JFC. Normally headed by the joint force chief planner, joint planning group responsibilities may include, but are not limited to, crisis action planning (to include course of action development and refinement), coordination of joint force operation order development, and planning for future operations (e.g., transition, termination, follow-on). Also called **JPG.** See also **course of action development; crisis action planning; joint operation planning.** (JP 5-00.2)

joint psychological operations task force — A joint special operations task force composed of headquarters and operational assets. It assists the joint force commander in developing strategic, operational, and tactical psychological operation plans for a theater campaign or other operations. Mission requirements will determine its composition and assigned or attached units to support the joint task force commander. Also called **JPOTF.** See also **joint special operations task force; psychological operations; special operations.** (JP 3-05.1)

joint publication — A publication containing joint doctrine and/or joint tactics, techniques, and procedures that involves the employment of forces prepared under the cognizance of Joint Staff directorates and applicable to the Military Departments, combatant commands, and other authorized agencies. It is approved by the Chairman of the Joint Chiefs of Staff, in coordination with the combatant commands and Services. Also called **JP.** See also **Chairman of the Joint Chiefs of Staff Instruction; Chairman of the Joint Chiefs of Staff Manual; joint doctrine; joint tactics, techniques, and procedures; joint test publication.** (JP 1-01)

joint readiness — See **readiness.**

joint rear area — A specific land area within a joint force commander's operational area designated to facilitate protection and operation of installations and forces supporting the joint force. Also called **JRA.** See also **joint force; joint force commander; rear area.** (JP 3-10)

joint rear area coordinator — The officer with responsibility for coordinating the overall security of the joint rear area in accordance with joint force commander directives and priorities in order to assist in providing a secure environment to facilitate sustainment, host-nation support, infrastructure development, and movements of the joint force. The joint rear area coordinator also coordinates intelligence support and ensures that area management is practiced with due consideration for security requirements. Also called **JRAC.** (JP 3-10)

joint rear area operations — Those operations in the joint rear area that facilitate protection or support of the joint force. See also **joint force; joint rear area; rear area.** (JP 3-10)

joint rear tactical operations center — A joint operations cell tailored to assist the joint rear area coordinator in meeting mission responsibilities. Also called **JRTOC.** (JP 3-10)

joint reception center — The center established in the operational area (per direction of the joint force commander), with responsibility for the reception, accountability, training, processing, of military and civilian individual augmentees upon their arrival in the operational area. Also the center where augmentees will normally be outprocessed through upon departure from the operational area. Also called **JRC.** (JP 4-01.8)

joint reception complex — The group of nodes (air and/or sea) designated by the supported combatant command, in coordination with the host nation and United States Transportation Command, that receives, processes, services, supports, and facilitates onward movement of personnel, equipment, materiel, and units deploying into, out of, or within a theater line of communications. See also **group; node**. (JP 4-01.8)

joint reception, staging, onward movement, and integration — A phase of joint force projection occurring in the operational area. This phase comprises the essential processes required to transition arriving personnel, equipment, and materiel into forces capable of meeting operational requirements. Also called **JRSOI.** See also **integration; joint force; reception; staging**. (JP 4-01.8)

joint regional defense command — A joint task force headquarters formed on order of the Commander in Chief, United States Atlantic Command within designated continental United States (CONUS) multistate regions, to command and control (1) execution of land defense of CONUS; and (2) military assistance to civil authority missions. (JP 3-57)

joint restricted frequency list — A time and geographically-oriented listing of TABOO, PROTECTED, and GUARDED functions, nets, and frequencies. It should be limited to the minimum number of frequencies necessary for friendly forces to accomplish objectives. Also called **JRFL.** See also **electronic warfare; guarded frequencies; protected frequencies; TABOO frequencies.** (JP 3-51)

joint search and rescue center — A primary search and rescue facility suitably staffed by supervisory personnel and equipped for planning, coordinating, and executing joint search and rescue and combat search and rescue operations within the geographical area assigned to the joint force. The facility is operated jointly by personnel from two or more Service or functional components or it may have a multinational staff of personnel from two or more allied or coalition nations (multinational search and rescue center). The joint search and rescue center should be staffed equitably by trained personnel drawn from each joint force component, including US Coast Guard participation where practical. Also called **JSRC.** See also **combat search and**

rescue; joint search and rescue center director; rescue coordination center; search and rescue. (JP 3-50.2)

joint search and rescue center director — The designated representative with overall responsibility for operation of the joint search and rescue center. See also **combat search and rescue; joint search and rescue center; search and rescue.** (JP 3-50.2)

joint servicing — That function performed by a jointly staffed and financed activity in support of two or more Military Services. See also **servicing.**

joint special operations air component commander — The commander within the joint force special operations command responsible for planning and executing joint special air operations and for coordinating and deconflicting such operations with conventional nonspecial operations air activities. The joint special operations air component commander normally will be the commander with the preponderance of assets and/or greatest ability to plan, coordinate, allocate, task, control, and support the assigned joint special operations aviation assets. The joint special operations air component commander may be directly subordinate to the joint force special operations component commander or to any nonspecial operations component or joint force commander as directed. Also called **JSOACC.** (JP 3-05.3)

joint special operations area — A restricted area of land, sea, and airspace assigned by a joint force commander to the commander of a joint special operations force to conduct special operations activities. The commander of joint special operations forces may further assign a specific area or sector within the joint special operations area to a subordinate commander for mission execution. The scope and duration of the special operations forces' mission, friendly and hostile situation, and politico-military considerations all influence the number, composition, and sequencing of special operations forces deployed into a joint special operations area. It may be limited in size to accommodate a discrete direct action mission or may be extensive enough to allow a continuing broad range of unconventional warfare operations. Also called **JSOA.** (JP 3-05.3)

joint special operations task force — A joint task force composed of special operations units from more than one Service, formed to carry out a specific special operation or prosecute special operations in support of a theater campaign or other operations. The joint special operations task force may have conventional non-special operations units assigned or attached to support the conduct of specific missions. Also called **JSOTF.** (JP 3-05)

joint specialty officer or joint specialist — An officer on the active duty list who is particularly trained in, and oriented toward, joint matters. Also called **JSO.**

joint staff — 1. The staff of a commander of a unified or specified command, subordinate unified command, joint task force, or subordinate functional component (when a functional component command will employ forces from more than one Military Department), that includes members from the several Services comprising the force. These members should be assigned in such a manner as to ensure that the commander understands the tactics, techniques, capabilities, needs, and limitations of the component parts of the force. Positions on the staff should be divided so that Service representation and influence generally reflect the Service composition of the force. 2. (capitalized as **Joint Staff**) The staff under the Chairman of the Joint Chiefs of Staff as provided for in the National Security Act of 1947, as amended by the Goldwater-Nichols Department of Defense Reorganization Act of 1986. The Joint Staff assists the Chairman and, subject to the authority, direction, and control of the Chairman and the other members of the Joint Chiefs of Staff in carrying out their responsibilities. Also called **JS.** See also **staff.** (JP 0-2)

Joint Staff doctrine sponsor — The sponsor for a joint doctrine or joint tactics, techniques, and procedures (JTTP) project. Each joint doctrine or JTTP project will be assigned a Joint Staff doctrine sponsor (JSDS). The JSDS will assist the lead agent and primary review authority as requested and directed. The JSDS will coordinate the draft document with the Joint Staff and provide Joint Staff comments and recommendations to the primary review authority. The JSDS will receive the revised draft from the lead agent, and process the preliminary coordination and final coordination (and test publications, if applicable) for approval. Also called **JSDS.** See also **joint doctrine; joint tactics, techniques, and procedures.** (JP 1-01)

joint state area command — A joint task force headquarters formed on order of Commander in Chief, United States Atlantic Command, within existing state boundaries to command and control United States and federalized state elements designated to execute land defense of the continental United States, military support to civil defense, and military assistance to civil authority missions. See also **joint regional defense command.** (JP 3-57)

Joint Strategic Capabilities Plan — The Joint Strategic Capabilities Plan (JSCP) provides guidance to the combatant commanders and the Joint Chiefs of Staff to accomplish task and missions based on current military capabilities. It apportions resources to combatant commanders based on military capabilities resulting from completed program and budget actions and intelligence assessments. The JSCP provides a coherent framework for capabilities-based military advice provided to the National Command Authorities. Also called **JSCP.** See also **combatant commander; joint.** (JP 1-0)

Joint Strategic Planning System — The primary means by which the Chairman of the Joint Chiefs of Staff, in consultation with the other members of the Joint Chiefs of Staff and the combatant commanders, carries out the statutory responsibilities to assist the President and Secretary of Defense in providing strategic direction to the Armed Forces; prepares strategic plans; prepares and reviews contingency plans; advises the President and Secretary of Defense on requirements, programs, and budgets; and provides net assessment on the capabilities of the Armed Forces of the United States and its allies as compared with those of their potential adversaries. Also called **JSPS.**

joint suppression of enemy air defenses — A broad term that includes all suppression of enemy air defense activities provided by one component of the joint force in support of another. Also called **J-SEAD.** See also **air defense suppression; suppression of enemy air defenses.** (JP 3-01.4)

joint table of allowances — A document that authorizes end-items of materiel for units operated jointly by two or more military assistance advisory groups and missions. Also called **JTA.**

joint table of distribution — A manpower document that identifies the positions and enumerates the spaces that have been approved for each organizational element of a joint activity for a specific fiscal year (authorization year), and those spaces which have been accepted for planning and programming purposes for the four subsequent fiscal years (program years). Also called **JTD.** See also **joint manpower program.**

Joint Tactical Air Reconnaissance/ Surveillance Mission Report — A preliminary report of information from tactical reconnaissance aircrews rendered by designated debriefing personnel immediately after landing and dispatched prior to compilation of the initial photo interpretation report. It provides a summary of the route conditions, observations, and aircrew actions and identifies sensor products. Also called **MISREP.**

joint tactics, techniques, and procedures — The actions and methods that implement joint doctrine and describe how forces will be employed in joint operations. They are authoritative; as such, joint tactics, techniques, and procedures will be followed except when, in the judgment of the commander, exceptional circumstances dictate otherwise. They will be promulgated by the Chairman of the Joint Chiefs of Staff, in coordination with the combatant commands and Services. Also called **JTTP.** See also **joint doctrine.** (JP 1-01)

joint targeting coordination board — A group formed by the joint force commander to accomplish broad targeting oversight functions that may include but are not limited to coordinating targeting information, providing targeting guidance and priorities, and refining the joint integrated prioritized target list. The board is normally comprised of representatives from the joint force staff, all components, and if required, component subordinate units. Also called **JTCB.** See also **joint integrated prioritized target list.** (JP 3-60)

joint targeting steering group — A group formed by a combatant commander to assist in developing targeting guidance and reconciling competing requests for assets from multiple joint task forces. Also called **JTSG.** See also **group; joint; targeting.** (JP 3-60)

joint target list — A consolidated list of selected targets considered to have military significance in the combatant commander's area of responsibility. Also called **JTL.** See also **joint; target.** (JP 3-60)

joint task force — A joint force that is constituted and so designated by the Secretary of Defense, a combatant commander, a subunified commander, or an existing joint task force commander. Also called **JTF.** (JP 0-2)

Joint Technical Coordinating Group for Munitions Effectiveness — A Joint Staff-level organization tasked to produce generic target vulnerability and weaponeering studies. The special operations working group is a subordinate organization specializing in studies for special operations. Also called **JTCG-ME.** (JP 3-05.5)

joint test publication — A proposed version of a joint doctrine or joint tactics, techniques, and procedures publication that normally contains contentious issues and is nominated for a test publication and evaluation stage. Joint test publications are approved for evaluation by the Director, Operational Plans and Interoperability (J-7), Joint Staff. Publication of a test publication does not constitute Chairman of the Joint Chiefs of Staff approval of the publication. Prior to final approval as joint doctrine, test publications are expected to be further refined based upon evaluation results. Test publications are automatically superseded upon completion of the evaluation and promulgation of the proposed publication. See also **Chairman of the Joint Chiefs of Staff Instruction; joint doctrine; joint**

publication; joint tactics, techniques, and procedures. (JP 1-01)

joint theater missile defense — The integration of joint force capabilities to destroy enemy theater missiles in flight or prior to launch or to otherwise disrupt the enemy's theater missile operations through an appropriate mix of mutually supportive passive missile defense; active missile defense; attack operations; and supporting command, control, communications, computers, and intelligence measures. Enemy theater missiles are those that are aimed at targets outside the continental United States. Also called **JTMD.** (JP 3-01.5)

joint total asset visibility — The capability designed to consolidate source data from a variety of joint and Service automated information systems to provide joint force commanders with visibility over assets in-storage, in-process, and in-transit. Also called **JTAV.** See also **total asset visibility.** (JP 4-01.8)

Joint Transportation Board — Responsible to the Chairman of the Joint Chiefs of Staff, the Joint Transportation Board assures that common-user transportation resources assigned or available to the Department of Defense (DOD) are allocated as to achieve maximum benefit in meeting DOD objectives. Also called **JTB.** See also **common-user transportation.** (JP 4-01)

joint urban operations — All joint operations planned and conducted across the range of military operations on or against objectives on a topographical complex and its adjacent natural terrain where manmade construction or the density of noncombatants are the dominant features. Also called **JUOs.** See also **joint operations.** (JP 3-0)

joint warfighting capabilities assessment — A team of warfighting and functional area experts from the Joint Staff, unified commands, Services, Office of the Secretary of Defense, and Defense agencies tasked by the Joint Requirements Oversight Council with completing assessments and providing military recommendations to improve joint warfighting capabilities. Also called **JWCA.**

Joint Worldwide Intelligence Communications System — The sensitive, compartmented information portion of the Defense Information Systems Network. It incorporates advanced networking technologies that permit point-to-point or multipoint information exchange involving voice, text, graphics, data, and video teleconferencing. Also called **JWICS.** (JP 2-0)

joint zone (air, land, or sea) — An area established for the purpose of permitting friendly surface, air, and subsurface forces to operate simultaneously.

join up — **(*)** To form separate aircraft or groups of aircraft into a specific formation.

jumpmaster — The assigned airborne qualified individual who controls paratroops from the time they enter the aircraft until they exit. See also **stick commander (air transport).**

jump speed — The airspeed at which paratroops can jump with comparative safety from an aircraft.

Intentionally Blank

K

K-day — The basic date for the introduction of a convoy system on any particular convoy lane. See also **D-day; M-day.**

key employee — Any Reservist identified by his or her employer, private or public, as filling a key position.

key facilities list — A register of selected command installations and industrial facilities of primary importance to the support of military operations or military production programs. It is prepared under the policy direction of the Joint Chiefs of Staff.

key point — **(*)** A concentrated site or installation, the destruction or capture of which would seriously affect the war effort or the success of operations.

key position — A civilian position, public or private (designated by the employer and approved by the Secretary concerned), that cannot be vacated during war or national emergency.

keystone publications — Joint doctrine publications that establish the doctrinal foundation for a series of joint publications in the hierarchy of joint publications. Keystone publications are provided for joint personnel, intelligence, operations, logistics, plans, and command, control, communications, and computer systems support series publications. See also **above-the-line publications; below-the-line publications; capstone publication; joint publication**. (JP 1-01)

key terrain — **(*)** Any locality, or area, the seizure or retention of which affords a marked advantage to either combatant. See also **vital ground.**

kill box — A three-dimensional area reference that enables timely, effective coordination and control and facilitates rapid attacks. (JP 3-60)

killed in action — A casualty category applicable to a hostile casualty, other than the victim of a terrorist activity, who is killed outright or who dies as a result of wounds or other injuries before reaching a medical treatment facility. Also called **KIA.** See also **casualty category.**

killing zone — An area in which a commander plans to force the enemy to concentrate so as to be destroyed with conventional weapons or the tactical employment of nuclear weapons.

kill probability — **(*)** A measure of the probability of destroying a target.

kiloton weapon — **(*)** A nuclear weapon, the yield of which is measured in terms of thousands of tons of trinitrotoluene explosive equivalents, producing yields from 1 to 999 kilotons. See also **megaton weapon; nominal weapon; subkiloton weapon.**

kite — **(*)** In naval mine warfare, a device which when towed, submerges and planes at a predetermined level without sideways displacement.

Intentionally Blank

L

land control operations — The employment of ground forces, supported by naval and air forces (as appropriate) to achieve military objectives in vital land areas. Such operations include destruction of opposing ground forces, securing key terrain, protection of vital land lines of communications, and establishment of local military superiority in areas of land operations. See also **sea control operations.**

land forces — Personnel, weapon systems, vehicles, and support elements operating on land to accomplish assigned missions and tasks.

landing aid — **(*)** Any illuminating light, radio beacon, radar device, communicating device, or any system of such devices for aiding aircraft in an approach and landing.

landing approach — **(*)** The continuously changing position of an aircraft in space directed toward effecting a landing on a predetermined area.

landing area — 1. That part of the operational area within which are conducted the landing operations of an amphibious force. It includes the beach, the approaches to the beach, the transport areas, the fire support areas, the airspace occupied by close supporting aircraft, and the land included in the advance inland to the initial objective. 2. (Airborne) The general area used for landing troops and materiel either by airdrop or air landing. This area includes one or more drop zones or landing strips. 3. Any specially prepared or selected surface of land, water, or deck designated or used for takeoff and landing of aircraft. See also **airfield; amphibious force; landing beach; landing force.** (JP 3-02)

landing attack — An attack against enemy defenses by troops landed from ships, aircraft, boats, or amphibious vehicles. See also **assault.**

landing beach — That portion of a shoreline usually required for the landing of a battalion landing team. However, it may also be that portion of a shoreline constituting a tactical locality (such as the shore of a bay) over which a force larger or smaller than a battalion landing team may be landed.

landing craft — **(*)** A craft employed in amphibious operations, specifically designed for carrying troops and their equipment and for beaching, unloading, and retracting. It is also used for resupply operations.

landing craft and amphibious vehicle assignment table — A table showing the assignment of personnel and materiel to each landing craft and amphibious vehicle and the assignment of the landing craft and amphibious vehicles to waves for the ship-to-shore movement.

landing craft availability table — A tabulation of the type and number of landing craft that will be available from each ship of the transport group. The table is the basis for the assignment of landing craft to the boat groups for the ship-to-shore movement.

landing diagram — **(*)** A graphic means of illustrating the plan for the ship-to-shore movement.

landing force — A Marine Corps or Army task organization formed to conduct amphibious operations. The landing force, together with the amphibious task force and other forces, constitute the amphibious force. Also called **LF.** See also **amphibious force; amphibious operation; amphibious task force; task organization.** (JP 3-02)

landing force supplies — Those supplies remaining in assault shipping after initial combat supplies and floating dumps have been unloaded. They are landed selectively in accordance with the requirements of the landing force until the situation ashore permits the inception of general unloading. (JP 3-02.2)

landing force support party — A temporary landing force organization composed of Navy and landing force elements, that facilitates the ship-to-shore movement and provides initial combat support and combat service support to the landing force. The landing force support party is brought into existence by a formal activation order issued by the commander, landing force. Also called **LFSP.** See also **combat service support; combat support; landing force; ship-to-shore movement.** (JP 3-02)

landing group — In amphibious operations, a subordinate task organization of the landing force capable of conducting landing operations, under a single tactical command, against a position or group of positions. (JP 3-02)

landing group commander — In amphibious operations, the officer designated by the commander, landing force as the single tactical commander of a subordinate task organization capable of conducting landing operations against a position or group of positions. See also **amphibious operation; commander, landing force.** (JP 3-02)

landing mat — **(*)** A prefabricated, portable mat so designed that any number of planks (sections) may be rapidly fastened together to form surfacing for emergency runways, landing beaches, etc.

landing plan — 1. In amphibious operations, a collective term referring to all individually prepared naval and landing force documents that, taken together, present in detail all instructions for execution of the ship-to-shore movement. 2. In airlift operations, the sequence, method of delivery, and place of arrival of troops and materiel. (JP 3-17)

landing point — **(*)** A point within a landing site where one helicopter or vertical takeoff and landing aircraft can land. See also **airfield.**

landing roll — **(*)** The movement of an aircraft from touchdown through deceleration to taxi speed or full stop.

landing schedule — In an amphibious operation, a schedule that shows the beach, hour, and priorities of landing of assault units, and which coordinates the movements of landing craft from the transports to the beach in order to execute the scheme of maneuver ashore.

landing sequence table — A document that incorporates the detailed plans for ship-to-shore movement of nonscheduled units. (JP 3-02.2)

landing ship — **(*)** An assault ship which is designed for long sea voyages and for rapid unloading over and on to a beach.

landing ship dock — **(*)** A ship designed to transport and launch loaded amphibious craft and/or amphibian vehicles with their crews and embarked personnel and/or equipment and to render limited docking and repair services to small ships and craft. Also called **LSD.** (JP 3-02.2)

landing signal officer — Officer responsible for the visual control of aircraft in the terminal phase of the approach immediately prior to landing. Also called **LSO.** See also **terminal phase.** (JP 3-04.1)

landing site — **(*)** 1. A site within a landing zone containing one or more landing points. See also airfield. 2. In amphibious operations, a continuous segment of coastline over which troops, equipment and supplies can be landed by surface means.

landing threshold — The beginning of that portion of a runway usable for landing.

landing zone — **(*)** Any specified zone used for the landing of aircraft. Also called **LZ.** See also **airfield.**

landing zone control — See **pathfinder drop zone control.**

landing zone control party — **(*)** Personnel specially trained and equipped to establish and operate communications devices from the ground for traffic control of aircraft/helicopters for a specific landing zone.

landmark — **(*)** A feature, either natural or artificial, that can be accurately determined on the ground from a grid reference.

land mine warfare — See **mine warfare.**

land search — The search of terrain by Earth-bound personnel.

lane marker — **(*)** In land mine warfare, sign used to mark a minefield lane. Lane markers, at the entrance to and exit from the lane, may be referenced to a landmark or intermediate marker. See also **marker; minefield lane.**

lap — **(*)** In naval mine warfare, that section or strip of an area assigned to a single sweeper or formation of sweepers for a run through the area.

lap course — **(*)** In naval mine warfare, the true course desired to be made good during a run along a lap.

lap track — **(*)** In naval mine warfare, the center line of a lap; ideally, the track to be followed by the sweep or detecting gear.

lap turn — **(*)** In naval mine warfare, the maneuver a minesweeper carries out during the period between the completion of one run and the commencement of the run immediately following.

lap width — **(*)** In naval mine warfare, the swept path of the ship or formation divided by the percentage coverage being swept to.

large-lot storage — A quantity of material that will require four or more pallet columns stored to maximum height. Usually accepted as stock stored in carload or greater quantities. See also **storage.**

large-scale map — A map having a scale of 1:75,000 or larger. See also **map.**

large spread — A report by an observer or a spotter to the ship to indicate that the distance between the bursts of a salvo is excessive.

laser — Any device that can produce or amplify optical radiation primarily by the process of controlled stimulated emission. A laser may emit electromagnetic radiation from the ultraviolet portion of the spectrum through the infrared portion. Also, an acronym for "light amplification by stimulated emission of radiation." (JP 3-09.1)

laser footprint — The projection of the laser beam and buffer zone on the ground or target area. The laser footprint may be part of the laser surface danger zone if that footprint lies within the nominal visual hazard distance of the laser. See also **buffer zone; laser.** (JP 3-09.1)

laser guidance unit — A device which incorporates a laser seeker to provide guidance commands to the control system of a missile, projectile or bomb.

laser guided weapon — **(*)** A weapon which uses a seeker to detect laser energy reflected from a laser marked/designated target and through signal processing provides guidance commands to a control system which guides the weapon to the point from which the laser energy is being reflected. Also called **LGW.** See also **laser.** (JP 3-09.1)

laser illuminator — A device for enhancing the illumination in a zone of action by irradiating with a laser beam.

laser intelligence — Technical and geo-location intelligence derived from laser systems; a subcategory of electro-optical intelligence. Also called **LASINT.** See also **electro-optical intelligence; intelligence.** (JP 2-0)

laser linescan system — **(*)** An active airborne imagery recording system which uses a laser as the primary source of illumination to scan the ground beneath the flight path, adding successive across-track lines to the record as the vehicle advances. See also **infrared linescan system.**

laser pulse duration — **(*)** The time during which the laser output pulse power remains continuously above half its maximum value.

laser rangefinder — **(*)** A device which uses laser energy for determining the distance from the device to a place or object.

laser seeker — **(*)** A device based on a direction sensitive receiver which detects the energy reflected from a laser designated target and defines the direction of the target relative to the receiver. See also **laser guided weapon.**

laser spot — The area on a surface illuminated by a laser. See also **laser; spot.** (JP 3-09.1)

laser spot tracker — A device that locks on to the reflected energy from a laser-marked or designated target and defines the direction of the target relative to itself. Also called **LST.**

laser target designating system — **(*)** A system which is used to direct (aim or point) laser energy at a target. The system consists of the laser designator or laser target marker with its display and control components necessary to acquire the target and direct the beam of laser energy thereon.

laser target designator — A device that emits a beam of laser energy which is used to mark a specific place or object. Also called **LTD.** See also **laser; target.** (JP 3-09.1)

laser-target/gun-target angle — The angle between the laser-to-target line and the laser guided weapon/gun-target line at the point where they cross the target. See also **laser; laser guided weapon; target.** (JP 3-09.1)

laser-target line — An imaginary straight line from the laser designator to the target with respect to magnetic north. See also **laser; laser target designator; target.** (JP 3-09.1)

laser target marker — See **laser designator.**

laser target marking system — See **laser target designating system.**

lashing — **(*)** See **tie down. (DOD only)** See **restraint of loads.**

lashing point — See **tie down point.**

late — **(*)** In artillery and naval gunfire support, a report made to the observer or spotter, whenever there is a delay in reporting "shot" by coupling a time in seconds with the report.

lateral gain — **(*)** The amount of new ground covered laterally by successive photographic runs over an area.

lateral route — **(*)** A route generally parallel to the forward edge of the battle area, which crosses, or feeds into, axial routes. See also **route.**

lateral spread — A technique used to place the mean point of impact of two or more units 100 meters apart on a line perpendicular to the gun-target line.

lateral tell — See **track telling.**

latest arrival date — A day, relative to C-Day, that is specified by the supported combatant commander as the latest date when a unit, a resupply shipment, or replacement personnel can arrive at the port of debarkation and support the concept of operations. Used with the earliest arrival date, it defines a delivery window for transportation planning. Also called **LAD.**

late time — See **span of detonation (atomic demolition munition employment), Part 3.**

latitude band — **(*)** Any latitudinal strip, designated by accepted units of linear or angular measurement, which circumscribes the Earth. Also called **latitudinal band.**

lattice — **(*)** A network of intersecting positional lines printed on a map or chart from which a fix may be obtained.

launch pad — **(*)** A concrete or other hard surface area on which a missile launcher is positioned.

launch time — The time at which an aircraft or missile is scheduled to be airborne. See also **airborne order.**

launch under attack — Execution by National Command Authorities of Single Integrated Operational Plan forces subsequent to tactical warning of strategic nuclear attack against the United States and prior to first impact. Also called **LUA.**

launch window — The earliest and latest time a rocket may launch.

laundering — In counterdrug operations, the process of transforming drug money into a more manageable form while concealing its illicit origin. Foreign bank accounts and dummy corporations are used as shelters. See also **counterdrug operations.** (JP 3-07.4)

law enforcement agency — Any of a number of agencies (outside the Department of Defense) chartered and empowered to enforce US laws in the following jurisdictions: The United States, a state (or political subdivision) of the United States, a territory or possession (or political subdivision) of the United States, or within the borders of a host nation. Also called **LEA.** (JP 3-07.4)

law of armed conflict — See **law of war.**

law of war — That part of international law that regulates the conduct of armed hostilities. Also called **the law of armed conflict.** See also **rules of engagement.**

lay — 1. Direct or adjust the aim of a weapon. 2. Setting of a weapon for a given range, a given direction, or both. 3. To drop one or more aerial bombs or aerial mines onto the surface from an aircraft. 4. To spread a smoke screen on the ground from an aircraft. 5. To calculate or project a course. 6. To lay on: a. to execute a bomber strike; b. to set up a mission.

laydown bombing — (*) A very low level bombing technique wherein delay fuzes and/or devices are used to allow the attacker to escape the effects of the bomb.

layer depth — The depth from the surface of the sea to the point above the first major negative thermocline at which sound velocity is maximum.

lay leader or lay reader — A volunteer ("lay leader" in Army and Air Force; "lay reader" in Navy and Marine Corps) appointed by the commanding officer and supervised and trained by the command chaplain to serve for a period of time to meet the needs of a particular religious faith group when their military chaplains are not available. The lay leader or lay reader may conduct services, but may not exercise any other activities usually reserved for the ordained clergy. See also **command chaplain; command chaplain of the combatant command; religious ministry support; religious ministry support plan; religious ministry support team; Service component command chaplain.** (JP 1-05)

lay reader — See **lay leader or lay reader.** (JP 1-05)

lay reference number — (*) In naval mine warfare, a number allocated to an individual mine by the minefield planning authority to provide a simple means of referring to it.

lead agency — Designated among US Government agencies to coordinate the interagency oversight of the day-to-day conduct of an ongoing operation. The lead agency is to chair the interagency working group established to coordinate policy related to a particular operation. The lead agency determines the agenda, ensures cohesion among the agencies, and is responsible for implementing decisions. (JP 3-08)

lead agent — Individual Services, combatant commands, or Joint Staff directorates may be assigned as lead agents for developing and maintaining joint doctrine, joint tactics, techniques, and procedures (JTTP) publications. The lead agent is responsible for developing, coordinating, reviewing, and maintaining an assigned doctrine or JTTP. Also called **LA.** See also **coordinating review authority; joint doctrine; joint publication; joint tactics, techniques, and procedures; joint test publication; primary review authority**. (JP 1-01)

lead aircraft — 1. The airborne aircraft designated to exercise command of other aircraft within the flight. 2. An aircraft in the van of two or more aircraft.

lead mobility wing — An Air Mobility Command unit designated to provide an on-call 32-member cross-functional initial response team (IRT) for short-notice deployment in response to humanitarian crises. When requested by a supported geographic combatant commander, this IRT arrives at an airfield in the disaster area to provide mobility expertise and leadership, assess the requirements for follow-on relief forces, and establish a reception base to serve as a conduit for relief supplies or the repatriation of noncombatants. The IRT is attached to the joint task force established by the supported geographic combatant commander. Also called **LMW.** See also **Air Mobility Command; mobility; wing.** (JP 3-57)

lead nation — One nation assumes the responsibility for procuring and providing a broad spectrum of logistic support for all or a part of the multinational force and/or headquarters. Compensation and/or reimbursement will then be subject to agreements between the parties involved. The lead nation may also assume the responsibility to coordinate logistics of the other nations within its functional and regional area of responsibility. See also **logistic support; multinational force.** (JP 4-0)

lead Service or agency for common-user logistics — A Service component or Department of Defense agency that is responsible for execution of common-user item or service support in a specific combatant command or multinational operation as defined in the combatant or subordinate joint force commander's operation plan, operation order, and/or directives. See also **common-user logistics.** (JP 4-07)

leapfrog — **(*)** Form of movement in which like supporting elements are moved successively through or by one another along the axis of movement of supported forces.

leaver — **(*)** A merchant ship which breaks off from a convoy to proceed to a different destination and becomes independent. Also called **convoy leaver.** See also **leaver convoy; leaver section.**

leaver convoy — **(*)** A convoy which has broken off from the main convoy and is proceeding to a different destination. See also **leaver; leaver section.**

leaver section — **(*)** A group of ships forming part of the main convoy which will subsequently break off to become leavers or a leaver convoy. See also **leaver; leaver convoy.**

left (or right) — **(*)** 1. Terms used to establish the relative position of a body of troops. The person using the terms "left" or "right" is assumed to be facing in the direction of the enemy regardless of whether the troops are advancing towards or withdrawing from the enemy. 2. Correction used in adjusting fire to indicate that a lateral shift of the mean point of impact perpendicular to the reference line or spotting line is desired.

left (right) bank — That bank of a stream or river on the left (right) of the observer when facing in the direction of flow or downstream.

letter of assist — A contractual document issued by the United Nations (UN) to a government authorizing it to provide goods or services to a peacekeeping operation; the UN agrees either to purchase the goods or services or authorizes the government to supply them subject to reimbursement by the UN. A letter of assist typically details specifically what is to be provided by the contributing government and establishes a funding limit that cannot be exceeded. Also called **LOA.** See also **peacekeeping.** (JP 1-06)

level of detail — Within the current joint planning and execution systems, movement characteristics are described at five distinct levels of detail. a. **level I**-aggregated level — Expressed as total number of passengers and total short tons, total measurement tons, total square feet, and/or total hundreds of barrels by unit line number (ULN), cargo increment number (CIN), and personnel increment number (PIN). b. **level II**-summary level — Expressed as total number of passengers by ULN and PIN and short tons, measurement tons (including barrels), total square feet of bulk, oversize, outsize, and non-air-transportable cargo by ULN and CIN. c. **level III**-detail by cargo category — Expressed as total number of passengers by ULN and PIN and short tons and/or measurement tons (including barrels) as well as total square feet of cargo as identified by the ULN or CIN three-position cargo category code. d. **level IV**-detail expressed as number of passengers and individual dimensional data (expressed in length, width, and height in number of inches) of cargo by equipment type by ULN. e. **level V**-detail by priority of shipment — Expressed as total number of passengers by Service specialty code in deployment sequence by ULN, individual weight (in pounds), and dimensional data (expressed in length, width, and height in number of inches) of equipment in deployment sequence by ULN.

level-of-effort munitions — **(*)** In stockpile planning, munitions stocked on the basis of expected daily expenditure rate, the number of combat days, and the attrition rate assumed, to counter targets the number of which is unknown. See also **threat-oriented munitions.**

level of effort-oriented items — Items for which requirements computations are based on such factors as equipment and personnel density and time and rate of use. See also **combination mission/level of effort-oriented items; mission-oriented items.**

level of supply — **(*)** The quantity of supplies or materiel authorized or directed to be held in anticipation of future demands. See also **operating level of supply; order and shipping time; procurement lead time; requisitioning objective; safety level of supply; stockage objective.**

leveraging — In information operations, the effective use of information, information systems, and technology to increase the means and synergy in accomplishing information operations strategy. See also **information; information operations; information system; operation.** (JP 3-13)

L-hour — See **times.**

liaison — That contact or intercommunication maintained between elements of military forces or other agencies to ensure mutual understanding and unity of purpose and action. (JP 3-08)

liberated territory — **(*)** Any area, domestic, neutral, or friendly, which, having been occupied by an enemy, is retaken by friendly forces.

licensed production — A direct commercial arrangement between a US company and a foreign government, international organization, or foreign company, providing for the transfer of production information which enables the foreign government, international organization, or commercial producer to manufacture, in whole or in part, an item of US defense equipment. A typical license production arrangement would include the functions of production engineering, controlling, quality assurance and determining of resource requirements. It may or may not include design engineering information and critical materials production and design information. A licensed production arrangement is accomplished under the provisions of a manufacturing license agreement per the US International Traffic in Arms Regulation.

life cycle — The total phases through which an item passes from the time it is initially developed until the time it is either consumed in use or disposed of as being excess to all known materiel requirements.

lifeguard submarine — **(*)** A submarine employed for rescue in an area which cannot be adequately covered by air or surface rescue facilities because of enemy opposition, distance from friendly bases, or other reasons. It is stationed near the objective and sometimes along the route to be flown by the strike aircraft.

life support equipment — Equipment designed to sustain aircrew members and passengers throughout the flight environment, optimizing their mission effectiveness and affording a means of safe and reliable escape, descent, survival, and recovery in emergency situations.

light artillery — See **field artillery.**

light damage — See **nuclear damage, Part 1.**

lightening — **(*)** The operation (normally carried out at anchor) of transferring crude oil cargo from a large tanker to a smaller tanker, so reducing the draft of the larger tanker to enable it to enter port.

lighterage — A small craft designed to transport cargo or personnel from ship to shore. Lighterage includes amphibians, landing craft, discharge lighters, causeways, and barges. (JP 3-02)

light filter — **(*)** An optical element such as a sheet of glass, gelatine, or plastic dyed in a specific manner to absorb selectively light of certain colors.

light line — **(*)** A designated line forward of which vehicles are required to use black-out lights at night.

lightweight amphibious container handler — A United States Marine Corps piece of equipment usually maneuvered by a bulldozer and used to retrieve 20-foot equivalent containers from landing craft in the surf and place them on flatbed truck trailers. See also **container.** (JP 4-01.6)

limited production-type item — An item under development, commercially available or available from other Government agencies, for which an urgent operational requirement exists and for which no other existing item is suitable. Such an item appears to fulfill an approved materiel requirement or other Military Department-approved requirements and to be promising enough operationally to warrant initiating procurement and/or production for service issue prior to completion of development and/or test or adoption as a standard item.

limited standard item — An item of supply determined by standardization action as authorized for procurement only to support in-service military materiel requirements.

limited war — Armed conflict short of general war, exclusive of incidents, involving the overt engagement of the military forces of two or more nations.

limiting factor — A factor or condition that, either temporarily or permanently, impedes mission accomplishment. Illustrative examples are transportation network deficiencies, lack of in-place facilities, malpositioned forces or materiel, extreme climatic conditions, distance, transit or overflight rights, political conditions, etc.

limit of fire — **(*)** 1. The boundary marking off the area on which gunfire can be delivered. 2. Safe angular limits for firing at aerial targets.

linear scale — See **graphic scale; scale.**

line of communications — A route, either land, water, and/or air, that connects an operating military force with a base of operations and along which supplies and military forces move. Also called **LOC.** See also **base of operations; route.**

line of demarcation — A line defining the boundary of a buffer zone or area of limitation. A line of demarcation may also be used to define the forward limits of disputing or belligerent forces after each phase of disengagement or withdrawal has been completed. See also **area of limitation; buffer zone; disengagement; peace operations.** (JP 3-07.3)

line of departure — **(*)** 1. In land warfare, a line designated to coordinate the departure of attack elements. 2. In amphibious warfare, a suitably marked offshore coordinating line to assist assault craft to land on designated beaches at scheduled times. Also called **LD.**

line overlap — See **overlap, Part 1.**

line-route map — A map or overlay for signal communications operations that shows the actual routes and types of construction of wire circuits in the field. It also gives the locations of switchboards and telegraph stations. See also **map.**

line search — **(*)** Reconnaissance along a specific line of communications, such as a road, railway or waterway, to detect fleeting targets and activities in general.

lines of operations — Lines that define the directional orientation of the force in time and space in relation to the enemy. They connect the force with its base of operations and its objectives. (JP 5-0)

link — **(*)** 1. In communications, a general term used to indicate the existence of communications facilities between two points. 2. A maritime route, other than a coastal or transit route, which links any two or more routes.

link encryption — The application of online crypto-operation to a link of a communications system so that all information passing over the link is encrypted in its entirety.

link-lift vehicle — The conveyance, together with its operating personnel, used to satisfy a movement requirement between nodes.

link-route segments — Route segments that connect nodes wherein link-lift vehicles perform the movement function.

liquid explosive — **(*)** Explosive which is fluid at normal temperatures.

liquid propellant — Any liquid combustible fed to the combustion chamber of a rocket engine.

listening watch — A continuous receiver watch established for the reception of traffic addressed to, or of interest to, the unit maintaining the watch, with complete log optional.

list of targets — A tabulation of confirmed or suspect targets maintained by any echelon for informational and fire support planning purposes. See also **target list.**

litter — A basket or frame utilized for the transport of injured persons.

litter patient — A patient requiring litter accommodations while in transit.

load — **(*)** The total weight of passengers and/or freight carried on board a ship, aircraft, train, road vehicle, or other means of conveyance. See also **airlift capability; airlift requirement; allowable load.**

load control group — **(*)** Personnel who are concerned with organization and control of loading within the pick-up zone.

loading — **(*)** The process of putting personnel, materiel, supplies and other freight on board ships, aircraft, trains, road vehicles, or other means of conveyance. See also **embarkation.**

loading chart (aircraft) — Any one of a series of charts carried in an aircraft that shows the proper location for loads to be transported and that pertains to check-lists, balance records, and clearances for weight and balance.

loading (ordnance) — An operation that installs airborne weapons and stores on or in an aircraft and may include fuzing of bombs and stray voltage checks. See also **loading; ordnance.** (JP 3-04.1)

loading plan — **(*)** All of the individually prepared documents which, taken together, present in detail all instructions for the arrangement of personnel, and the loading of equipment for one or more units or other special grouping of personnel or material moving by highway, water, rail, or air transportation. See also **ocean manifest.**

loading point — **(*)** A point where one aircraft can be loaded or unloaded.

loading site — **(*)** An area containing a number of loading points.

loading time — In airlift operations, a specified time, established jointly by the airlift and airborne commanders concerned, when aircraft and loads are available and loading is to begin. (JP 3-17)

loadmaster — An Air Force technician qualified to plan loads, to operate auxiliary materials handling equipment, and to supervise loading and unloading of aircraft. (JP 3-17)

load signal — In evasion and recovery operations, a visual signal displayed in a covert manner to indicate the presence of an individual or object at a given location. See also **evasion; evasion and recovery; recovery operations; signal.** (JP 3-50.3)

load spreader — **(*)** Material used to distribute the weight of a load over a given area to avoid exceeding designed stress.

localizer — **(*)** A directional radio beacon which provides to an aircraft an indication of its lateral position relative to a predetermined final approach course. See also **instrument landing system.**

local mean time — **(*)** The time interval elapsed since the mean sun's transit of the observer's anti-meridian.

local procurement — The process of obtaining personnel, services, supplies, and equipment from local or indigenous sources.

local purchase — The function of acquiring a decentralized item of supply from sources outside the Department of Defense.

lock on — Signifies that a tracking or target-seeking system is continuously and automatically tracking a target in one or more coordinates (e.g., range, bearing, elevation).

lodgment — A designated area in a hostile or potentially hostile territory that, when seized and held, makes the continuous landing of troops and materiel possible and provides maneuver space for subsequent operations. See also **hostile.** (JP 3-18)

lodgment area — See **airhead, Part 1; beachhead.**

loft bombing — A method of bombing in which the delivery plane approaches the target at a very low altitude, makes a definite pullup at a given point, releases the bomb at a predetermined point during the pullup, and tosses the bomb onto the target. See also **toss bombing.**

logistic assessment — An evaluation of: a. the logistic support required to support particular military operations in a theater, country, or area; and b. the actual and/or potential logistic support available for the conduct of military operations either within the theater, country, or area, or located elsewhere.

logistic estimate of the situation — An appraisal resulting from an orderly examination of the logistic factors influencing contemplated courses of action in order to provide conclusions concerning the degree and manner of that influence.

logistic implications test — An analysis of the major logistic aspects of a joint strategic war plan and the consideration of the logistic implications resultant therefrom as they may limit the acceptability of the plan. The logistic analysis and consideration are conducted concurrently with the development of the strategic plan. The objective is to establish whether the logistic requirements generated by the plan are in balance with availabilities, and to set forth those logistic implications that should be weighed by the Joint Chiefs of Staff in their consideration of the plan. See also **feasibility test.**

logistic and movement control center — A center organized from service support elements (or the supporting establishment) in the geographic proximity of the marshaling units. It is tasked by the force movement control center to provide organic and commercial transportation, transportation scheduling, materials handling equipment, and all other logistic support required by parent commands during marshaling and embarkation. Also called **LMCC.** See also **control center; embarkation; force movement; marshalling.** (JP 4-01.8)

logistic marking and reading symbology — A system designed to improve the flow of cargo through the seaport of embarkation and debarkation using bar code technology. See also **logistics.** (JP 4-01.6)

logistic routes — See **line of communications.**

logistics — The science of planning and carrying out the movement and maintenance of forces. In its most comprehensive sense, those aspects of military operations which deal with: a. design and development, acquisition, storage, movement, distribution, maintenance, evacuation, and disposition of materiel; b. movement, evacuation, and hospitalization of personnel; c. acquisition or construction, maintenance, operation, and disposition of facilities; and d. acquisition or furnishing of services.

logistics over-the-shore operation area — That geographic area required to successfully conduct a logistics over-the-shore operation. Also called **LOA.** See also **logistics over-the-shore operations.** (JP 4-01.6)

logistics over-the-shore operations — The loading and unloading of ships without the benefit of deep draft-capable, fixed port facilities in friendly or nondefended territory and, in time of war, during phases of theater development in which there is no opposition by the enemy; or as a means of moving forces closer to tactical assembly areas dependent on threat force capabilities. Also called **LOTS operations.** See also **joint logistics over-the-shore operations.** (JP 4-01.8)

logistic sourcing — The identification of the origin and determination of the availability of the time-phased force and deployment data nonunit logistic requirements.

logistic support — Logistic support encompasses the logistic services, materiel, and transportation required to support the continental United States-based and worldwide deployed forces.

logistic support (medical) — Medical care, treatment, hospitalization, and evacuation as well as the furnishing of medical services, supplies, materiel, and adjuncts thereto.

long-range bomber aircraft — A bomber designed for a tactical operating radius over 2,500 nautical miles at design gross weight and design bomb load.

long-range transport aircraft — See **transport aircraft.**

long ton — 2,240 pounds. Also called **LT; L/T; or LTON.** (JP 4-01.7)

look — **(*)** In mine warfare, a period during which a mine circuit is receptive of an influence.

loran — **(*)** A long-range radio navigation position fixing system using the time difference of reception of pulse type transmissions from two or more fixed stations. This term is derived from the words long-range electronic navigation.

lot — Specifically, a quantity of material all of which was manufactured under identical conditions and assigned an identifying lot number.

low airburst — **(*)** The fallout safe height of burst for a nuclear weapon which maximizes damage to or casualties on surface targets. See also **types of burst.**

low-altitude missile engagement zone — See **weapon engagement zone.** (JP 3-52)

low-altitude parachute extraction system — A low-level, self-contained system capable of delivering heavy loads into an area where air landing is not feasible from an optimum aircraft wheel altitude of 5 to 10 feet above ground level. One or more platforms may be dropped. Also called **LAPES.** (JP 3-17)

low angle — **(*)** In artillery and naval gunfire support, an order or request to obtain low angle fire.

low angle fire — **(*)** Fire delivered at angles of elevation below the elevation that corresponds to the maximum range of the gun and ammunition concerned.

low angle loft bombing — **(*)** Type of loft bombing of free fall bombs wherein weapon release occurs at an angle less than 35 degrees above the horizontal. See also **loft bombing.**

low dollar value item — An item that normally requires considerably less management effort than those in the other management intensity groupings.

low level flight — See **terrain flight.**

low level transit route — **(*)** A temporary corridor of defined dimensions established in the forward area to minimize the risk to friendly aircraft from friendly air defenses or surface forces. Also called **LLTR.**

low oblique — See **oblique air photograph.**

low velocity drop — **(*)** A drop procedure in which the drop velocity does not exceed 30 feet per second.

low visibility operations — Sensitive operations wherein the political-military restrictions inherent in covert and clandestine operations are either not necessary or not feasible; actions are taken as required to limit exposure of those involved and/or their activities. Execution of these operations is undertaken with the knowledge that the action and/or sponsorship of the operation may preclude plausible denial by the initiating power.

M

mach number — The ratio of the velocity of a body to that of sound in the surrounding medium.

magnetic bearing — See **bearing.**

magnetic circuit — See **magnetic mine.**

magnetic compass — **(*)** An instrument containing a freely suspended magnetic element which displays the direction of the horizontal component of the Earth's magnetic field at the point of observation.

magnetic declination — **(*)** The angle between the magnetic and geographical meridians at any place, expressed in degrees east or west to indicate the direction of magnetic north from true north. In nautical and aeronautical navigation, the term magnetic variation is used instead of magnetic declination and the angle is termed variation of the compass or magnetic variation. Magnetic declination is not otherwise synonymous with magnetic variation which refers to regular or irregular change with time of the magnetic declination, dip, or intensity. See also **magnetic variation.**

magnetic equator — **(*)** A line drawn on a map or chart connecting all points at which the magnetic inclination (dip) is zero for a specified epoch. Also called **aclinic line.**

magnetic mine — **(*)** A mine which responds to the magnetic field of a target.

magnetic minehunting — The process of using magnetic detectors to determine the presence of mines or minelike objects.

magnetic north — **(*)** The direction indicated by the north seeking pole of a freely suspended magnetic needle, influenced only by the Earth's magnetic field.

magnetic tape — A tape or ribbon of any material impregnated or coated with magnetic or other material on which information may be placed in the form of magnetically polarized spots.

magnetic variation — **(*)** 1. In navigation, at a given place and time, the horizontal angle between the true north and magnetic north measured east or west according to whether magnetic north lies east or west of true north. See also **magnetic declination.** 2. In cartography, the annual change in direction of the horizontal component of the Earth's magnetic field.

mail embargo — A temporary shutdown or redirection of mail flow to or from a specific location. (JP 1-0)

main airfield — **(*)** An airfield planned for permanent occupation in peacetime, also suitable for use in wartime and having sufficient operational facilities for full use of its combat potential. See also **airfield; departure airfield; diversion airfield; redeployment airfield.**

main armament — The request of the observer or spotter to obtain fire from the largest guns installed on the fire support ship.

main attack — **(*)** The principal attack or effort into which the commander throws the full weight of the offensive power at his disposal. An attack directed against the chief objective of the campaign, major operation, or battle.

main battle area — That portion of the battlefield in which the decisive battle is fought to defeat the enemy. For any particular command, the main battle area extends rearward from the forward edge of the battle area to the rear boundary of the command's subordinate units.

main convoy — (*) The convoy as a whole which sails from the convoy assembly port/anchorage to its destination. It may be supplemented by joiners or joiner convoys, and leavers or leaver convoys may break off.

main deck — The highest deck running the full length of a vessel (except for an aircraft carrier's hanger deck). See also **watercraft.** (JP 4-01.6)

main detonating line — (*) In demolition, a line of detonating cord used to transmit the detonation wave to two or more branches.

main line of resistance — A line at the forward edge of the battle position, designated for the purpose of coordinating the fire of all units and supporting weapons, including air and naval gunfire. It defines the forward limits of a series of mutually supporting defensive areas, but it does not include the areas occupied or used by covering or screening forces.

main operations base — In special operations, a base established by a joint force special operations component commander or a subordinate special operations component commander in friendly territory to provide sustained command and control, administration, and logistical support to special operations activities in designated areas. Also called **MOB.** See also **advanced operations base; forward operations base.** (JP 3-05.3)

main supply route — The route or routes designated within an operational area upon which the bulk of traffic flows in support of military operations. Also called **MSR.**

maintain — When used in the context of deliberate planning, the directed command will keep the referenced operation plan, operation plan in concept format, or concept summary, and any associated Joint Operation Planning and Execution System (JOPES) automated data processing files active in accordance with applicable tasking documents describing the type and level of update or maintenance to be performed. General guidance is contained in JOPES, Volumes I and II. See also **archive; retain.**

maintenance area — A general locality in which are grouped a number of maintenance activities for the purpose of retaining or restoring materiel to a serviceable condition.

maintenance engineering — The application of techniques, engineering skills, and effort, organized to ensure that the design and development of weapon systems and equipment provide adequately for their effective and economical maintenance.

maintenance (materiel) — 1. All action taken to retain materiel in a serviceable condition or to restore it to serviceability. It includes inspection, testing, servicing, classification as to serviceability, repair, rebuilding, and reclamation. 2. All supply and repair action taken to keep a force in condition to carry out its mission. 3. The routine recurring work required to keep a facility (plant, building, structure, ground facility, utility system, or other real property) in such condition that it may be continuously used at its original or designed capacity and efficiency for its intended purpose.

maintenance status — 1. A nonoperating condition, deliberately imposed, with adequate personnel to maintain and preserve installations, materiel, and facilities in such a condition that they may be readily restored to operable condition in a minimum time by the assignment of additional personnel and without extensive repair or overhaul. 2. That condition of materiel that is in fact, or is administratively classified as, unserviceable, pending completion of required servicing or repairs. 3. A condition of materiel readiness that reports the level of operational readiness for a piece of equipment.

major combat element — Those organizations and units described in the Joint Strategic Capabilities Plan that directly produce combat capability. The size of the element varies by Service, force capability, and the total number of such elements available. Examples are Army divisions and separate brigades, Air Force squadrons, Navy task forces, and Marine expeditionary forces. See also **major force.**

major disaster — See **domestic emergencies.**

major fleet — A principal, permanent subdivision of the operating forces of the Navy with certain supporting shore activities. Presently there are two such fleets: the Pacific Fleet and the Atlantic Fleet. See also **fleet.**

major force — A military organization comprised of major combat elements and associated combat support, combat service support, and sustainment increments. The major force is capable of sustained military operations in response to plan employment requirements. See also **major combat element.**

major nuclear power — **(*)** Any nation that possesses a nuclear striking force capable of posing a serious threat to every other nation.

major operation — A series of tactical actions (battles, engagements, strikes) conducted by various combat forces of a single or several Services, coordinated in time and place, to accomplish operational and, sometimes, strategic objectives in an operational area. These actions are conducted simultaneously or sequentially in accordance with a common plan and are controlled by a single commander. See also **operation.** (JP 3-0)

major weapon system — One of a limited number of systems or subsystems that for reasons of military urgency, criticality, or resource requirements, is determined by the Department of Defense as being vital to the national interest.

make safe — One or more actions necessary to prevent or interrupt complete function of the system (traditionally synonymous with "dearm," "disarm," and "disable"). Among the necessary actions are: (1) install (safety devices such as pins or locks); (2) disconnect (hoses, linkages, batteries); (3) bleed (accumulators, reservoirs); (4) remove (explosive devices such as initiators, fuzes, detonators); and (5) intervene (as in welding, lockwiring).

management and control system (mobility) — Those elements of organizations and/or activities that are part of, or are closely related to, the mobility system, and which authorize requirements to be moved, to obtain and allocate lift resources, or to direct the operation of linklift vehicles.

maneuver — 1. A movement to place ships, aircraft, or land forces in a position of advantage over the enemy. 2. A tactical exercise carried out at sea, in the air, on the ground, or on a map in imitation of war. 3. The operation of a ship, aircraft, or vehicle, to cause it to perform desired movements. 4. Employment of forces in the battlespace through movement in combination with fires to achieve a position of advantage in respect to the enemy in order to accomplish the mission. See also **mission; operation.** (JP 3-0)

maneuverable reentry vehicle — A reentry vehicle capable of performing preplanned flight maneuvers during the reentry phase. See also **multiple independently targetable reentry vehicle; multiple reentry vehicle; reentry vehicle.**

manifest — A document specifying in detail the passengers or items carried for a specific destination.

manipulative electromagnetic deception — See **electromagnetic deception.**

man portable — Capable of being carried by one man. Specifically, the term may be used to qualify: 1. Items designed to be carried as an integral part of individual, crew-served, or team equipment of the dismounted soldier in conjunction with assigned duties. Upper weight limit: approximately 14 kilograms (31 pounds.) 2. In land warfare, equipment which can be carried by one man over long distance without serious degradation of the performance of normal duties.

manpower — See **manpower requirements; manpower resources.**

manpower management — **(*)** The means of manpower control to ensure the most efficient and economical use of available manpower.

manpower management survey — **(*)** Systematic evaluation of a functional area, utilizing expert knowledge, manpower scaling guides, experience, and other practical considerations in determining the validity and managerial efficiency of the function's present or proposed manpower establishment.

manpower requirements — Human resources needed to accomplish specified work loads of organizations.

manpower resources — Human resources available to the Services that can be applied against manpower requirements.

man space — The space and weight factor used to determine the combat capacity of vehicles, craft, and transport aircraft, based on the requirements of one person with individual equipment. The person is assumed to weigh between 222-250 pounds and to occupy 13.5 cubic feet of space. See also **boat space.**

man transportable — Items that are usually transported on wheeled, tracked, or air vehicles, but have integral provisions to allow periodic handling by one or more individuals for limited distances (100-500 meters). Upper weight limit: approximately 65 pounds per individual.

map — **(*)** A graphic representation, usually on a plane surface and at an established scale, of natural or artificial features on the surface of a part or the whole of the Earth or other planetary body. The features are positioned relative to a coordinate reference system. See also **administrative map; chart index; chart series; chart sheet; controlled map; general map; large-scale map; line-route map; map chart; map index; map series; map sheet; medium-scale map; operation map; planimetric map; situation map; small-scale map; strategic map; tactical map; topographic map; traffic circulation map.**

map chart — A representation of a land-sea area, using the characteristics of a map to represent the land area and the characteristics of a chart to represent the sea area, with such special characteristics as to make the map-chart most useful in military operations, particularly amphibious operations. See also **map.**

map convergence — (*) The angle at which one meridian is inclined to another on a map or chart. See also **convergence.**

map exercise — An exercise in which a series of military situations is stated and solved on a map.

map index — (*) Graphic key primarily designed to give the relationship between sheets of a series, their coverage, availability, and further information on the series. See also **map.**

mapping camera — See **air cartographic camera.**

map reference — (*) A means of identifying a point on the surface of the Earth by relating it to information appearing on a map, generally the graticule or grid.

map reference code — (*) A code used primarily for encoding grid coordinates and other information pertaining to maps. This code may be used for other purposes where the encryption of numerals is required.

map series — (*) A group of maps or charts usually having the same scale and cartographic specifications, and with each sheet appropriately identified by producing agency as belonging to the same series.

map sheet — (*) An individual map or chart either complete in itself or part of a series. See also **map.**

margin — (*) In cartography, the area of a map or chart lying outside the border.

marginal data — (*) All explanatory information given in the margin of a map or chart which clarifies, defines, illustrates, and/or supplements the graphic portion of the sheet.

marginal information — See **marginal data.**

marginal weather — Weather that is sufficiently adverse to a military operation so as to require the imposition of procedural limitations. See also **adverse weather.**

Marine air command and control system — A system that provides the aviation combat element commander with the means to command, coordinate, and control all air operations within an assigned sector and to coordinate air operations with other Services. It is composed of command and control agencies with communications-electronics equipment that incorporates a capability from manual through semiautomatic control. Also called **MACCS.** See also **direct air support center; tactical air operations center.** (JP 3-09.3)

Marine air-ground task force — The Marine Corps principal organization for all missions across the range of military operations, composed of forces task-organized under a single commander capable of responding rapidly to a contingency anywhere in the world. The types of forces in the Marine air-ground task force (MAGTF) are functionally grouped into four core elements: a command element, an aviation combat element, a ground combat element, and a combat service support element. The four core elements are categories of forces, not formal commands. The basic structure of the MAGTF never varies, though the number, size, and type of Marine Corps units comprising each of its four elements will always be mission dependent. The flexibility of the organizational structure allows for one or more subordinate MAGTFs to be assigned. Also called **MAGTF.** See also **aviation combat element; combat service support element; command element; ground combat element; Marine expeditionary force; Marine expeditionary force (forward); Marine expeditionary unit; special purpose Marine air-ground task force; task force.**

Marine base — A base for support of Marine ground forces, consisting of activities or facilities for which the Marine Corps has operating responsibilities, together with interior lines of communications and the minimum surrounding area necessary for local security. (Normally, not greater than an area of 20 square miles.) See also **base complex.**

Marine division and wing team — A Marine Corps air-ground team consisting of one division and one aircraft wing, together with their normal reinforcements.

marine environment — The oceans, seas, bays, estuaries, and other major water bodies, including their surface interface and interaction, with the atmosphere and with the land seaward of the mean high water mark.

Marine expeditionary brigade — A Marine air-ground task force that is constructed around a reinforced infantry regiment, a composite Marine aircraft group, and a brigade service support group. The Marine expeditionary brigade (MEB), commanded by a general officer, is task-organized to meet the requirements of a specific situation. It can function as part of a joint task force, as the lead echelon of the Marine expeditionary force (MEF), or alone. It varies in size and composition, and is larger than a Marine expeditionary unit but smaller than a MEF. The MEB is capable of conducting missions across the full range of military operations. Also called **MEB.** See also **brigade; Marine air-ground task force; Marine expeditionary force.** (JP 3-18)

Marine expeditionary force — The largest Marine air-ground task force (MAGTF) and the Marine Corps principal warfighting organization, particularly for larger crises or contingencies. It is task-organized around a permanent command element and normally contains one or more Marine divisions, Marine aircraft wings, and Marine force service support groups. The Marine expeditionary force is capable of missions across the range of military operations, including amphibious assault and sustained operations ashore in any environment. It can operate from a sea base, a land base, or both. Also called **MEF.** See also **aviation combat element; combat service support element; command element; ground combat element; Marine air-ground task force; Marine expeditionary force (forward); Marine expeditionary unit; special purpose Marine air-ground task force; task force.**

Marine expeditionary force (forward) — A designated lead echelon of a Marine expeditionary force (MEF), task-organized to meet the requirements of a specific situation. A Marine expeditionary force (forward) varies in size and composition, and may be commanded by the MEF commander personally or by another designated commander. It may be tasked with preparing for the subsequent arrival of the rest of the MEF/joint/multinational forces, and/or the conduct of other specified tasks, at the discretion of the MEF commander. A Marine expeditionary force (forward) may also be a stand-alone Marine air-ground task force (MAGTF), task-organized for a mission in which an MEF is not required. Also called **MEF (FWD).** See also **aviation combat element; combat service support element; command element; ground combat element; Marine air-ground task force; Marine expeditionary force; Marine expeditionary unit; Marine expeditionary unit (special operations capable); special purpose Marine air-ground task force; task force.**

Marine expeditionary unit — A Marine air-ground task force (MAGTF) that is constructed around an infantry battalion reinforced, a helicopter squadron reinforced, and a task-organized combat service support element. It normally fulfills Marine Corps forward sea-based deployment requirements. The Marine expeditionary unit provides an immediate reaction capability for crisis response and is capable of limited combat operations. Also called **MEU.** See also **aviation combat element; combat service support element; command element; ground combat element; Marine air-ground task force; Marine expeditionary force; Marine expeditionary force (forward); Marine expeditionary unit (special operations capable); special purpose Marine air-ground task force; task force.**

Marine expeditionary unit (special operations capable) — The Marine Corps standard, forward-deployed, sea-based expeditionary organization. The Marine expeditionary unit (special operations capable) (MEU[SOC]) is a Marine expeditionary unit, augmented with selected personnel and equipment, that is trained and equipped with an enhanced capability to conduct amphibious operations and a variety of specialized missions of limited scope and duration. These capabilities include specialized demolition, clandestine reconnaissance and surveillance, raids, *in-extremis* hostage recovery, and enabling operations for follow-on forces. The MEU(SOC) is not a special operations force but, when directed by the National Command Authorities, the combatant commander, and/or other operational commander, may conduct limited special operations *in extremis*, when other forces are inappropriate or unavailable. Also called **MEU(SOC)**. See also **aviation combat element; combat service support element; command element; ground combat element; Marine air-ground task force; Marine expeditionary force; Marine expeditionary force (forward); Marine expeditionary unit; special purpose Marine air-ground task force; task force.**

Marine Logistics Command — The US Marines may employ the concept of the Marine Logistics Command (MLC) in major regional contingencies to provide operational logistic support, which will include arrival and assembly operations. The combat service support operations center will be the MLC's primary combat service support coordination center for units undergoing arrival and assembly. Also called **MLC.** See also **combat service support operations center.** (JP 4-01.8)

maritime control area — An area generally similar to a defensive sea area in purpose except that it may be established any place on the high seas. Maritime control areas are normally established only in time of war. See also **defensive sea area.**

maritime defense sector — **(*)** One of the subdivisions of a coastal area.

maritime environment — The oceans, seas, bays, estuaries, islands, coastal areas, and the airspace above these, including the littorals.

maritime power projection — Power projection in and from the maritime environment, including a broad spectrum of offensive military operations to destroy enemy forces or logistic support or to prevent enemy forces from approaching within enemy weapons' range of friendly forces. Maritime power projection may be accomplished by amphibious assault operations, attack of targets ashore, or support of sea control operations.

maritime pre-positioning force operation — A rapid deployment and assembly of a Marine expeditionary force in a secure area using a combination of strategic airlift and forward-deployed maritime pre-positioning ships. See also **Marine expeditionary force; maritime pre-positioning ships.** (JP 4-01.6)

maritime pre-positioning ships — Civilian-crewed, Military Sealift Command-chartered ships that are organized into three squadrons and are usually forward-deployed. These ships are loaded with pre-positioned equipment and 30 days of supplies to support three Marine expeditionary brigades. Also called **MPS.** See also **Navy cargo handling battalion.**

maritime search and rescue region — The waters subject to the jurisdiction of the United States; the territories and possessions of the United States (except Canal Zone and the inland area of Alaska), and designated areas of the high seas. See also **search and rescue region.**

maritime special purpose force — A task-organized force formed from elements of a Marine expeditionary unit (special operations capable) and naval special warfare forces that can be quickly tailored to a specific mission. The maritime special purpose force can execute on short notice a wide variety of missions in a supporting, supported, or unilateral role. It focuses on operations in a maritime environment and is capable of operations in conjunction with or in support of special operations forces. The maritime special purpose force is integral to and directly relies upon the Marine expeditionary unit (special operations capable) for all combat and combat service support. Also called **MSPF.** (JP 3-05)

maritime superiority — That degree of dominance of one force over another that permits the conduct of maritime operations by the former and its related land, sea, and air forces at a given time and place without prohibitive interference by the opposing force.

maritime supremacy — That degree of maritime superiority wherein the opposing force is incapable of effective interference.

marker — **(*)** 1. A visual or electronic aid used to mark a designated point. 2. In land mine warfare: See **gap marker; intermediate marker; lane marker; row marker; strip marker.** 3. In naval operations, a maritime unit which maintains an immediate offensive or obstructive capability against a specified target.

marker ship — **(*)** In an amphibious operation, a ship which takes accurate station on a designated control point. It may fly identifying flags by day and show lights to seaward by night.

marking — To maintain contact on a target from such a position that the marking unit has an immediate offensive capability.

marking error — **(*)** In naval mine warfare, the distance and bearing of a marker from a target.

marking fire — **(*)** Fire placed on a target for the purpose of identification.

marking panel — **(*)** A sheet of material displayed for visual communication, usually between friendly units. See also **panel code.**

married failure — (*) In naval mine warfare, a moored mine lying on the seabed connected to its sinker from which it has failed to release owing to defective mechanism.

marshal — A bearing, distance, and altitude fix designated by an air operations center, helicopter direction center, or carrier air traffic control center on which the pilot will orientate holding, and from which initial approach will commence during an instrument approach. See also **air operations center; helicopter directions center.** (JP 3-04.1)

marshalling — (*) 1. The process by which units participating in an amphibious or airborne operation group together or assemble when feasible or move to temporary camps in the vicinity of embarkation points, complete preparations for combat, or prepare for loading. 2. The process of assembling, holding, and organizing supplies and/or equipment, especially vehicles of transportation, for onward movement. See also **stage; staging area.**

marshalling area — A location in the vicinity of a reception terminal or pre-positioned equipment storage site where arriving unit personnel, equipment, materiel, and accompanying supplies are reassembled, returned to the control of the unit commander, and prepared for onward movement. The joint complex commander designating the location will coordinate the use of the facilities with other allied commands and the host nation, and will provide life support to the units while in the marshalling area. See also **marshalling.** (JP 4-01.8)

mass — (*) 1. The concentration of combat power. 2. The military formation in which units are spaced at less than the normal distances and intervals.

mass casualty — Any large number of casualties produced in a relatively short period of time, usually as the result of a single incident such as a military aircraft accident, hurricane, flood, earthquake, or armed attack that exceeds local logistic support capabilities. See also **casualty.**

massed fire — 1. The fire of the batteries of two or more ships directed against a single target. 2. Fire from a number of weapons directed at a single point or small area.

master — The commanding officer of a United States Naval Ship, a commercial ship, or a government-owned general agency agreement ship operated for the Military Sealift Command by a civilian company to transport Department of Defense cargo. Also called **MA.** (JP 3-02.2)

master air attack plan — A plan that contains key information that forms the foundation of the joint air tasking order. Sometimes referred to as the air employment plan or joint air tasking order shell. Information that may be found in the plan includes joint force commander guidance, joint force air component commander guidance, support plans, component requests, target update requests, availability of capabilities and forces, target information from target lists, aircraft allocation, etc. Also called **MAAP.** See also **air attack; target.** (JP 3-60)

master film — (*) The earliest generation of imagery (negative or positive) from which subsequent copies are produced.

master plot — **(*)** A portion of a map or overlay on which are drawn the outlines of the areas covered by an air photographic sortie. Latitude and longitude, map, and sortie information are shown. See also **sortie plot.**

materials handling — **(*)** The movement of materials (raw materials, scrap, semifinished, and finished) to, through, and from productive processes; in warehouses and storage; and in receiving and shipping areas.

materials handling equipment — Mechanical devices for handling of supplies with greater ease and economy. Also called **MHE.** See also **materials handling.** (JP 4-01.8)

materiel — All items (including ships, tanks, self-propelled weapons, aircraft, etc., and related spares, repair parts, and support equipment, but excluding real property, installations, and utilities) necessary to equip, operate, maintain, and support military activities without distinction as to its application for administrative or combat purposes. See also **equipment; personal property.**

materiel cognizance — Denotes responsibility for exercising supply management over items or categories of materiel.

materiel control — See **inventory control.**

materiel inventory objective — The quantity of an item required to be on hand and on order on M-day in order to equip, provide a materiel pipeline, and sustain the approved US force structure (active and reserve) and those Allied forces designated for US materiel support, through the period prescribed for war materiel planning purposes. It is the quantity by which the war materiel requirement exceeds the war materiel procurement capability and the war materiel requirement adjustment. It includes the M-day force materiel requirement and the war reserve materiel requirement.

materiel management — See **inventory control.**

materiel pipeline — The quantity of an item required in the worldwide supply system to maintain an uninterrupted replacement flow.

materiel planning — A subset of logistic planning consisting of a four-step process. a. **requirements definition.** Requirements for significant items must be calculated at item level detail (i.e., National Stock Number) to support sustainability planning and analysis. Requirements include unit roundout, consumption and attrition replacement, safety stock, and the needs of allies. b. **apportionment.** Items are apportioned to the combatant commanders based on a global scenario to avoid sourcing of items to multiple theaters. The basis for apportionment is the capability provided by unit stocks, host-nation support, theater pre-positioned war reserve stocks and industrial base, and continental United States Department of Defense stockpiles and available production. Item apportionment cannot exceed total capabilities. c. **sourcing.** Sourcing is the matching of available capabilities on a given date against item requirements to support sustainability analysis and the identification of locations to support transportation planning. Sourcing of any item is done within the combatant commander's apportionment. d. **documentation.** Sourced item requirements and corresponding shortfalls are major inputs to the combatant commander's sustainability analysis. Sourced item requirements are translated into movement requirements and documented in the Joint Operation Planning and Execution System database for transportation feasibility analysis. Movement requirements for nonsignificant items are estimated in tonnage.

materiel readiness — The availability of materiel required by a military organization to support its wartime activities or contingencies, disaster relief (flood, earthquake, etc.), or other emergencies.

materiel release confirmation — A notification from a shipping or storage activity advising the originator of a materiel release order of the positive action taken on the order. It will also be used with appropriate shipment status document identifier codes as a reply to a followup initiated by the inventory control point.

materiel release order — An order issued by an accountable supply system manager (usually an inventory control point or accountable depot or stock point) directing a non-accountable activity (usually a storage site or materiel drop point) within the same supply distribution complex to release and ship materiel.

materiel requirements — Those quantities of items of equipment and supplies necessary to equip, provide a materiel pipeline, and sustain a Service, formation, organization, or unit in the fulfillment of its purposes or tasks during a specified period.

maximum effective range — The maximum distance at which a weapon may be expected to be accurate and achieve the desired effect.

maximum elevation figure — **(*)** A figure, shown in each quadrangle bounded by ticked graticule lines on aeronautical charts, which represents the height in thousands and hundreds of feet, above mean sea level, of the highest known natural or manmade feature in that quadrangle, plus suitable factors to allow for inaccuracy and incompleteness of the topographical heighting information.

maximum enlisted amount — For any month, the sum of: a. the highest rate of basic pay payable for such month to any enlisted member of the Armed Forces of the United States at the highest pay grade applicable to enlisted members; and b. in the case of officers entitled to special pay under Title 37, United States Code, for such month, the amount of such special pay payable to such officers for such month. (JP 1-0)

maximum landing weight — **(*)** The maximum gross weight due to design or operational limitations at which an aircraft is permitted to land.

maximum operating depth — The keel depth that a submarine is not to exceed during operations. This depth is determined by the submarine's national naval authority. See also **test depth**.

maximum ordinate — **(*)** In artillery and naval gunfire support, the height of the highest point in the trajectory of a projectile above the horizontal plane passing through its origin. Also called **vertex height.**

maximum permissible concentration — See **radioactivity concentration guide.**

maximum permissible dose — **(*)** That radiation dose which a military commander or other appropriate authority may prescribe as the limiting cumulative radiation dose to be received over a specific period of time by members of the command, consistent with current operational military considerations.

maximum range — **(*)** The greatest distance a weapon can fire without consideration of dispersion.

maximum sustained speed — **(*)** In road transport, the highest speed at which a vehicle, with its rated payload, can be driven for an extended period on a level first-class highway without sustaining damage.

maximum take-off weight — **(*)** The maximum gross weight due to design or operational limitations at which an aircraft is permitted to take off.

mayday — Distress call.

M-day — See **times.**

M-day force materiel requirement — The quantity of an item required to be on hand and on order (on M-day minus one day) to equip and provide a materiel pipeline for the approved peacetime US force structure, both active and reserve.

meaconing — **(*)** A system of receiving radio beacon signals and rebroadcasting them on the same frequency to confuse navigation. The meaconing stations cause inaccurate bearings to be obtained by aircraft or ground stations.

mean lethal dose — **(*)** 1. The amount of nuclear irradiation of the whole body which would be fatal to 50 percent of the exposed personnel in a given period of time. 2. The dose of chemical agent that would kill 50 percent of exposed, unprotected, and untreated personnel.

mean line of advance — In naval usage, the direction expected to be made good over a sustained period.

mean point of burst — See **mean point of impact.**

mean point of impact — **(*)** The point whose coordinates are the arithmetic means of the coordinates of the separate points of impact/burst of a finite number of projectiles fired or released at the same aiming point under a given set of conditions.

mean sea level — The average height of the surface of the sea for all stages of the tide; used as a reference for elevations. Also called **MSL.**

means of transport — See **mode of transport.**

measured mile — **(*)** In maritime navigation, distance precisely measured and marked, used by a vessel to calibrate its log.

measurement and signature intelligence — Scientific and technical intelligence obtained by quantitative and qualitative analysis of data (metric, angle, spatial, wavelength, time dependence, modulation, plasma, and hydromagnetic) derived from specific technical sensors for the purpose of identifying any distinctive features associated with the target, source, emitter, or sender measurement of the same. The detected feature may be either reflected or emitted. Also called **MASINT.** See also **intelligence; scientific and technical intelligence.** (JP 2-01.3)

measurement ton — The unit of volumetric measurement of equipment associated with surface-delivered cargo. Measurement tons equal total cubic feet divided by 40 (1MTON = 40 cubic feet). Also called **M/T, MT, MTON.**

measures of effectiveness — Tools used to measure results achieved in the overall mission and execution of assigned tasks. Measures of effectiveness are a prerequisite to the performance of combat assessment.

Also called **MOEs.** See also **combat assessment; mission.** (JP 3-60)

mechanical sweep — **(*)** In naval mine warfare, any sweep used with the object of physically contacting the mine or its appendages.

median incapacitating dose — **(*)** The amount or quantity of chemical agent which when introduced into the body will incapacitate 50 percent of exposed, unprotected personnel.

media pool — A limited number of news media who represent a larger number of news media organizations for purposes of news gathering and sharing of material during a specified activity. Pooling is typically used when news media support resources cannot accommodate a large number of journalists. The Department of Defense National Media Pool is available for coverage of the earliest stages of a contingency. Additionally, the combatant commanders may also find it necessary to form limited local pools to report on specific missions. See also **news media representative; public affairs.** (JP 3-61)

medical evacuees — Personnel who are wounded, injured, or ill and must be moved to or between medical facilities.

medical intelligence — That category of intelligence resulting from collection, evaluation, analysis, and interpretation of foreign medical, bio-scientific, and environmental information that is of interest to strategic planning and to military medical planning and operations for the conservation of the fighting strength of friendly forces and the formation of assessments of foreign medical capabilities in both military and civilian sectors. Also called **MEDINT.** See also **intelligence.** (JP 2-02)

medical officer — **(*)** Physician with officer rank. Also called **MO.**

medical regulating — The actions and coordination necessary to arrange for the movement of patients through the levels of care. This process matches patients with a medical treatment facility that has the necessary health service support capabilities and available bed space. See also **health service support; medical treatment facility.** (JP 4-02)

medical surveillance — The ongoing, systematic collection of health data essential to the evaluation, planning, and implementation of public health practice, closely integrated with timely dissemination of data as required by higher authority. See also **surveillance.** (JP 4-02)

medical treatment facility — A facility established for the purpose of furnishing medical and/or dental care to eligible individuals.

medium-angle loft bombing — Type of loft bombing wherein weapon release occurs at an angle between 35 and 75 degrees above the horizontal.

medium artillery — See **field artillery.**

medium-lot storage — Generally defined as a quantity of material that will require one to three pallet stacks, stored to maximum height. Thus, the term refers to relatively small lots as distinguished from definitely large or small lots. See also **storage.**

medium-range ballistic missile — A ballistic missile with a range capability from about 600 to 1,500 nautical miles.

medium-range bomber aircraft — A bomber designed for a tactical operating radius of under 1,000 nautical miles at design gross weight and design bomb load.

medium-range transport aircraft — See **transport aircraft.**

medium-scale map — A map having a scale larger than 1:600,000 and smaller than 1:75,000. See also **map.**

meeting engagement — **(*)** A combat action that occurs when a moving force, incompletely deployed for battle, engages an enemy at an unexpected time and place.

megaton weapon — **(*)** A nuclear weapon, the yield of which is measured in terms of millions of tons of trinitrotoluene explosive equivalents. See also **kiloton weapon; nominal weapon; subkiloton weapon.**

merchant convoy — **(*)** A convoy consisting primarily of merchant ships controlled by the naval control of shipping organization.

merchant intelligence — In intelligence handling, communication instructions for reporting by merchant vessels of vital intelligence sightings. Also called **MERINT.**

merchant ship — **(*)** A vessel engaged in mercantile trade except river craft, estuarial craft, or craft which operate solely within harbor limits.

merchant ship casualty report — A report by message, or other means, of a casualty to a merchant ship at sea or in port. Merchant ship casualty reports are sent by the escort force commander or other appropriate authority to the operational control authority in whose area the casualty occurred.

merchant ship communications system — **(*)** A worldwide system of communications to and from merchant ships using the peacetime commercial organization as a basis but under operational control authority, with the ability to employ the broadcast mode to ships when the situation makes radio silence necessary. Also called **mercomms system.**

merchant ship control zone — **(*)** A defined area of sea or ocean inside which it may be necessary to offer guidance, control, and protection to Allied shipping.

merchant ship reporting and control message system — **(*)** A worldwide message system for reporting the movements of and information relating to the control of merchant ships.

mercomms system — See **merchant ship communications system.**

message — Any thought or idea expressed briefly in a plain or secret language and prepared in a form suitable for transmission by any means of communication.

message center — See **telecommunications center.**

message (telecommunications) — Record information expressed in plain or encrypted language and prepared in a format specified for intended transmission by a telecommunications system.

metadata — Information about information; more specifically, information about the meaning of other data. See also **data; information.** (JP 2-03)

meteorological and oceanographic — A term used to convey all meteorological (weather) and oceanographic (physical oceanography) factors as provided by Service components. These factors include the whole range of atmospheric and oceanographic phenomena, from the sub-bottom of the earth's oceans up to the space environment (space weather). Also called **METOC.** (JP 3-59)

Meteorological and Oceanographic Forecast Center — The collective of electronically connected, shore-based meteorological and oceanographic (METOC) production facilities that includes centers such as Air Force Weather Agency, Navy Fleet Numerical METOC Center, 55th Space Weather Squadron, Naval Oceanographic Office, Warfighting Support Center, Air Force Combat Climatology Center, Fleet Numerical METOC Center Detachment, Asheville, North Carolina, and the Air Force and Navy theater and/or regional METOC production activities. Also called **MFC.** See also **meteorological and oceanographic.** (JP 3-59)

meteorological data — Meteorological facts pertaining to the atmosphere, such as wind, temperature, air density, and other phenomena that affect military operations.

meteorology — The study dealing with the phenomena of the atmosphere including the physics, chemistry, and dynamics extending to the effects of the atmosphere on the earth's surface and the oceans. (JP 3-59)

microform — **(*)** A generic term for any form, whether film, video tape, paper, or other medium, containing miniaturized or otherwise compressed images which cannot be read without special display devices.

midcourse guidance — The guidance applied to a missile between termination of the boost phase and the start of the terminal phase of flight.

midcourse phase — That portion of the trajectory of a ballistic missile between the boost phase and the reentry phase. See also **ballistic trajectory; boost phase; reentry phase; terminal phase.**

militarily significant fallout — Radioactive contamination capable of inflicting radiation doses on personnel which may result in a reduction of their combat effectiveness.

Military Affiliate Radio System — A program conducted by the Departments of the Army, Navy, and Air Force in which amateur radio stations and operators participate in and contribute to the mission of providing auxiliary and emergency communications on a local, national, or international basis as an adjunct to normal military communications. Also called **MARS.**

military assistance advisory group — A joint Service group, normally under the military command of a commander of a unified command and representing the Secretary of Defense, which primarily administers the US military assistance planning and programming in the host country. Also called **MAAG.**

Military Assistance Articles and Services List — A Department of Defense publication listing source, availability, and price of items and services for use by the unified commands and Military Departments in preparing military assistance plans and programs.

Military Assistance Program — That portion of the US security assistance authorized by the Foreign Assistance Act of 1961, as amended, which provides defense articles and services to recipients on a nonreimbursable (grant) basis. Also called **MAP.**

Military Assistance Program training — See **international military education and training.**

military capability — The ability to achieve a specified wartime objective (win a war or battle, destroy a target set). It includes four major components: force structure, modernization, readiness, and sustainability. a. **force structure** — Numbers, size, and composition of the units that comprise US defense forces; e.g., divisions, ships, air wings. b. **modernization** — Technical sophistication of forces, units, weapon systems, and equipments. c. **unit readiness** — The ability to provide capabilities required by the combatant commanders to execute their assigned missions. This is derived from the ability of each unit to deliver the outputs for which it was designed. d. **sustainability** — The ability to maintain the necessary level and duration of operational activity to achieve military objectives. Sustainability is a function of providing for and maintaining those levels of ready forces, materiel, and consumables necessary to support military effort. See also **readiness.**

military characteristics — Those characteristics of equipment upon which depends its ability to perform desired military functions. Military characteristics include physical and operational characteristics but not technical characteristics.

military civic action — The use of preponderantly indigenous military forces on projects useful to the local population at all levels in such fields as education, training, public works, agriculture, transportation, communications, health, sanitation, and others contributing to economic and social development, which would also serve to improve the standing of the military forces with the population. (US forces may at times advise or engage in military civic actions in overseas areas.)

military construction — Any construction, alteration, development, conversion, or extension of any kind carried out with respect to a military installation. Also called **MILCON.** (JP 4-04)

military container moved via ocean — Commercial or Government owned (or leased) shipping containers that are moved via ocean transportation without bogey wheels attached, i.e., lifted on and off the ship. Also called **SEAVAN.**

military convoy — (*) A land or maritime convoy that is controlled and reported as a military unit. A maritime convoy can consist of any combination of merchant ships, auxiliaries, or other military units.

military currency — (*) Currency prepared by a power and declared by its military commander to be legal tender for use by civilian and/or military personnel as prescribed in the areas occupied by its forces. It should be of distinctive design to distinguish it from the official currency of the countries concerned, but may be denominated in the monetary unit of either.

military damage assessment — An appraisal of the effects of an attack on a nation's military forces to determine residual military capability and to support planning for recovery and reconstitution. See also **damage assessment.**

military deception — Actions executed to deliberately mislead adversary military decision makers as to friendly military capabilities, intentions, and operations, thereby causing the adversary to take specific actions (or inactions) that will contribute to the accomplishment of the friendly mission. The five categories of military deception are as follows. a. **strategic military deception** — Military deception planned and executed by and in support of senior military commanders to result in adversary military policies and actions that support the originator's strategic military objectives, policies, and operations. b. **operational military deception** — Military deception planned and executed by and in support of operational-level commanders to result in adversary actions that are favorable to the originator's objectives and operations. Operational military deception is planned and conducted in a theater to support campaigns and major operations. c. **tactical military deception** — Military deception planned and executed by and in support of tactical commanders to result in adversary actions that are favorable to the originator's objectives and operations. Tactical military deception is planned and conducted to support battles and engagements. d. **Service military deception** — Military deception planned and executed by the Services that pertain to Service support to joint operations. Service military deception is designed to protect and enhance the combat capabilities of Service forces and systems. e. **military deception in support of operations security (OPSEC)** — Military deception planned and executed by and in support of all levels of command to support the prevention of the inadvertent compromise of sensitive or classified activities, capabilities, or intentions. Deceptive OPSEC measures are designed to distract foreign intelligence away from, or provide cover for, military operations and activities. See also **deception.** (JP 3-58)

Military Department — One of the departments within the Department of Defense created by the National Security Act of 1947, as amended. Also called **MILDEP.** See also **Department of the Air Force; Department of the Army; Department of the Navy.**

military designed vehicle — A vehicle having military characteristics resulting from military research and development processes, designed primarily for use by forces in the field in direct connection with, or support of, combat or tactical operations.

military education — The systematic instruction of individuals in subjects that will enhance their knowledge of the science and art of war. See also **military training.**

military geographic documentation — Military geographic information that has been evaluated, processed, summarized, and published.

military geographic information — Information concerning physical aspects, resources, and artificial features of the terrain that is necessary for planning and operations.

military geography — The specialized field of geography dealing with natural and manmade physical features that may affect the planning and conduct of military operations.

military government — See **civil affairs.**

military government ordinance — An enactment on the authority of a military governor promulgating laws or rules regulating the occupied territory under such control.

military governor — **(*)** The military commander or other designated person who, in an occupied territory, exercises supreme authority over the civil population subject to the laws and usages of war and to any directive received from the commander's government or superior.

military grid — **(*)** Two sets of parallel lines intersecting at right angles and forming squares; the grid is superimposed on maps, charts, and other similar representations of the surface of the Earth in an accurate and consistent manner to permit identification of ground locations with respect to other locations and the computation of direction and distance to other points. See also **military grid reference system.**

military grid reference system — **(*)** A system which uses a standard-scaled grid square, based on a point of origin on a map projection of the surface of the Earth in an accurate and consistent manner to permit either position referencing or the computation of direction and distance between grid positions. Also called **MGRS.** See also **military grid.**

military independent — **(*)** A merchant ship or auxiliary sailed singly but controlled and reported as a military unit. See also **independent.**

military installation — A base, camp, post, station, yard, center, or other activity under the jurisdiction of the Secretary of a Military Department or, in the case of an activity in a foreign country, under the operational control of the Secretary of a Military Department or the Secretary of Defense. See also **base; station.** (JP 4-04)

military intelligence — Intelligence on any foreign military or military-related situation or activity which is significant to military policymaking or the planning and conduct of military operations and activities. Also called **MI.**

Military Intelligence Board — A decisionmaking forum which formulates Defense intelligence policy and programming priorities. The Military Intelligence Board, chaired by the Director, Defense Intelligence Agency, who is dual-hatted as Director of Military Intelligence, consists of senior military and civilian intelligence officials of each Service, US Coast Guard, each Combat Support Agency, the Joint Staff/J-2/J-6, Deputy Assistant Secretary of Defense (Intelligence), Intelligence Program Support Group, DIA's Directorates for Intelligence Production, Intelligence Operations, and Information and Services, and the combatant command J-2s. Also called **MIB.** See also **intelligence; military intelligence.** (JP 2-0)

Military Intelligence Integrated Data System/Integrated Database — An architecture for improving the manner in which military intelligence is analyzed, stored, and disseminated. The Integrated Database (IDB) forms the core automated database for the Military Intelligence Integrated Data System (MIIDS) program and integrates the data in the installation, order of battle, equipment, and selected electronic warfare and command, control, and communications files. The IDB is the national-level repository for the general military intelligence information available to the entire Department of Defense Intelligence Information System community and maintained by DIA and the commands. The IDB is kept synchronized by system transactions to disseminate updates. Also called **MIIDS/IDB.** See also **architecture; military intelligence.** (JP 2-01)

military intervention — The deliberate act of a nation or a group of nations to introduce its military forces into the course of an existing controversy.

military journalist — A US Service member or Department of Defense civilian employee providing photographic, print, radio, or television command information for military internal audiences. See also **command information.** (JP 3-61)

military land transportation resources — All military-owned transportation resources, designated for common-user, over the ground, point-to-point use.

military load classification — (*) A standard system in which a route, bridge, or raft is assigned class number(s) representing the load it can carry. Vehicles are also assigned number(s) indicating the minimum class of route, bridge, or raft they are authorized to use. See also **route classification.**

military necessity — (*) The principle whereby a belligerent has the right to apply any measures which are required to bring about the successful conclusion of a military operation and which are not forbidden by the laws of war.

military nuclear power — (*) A nation which has nuclear weapons and the capability for their employment.

military objective — A derived set of military actions to be taken to implement National Command Authorities guidance in support of national objectives. A military objective defines the results to be achieved by the military and assign tasks to commanders. See also **national objectives.**

military occupation — A condition in which territory is under the effective control of a foreign armed force. See also **occupied territory; phases of military government.**

military operations other than war — Operations that encompass the use of military capabilities across the range of military operations short of war. These military actions can be applied to complement any combination of the other instruments of national power and occur before, during, and after war. Also called **MOOTW.** (JP 3-07)

military options — A range of military force responses that can be projected to accomplish assigned tasks. Options include one or a combination of the following: civic action, humanitarian assistance, civil affairs, and other military activities to develop positive relationships with other countries; confidence building and other measures to reduce military tensions; military presence; activities to convey threats to adversaries as well as truth projections; military deceptions and psychological operations; quarantines, blockades, and harassment operations; raids; intervention operations; armed conflict involving air, land, maritime, and strategic warfare operations; support for law enforcement authorities to counter international criminal activities (terrorism, narcotics trafficking, slavery, and piracy); support for law enforcement authorities to suppress domestic rebellion; and support for insurgency, counterinsurgency, and civil war in foreign countries. See also **civil affairs; foreign humanitarian assistance; military civic action.** (JP 5-01.3)

military ordinary mail — A special military airlift service for ordinary official mail being sent to, from, or between overseas areas. Also called **MOM.**

military payment certificate — An instrument (scrip) denominated in US dollars that is used as the official medium of exchange in US military operations designated as military payment certificate areas. Also called **MPC.** (JP 1-06)

military performance specification container — A container that meets specific written standards. Aviation and Troop Command, US Army, procures military performance specification containers for the Army and will perform like services for other Department of Defense components on request. Also called **MILSPEC container.** (JP 4-01.7)

military post office — A branch of a designated US-based post office such as New York, San Francisco, Miami, or Seattle established by US Postal Service authority and operated by one of the Military Services. The term includes Army,

Air Force, Navy, Marine Corps, and established Coast Guard post offices Also called **MPO.**

military postal clerk — A person of the US Armed Forces officially designated to perform postal duties.

Military Postal Service — The command, organization, personnel, and facilities established to provide, through military post offices, a means for the transmission of mail to and from the Department of Defense, members of the US Armed Forces, and other authorized agencies and individuals. Also called **MPS.**

Military Postal Service Agency — The single manager operating agency established to manage the Military Postal Service. Also called **MPSA.**

military posture — The military disposition, strength, and condition of readiness as it affects capabilities.

military requirement — (*) An established need justifying the timely allocation of resources to achieve a capability to accomplish approved military objectives, missions, or tasks. Also called **operational requirement.** See also **objective force level.**

military resources — Military and civilian personnel, facilities, equipment, and supplies under the control of a Department of Defense component.

Military Sealift Command — A major command of the US Navy, and the US Transportation Command's component command responsible for designated common-user sealift transportation services to deploy, employ, sustain, and redeploy US forces on a global basis. Also called **MSC.** See also **transportation component command.**

Military Sealift Command-controlled ships — Those ships assigned by the Military Sealift Command (MSC) for a specific operation. They may be MSC nucleus fleet ships, contract-operated MSC ships, MSC-controlled time or voyage-chartered commercial ships, or MSC-controlled ships allocated by the Maritime Administration to MSC to carry out Department of Defense objectives. (JP 3-02)

Military Sealift Command force — The Military Sealift Command (MSC) force common-user sealift consists of three subsets: the Naval Fleet Auxiliary Force, common-user ocean transportation, and the special mission support force. These ship classes include government-owned ships (normally civilian manned) and ships acquired by MSC charter or allocated from other government agencies. See also **common-user sealift; Military Sealift Command.** (JP 4-01.6)

Military Service — A branch of the Armed Forces of the United States, established by act of Congress, in which persons are appointed, enlisted, or inducted for military service, and which operates and is administered within a military or executive department. The Military Services are: the United States Army, the United States Navy, the United States Air Force, the United States Marine Corps, and the United States Coast Guard.

military standard requisitioning and issue procedure — A uniform procedure established by the Department of Defense for use within the Department of Defense to govern requisition and issue of materiel within standardized priorities. Also called **MILSTRIP.**

military standard transportation and movement procedures — Uniform and standard transportation data, documentation, and control procedures applicable to all cargo movements in the Department of Defense transportation system. Also called **MILSTAMP.**

military strategy — The art and science of employing the armed forces of a nation to secure the objectives of national policy by the application of force or the threat of force. See also **strategy.**

military symbol — (*) A graphic sign used, usually on map, display or diagram, to represent a particular military unit, installation, activity, or other item of military interest.

military technician — A Federal civilian employee providing full-time support to a National Guard, Reserve, or Active Component organization for administration, training, and maintenance of the Selected Reserve. Also called **MILTECH.** (JP 1-03.17)

military traffic — Department of Defense personnel, mail, and cargo to be, or being, transported.

Military Traffic Management Command — A major command of the US Army, and the US Transportation Command's component command responsible for designated continental United States land transportation as well as common-user water terminal and traffic management service to deploy, employ, sustain, and redeploy US forces on a global basis. Also called **MTMC.** See also **transportation component command.**

military training — 1. The instruction of personnel to enhance their capacity to perform specific military functions and tasks. 2. The exercise of one or more military units conducted to enhance their combat readiness. See also **military education.**

military van (container) — Military-owned, demountable container, conforming to US and international standards, operated in a centrally controlled fleet for movement of military cargo. Also called **MILVAN.**

MILSPEC container — See **military performance specification containers.** (JP 4-01.7)

MILVAN — See **military van (container).**

MILVAN chassis — The compatible chassis to which the military van (container) is attached by coupling the lower four standard corner fittings of the container to compatible mounting blocks in the chassis to permit road movement.

mine — (*) 1. In land mine warfare, an explosive or material, normally encased, designed to destroy or damage ground vehicles, boats, or aircraft, or designed to wound, kill, or otherwise incapacitate personnel. It may be detonated by the action of its victim, by the passage of time, or by controlled means. 2. In naval mine warfare, an explosive device laid in the water with the intention of damaging or sinking ships or of deterring shipping from entering an area. The term does not include devices attached to the bottoms of ships or to harbor installations by personnel operating underwater, nor does it include devices which explode immediately on expiration of a predetermined time after laying. See also **land mine warfare; mine warfare.** (JP 3-15)

mineable waters — (*) Waters where naval mines of any given type may be effective against any given target.

mine clearance — (*) The process of removing all mines from a route or area.

mine-cluster — A number of mines (not to exceed five) laid within a two-meter semicircle of the central mine.

mine countermeasures — All methods for preventing or reducing damage or danger from mines. Also called **MCM.** (JP 3-15)

mined area — **(*)** An area declared dangerous due to the presence or suspected presence of mines.

mine defense — **(*)** The defense of a position, area, etc., by land or underwater mines. A mine defense system includes the personnel and equipment needed to plant, operate, maintain, and protect the minefields that are laid.

mine disposal — The operation by suitably qualified personnel designed to render safe, neutralize, recover, remove, or destroy mines.

minefield — 1. In land warfare, an area of ground containing mines emplaced with or without a pattern. 2. In naval warfare, an area of water containing mines laid with or without a pattern. See also **land mine warfare; mine; mine warfare.** (JP 3-15)

minefield breaching — **(*)** In land mine warfare, the process of clearing a lane through a minefield under tactical conditions. See also **minefield lane.**

minefield density — In land mine warfare, the average number of mines per meter of minefield front, or the average number of mines per square meter of minefield. In naval warfare, the average number of mines per nautical mile.

minefield lane — A marked lane, unmined, or cleared of mines, leading through a minefield.

minefield marking — Visible marking of all points required in laying a minefield and indicating the extent of such minefields.

minefield record — **(*)** A complete written record of all pertinent information concerning a minefield, submitted on a standard form by the officer in charge of the laying operations.

minefield report — An oral, electronic, or written communication concerning mining activities (friendly or enemy) submitted in a standard format by the fastest secure means available. (JP 3-15)

minehunting — Employment of sensor and neutralization systems, whether air, surface, or subsurface, to locate and dispose of individual mines. Minehunting is conducted to eliminate mines in a known field when sweeping is not feasible or desirable, or to verify the presence or absence of mines in a given area. See also **minesweeping.** (JP 3-15)

mine row — **(*)** A single row of mines or clusters of mines. See also **mine strip.**

mine spotting — **(*)** In naval mine warfare, the process of visually observing a mine or minefield.

mine strip — **(*)** In land mine warfare, two parallel mine rows laid simultaneously six meters or six paces apart. See also **mine row.**

minesweeping — The technique of clearing mines using either mechanical, explosive, or influence sweep equipment. Mechanical sweeping removes, disturbs, or otherwise neutralizes the mine; explosive sweeping causes sympathetic detonations in, damages, or displaces the mine; and influence sweeping produces either the acoustic and/or magnetic influence required to detonate the mine. See also **minehunting.** (JP 3-15)

mine warfare — The strategic, operational, and tactical use of mines and mine countermeasures. Mine warfare is divided into two basic subdivisions: the laying of mines to degrade the enemy's capabilities to wage land, air, and maritime warfare; and the countering of enemy-laid mines to permit friendly maneuver or use of selected land or sea areas. Also called **MIW.** (JP 3-15)

mine warfare chart — **(*)** A special naval chart, at a scale of 1:50,000 or larger (preferably 1:25,000 or larger) designed for planning and executing mine warfare operations, either based on an existing standard nautical chart, or produced to special specifications.

mine warfare forces (naval) — Navy forces charged with the strategic, operational, and tactical use of naval mines and their countermeasures. Such forces are capable of offensive and defensive measures in connection with laying and clearing mines.

mine warfare group — **(*)** A task organization of mine warfare units for the conduct of minelaying and/or mine countermeasures in maritime operations.

minewatching — **(*)** In naval mine warfare, the mine countermeasures procedure to detect, record and, if possible, track potential minelayers and to detect, find the position of, and/or identify mines during the actual minelaying.

mine weapons — **(*)** The collective term for all weapons which may be used in mine warfare.

minimize — A condition wherein normal message and telephone traffic is drastically reduced in order that messages connected with an actual or simulated emergency shall not be delayed.

minimum aircraft operating surface — **(*)** The minimum surface on an airfield which is essential for the movement of aircraft. It includes the aircraft dispersal areas, the minimum operating strip, and the taxiways between them. See also **minimum operating strip.**

minimum attack altitude — The lowest altitude determined by the tactical use of weapons, terrain consideration, and weapons effects that permits the safe conduct of an air attack and/or minimizes effective enemy counteraction.

minimum crossing altitude — The lowest altitude at certain radio fixes at which an aircraft must cross when proceeding in the direction of a higher minimum *en route* instrument flight rules altitude.

minimum essential equipment — That part of authorized allowances of Army equipment, clothing, and supplies needed to preserve the integrity of a unit during movement without regard to the performance of its combat or service mission. Items common within this category will normally be carried by or accompany troops to the port and will be placed aboard the same ships with the troops. As used in movement directives, minimum essential equipment refers to specific items of both organizational and individual clothing and equipment.

minimum force — Those minimum actions, including the use of armed force, sufficient to bring a situation under control or to defend against hostile act or hostile intent. All actions must cease as soon as the target complies with instructions or ceases hostile action. The firing of weapons is to be considered as a means of last resort.

minimum nuclear safe distance — (*) The sum of the radius of safety and the buffer distance.

minimum nuclear warning time — (*) The sum of system reaction time and personnel reaction time.

minimum obstruction clearance altitude — The specified altitude in effect between radio fixes on very high frequency omnirange airways, off-airway routes, or route segments, which meets obstruction clearance requirements for the entire route segment, and that assures acceptable navigational signal coverage only within 22 miles of a very high frequency omnirange.

minimum operating strip — (*) A runway which meets the minimum requirements for operating assigned and/or allocated aircraft types on a particular airfield at maximum or combat gross weight. See also **minimum aircraft operating surface.**

minimum range — 1. Least range setting of a gun at which the projectile will clear an obstacle or friendly troops between the gun and the target. 2. Shortest distance to which a gun can fire from a given position. 3. The range at which a projectile or fuse will be armed.

minimum reception altitude — The lowest altitude required to receive adequate signals to determine specific very high frequency omnirange and tactical air navigation fixes.

minimum residual radioactivity weapon — (*) A nuclear weapon designed to have optimum reduction of unwanted effects from fallout, rainout, and burst site radioactivity. See also **salted weapon.**

minimum-risk level — A specific altitude or altitude block that allows homebound aircraft to return in a homebound direction without lateral restrictions. Also called **MRL.** (JP 3-52)

minimum-risk route — A temporary corridor of defined dimensions recommended for use by high-speed, fixed-wing aircraft that presents the minimum known hazards to low-flying aircraft transiting the combat zone. Also called **MRR.** (JP 3-52)

minimum safe altitude — (*) The altitude below which it is hazardous to fly owing to presence of high ground or other obstacles.

minor control — See **photogrammetric control.**

minor installation — In the Air Force, a facility operated by an Active, Reserve, or Guard unit of at least squadron size that does not otherwise satisfy all the criteria for a major installation. This category includes Air Force stations; air stations; Air Reserve stations; and Air Guard stations. Examples of minor installations are Active, Reserve, or Guard flying operations that are located at civilian-owned airports. See also **installation complex.**

minor port — (*) A port having facilities for the discharge of cargo from coasters or lighters only.

misfire — (*) 1. Failure to fire or explode properly. 2. Failure of a primer or the propelling charge of a round or projectile to function wholly or in part.

missed approach — (*) An approach which is not completed by landing.

missile assembly-checkout facility — A building, van, or other type structure located near the operational missile launching location and designed for the final assembly and checkout of the missile system.

missile control system — (*) A system that serves to maintain attitude stability and to correct deflections. See also **missile guidance system.**

missile destruct — (*) Intentional destruction of a missile or similar vehicle for safety or other reasons.

missile destruct system — (*) A system which, when operated by external command or preset internal means, destroys the missile or similar vehicle.

missile guidance system — (*) A system which evaluates flight information, correlates it with target data, determines the desired flight path of a missile, and communicates the necessary commands to the missile flight control system. See also **missile control system.**

missile release line — The line at which an attacking aircraft could launch an air-to-surface missile against a specific target.

missing — A casualty status for which the United States Code provides statutory guidance concerning missing members of the Military Services. Excluded are personnel who are in an absent without leave, deserter, or dropped-from-rolls status. A person declared missing is categorized as follows. a. **beleaguered —** The casualty is a member of an organized element that has been surrounded by a hostile force to prevent escape of its members. b. **besieged —** The casualty is a member of an organized element that has been surrounded by a hostile force, compelling it to surrender. c. **captured —** The casualty has been seized as the result of action of an unfriendly military or paramilitary force in a foreign country. d. **detained —** The casualty is prevented from proceeding or is restrained in custody for alleged violation of international law or other reason claimed by the government or group under which the person is being held. e. **interned —** The casualty is definitely known to have been taken into custody of a nonbelligerent foreign power as the result of and for reasons arising out of any armed conflict in which the Armed Forces of the United States are engaged. f. **missing —** The casualty is not present at his or her duty location due to apparent involuntary reasons and whose location is unknown. g. **missing in action —** The casualty is a hostile casualty, other than the victim of a terrorist activity, who is not present at his or her duty location due to apparent involuntary reasons and whose location is unknown. Also called **MIA.** See also **casualty category; casualty status.**

missing in action — See **missing.**

mission — 1. The task, together with the purpose, that clearly indicates the action to be taken and the reason therefore. 2. In common usage, especially when applied to lower military units, a duty assigned to an individual or unit; a task. 3. The dispatching of one or more aircraft to accomplish one particular task.

mission-capable — Material condition of an aircraft indicating it can perform at least one and potentially all of its designated missions. Mission-capable is further defined as the sum of full mission-capable and partial mission-capable. Also called **MC.** See also **full mission-capable; partial mission-capable; partial mission-capable, maintenance; partial mission-capable, supply.**

mission cycle — The mission cycle, as it pertains to targeting, is a decisionmaking process used by commanders to employ forces. Within the cycle there are six general mission steps: detection, location, identification, decision, execution, and assessment. (JP 3-55)

mission-essential materiel — 1. That materiel authorized and available to combat, combat support, combat service support, and combat readiness training forces in order to accomplish their assigned missions. 2. For the purpose of sizing organic industrial facilities, that Service-designated materiel authorized to combat, combat support, combat service support, and combat readiness training forces and activities, including Reserve and National Guard activities, that is required to support approved emergency and/or war plans, and where the materiel is used to: a. destroy the enemy or the enemy's capacity to continue war; b. provide battlefield protection of personnel; c. communicate under war conditions; d. detect, locate, or maintain surveillance over the enemy; e. provide combat transportation and support of men and materiel; and f. support training functions. Mission-essential materiel should also be suitable for employment under emergency plans to meet the purposes enumerated above.

mission needs statement — A formatted non-system-specific statement containing operational capability needs and written in broad operational terms. It describes required operational capabilities and constraints to be studied during the Concept Exploration and Definition Phase of the Requirements Generation Process. Also called **MNS.**

mission-oriented items — Items for which requirements computations are based upon the assessment of enemy capabilities expressed as a known or estimated quantity of total targets to be destroyed. See also **combination mission/level of effort-oriented items; level of effort-oriented items.**

mission-oriented protective posture — A flexible system of protection against nuclear, biological, and chemical contamination. This posture requires personnel to wear only that protective clothing and equipment (mission-oriented protective posture gear) appropriate to the threat level, work rate imposed by the mission, temperature, and humidity. Also called **MOPP.** See also **mission-oriented protective posture gear**. (JP 3-11)

mission-oriented protective posture gear — Military term for individual protective equipment including suit, boots, gloves, mask with hood, first aid treatments, and decontamination kits issued to soldiers. Also called **MOPP gear.** See also **decontamination; mission-oriented protective posture.** (JP 3-11)

mission review report (photographic interpretation) — An intelligence report containing information on all targets covered by one photographic sortie.

mission specific data sets — Further densification of global geospatial foundation data. Information created to support specific operations, operation plans, training, or system development. Information conforms to established Department of Defense data specifications. Also called **MSDS.** See also **geospatial information and services.** (JP 2-03)

mission type order — 1. Order issued to a lower unit that includes the accomplishment of the total mission assigned to the higher headquarters. 2. Order to a unit to perform a mission without specifying how it is to be accomplished.

mixed — **(*)** In artillery and naval gunfire support, a spotting, or an observation, by a spotter or an observer to indicate that the rounds fired resulted in an equal number of air and impact bursts.

mixed bag — **(*)** In naval mine warfare, a collection of mines of various types, firing systems, sensitivities, arming delays and ship counter settings.

mixed minefield — **(*)** A minefield containing both antitank and antipersonnel mines. See also **minefield.**

mobile defense — Defense of an area or position in which maneuver is used with organization of fire and utilization of terrain to seize the initiative from the enemy.

mobile inshore undersea warfare unit — A Navy surveillance unit that provides seaward security to joint logistics over-the-shore operations from either a port or harbor complex or unimproved beach sites. The mobile inshore undersea warfare unit is equipped with mobile radar, sonar, and communications equipment located within a mobile van. Also called **MIUWU.** See also **joint logistics over-the-shore operations.** (JP 4-01.6)

mobile mine — **(*)** In naval mine warfare, a mine designed to be propelled to its proposed laying position by propulsion equipment like a torpedo. It sinks at the end of its run and then operates like a mine. See also **mine.**

mobile support group (naval) — Provides logistic support to ships at an anchorage; in effect a naval base afloat, although certain of its supporting elements may be located ashore.

mobile training team — A team consisting of one or more US military or civilian personnel sent on temporary duty, often to a foreign nation, to give instruction. The mission of the team is to train indigenous personnel to operate, maintain, and employ weapons and support systems, or to develop a self-training capability in a particular skill. The National Command Authorities may direct a team to train either military or civilian indigenous personnel, depending upon host-nation requests. Also called **MTT.** (JP 3-05)

mobility — **(*)** A quality or capability of military forces which permits them to move from place to place while retaining the ability to fulfill their primary mission.

Mobility Air Forces — The Mobility Air Forces are comprised of those air components and Service components that are assigned air mobility forces and/or that routinely exercise command authority over their operations. Also called **MAF.**

mobility analysis — An in-depth examination of all aspects of transportation planning in support of operation plan and operation order development.

mobility corridor — Areas where a force will be canalized due to terrain restrictions. They allow military forces to capitalize on the principles of mass and speed and are therefore relatively free of obstacles. (JP 2-01.3)

mobility echelon — A subordinate element of a unit that is scheduled for deployment separately from the parent unit.

mobility system support resources — Those resources that are required to: a. complement the airlift and sealift forces; and/or b. perform those work functions directly related to the origination, processing, or termination of a movement requirement.

mobilization — 1. The act of assembling and organizing national resources to support national objectives in time of war or other emergencies. See also **industrial mobilization.** 2. The process by which the Armed Forces or part of them are brought to a state of readiness for war or other national emergency. This includes activating all or part of the Reserve

Components as well as assembling and organizing personnel, supplies, and materiel. Mobilization of the Armed Forces includes but is not limited to the following categories. a. **selective mobilization** — Expansion of the active Armed Forces resulting from action by Congress and/or the President to mobilize Reserve Component units, Individual Ready Reservists, and the resources needed for their support to meet the requirements of a domestic emergency that is not the result of an enemy attack. b. **partial mobilization** — Expansion of the active Armed Forces resulting from action by Congress (up to full mobilization) or by the President (not more than 1,000,000 for not more than 24 consecutive months) to mobilize Ready Reserve Component units, individual reservists, and the resources needed for their support to meet the requirements of a war or other national emergency involving an external threat to the national security. c. **full mobilization** — Expansion of the active Armed Forces resulting from action by Congress and the President to mobilize all Reserve Component units in the existing approved force structure, as well as all individual reservists, retired military personnel, and the resources needed for their support to meet the requirements of a war or other national emergency involving an external threat to the national security. Reserve personnel can be placed on active duty for the duration of the emergency plus six months. d. **total mobilization** — Expansion of the active Armed Forces resulting from action by Congress and the President to organize and/or generate additional units or personnel beyond the existing force structure, and the resources needed for their support, to meet the total requirements of a war or other national emergency involving an external threat to the national security. Also called **MOB.** (JP 4-05)

mobilization base — The total of all resources available, or that can be made available, to meet foreseeable wartime needs. Such resources include the manpower and materiel resources and services required for the support of essential military, civilian, and survival activities, as well as the elements affecting their state of readiness, such as (but not limited to) the following: manning levels, state of training, modernization of equipment, mobilization materiel reserves and facilities, continuity of government, civil defense plans and preparedness measures, psychological preparedness of the people, international agreements, planning with industry, dispersion, and standby legislation and controls.

mobilization exercise — An exercise involving, either completely or in part, the implementation of mobilization plans.

mobilization reserves — Not to be used. See **war reserves.**

mobilization site — The designated location where a Reserve Component unit or individual mobilizes or moves after mobilization for further processing, training, and employment. This differs from a mobilization station in that it is not necessarily a military installation. See also **mobilization; mobilization station; Reserve Components.** (JP 4-05)

mobilization staff officer — The action officer assigned the principle responsibility or additional duties related to Reserve Component mobilization actions. See also **mobilization; Reserve Components.** (JP 4-05.1)

mobilization station — The designated military installation to which a Reserve Component unit or individual is moved for further processing, organizing, equipping, training, and employment and from which the unit or individual may move to an aerial port of embarkation or seaport of embarkation. See also **mobilization; mobilization site; Reserve Components.** (JP 4-05)

mock-up — (*) A model, built to scale, of a machine, apparatus, or weapon, used in studying the construction of, and in testing a new development, or in teaching personnel how to operate the actual machine, apparatus, or weapon.

mode (identification, friend or foe) — The number or letter referring to the specific pulse spacing of the signals transmitted by an interrogator or transponder.

mode of transport — The various modes used for a movement. For each mode, there are several means of transport. They are: a. inland surface transportation (rail, road, and inland waterway); b. sea transport (coastal and ocean); c. air transportation; and d. pipelines.

modernization — See **military capability.**

Modernized Integrated Database — The national level repository for the general military intelligence available to the entire Department of Defense Intelligence Information System community and, through Global Command and Control System integrated imagery and intelligence, to tactical units. This data is maintained and updated by the Defense Intelligence Agency. Commands and Services are delegated responsibility to maintain their portion of the database. Also called **MIDB.** See also **database.** (JP 3-51)

modified combined obstacle overlay — A joint intelligence preparation of the battlespace product used to portray the effects of each battlespace dimension on military operations. It normally depicts militarily significant aspects of the battlespace environment, such as obstacles restricting military movement, key geography, and military objectives. Also called **MCOO.** See also **joint intelligence preparation of a battlespace**. (JP 2-01.3)

moment — (*) In air transport, the weight of a load multiplied by its distance from a reference point in the aircraft.

monitoring — (*) 1. The act of listening, carrying out surveillance on, and/or recording the emissions of one's own or allied forces for the purposes of maintaining and improving procedural standards and security, or for reference, as applicable. 2. The act of listening, carrying out surveillance on, and/or recording of enemy emissions for intelligence purposes. 3. The act of detecting the presence of radiation and the measurement thereof with radiation measuring instruments. Also called **radiological monitoring.**

monitoring service — The general surveillance of known air traffic movements by reference to a radar scope presentation or other means, for the purpose of passing advisory information concerning conflicting traffic or providing navigational assistance. Direct supervision or control is not exercised, nor is positive separation provided.

moored — Lying with both anchors down or tied to a pier, anchor buoy, or mooring buoy. (JP 4-01.6)

moored mine — (*) A contact or influence-operated mine of positive buoyancy held below the surface by a mooring attached to a sinker or anchor on the bottom. See also **mine.**

mopping up — (*) The liquidation of remnants of enemy resistance in an area that has been surrounded or isolated, or through which other units have passed without eliminating all active resistance.

mortar — A muzzle-loading, indirect fire weapon with either a rifled or smooth bore. It usually has a shorter range than a howitzer, employs a higher angle of fire, and has a tube with a length of 10 to 20 calibers. See also **gun; howitzer.**

mortuary affairs — Covers the search for, recovery, identification, preparation, and disposition of remains of persons for whom the Services are responsible by status and Executive Order. See also **joint mortuary affairs office.** (JP 4-06)

mosaic — (*) An assembly of overlapping photographs that have been matched to form a continuous photographic representation of a portion of the surface of the Earth. See also **controlled mosaic; semi-controlled mosaic.**

most capable Service or agency — The organization that is best suited to provide common supply commodity or logistic service support within a specific joint operation. In this context, "best suited" could mean the Service or agency that has required or readily available resources and/or expertise. The most capable Service may or may not be the dominant user in any particular operation. See also **agency.** (JP 4-07)

motorized unit — (*) A unit equipped with complete motor transportation that enables all of its personnel, weapons, and equipment to be moved at the same time without assistance from other sources.

mounting — (*) 1. All preparations made in areas designated for the purpose, in anticipation of an operation. It includes the assembly in the mounting area, preparation and maintenance within the mounting area, movement to loading points, and subsequent embarkation into ships, craft, or aircraft if applicable. 2. **(DOD only)** A carriage or stand upon which a weapon is placed.

mounting area — A general locality where assigned forces of an amphibious or airborne operation, with their equipment, are assembled, prepared, and loaded in shipping and/or aircraft preparatory to an assault. See also **embarkation area.**

movement control — 1. The planning, routing, scheduling, and control of personnel and cargo movements over lines of communications. 2. An organization responsible for the planning, routing, scheduling, and control of personnel and cargo movements over lines of communications. Also called **movement control center or MCC.** See also **line of communications; non-unit-related cargo; non-unit-related personnel**. (JP 4-01.8)

movement control center — See **movement control.**

movement control post — (*) The post through which the control of movement is exercised by the commander, depending on operational requirements.

movement control team — Movement control teams (MCTs) are Army units that decentralize the execution of movement responsibilities on an area basis or at key transportation nodes. The mission of the MCTs is movement control of personnel and materiel as well as the coordination of bulk fuel and water transportation at pipeline and production take-off points. To this end, the MCTs contribute to the development of procedures, documents, and practices to facilitate local movement. Their role is to expedite, coordinate, and monitor traffic moving through the transportation system. MCTs are tailored to meet the anticipated workload. Other Service movement requirements that exceed organic capability will be requested through the Army MCTs. The movement control center is the higher headquarters for the MCTs and is located at Corps level. Also called **MCT.** (JP 4-01.7)

movement credit — **(*)** The allocation granted to one or more vehicles in order to move over a controlled route in a fixed time according to movement instructions.

movement directive — The basic document published by the Department of the Army or the Department of the Air Force (or jointly) that authorizes a command to take action to move a designated unit from one location to another.

movement group — Those ships and embarked units that load out and proceed to rendezvous in the objective area. (JP 3-02.2)

movement order — An order issued by a commander covering the details for a move of the command.

movement phase — In amphibious operations, the period during which various elements of the amphibious force move from points of embarkation to the operational area. This move may be via rehearsal, staging, or rendezvous areas. The movement phase is completed when the various elements of the amphibious force arrive at their assigned positions in the operational area. See also **amphibious force; amphibious operation.** (JP 3-02)

movement plan — In amphibious operations, the naval plan providing for the movement of the amphibious task force to the objective area. It includes information and instructions concerning departure of ships from embarkation points, the passage at sea, and the approach to and arrival in assigned positions in the objective area. See also **amphibious operation; amphibious task force.** (JP 3-02)

movement report control center — The controlling agency for the entire movement report system. It has available all information relative to the movements of naval ships and other ships under naval control.

movement report system — A system established to collect and make available to certain commands vital information on the status, location, and movement of flag commands, commissioned fleet units, and ships under operational control of the Navy.

movement requirement — A stated movement mode and time-phased need for the transport of units, personnel, and/or materiel from a specified origin to a specified destination.

movement restriction — **(*)** A restriction temporarily placed on traffic into and/or out of areas to permit clearance of or prevention of congestion.

movement schedule — A schedule developed to monitor or track a separate entity, whether it is a force requirement, cargo or personnel increment, or lift asset. The schedule reflects the assignment of specific lift resources (such as an aircraft or ship) that will be used to move the personnel and cargo included in a specific movement increment. Arrival and departure times at ports of embarkation, etc., are detailed to show a flow and workload at each location. Movement schedules are detailed enough to support plan implementation.

movement table — **(*)** A table giving detailed instructions or data for a move. When necessary it will be qualified by the words road, rail, sea, air, etc., to signify the type of movement. Normally issued as an annex to a movement order or instruction.

movement to contact — A form of the offense designed to develop the situation and to establish or regain contact. See also **meeting engagement; reconnaissance in force.**

moving havens — Restricted areas established to provide a measure of security to submarines and surface ships in transit through areas in which the existing attack restrictions would be inadequate to prevent attack by friendly forces. See also **moving submarine haven; moving surface ship haven.**

moving map display — **(*)** A display in which a symbol, representing the vehicle, remains stationary while the map or chart image moves beneath the symbol so that the display simulates the horizontal movement of the vehicle in which it is installed. Occasionally the design of the display is such that the map or chart image remains stationary while the symbol moves across a screen. See also **projected map display.**

moving mine — **(*)** The collective description of mines, such as drifting, oscillating, creeping, mobile, rising, homing, and bouquet mines.

moving submarine haven — An area established by a submarine operating authority to prevent mutual interference among friendly submarines, or between friendly submarines and ships operating with towed bodies or arrays. See also **moving havens.**

moving surface ship haven — Established by surface ship notices, a moving surface ship haven will normally be a circle with a specified radius centered on the estimated position of the ship or the guide of a group of ships. See also **moving havens.**

moving target indicator — **(*)** A radar presentation which shows only targets which are in motion. Signals from stationary targets are subtracted out of the return signal by the output of a suitable memory circuit.

multichannel — Pertaining to communications, usually full duplex, on more than one channel simultaneously. Multichannel transmission may be accomplished by either time-, frequency-, code-, and phase-division multiplexing or space diversity.

multi-modal — **(*)** In transport operations, a term applied to the movement of passengers and cargo by more than one method of transport.

multinational — Between two or more forces or agencies of two or more nations or coalition partners. See also **alliance**; **coalition.** (JP 5-0)

multinational doctrine — Fundamental principles that guide the employment of forces of two or more nations in coordinated action toward a common objective. It is ratified by participating nations. See also **doctrine; joint doctrine; multi-Service doctrine.**

multinational exercise — An exercise containing one or more non-US participating force(s). See also **exercise.**

multinational force — A force composed of military elements of nations who have formed an alliance or coalition for some specific purpose. Also called **MNF.** See also **multinational force commander; multinational operations.** (JP 0-2)

multinational force commander — A general term applied to a commander who exercises command authority over a military force composed of elements from two or more nations. The extent of the multinational force commander's command authority is determined by the participating nations. Also called **MNFC.** See also **multinational force.** (JP 3-16)

multinational joint special operations task force — A task force composed of special operations units from one or more foreign countries and more than one US Military Department formed to carry out a specific special operation or prosecute special operations in support of a theater campaign or other operations. The multinational joint special operations task force may have conventional nonspecial operations units assigned or attached to support the conduct of specific missions. Also called **MJSOTF.** See also **joint special operations task force; special operations; task force.** (JP 3-05)

multinational operations — A collective term to describe military actions conducted by forces of two or more nations, usually undertaken within the structure of a coalition or alliance. See also **alliance; coalition; coalition action.** (JP 3-16)

multinational staff — A staff composed of personnel of two or more nations within the structure of a coalition or alliance. See also **integrated staff; joint staff; parallel staff.**

multinational warfare — Warfare conducted by forces of two or more nations, usually undertaken within the structure of a coalition or alliance. (JP 3-05)

multiple drill — See **multiple unit training assemblies.**

multiple inactive duty training periods — Two scheduled inactive duty training periods performed in one calendar day, each at least four hours in duration. No more than two inactive duty training periods may be performed in one day.

multiple independently targetable reentry vehicle — A reentry vehicle carried by a delivery system that can place one or more reentry vehicles over each of several separate targets. See also **maneuverable reentry vehicle; multiple reentry vehicle; reentry vehicle.**

multiple reentry vehicle — The reentry vehicle of a delivery system that places more than one reentry vehicle over an individual target. See also **maneuverable reentry vehicle; multiple independently targetable reentry vehicle; reentry vehicle.**

multiple unit training assemblies — Two or more unit training assemblies executed during one or more consecutive days. No more than two unit training assemblies may be performed in one calendar day.

multiple warning phenomenology — Deriving warning information from two or more systems observing separate physical phenomena associated with the same events to attain high credibility while being less susceptible to false reports or spoofing.

multiplexer — A device that combines (multiplexes) multiple input signals (information channels) into an aggregate signal (common channel) for transmission.

multi-Service doctrine — Fundamental principles that guide the employment of forces of two or more Services in coordinated action toward a common objective. It is ratified by two or more

Services, and is promulgated in multi-Service publications that identify the participating Services, e.g., Army-Navy doctrine. See also **joint doctrine; joint tactics, techniques, and procedures; multinational doctrine.**

multi-spectral imagery — (*) The image of an object obtained simultaneously in a number of discrete spectral bands. Also called **MSI.**

multi-spot ship — Those ships certified to have three or more adjacent landing areas. See also **spot.** (JP 3-04.1)

munition — (*) A complete device charged with explosives, propellants, pyrotechnics, initiating composition, or nuclear, biological, or chemical material for use in military operations, including demolitions. Certain suitably modified munitions can be used for training, ceremonial, or nonoperational purposes. Also called **ammunition.** (Note: In common usage, "munitions" [plural] can be military weapons, ammunition, and equipment.) See also **explosive ordnance.** (JP 3-11)

munitions effectiveness assessment — Conducted concurrently and interactively with battle damage assessment, the assessment of the military force applied in terms of the weapon system and munitions effectiveness to determine and recommend any required changes to the methodology, tactics, weapon system, munitions, fusing, and/or weapon delivery parameters to increase force effectiveness. Munitions effectiveness assessment is primarily the responsibility of operations with required inputs and coordination from the intelligence community. Also called **MEA.** See also **assessment; battle damage assessment; munition.** (JP 3-60)

mutual support — (*) That support which units render each other against an enemy, because of their assigned tasks, their position relative to each other and to the enemy, and their inherent capabilities. See also **close support; direct support; support.**

muzzle brake — A device attached to the muzzle of a weapon that utilizes escaping gas to reduce recoil.

muzzle compensator — A device attached to the muzzle of a weapon that utilizes escaping gas to control muzzle movement.

muzzle velocity — The velocity of a projectile with respect to the muzzle at the instant the projectile leaves the weapon.

N

named area of interest — The geographical area where information that will satisfy a specific information requirement can be collected. Named areas of interest are usually selected to capture indications of adversary courses of action, but also may be related to conditions of the battlespace. Also called **NAI.** See also **area of interest.** (JP 2-01.3)

napalm — 1. Powdered aluminum soap or similar compound used to gelatinize oil or gasoline for use in napalm bombs or flame throwers. 2. The resultant gelatinized substance.

nap-of-the-earth flight — See **terrain flight.**

narco-terrorism — Terrorism conducted to further the aims of drug traffickers. It may include assassinations, extortion, hijackings, bombings, and kidnappings directed against judges, prosecutors, elected officials, or law enforcement agents, and general disruption of a legitimate government to divert attention from drug operations. (JP 3-07.4)

national censorship — The examination and control under civil authority of communications entering, leaving, or transiting the borders of the United States, its territories, or its possessions. See also **censorship.**

National Command Authorities — The President and the Secretary of Defense or their duly deputized alternates or successors. Also called **NCA.**

National Communications System — The telecommunications system that results from the technical and operational integration of the separate telecommunications systems of the several executive branch departments and agencies having a significant telecommunications capability. Also called **NCS.**

national defense area — An area established on non-Federal lands located within the United States or its possessions or territories for the purpose of safeguarding classified defense information or protecting Department of Defense (DOD) equipment and/or materiel. Establishment of a national defense area temporarily places such non-Federal lands under the effective control of the Department of Defense and results only from an emergency event. The senior DOD representative at the scene will define the boundary, mark it with a physical barrier, and post warning signs. The landowner's consent and cooperation will be obtained whenever possible; however, military necessity will dictate the final decision regarding location, shape, and size of the national defense area. Also called **NDA.**

National Defense Reserve Fleet — 1. Including the Ready Reserve Force, a fleet composed of ships acquired and maintained by the Maritime Administration (MARAD) for use in mobilization or emergency. 2. Less the Ready Reserve Force, a fleet composed of the older dry cargo ships, tankers, troop transports, and other assets in MARAD's custody that are maintained at a relatively low level of readiness. They are acquired by MARAD from commercial ship operators under the provisions of the Merchant Marine Act of 1936 and are available only on mobilization or congressional declaration of an emergency. Because the ships are maintained in a state of minimum preservation, activation requires 30 to 90 days and extensive shipyard work, for many. Also called **NDRF.** See also **Ready Reserve Force.** (JP 3-02.2)

national emergency — A condition declared by the President or the Congress by virtue of powers previously vested in them that authorize certain emergency actions to be undertaken in the national interest. Action to be taken may include partial, full, or total mobilization of national resources. See also **mobilization.**

national information infrastructure — The nationwide interconnection of communications networks, computers, databases, and consumer electronics that make vast amounts of information available to users. The national information infrastructure encompasses a wide range of equipment, including cameras, scanners, keyboards, facsimile machines, computers, switches, compact disks, video and audio tape, cable, wire, satellites, fiber-optic transmission lines, networks of all types, televisions, monitors, printers, and much more. The friendly and adversary personnel who make decisions and handle the transmitted information constitute a critical component of the national information infrastructure. Also called **NII.** See also **defense information infrastructure; global information infrastructure; information.** (JP 3-13)

national infrastructure — **(*)** Infrastructure provided and financed by a NATO member in its own territory solely for its own forces (including those forces assigned to or designated for NATO). See also **infrastructure.**

national intelligence — Integrated departmental intelligence that covers the broad aspects of national policy and national security, is of concern to more than one department or agency, and transcends the exclusive competence of a single department or agency.

national intelligence estimate — A strategic estimate of the capabilities, vulnerabilities, and probable courses of action of foreign nations produced at the national level as a composite of the views of the intelligence community. Also called **NIE.**

national intelligence support team — A nationally sourced team composed of intelligence and communications experts from either Defense Intelligence Agency, Central Intelligence Agency, National Security Agency, or any combination of these agencies. Also called **NIST.** See also **intelligence; national intelligence.** (JP 2-0)

national intelligence surveys — Basic intelligence studies produced on a coordinated interdepartmental basis and concerned with characteristics, basic resources, and relatively unchanging natural features of a foreign country or other area.

National Military Command System — The priority component of the Global Command and Control System designed to support the National Command Authorities and Joint Chiefs of Staff in the exercise of their responsibilities. Also called **NMCS.**

National Military Joint Intelligence Center — National-level focal point for all defense intelligence activities in support of joint operations. Also called **NMJIC.** See also **intelligence; joint intelligence center.** (JP 3-33)

national military strategy — The art and science of distributing and applying military power to attain national objectives in peace and war. Also called **NMS.** See also **military strategy; national security strategy; strategy; theater strategy.** (JP 3-0)

national objectives — The aims, derived from national goals and interests, toward which a national policy or strategy is directed and efforts and resources of the nation are applied. See also **military objective.**

national policy — A broad course of action or statements of guidance adopted by the government at the national level in pursuit of national objectives.

National Reconnaissance Office — A Department of Defense agency tasked to ensure that the United States has the technology and spaceborne and airborne assets needed to acquire intelligence worldwide, including support to such functions as monitoring of arms control agreements, indications and warning, and the planning and conducting of military operations. This mission is accomplished through research and development, acquisition, and operation of spaceborne and airborne intelligence data collection systems. Also called **NRO.** (JP 2-0)

national security — A collective term encompassing both national defense and foreign relations of the United States. Specifically, the condition provided by: a. a military or defense advantage over any foreign nation or group of nations; b. a favorable foreign relations position; or c. a defense posture capable of successfully resisting hostile or destructive action from within or without, overt or covert. See also **security.**

National Security Agency/Central Security Service Representative — The senior theater or military command representative of the Director, National Security Agency/Chief, Central Security Service in a specific country or military command headquarters who provides the Director, National Security Agency, with information on command plans requiring cryptologic support. The National Security Agency/Central Security Service representative serves as a special advisor to the combatant commander for cryptologic matters, to include signals intelligence, communications security, and computer security. Also called **NCR.** See also **counterintelligence.** (JP 2-01.2)

National Security Council — A governmental body specifically designed to assist the President in integrating all spheres of national security policy. The President, Vice President, Secretary of State, and Secretary of Defense are statutory members. The Chairman of the Joint Chiefs of Staff; Director, Central Intelligence Agency; and the Assistant to the President for National Security Affairs serve as advisers. Also called **NSC.**

national security interests — The foundation for the development of valid national objectives that define US goals or purposes. National security interests include preserving US political identity, framework, and institutions; fostering economic well-being; and bolstering international order supporting the vital interests of the United States and its allies.

national security strategy — The art and science of developing, applying, and coordinating the instruments of national power (diplomatic, economic, military, and informational) to achieve objectives that contribute to national security. Also called **national strategy or grand strategy.** See also **military strategy; national military strategy; strategy; theater strategy.** (JP 3-0)

national shipping authority — **(*)** The organization within each Allied government responsible in time of war for the direction of its own merchant shipping. Also called **NSA.** (JP 4-01.2)

National Stock Number — The 13-digit stock number replacing the 11-digit Federal Stock Number. It consists of the 4-digit Federal Supply Classification code and the

9-digit National Item Identification Number. The National Item Identification Number consists of a 2-digit National Codification Bureau number designating the central cataloging office (whether North Atlantic Treaty Organization or other friendly country) that assigned the number and a 7-digit (xxx-xxxx) nonsignificant number. The number shall be arranged as follows: 9999-00-999-9999. Also called **NSN.**

national strategy — The art and science of developing and using the diplomatic, economic, and informational powers of a nation, together with its armed forces, during peace and war to secure national objectives. Also called **national security strategy or grand strategy.** See also **strategy.**

national support element — Any national organization or activity that supports national forces that are a part of a multinational force. Their mission is nation-specific support to units and common support that is retained by the nation. Also called **NSE.** See also **multinational force; support.** (JP 0-2)

nation assistance — Civil and/or military assistance rendered to a nation by foreign forces within that nation's territory during peacetime, crises or emergencies, or war based on agreements mutually concluded between nations. Nation assistance programs include, but are not limited to, security assistance, foreign internal defense, other US Code title 10 (DOD) programs, and activities performed on a reimbursable basis by Federal agencies or international organizations. (JP 3-57)

natural disaster — An emergency situation posing significant danger to life and property that results from a natural cause. See also **domestic emergencies.** (JP 3-07.6)

nautical chart — See **hydrographic chart.**

nautical mile — A measure of distance equal to one minute of arc on the Earth's surface. The United States has adopted the international nautical mile equal to 1,852 meters or 6,076.11549 feet. Also called **nm.**

nautical plotting chart — (*) An outline chart, devoid of hydrographic information, of a specific scale and projection, usually portraying a graticule and compass rose, designed to be ancillary to standard nautical charts, and produced either as an individual chart or a part of a coordinated series.

naval advanced logistic support site — An overseas location used as the primary transshipment point in the theater of operations for logistic support. A naval advanced logistic support site possesses full capabilities for storage, consolidation, and transfer of supplies and for support of forward-deployed units (including replacements units) during major contingency and wartime periods. Naval advanced logistic support sites, with port and airfield facilities in close proximity, are located within the theater of operations but not near the main battle areas, and must possess the throughput capacity required to accommodate incoming and outgoing intertheater airlift and sealift. When fully activated, the naval advanced logistic support site should consist of facilities and services provided by the host nation, augmented by support personnel located in the theater of operations, or both. Also called **NALSS or Naval ALSS.** See also **logistic support; naval forward logistic site; support; theater of operations.** (JP 3-35)

naval base — A naval base primarily for support of the forces afloat, contiguous to a port or anchorage, consisting of activities or facilities for which the Navy has operating responsibilities, together with interior lines of communications and the minimum surrounding area necessary for local security. (Normally, not greater than an area of 40 square miles.) See also **base complex.**

naval beach group — A permanently organized naval command within an amphibious force comprised of a commander and staff, a beachmaster unit, an amphibious construction battalion, and assault craft units, designed to provide an administrative group from which required naval tactical components may be made available to the attack force commander and to the amphibious landing force commander. Also called **NBG.** See also **shore party.**

naval coastal warfare — Coastal sea control, harbor defense, and port security, executed both in coastal areas outside the United States in support of national policy and in the United States as part of this Nation's defense. Also called **NCW.** (JP 3-10)

naval coastal warfare area — An assigned geographic area of operations which includes offshore waters, harbor approaches, harbors, ports, waterfront facilities, and those internal waters and rivers which provide access to port facilities. See also **area of operations; naval coastal warfare.** (JP 3-10)

naval coastal warfare commander — An officer designated to conduct naval coastal warfare missions within a designated naval coastal geographic area. Also called **NCWC.** (JP 3-10)

naval construction force — The combined construction units of the Navy, including primarily the mobile construction battalions and the amphibious construction battalions. These units are part of the operating forces and represent the Navy's capability for advanced base construction. Also called **NCF.**

naval control of shipping — **(*)** Control exercised by naval authorities of movement, routing, reporting, convoy organization, and tactical diversion of allied merchant shipping. It does not include the employment or active protection of such shipping.

naval control of shipping officer — A naval officer appointed to form merchant convoys and control and coordinate the routing and movements of such convoys, independently sailed merchant ships, and hospital ships in and out of a port or base, subject to the directions of the operational control authority.

naval control of shipping organization — The organization within the Navy that carries out the specific responsibilities of the Chief of Naval Operations to provide for the control and protection of movements of merchant ships in time of war. Also called **NCSORG.**

naval expeditionary warfare — Military operations mounted from the sea, usually on short notice, consisting of forward deployed, or rapidly deployable, self-sustaining naval forces tailored to achieve a clearly stated objective. Also called **NEW.** See also **expedition.** (JP 3-33)

naval forward logistic site — An overseas location, with port and airfield facilities nearby, which provides logistic support to naval forces within the theater of operations during major contingency and wartime periods. Naval forward logistic sites may be located in close proximity to main battle areas to permit forward staging of services, throughput of high priority cargo, advanced maintenance, and battle damage repair. Naval forward logistic sites are linked to in-theater naval advanced logistic support sites by intratheater airlift and sealift, but may also serve as transshipment points for intertheater movement of high-priority cargo into areas of direct combat. In providing fleet logistic support, naval forward logistic site capabilities may range from very austere to near those of a naval advanced logistic support site. Also called **NFLS or Naval FLS.** See also **logistic support; naval advanced logistic support site; staging.** (JP 3-35)

naval gunfire operations center — (*) The agency established in a ship to control the execution of plans for the employment of naval gunfire, process requests for naval gunfire support, and to allot ships to forward observers. Ideally located in the same ship as the supporting arms coordination center.

naval gunfire spotting team — The unit of a shore fire control party that designates targets; controls commencement, cessation, rate, and types of fire; and spots fire on the target. See also **field artillery observer; spotter.**

naval gunfire support — Fire provided by Navy surface gun systems in support of a unit or units tasked with achieving the commander's objectives. A subset of naval surface fire support. Also called **NGFS.** See also **naval surface fire support.** (JP 3-09.2)

naval mobile environmental team — A team of naval personnel organized, trained, and equipped to support maritime special operations by providing weather, oceanography, mapping, charting, and geodesy support. Also called **NMET.** (JP 3-05)

naval operation — 1. A naval action (or the performance of a naval mission) that may be strategic, operational, tactical, logistic, or training. 2. The process of carrying on or training for naval combat in order to gain the objectives of any battle or campaign.

naval or Marine (air) base — An air base for support of naval or Marine air units, consisting of landing strips, seaplane alighting areas, and all components of related facilities for which the Navy or Marine Corps has operating responsibilities, together with interior lines of communications and the minimum surrounding area necessary for local security. (Normally, not greater than an area of 20 square miles.) See also **base complex.**

naval port control office — The authority established at a port or port complex to coordinate arrangements for logistic support and harbor services to ships under naval control and to otherwise support the naval control of shipping organization.

naval special warfare — A designated naval warfare specialty that conducts operations in the coastal, riverine, and maritime environments. Naval special warfare emphasizes small, flexible, mobile units operating under, on, and from the sea. These operations are characterized by stealth, speed, and precise, violent application of force. Also called **NSW.** (JP 3-05)

naval special warfare forces — Those Active and Reserve Component Navy forces designated by the Secretary of Defense that are specifically organized, trained, and equipped to conduct and support special operations. Also called **NSW forces or NAVSOF.** (JP 3-05.5)

naval special warfare group — A permanent Navy echelon III major command to which most naval special warfare forces are assigned for some operational and all administrative purposes. It consists of a group headquarters with command and control, communications, and support staff; sea-air-land teams; and sea-air-land team delivery vehicle teams. Also called **NSWG.** (JP 3-05.3)

naval special warfare special operations component — The Navy special operations component of a unified or subordinate unified command or joint special operations task force. Also called **NAVSOC.** (JP 3-05)

naval special warfare task element — A provisional subordinate element of a naval special warfare task unit, employed to extend the command and control and support capabilities of its parent task unit. Also called **NSWTE.** See also **naval special warfare task unit.** (JP 3-05.3)

naval special warfare task group — A provisional naval special warfare organization that plans, conducts, and supports special operations in support of fleet commanders and joint force special operations component commanders. Also called **NSWTG.** (JP 3-05.3)

naval special warfare task unit — A provisional subordinate unit of a naval special warfare task group. Also called **NSWTU.** See also **naval special warfare task group.** (JP 3-05.3)

naval special warfare unit — A permanent Navy organization forward based to control and support attached naval special warfare forces. Also called **NSWU.** (JP 3-05.3)

naval stores — **(*)** Any articles or commodities used by a naval ship or station, such as equipment; consumable supplies; clothing; petroleum, oils, and lubricants; medical supplies; and ammunition.

naval support area — **(*)** A sea area assigned to naval ships detailed to support an amphibious operation. See also **fire support area.**

naval surface fire support — Fire provided by Navy surface gun and missile systems in support of a unit or units tasked with achieving the commander's objectives. Also called **NSFS.** See also **fire support.** (JP 3-02)

naval tactical data system — A complex of data inputs, user consoles, converters, adapters, and radio terminals interconnected with high-speed, general-purpose computers and its stored programs. Combat data is collected, processed, and composed into a picture of the overall tactical situation that enables the force commander to make rapid, accurate evaluations and decisions.

navigational grid — **(*)** A series of straight lines, superimposed over a conformal projection and indicating grid north, used as an aid to navigation. The interval of the grid lines is generally a multiple of 60 or 100 nautical miles. See also **military grid.**

navigation head — A transshipment point on a waterway where loads are transferred between water carriers and land carriers. A navigation head is similar in function to a railhead or truckhead.

navigation mode — In a flight control system, a control mode in which the flight path of an aircraft is automatically maintained by signals from navigation equipment.

Navy cargo handling battalion — A mobile logistic support unit capable of worldwide deployment in its entirety or in specialized detachments. It is organized, trained, and equipped to: a. load and off-load Navy and Marine Corps cargo carried in maritime pre-positioning ships and merchant breakbulk or container ships in all environments; b.

operate an associated temporary ocean cargo terminal; c. load and offload Navy and Marine Corps cargo carried in military-controlled aircraft; and d. operate an associated expeditionary air cargo terminal. Also called **NCHB or Navy CHB.** Three sources of Navy cargo handling battalions are: a. **Navy cargo handling and port group** — The active duty, cargo handling, battalion-sized unit composed solely of active duty personnel. Also called **NAVCHAPGRU.** b. **Naval Reserve cargo handling training battalion** — The active duty, cargo handling training battalion composed of both active duty and reserve personnel. Also called **NRCHTB.** c. **Naval Reserve cargo handling battalion** — A reserve cargo handling battalion composed solely of selected reserve personnel. Also called **NRCHB.** See also **maritime pre-positioning ships.**

Navy cargo handling force — The combined cargo handling units of the Navy, including primarily the Navy cargo handling and port group, the Naval Reserve cargo handling training battalion, and the Naval Reserve cargo handling battalion. These units are part of the operating forces and represent the Navy's capability for open ocean cargo handling. Also called **NCHF.** See also **Navy cargo handling battalion.**

Navy special operations component — The Navy component of a joint force special operations component. Also called **NAVSOC.** See also **Air Force special operations component; Army special operations component.** (JP 3-05.3)

Navy support element — The maritime pre-positioning force element that is composed of naval beach group staff and subordinate unit personnel, a detachment of Navy cargo handling force personnel, and other Navy components, as required. It is tasked with conducting the off-load and ship-to-shore movement of maritime pre-positioned equipment and/or supplies. Also called **NSE.** (JP 3-02.2)

Navy tactical air control center — See **tactical air control center.** (JP 3-09.3)

N-day — See **times.**

near miss (aircraft) — Any circumstance in flight where the degree of separation between two aircraft is considered by either pilot to have constituted a hazardous situation involving potential risk of collision.

near real time — (*) Pertaining to the timeliness of data or information which has been delayed by the time required for electronic communication and automatic data processing. This implies that there are no significant delays. Also called **NRT.** See also **real time.**

neatlines — (*) The lines that bound the body of a map, usually parallels and meridians. See also **graticule.**

need to know — A criterion used in security procedures that requires the custodians of classified information to establish, prior to disclosure, that the intended recipient must have access to the information to perform his or her official duties.

negative phase of the shock wave — The period during which the pressure falls below ambient and then returns to the ambient value. See also **positive phase of the shock wave.**

negative photo plane — (*) The plane in which a film or plate lies at the moment of exposure.

negligible risk (nuclear) — A degree of risk where personnel are reasonably safe, with the exceptions of dazzle or temporary loss of night vision. See also **emergency risk (nuclear).**

negotiations — A discussion between authorities and a barricaded offender or terrorist to effect hostage release and terrorist surrender. See also **antiterrorism.** (JP 3-07.2)

NEOPACK — Preassembled package of selected maps, charts, and other geographic materials of various scales to support the planning and conduct of noncombatant evacuation operations in selected countries or areas. See also **noncombatant evacuation operations; noncombatant evacuees.** (JP 3-07.5)

nerve agent — **(*)** A potentially lethal chemical agent which interferes with the transmission of nerve impulses.

net call sign — **(*)** A call sign which represents all stations within a net. See also **call sign.**

net, chain, cell system — Patterns of clandestine organization, especially for operational purposes. Net is the broadest of the three; it usually involves: a. a succession of echelons; and b. such functional specialists as may be required to accomplish its mission. When it consists largely or entirely of nonstaff employees, it may be called an agent net. Chain focuses attention upon the first of these elements; it is commonly defined as a series of agents and informants who receive instructions from and pass information to a principal agent by means of cutouts and couriers. Cell system emphasizes a variant of the first element of net; its distinctive feature is the grouping of personnel into small units that are relatively isolated and self-contained. In the interest of maximum security for the organization as a whole, each cell has contact with the rest of the organization only through an agent of the organization and a single member of the cell. Others in the cell do not know the agent, and nobody in the cell knows the identities or activities of members of other cells.

net (communications) — An organization of stations capable of direct communications on a common channel or frequency.

net control station — A communications station designated to control traffic and enforce circuit discipline within a given net. Also called **NCS.**

net explosive weight — The actual weight in pounds of explosive mixtures or compounds, including the trinitrotoluene equivalent of energetic material, that is used in determination of explosive limits and explosive quantity data arcs. Also called **NEW.** (JP 4-09)

net inventory assets — That portion of the total materiel assets that is designated to meet the materiel inventory objective. It consists of the total materiel assets less the peacetime materiel consumption and losses through normal appropriation and procurement leadtime periods.

net sweep — **(*)** In naval mine warfare, a two-ship sweep, using a netlike device, designed to collect drifting mines or scoop them up from the sea bottom.

net weight — Weight of a ground vehicle without fuel, engine oil, coolant, on-vehicle materiel, cargo, or operating personnel.

neutral — In combat and combat support operations, an identity applied to a track whose characteristics, behavior, origin, or nationality indicate that it is neither supporting nor opposing friendly forces. See also **hostile; suspect; unknown.**

neutrality — In international law, the attitude of impartiality during periods of war adopted by third states toward a belligerent and subsequently recognized by the belligerent, which creates rights and duties between the impartial states and the belligerent. In a United Nations enforcement action, the rules of neutrality apply to impartial members of the United Nations except so far as they are excluded by the obligation of such members under the United Nations Charter.

neutralization — **(*)** In mine warfare, a mine is said to be neutralized when it has been rendered, by external means, incapable of firing on passage of a target, although it may remain dangerous to handle.

neutralization fire — Fire which is delivered to render the target ineffective or unusable. See also **fire.**

neutralize — 1. As pertains to military operations, to render ineffective or unusable. 2. To render enemy personnel or material incapable of interfering with a particular operation. 3. To render safe mines, bombs, missiles, and boobytraps. 4. To make harmless anything contaminated with a chemical agent.

neutral state — In international law, a state that pursues a policy of neutrality during war. See also **neutrality.**

neutron induced activity — **(*)** Radioactivity induced in the ground or an object as a result of direct irradiation by neutrons.

news media representative — An individual employed by a civilian radio or television station, newspaper, newsmagazine, periodical, or news agency to gather and report on a newsworthy event. Also called **NMR.** See also **public affairs.** (JP 3-61)

nickname — A combination of two separate unclassified words that is assigned an unclassified meaning and is employed only for unclassified administrative, morale, or public information purposes.

night effect — **(*)** An effect mainly caused by variations in the state of polarization of reflected waves, which sometimes result in errors in direction finding bearings. The effect is most frequent at nightfall.

night vision device — Any electro-optical device that is used to detect visible and infrared energy and provide a visible image. Night vision goggles, forward-looking infrared, thermal sights, and low-light level television are night vision devices. Also called **NVD.** See also **forward-looking infrared; night vision goggles(s).** (JP 3-09.3)

night vision goggle(s) — An electro-optical image intensifying device that detects visible and near-infrared energy, intensifies the energy, and provides a visible image for night viewing. Night vision goggles can be either hand-held or helmet-mounted. Also called **NVG.** See also **night vision device.** (JP 3-09.3)

node — A location in a mobility system where a movement requirement is originated, processed for onward movement, or terminated.

node/command, control, communications, and computers node — The physical and functional grouping of communications and computer systems that provide terminating, switching, and gateway access services to support information exchange. See also **common operating environment; global grid.** (JP 6-02)

no-fire area — A land area designated by the appropriate commander into which fires or their effects are prohibited. Also called **NFA.** See also **fires.** (JP 3-09)

no-fire line — **(*)** A line short of which artillery or ships do not fire except on request or approval of the supported commander, but beyond which they may fire at any time without danger to friendly troops.

nominal filter — **(*)** A filter capable of cutting off a nominated minimum percentage by weight of solid particles greater than a stated micron size.

nominal focal length — **(*)** An approximate value of the focal length, rounded off to some standard figure, used for the classification of lenses, mirrors, or cameras.

nominal scale — See **principal scale; scale.**

nominal weapon — **(*)** A nuclear weapon producing a yield of approximately 20 kilotons. See also **kiloton weapon; megaton weapon; subkiloton weapon.**

nonair transportable — That which is not transportable by air by virtue of dimension, weight, or special characteristics or restrictions.

nonaligned state — A state that pursues a policy of nonalignment.

nonalignment — The political attitude of a state that does not associate or identify itself with the political ideology or objective espoused by other states, groups of states, or international causes, or with the foreign policies stemming therefrom. It does not preclude involvement, but expresses the attitude of no precommitment to a particular state (or block) or policy before a situation arises.

nonappropriated funds — Funds generated by DOD military and civilian personnel and their dependents and used to augment funds appropriated by the Congress to provide a comprehensive, morale-building welfare, religious, educational, and recreational program, designed to improve the well-being of military and civilian personnel and their dependents.

nonbattle injury — A person who becomes a casualty due to circumstances not directly attributable to hostile action or terrorist activity. Also called **NBI.**

noncombatant evacuation operations — Operations directed by the Department of State, the Department of Defense, or other appropriate authority whereby noncombatants are evacuated from foreign countries when their lives are endangered by war, civil unrest, or natural disaster to safe havens or to the United States. Also called **NEOs.** See also **evacuation; NEOPACK; noncombatant evacuees; operation; safe haven.** (JP 3-07)

noncombatant evacuees — 1. US citizens who may be ordered to evacuate by competent authority include: a. civilian employees of all agencies of the US Government and their dependents, except as noted in 2a below; b. military personnel of the US Armed Forces specifically designated for evacuation as noncombatants; and c. dependents of members of the US Armed Forces. 2. US (and non-US) citizens who may be authorized or assisted (but not necessarily ordered to evacuate) by competent authority include: a. civilian employees of US Government agencies and their dependents, who are residents in the country concerned on their own volition, but express the willingness to be evacuated; b. private US citizens and their dependents; c. military personnel and dependents of members of the US Armed Forces outlined in 1c above, short of an ordered evacuation; and d. designated aliens, including dependents of persons listed in 1a through 1c above, as prescribed by the Department of State. See also **noncombatant evacuation operations.** (JP 3-07.5)

noncontiguous facility — A facility for which the Service indicated has operating responsibility, but which is not located on, or in the immediate vicinity of, a base complex of that Service. Its area includes only that actually occupied by the facility, plus the minimum surrounding area necessary for close-in security. See also **base complex.**

nondeferrable issue demand — Issue demand related to specific periods of time that will not exist after the close of those periods, even though not satisfied during the period.

nondeployable account — An account where Reservists (officer and enlisted), either in units or individually, are assigned to a reserve component category or a training/retired category when the individual has not completed initial active duty for training or its equivalent. Reservists in a nondeployable account are not considered as trained strength assigned to units or mobilization positions and are not deployable overseas on land with those units or mobilization positions. See also **training pipeline.**

nondestructive electronic warfare — Those electronic warfare actions, not including employment of wartime reserve modes, that deny, disrupt, or deceive rather than damage or destroy. See also **electronic warfare.** (JP 3-51)

nonexpendable supplies and materiel — Supplies not consumed in use that retain their original identity during the period of use, such as weapons, machines, tools, and equipment.

nonfixed medical treatment facility — A medical treatment facility designed to be moved from place to place, including medical treatment facilities afloat.

nongovernmental organizations — Transnational organizations of private citizens that maintain a consultative status with the Economic and Social Council of the United Nations. Nongovernmental organizations may be professional associations, foundations, multinational businesses, or simply groups with a common interest in humanitarian assistance activities (development and relief). "Nongovernmental organizations" is a term normally used by non-United States organizations. Also called **NGOs.** (JP 1-06)

nonhostile casualty — A person who becomes a casualty due to circumstances not directly attributable to hostile action or terrorist activity. Casualties due to the elements, self-inflicted wounds, and combat fatigue are nonhostile casualties. Also called **NHCS.** See also **casualty; casualty type; hostile casualty.**

nonlethal weapons — Weapons that are explicitly designed and primarily employed so as to incapacitate personnel or material, while minimizing fatalities, permanent injury to personnel, and undesired damage to property and the environment. a. Unlike conventional lethal weapons that destroy their targets through blast, penetration, and fragmentation, nonlethal weapons employ means other than gross physical destruction to prevent the target from functioning. b. Nonlethal weapons are intended to have one, or both, of the following characteristics: (1) They have relatively reversible effects on personnel or materiel. (2) They affect objects differently within their area of influence.

non-linear approach — (*) In approach and landing systems, a final approach in which the nominal flight path is not a straight line.

nonorganic transportation requirement — Unit personnel and cargo for which the transportation source must be an outside agency, normally a component of US Transportation Command.

nonpersistent agent — A chemical agent that when released dissipates and/or loses its ability to cause casualties after 10 to 15 minutes. (JP 3-11)

nonprecision approach — Radar-controlled approach or an approach flown by reference to navigation aids in which glide slope information is not available. See also **final approach; precision approach.** (JP 3-04.1)

nonprior service personnel — Individuals without any prior military service, who have not completed basic inactive duty training, and who receive a commission in or enlist directly into an Armed Force of the United States.

nonrecurring demand — A request by an authorized requisitioner to satisfy a materiel requirement known to be a one-time occurrence. This materiel is required to provide initial stockage allowances, to meet planned program requirements, or to satisfy a one-time project or maintenance requirement. Nonrecurring demands normally will not be considered by the supporting supply system in the development of demand-based elements of the requirements computation.

non-registered publication — **(*)** A publication which bears no register number and for which periodic accounting is not required.

nonscheduled units — Units of the landing force held in readiness for landing during the initial unloading period, but not included in either scheduled or on-call waves. This category usually includes certain of the combat support units and most of the combat service support units with higher echelon (division and above) reserve units of the landing force. Their landing is directed when the need ashore can be predicted with a reasonable degree of accuracy.

non-self-sustaining containership — A containership that does not have a built-in capability to load or off-load containers, and requires a port crane or craneship service. Also called **NSSCS**. See also **containership; self-sustaining containership.** (JP 4-01.7)

nonstandard item — An item of supply determined by standardization action as not authorized for procurement.

nonstandard unit — A force requirement identified in a time-phased force and deployment data for which movement characteristics have not been described in the type unit characteristics file. The planner is required to submit detailed movement characteristics for these units.

nonstocked item — An item that does not meet the stockage criteria for a given activity, and therefore is not stocked at the particular activity.

nonstrategic nuclear forces — Those nuclear-capable forces located in an operational area with a capability to employ nuclear weapons by land, sea, or air forces against opposing forces, supporting installations, or facilities. Such forces may be employed, when authorized by competent authority, to support operations that contribute to the accomplishment of the commander's mission within the theater of operations.

non-submarine contact chart — **(*)** A special naval chart, at a scale of 1:100,000 to 1:1,000,000, showing bathymetry, bottom characteristics, wreck data, and non-submarine contact data for coastal and off-shore waters. It is designed for use in conducting submarine and antisubmarine warfare operations. Also called non-sub contact chart.

non-unit record — A time-phased force and deployment data file entry for non-unit-related cargo and personnel. Characteristics include using and providing organization, type of movement, routing data, cargo category, weight, volume, area required, and number of personnel requiring transportation.

non-unit-related cargo — All equipment and supplies requiring transportation to an operational area, other than those identified as the equipment or accompanying supplies of a specific unit (e.g., resupply, military support for allies, and support for nonmilitary programs, such as civil relief). Also called **NURC.**

non-unit-related personnel — All personnel requiring transportation to or from an operational area, other than those assigned to a specific unit (e.g., filler personnel; replacements; temporary duty/temporary additional duty personnel; civilians; medical evacuees; and retrograde personnel). Also called **NRP** or **NUP.**

non-US forces — Includes all armed forces of states other than US forces. US forces may act in defense of non-US forces when so designated by US National Command Authorities.

normal charge — Charge employing a standard amount of propellant to fire a gun under ordinary conditions, as compared with a reduced charge. See also **reduced charge.**

normal impact effect — See **cardinal point effect.**

normal intelligence reports — A category of reports used in the dissemination of intelligence, conventionally used for the immediate dissemination of individual items of intelligence. See also **intelligence reporting; specialist intelligence report.**

normal lighting — **(*)** Lighting of vehicles as prescribed or authorized by the law of a given country without restrictions for military reasons. See also **reduced lighting.**

normal operations — Generally and collectively, the broad functions that a combatant commander undertakes when assigned responsibility for a given geographic or functional area. Except as otherwise qualified in certain unified command plan paragraphs that relate to particular commands, "normal operations" of a combatant commander include: planning and execution of operations throughout the range of military operations; planning and conduct of cold war activities; planning and administration of military assistance; and maintaining the relationships and exercising the directive or coordinating authority prescribed in JP 0-2 and JP 4-01.

North American Aerospace Defense Command — A bi-national command of the US and Canada that provides aerospace surveillance, warning and assessment of aerospace attack, and maintains the sovereignty of US and Canadian airspace. Also called **NORAD.**

no-strike list — A list of geographic areas, complexes, or installations not planned for capture or destruction. Attacking these may violate the law of armed conflict or interfere with friendly relations with indigenous personnel or governments. Also called **NSL.** See also **law of armed conflict.** (JP 3-60)

notice to airmen — A notice containing information concerning the establishment, condition, or change in any aeronautical facility, service, procedures, or hazard, the timely knowledge of which is essential to personnel concerned with flight operations. Also called **NOTAM.**

notional ship — A theoretical or average ship of any one category used in transportation planning (e.g., a Liberty ship for dry cargo; a T-2 tanker for bulk petroleum, oils, and lubricants; a personnel transport of 2,400 troop spaces).

not mission capable, maintenance — Material condition indicating that systems and equipment are not capable of performing any of their assigned missions because of maintenance requirements. Also called **NMCM.** See also **not mission capable, supply.**

not mission capable, supply — Material condition indicating that systems and equipment are not capable of performing any of their assigned missions because of maintenance work stoppage due to a supply shortage. Also called **NMCS.** See also **not mission capable, maintenance.**

not seriously injured — The casualty status of a person whose injury may or may not require hospitalization; medical authority does not classify as very seriously injured, seriously injured, or incapacitating illness or injury; and the person can communicate with the next of kin. Also called **NSI.** See also **casualty status.**

no-wind position — See **air position.**

nuclear accident — See **nuclear weapon(s) accident.**

nuclear airburst — **(*)** The explosion of a nuclear weapon in the air, at a height greater than the maximum radius of the fireball. See also **types of burst.**

nuclear, biological, and chemical-capable nation — A nation that has the capability to produce and employ one or more types of nuclear, biological, and chemical weapons across the full range of military operations and at any level of war in order to achieve political and military objectives. (JP 3-11)

nuclear, biological, and chemical conditions — See **nuclear, biological, and chemical environment.** (JP 3-11)

nuclear, biological, and chemical defense — Defensive measures that enable friendly forces to survive, fight, and win against enemy use of nuclear, biological, or chemical (NBC) weapons and agents. US forces apply NBC defensive measures before and during integrated warfare. In integrated warfare, opposing forces employ nonconventional weapons along with conventional weapons (NBC weapons are nonconventional). See also **integrated warfare.** (JP 3-11)

nuclear, biological, and chemical environment — Environments in which there is deliberate or accidental employment, or threat of employment, of nuclear, biological, or chemical weapons; deliberate or accidental attacks or contamination with toxic industrial materials, including toxic industrial chemicals; or deliberate or accidental attacks or contamination with radiological (radioactive) materials. See also **contamination.** (JP 3-11)

nuclear bonus effects — **(*)** Desirable damage or casualties produced by the effects from friendly nuclear weapons that cannot be accurately calculated in targeting as the uncertainties involved preclude depending on them for a militarily significant result.

nuclear burst — See **types of burst.**

nuclear certifiable — **(*)** Indicates a unit or vehicle possessing the potential of passing functional tests and inspections of all normal and emergency systems affecting the nuclear weapons.

nuclear certified — See **nuclear certified delivery unit; nuclear certified delivery vehicle.**

nuclear certified delivery unit — **(*)** Any level of organization and support elements which are capable of executing nuclear missions in accordance with appropriate bilateral arrangements and NATO directives. See also **nuclear delivery unit.**

nuclear certified delivery vehicle — **(*)** A delivery vehicle whose compatibility with a nuclear weapon has been certified by the applicable nuclear power through formal procedures. See also **nuclear delivery vehicle.**

nuclear cloud — **(*)** An all-inclusive term for the volume of hot gases, smoke, dust, and other particulate matter from the nuclear bomb itself and from its environment, which is carried aloft in conjunction with the rise of the fireball produced by the detonation of the nuclear weapon.

nuclear collateral damage — Undesired damage or casualties produced by the effects from friendly nuclear weapons.

nuclear column — **(*)** A hollow cylinder of water and spray thrown up from an underwater burst of a nuclear weapon, through which the hot, high-pressure gases formed in the explosion are vented to the atmosphere. A somewhat similar column of dirt is formed in an underground explosion.

nuclear commitment — **(*)** A statement by a NATO member that specific forces have been committed or will be committed to NATO in a nuclear only or dual capable role.

nuclear contact surface burst — An explosion of a nuclear weapon whose center of energy is at the surface of land or water.

nuclear coordination — A broad term encompassing all the actions involved with planning nuclear strikes, including liaison between commanders, for the purpose of satisfying support requirements or because of the extension of weapons effects into the territory of another.

nuclear damage — **(*)** 1. **Light Damage** — Damage which does not prevent the immediate use of equipment or installations for which it was intended. Some repair by the user may be required to make full use of the equipment or installations. 2. **Moderate Damage** — Damage which prevents the use of equipment or installations until extensive repairs are made. 3. **Severe Damage** — Damage which prevents use of equipment or installations permanently.

nuclear damage assessment — **(*)** The determination of the damage effect to the population, forces, and resources resulting from actual nuclear attack. It is performed during and after an attack. The operational significance of the damage is not evaluated in this assessment.

nuclear defense — **(*)** The methods, plans, and procedures involved in establishing and exercising defensive measures against the effects of an attack by nuclear weapons or radiological warfare agents. It encompasses both the training for, and the implementation of, these methods, plans, and procedures. See also **nuclear, biological, and chemical defense; radiological defense.**

nuclear delivery unit — **(*)** Any level of organization capable of employing a nuclear weapon system or systems when the weapon or weapons have been released by proper authority.

nuclear delivery vehicle — **(*)** That portion of the weapon system which provides the means of delivery of a nuclear weapon to the target.

nuclear detonation detection and reporting system — **(*)** A system deployed to provide surveillance coverage of critical friendly target areas, and indicate place, height of burst, yield, and ground zero of nuclear detonations. Also called **NUDETS.**

nuclear dud — A nuclear weapon that, when launched at or emplaced on a target, fails to provide any explosion of that part of the weapon designed to produce the nuclear yield.

nuclear energy — All forms of energy released in the course of a nuclear fission or nuclear transformation.

nuclear exoatmospheric burst — The explosion of a nuclear weapon above the sensible atmosphere (above 120 kilometers) where atmospheric interaction is minimal. See also **types of burst.**

nuclear incident — An unexpected event involving a nuclear weapon, facility, or component, resulting in any of the following, but not constituting a nuclear weapon(s) accident: a. an increase in the possibility of explosion or radioactive contamination; b. errors committed in the assembly, testing, loading, or transportation of equipment, and/or the malfunctioning of equipment and materiel which could lead to an unintentional operation of all or part of the weapon arming and/or firing sequence, or which could lead to a substantial change in yield, or increased dud probability; and c. any act of God, unfavorable environment, or condition resulting in damage to the weapon, facility, or component.

nuclear intelligence — Intelligence derived from the collection and analysis of radiation and other effects resulting from radioactive sources. Also called **NUCINT.** See also **intelligence.** (JP 2-0)

nuclear logistic movement — The transport of nuclear weapons in connection with supply or maintenance operations. Under certain specified conditions, combat aircraft may be used for such movements.

nuclear nation — **(*)** Military nuclear powers and civil nuclear powers.

nuclear parity — A condition at a given point in time when opposing forces possess nuclear offensive and defensive systems approximately equal in overall combat effectiveness.

nuclear planning system — A system composed of personnel, directives, and electronic data processing systems to directly support geographic nuclear combatant commanders in developing, maintaining, and disseminating nuclear operation plans.

nuclear proximity-surface burst — An explosion of a nuclear weapon at a height less than the maximum radius of its fireball, but low enough to facilitate cratering and/or the propagation of a shock wave into the ground.

nuclear radiation — **(*)** Particulate and electromagnetic radiation emitted from atomic nuclei in various nuclear processes. The important nuclear radiations, from the weapon standpoint, are alpha and beta particles, gamma rays, and neutrons. All nuclear radiations are ionizing radiations, but the reverse is not true; X-rays for example, are included among ionizing radiations, but they are not nuclear radiations since they do not originate from atomic nuclei.

nuclear reactor — A facility in which fissile material is used in a self-supporting chain reaction (nuclear fission) to produce heat and/or radiation for both practical application and research and development.

nuclear round — See **complete round.**

nuclear safety line — **(*)** A line selected, if possible, to follow well-defined topographical features and used to delineate levels of protective measures, degrees of damage or risk to friendly troops, and/or to prescribe limits to which the effects of friendly weapons may be permitted to extend.

nuclear stalemate — A concept that postulates a situation wherein the relative strength of opposing nuclear forces results in mutual deterrence against employment of nuclear forces.

nuclear strike warning — **(*)** A warning of impending friendly or suspected enemy nuclear attack.

nuclear support — The use of nuclear weapons against hostile forces in support of friendly air, land, and naval operations. See also **immediate nuclear support; preplanned nuclear support.**

nuclear transmutation — Artificially induced modification (nuclear reaction) of the constituents of certain nuclei, thus giving rise to different nuclides.

nuclear underground burst — **(*)** The explosion of a nuclear weapon in which the center of the detonation lies at a point beneath the surface of the ground. See also **types of burst.**

nuclear underwater burst — **(*)** The explosion of a nuclear weapon in which the center of the detonation lies at a point beneath the surface of the water. See also **types of burst.**

nuclear vulnerability assessment — **(*)** The estimation of the probable effect on population, forces, and resources from a hypothetical nuclear attack. It is performed predominantly in the preattack period; however, it may be extended to the transattack or postattack periods.

nuclear warfare — **(*)** Warfare involving the employment of nuclear weapons. See also **postattack period; transattack period.**

nuclear warning message — A warning message that must be disseminated to all affected friendly forces any time a nuclear weapon is to be detonated if effects of the weapon will have impact upon those forces.

nuclear weapon — **(*)** A complete assembly (i.e., implosion type, gun type, or thermonuclear type), in its intended ultimate configuration which, upon completion of the prescribed arming, fusing, and firing sequence, is capable of producing the intended nuclear reaction and release of energy.

nuclear weapon degradation — The degeneration of a nuclear warhead to such an extent that the anticipated nuclear yield is lessened.

nuclear weapon employment time — (*) The time required for delivery of a nuclear weapon after the decision to fire has been made.

nuclear weapon exercise — (*) An operation not directly related to immediate operational readiness. It includes removal of a weapon from its normal storage location, preparing for use, delivery to an employment unit, and the movement in a ground training exercise, to include loading aboard an aircraft or missile and return to storage. It may include any or all of the operations listed above, but does not include launching or flying operations. Typical exercises include aircraft generation exercises, ground readiness exercises, ground tactical exercises, and various categories of inspections designed to evaluate the capability of the unit to perform its prescribed mission. See also **immediate operational readiness; nuclear weapon maneuver.**

nuclear weapon maneuver — (*) An operation not directly related to immediate operational readiness. It may consist of all those operations listed for a nuclear weapon exercise and is extended to include flyaway in combat aircraft, but does not include expenditure of the weapon. Typical maneuvers include nuclear operational readiness maneuvers and tactical air operations. See also **immediate operational readiness; nuclear weapon exercise.**

nuclear weapon(s) accident — An unexpected event involving nuclear weapons or radiological nuclear weapon components that results in any of the following; a. accidental or unauthorized launching, firing, or use by United States forces or United States supported allied forces of a nuclear-capable weapon system that could create the risk of an outbreak of war; b. nuclear detonation; c. nonnuclear detonation or burning of a nuclear weapon or radiological nuclear weapon component; d. radioactive contamination; e. seizure, theft, loss, or destruction of a nuclear weapon or radiological nuclear weapon component, including jettisoning; and f. public hazard, actual or implied.

nuclear weapons state — See **military nuclear power.**

nuclear weapons surety — Materiel, personnel, and procedures that contribute to the security, safety, and reliability of nuclear weapons and to the assurance that there will be no nuclear weapon accidents, incidents, unauthorized weapon detonations, or degradation in performance at the target.

nuclear yields — The energy released in the detonation of a nuclear weapon, measured in terms of the kilotons or megatons of trinitrotoluene required to produce the same energy release. Yields are categorized as follows:

very low — less than 1 kiloton;
low — 1 kiloton to 10 kilotons;
medium — over 10 kilotons to 50 kilotons;
high — over 50 kilotons to 500 kilotons;
very high — over 500 kilotons.

See also **nominal weapon; subkiloton weapon**.

nuisance minefield — (*) A minefield laid to delay and disorganize the enemy and to hinder the use of an area or route. See also **minefield.**

number . . . in (out) — (*) In artillery, term used to indicate a change in status of weapon number _______________.

numbered beach — In amphibious operations, a subdivision of a colored beach, designated for the assault landing of a battalion landing team or similarly sized unit, when landed as part of a larger force. (JP 3-02)

numbered fleet — A major tactical unit of the Navy immediately subordinate to a major fleet command and comprising various task forces, elements, groups, and units for the purpose of prosecuting specific naval operations. See also **fleet.**

numbered wave — See **wave.**

numerical scale — See **scale.**

O

objective — 1. The clearly defined, decisive, and attainable goals towards which every military operation should be directed. 2. The specific target of the action taken (for example, a definite terrain feature, the seizure or holding of which is essential to the commander's plan, or, an enemy force or capability without regard to terrain features). See also **target.** (JP 3-0)

objective area — **(*)** A defined geographical area within which is located an objective to be captured or reached by the military forces. This area is defined by competent authority for purposes of command and control. Also called **OA.**

objective force level — The level of military forces that needs to be attained within a finite time frame and resource level to accomplish approved military objectives, missions, or tasks. See also **military requirement.**

obligated reservist — An individual who has a statutory requirement imposed by the Military Selective Service Act of 1967 or Section 651, Title 10, United States Code, to serve on active duty in the armed forces or to serve while not on active duty in a Reserve Component for a period not to exceed that prescribed by the applicable statute.

oblique air photograph — **(*)** An air photograph taken with the camera axis directed between the horizontal and vertical planes. Commonly referred to as an "oblique." a. **High Oblique.** One in which the apparent horizon appears. b. **Low Oblique.** One in which the apparent horizon does not appear.

oblique air photograph strip — Photographic strip composed of oblique air photographs.

obliquity — The characteristic in wide-angle or oblique photography that portrays the terrain and objects at such an angle and range that details necessary for interpretation are seriously masked or are at a very small scale, rendering interpretation difficult or impossible.

observation helicopter — **(*)** Helicopter used primarily for observation and reconnaissance, but which may be used for other roles.

observation post — **(*)** A position from which military observations are made, or fire directed and adjusted, and which possesses appropriate communications; may be airborne. Also called **OP.**

observed fire — **(*)** Fire for which the point of impact or burst can be seen by an observer. The fire can be controlled and adjusted on the basis of observation. See also **fire.**

observed fire procedure — **(*)** A standardized procedure for use in adjusting indirect fire on a target.

observer-target line — **(*)** An imaginary straight line from the observer/spotter to the target. See also **spotting line.**

observer-target range — The distance along an imaginary straight line from the observer or spotter to the target.

obstacle — Any obstruction designed or employed to disrupt, fix, turn, or block the movement of an opposing force, and to impose additional losses in personnel, time, and equipment on the opposing force. Obstacles can be natural, manmade, or a combination of both. (JP 3-15)

obstacle belt — A brigade-level command and control measure, normally given graphically, to show where within an obstacle zone the ground tactical commander plans to limit friendly obstacle employment and focus the defense. It assigns an intent to the obstacle plan and provides the necessary guidance on the overall effect of obstacles within a belt. See also **obstacle.** (JP 3-15)

obstacle clearing — The total elimination or neutralization of obstacles.

obstacle restricted areas — A command and control measure used to limit the type or number of obstacles within an area. See also **obstacle.** (JP 3-15)

obstacle zone — A division-level command and control measure, normally done graphically, to designate specific land areas where lower echelons are allowed to employ tactical obstacles. See also **obstacle.** (JP 3-15)

obstructor — **(*)** In naval mine warfare, a device laid with the sole object of obstructing or damaging mechanical minesweeping equipment.

occupational and environmental health threats — Threats to the health of military personnel and to military readiness created by exposure to hazardous agents, environmental contamination, or toxic industrial materials. See also **health threat.** (JP 4-02)

occupation currency — See **military currency.**

occupied territory — Territory under the authority and effective control of a belligerent armed force. The term is not applicable to territory being administered pursuant to peace terms, treaty, or other agreement, express or implied, with the civil authority of the territory. See also **civil affairs agreement.**

Ocean Cargo Clearance Authority — The Military Traffic Management Command activity that books Department of Defense (DOD) sponsored cargo and passengers for surface movement, performs related contract administration, and accomplishes export and import surface traffic management functions for DOD cargo moving within the Defense Transportation System. Also called **OCCA.**

ocean convoy — **(*)** A convoy whose voyage lies, in general, outside the continental shelf. See also **convoy.**

ocean manifest — **(*)** A detailed listing of the entire cargo loaded into any one ship showing all pertinent data which will readily identify such cargo and where and how the cargo is stowed.

oceanography — The study of the sea, embracing and integrating all knowledge pertaining to the sea and its physical boundaries, the chemistry and physics of seawater, and marine biology.

ocean station ship — **(*)** A ship assigned to operate within a specified area to provide several services, including search and rescue, meteorological information, navigational aid, and communications facilities.

offensive counterair — Offensive operations to destroy, disrupt, or neutralize enemy aircraft, missiles, launch platforms, and their supporting structures and systems both before and after launch, but as close to their source as possible. Offensive counterair operations range throughout enemy territory and are generally conducted at the initiative of friendly forces. These operations include attack operations, fighter sweep, escort, and suppression of enemy air defenses. Also called **OCA.** See also **counterair; defensive counterair; operation.** (JP 3-01)

offensive counterair attack operations — Offensive action in support of the offensive counterair mission against surface targets that contribute to the enemy's air power capabilities. The objective of attack operations is to prevent the hostile use of aircraft and missile forces by attacking targets such as missile launch sites, airfields, naval vessels, command and control nodes, munitions stockpiles, and supporting infrastructure. Attack operations may be performed by fixed- or rotary-wing aircraft, surface-to-surface weapons, special operations forces, or ground forces. Also called **OCA attack ops.** See also **counterair; offensive counterair.**

offensive information operations — The integrated use of assigned and supporting capabilities and activities, mutually supported by intelligence, to affect adversary decision makers to achieve or promote specific objectives. These capabilities and activities include but are not limited to operations security, military deception, psychological operations, electronic warfare, physical attack and/or destruction, and special information operations, and could also include computer network attack. See also **computer network attack; defensive information operations; electronic warfare; information operations; intelligence; military deception; operations security; psychological operations; special information operations.** (JP 3-13)

offensive minefield — **(*)** In naval mine warfare, a minefield laid in enemy territorial water or waters under enemy control.

officer in tactical command — In maritime usage, the senior officer present eligible to assume command, or the officer to whom the senior officer has delegated tactical command. Also called **OTC.**

officer of the deck — The officer of the deck under way has been designated by the commanding officer to be in charge of the ship, including its safe and proper operation. The officer of the deck reports directly to the commanding officer for the safe navigation and general operation of the ship, to the executive officer (and command duty officer if appointed) for carrying out the ship's routine, and to the navigator on sighting navigational landmarks and making course and speed changes. Also called **OOD.** (JP 3-04.1)

official information — Information that is owned by, produced for or by, or is subject to the control of the United States Government.

off-load preparation party — A temporary task organization of Navy and Marine maintenance, embarkation, equipment operators, and cargo-handling personnel deployed to the maritime pre-positioning ship before or during its transit to the objective area to prepare the ship's off-load systems and embarked equipment for off-load. Also called **OPP.** See also **task organization.** (JP 4-01.8)

offset bombing — **(*)** Any bombing procedure which employs a reference or aiming point other than the actual target.

offset costs — Costs for which funds have been appropriated but will not be obligated because of a contingency operation. See also **contingency operation.** (JP 1-06)

offset distance (nuclear) — The distance the desired ground zero or actual ground zero is offset from the center of an area target or from a point target.

offset lasing — The technique of aiming a laser designator at a point other than the target and, after laser acquisition, moving the laser to designate the target for terminal attack guidance. See also **laser target designator.** (JP 3-09.1)

offshore assets — Oil and gas facilities, mining and industrial installations, ocean thermal energy conversion facilities, deep water ports, aids to navigation, and nuclear power plants located or in operation seaward of the coastline.

offshore bulk fuel system — The system used for transferring fuel from points offshore to reception facilities on the beach. It consists of two subsystems: amphibious assault bulk fuel system and the offshore petroleum discharge system. See also **amphibious assault bulk fuel system; offshore petroleum discharge system.** (JP 4-01.6)

offshore patrol — **(*)** A naval defense patrol operating in the outer areas of navigable coastal waters. It is a part of the naval local defense forces consisting of naval ships and aircraft and operates outside those areas assigned to the inshore patrol.

offshore petroleum discharge system — Provides a semipermanent, all-weather facility for bulk transfer of petroleum, oils, and lubricants (POL) directly from an offshore tanker to a beach termination unit (BTU) located immediately inland from the high watermark. POL then is either transported inland or stored in the beach support area. Major offshore petroleum discharge systems (OPDS) components are: the OPDS tanker with booster pumps and spread mooring winches; a recoverable single anchor leg mooring (SALM) to accommodate tankers of up to 70,000 deadweight tons; ship to SALM hose lines; up to 4 miles of 6-inch (internal diameter) conduit for pumping to the beach; and two BTUs to interface with the shoreside systems. OPDS can support a two-line system for multiproduct discharge, but ship standoff distance is reduced from 4 to 2 miles. Amphibious construction battalions install the OPDS with underwater construction team assistance. OPDS are embarked on selected Ready Reserve Force tankers modified to support the system. Also called **OPDS.** See also **facility; petroleum, oils, and lubricants; single-anchor leg mooring.** (JP 4-01.6)

off-the-shelf item — An item that has been developed and produced to military or commercial standards and specifications, is readily available for delivery from an industrial source, and may be procured without change to satisfy a military requirement.

oiler — **(*)** A naval or merchant tanker specially equipped and rigged for replenishing other ships at sea.

on berth — Said of a ship when it is properly moored to a quay, wharf, jetty, pier, or buoy or when it is at anchor and available for loading or discharging passengers and cargo.

on-call — 1. A term used to signify that a prearranged concentration, air strike, or final protective fire may be called for. 2. Preplanned, identified force or materiel requirements without designated time-phase and destination information. Such requirements will be called forward upon order of competent authority. See also **call for fire.**

on-call resupply — A resupply mission planned before insertion of a special operations team into the operations area but not executed until requested by the operating team. See also **automatic resupply; emergency resupply.** (JP 3-05.3)

on-call targets — Planned targets that are known to exist in an operational area and are located in sufficient time for deliberate planning to meet emerging situations specific to campaign objectives. See also **on-call; operational area; target.** (JP 3-60)

on-call target (nuclear) — A planned nuclear target other than a scheduled nuclear target for which a need can be anticipated but which will be delivered upon request rather than at a specific time. Coordination and warning of friendly troops and aircraft are mandatory.

on-call wave — See **wave.**

one day's supply — **(*)** A unit or quantity of supplies adopted as a standard of measurement, used in estimating the average daily expenditure under stated conditions. It may also be expressed in terms of a factor, e.g., rounds of ammunition per weapon per day.

one-look circuit — **(*)** A mine circuit which requires actuation by a given influence once only.

on hand — The quantity of an item that is physically available in a storage location and contained in the accountable property book records of an issuing activity.

on-scene commander — 1. The person designated to coordinate the rescue efforts at the rescue site. 2. Federal officer designated to direct federal crisis and consequence management efforts at the scene of a terrorist or weapons of mass destruction incident. Also called **OSC.**

on-station time — The time an aircraft can remain on station. May be determined by endurance or orders.

open improved storage space — Open area that has been graded and hard surfaced or prepared with topping of some suitable material so as to permit effective materials handling operations. See also **storage.**

open ocean — Ocean limit defined as greater than 12 nautical miles (nm) from shore, as compared with high seas that are over 200 nm from shore. See also **contiguous zone.**

open route — **(*)** A route not subject to traffic or movement control restrictions.

open-source intelligence — Information of potential intelligence value that is available to the general public. Also called **OSINT.** See also **intelligence.** (JP 2-0)

open unimproved wet space — That water area specifically allotted to and usable for storage of floating equipment. See also **storage.**

operating forces — Those forces whose primary missions are to participate in combat and the integral supporting elements thereof. See also **combat forces; combat service support element; combat support elements.**

operating level of supply — The quantities of materiel required to sustain operations in the interval between requisitions or the arrival of successive shipments. These quantities should be based on the established replenishment period (monthly, quarterly, etc.) See also **level of supply.**

operation — 1. A military action or the carrying out of a strategic, operational, tactical, service, training, or administrative military mission. 2. The process of carrying on combat, including movement, supply, attack, defense, and maneuvers needed to gain the objectives of any battle or campaign.

operational architecture — Descriptions of the tasks, operational elements, and information flows required to accomplish or support a warfighting function.

operational area — An overarching term encompassing more descriptive terms for geographic areas in which military operations are conducted. Operational areas include, but are not limited to, such descriptors as area of responsibility, theater of war, theater of operations, joint operations area, amphibious objective area, joint special operations area, and area of operations. See also **amphibious objective area; area of operations; area of responsibility; joint operations area; joint special operations area; theater of operations; theater of war.** (JP 3-0)

operational art — The employment of military forces to attain strategic and/or operational objectives through the design, organization, integration, and conduct of strategies, campaigns, major operations, and battles. Operational art translates the joint force commander's strategy into operational design and, ultimately, tactical action, by integrating the key activities at all levels of war. (JP 3-0)

operational authority — That authority exercised by a commander in the chain of command, defined further as combatant command (command authority), operational control, tactical control, or a support relationship. See also **combatant command (command authority); in support of; operational control; support; tactical control.** (JP 0-2)

operational characteristics — Those military characteristics that pertain primarily to the functions to be performed by equipment, either alone or in conjunction with other equipment; e.g., for electronic equipment, operational characteristics include such items as frequency coverage, channeling, type of modulation, and character of emission.

operational control — Command authority that may be exercised by commanders at any echelon at or below the level of combatant command. Operational control is inherent in combatant command (command authority) and may be delegated within the command. When forces are transferred between combatant commands, the command relationship the gaining commander will exercise (and the losing commander will relinquish) over these forces must be specified by the Secretary of Defense. Operational control is the authority to perform those functions of command over subordinate forces involving organizing and employing commands and forces, assigning tasks, designating objectives, and giving authoritative direction necessary to accomplish the mission. Operational control includes authoritative direction over all aspects of military operations and joint training necessary to accomplish missions assigned to the command. Operational control should be exercised through the commanders of subordinate organizations. Normally this authority is exercised through subordinate joint force commanders and Service and/or functional component commanders. Operational control normally provides full authority to organize commands and forces and to employ those forces as the commander in operational control considers necessary to accomplish assigned missions; it does not, in and of itself, include authoritative direction for logistics or matters of administration, discipline, internal organization, or unit training. Also called **OPCON.** See also **combatant command; combatant command (command authority); tactical control.** (JP 0-2)

operational control authority — **(*)** The naval commander responsible within a specified geographical area for the naval control of all merchant shipping under Allied naval control. Also called **OCA.**

operational decontamination — (*) Decontamination carried out by an individual and/or a unit, restricted to specific parts of operationally essential equipment, materiel and/or working areas, in order to minimize contact and transfer hazards and to sustain operations. This may include decontamination of the individual beyond the scope of immediate decontamination, as well as decontamination of mission-essential spares and limited terrain decontamination. See also **decontamination; immediate decontamination; thorough decontamination.**

operational design — The key considerations used as a framework in the course of planning for a campaign or major operation. See also **campaign; major operation.** (JP 5-00.1)

operational documentation — Visual information documentation of activities to convey information about people, places, and things. It is general purpose documentation normally accomplished in peacetime. Also called **OPDOC.** See also **visual information documentation.**

operational environment — A composite of the conditions, circumstances, and influences that affect the employment of military forces and bear on the decisions of the unit commander. Some examples are as follows. a. **permissive environment** — Operational environment in which host country military and law enforcement agencies have control as well as the intent and capability to assist operations that a unit intends to conduct. b. **uncertain environment** — Operational environment in which host government forces, whether opposed to or receptive to operations that a unit intends to conduct, do not have totally effective control of the territory and population in the intended operational area. c. **hostile environment** — Operational environment in which hostile forces have control as well as the intent and capability to effectively oppose or react to the operations a unit intends to conduct. (JP 3-05.3)

operational evaluation — The test and analysis of a specific end item or system, insofar as practicable under Service operating conditions, in order to determine if quantity production is warranted considering: a. the increase in military effectiveness to be gained; and b. its effectiveness as compared with currently available items or systems, consideration being given to: (1) personnel capabilities to maintain and operate the equipment; (2) size, weight, and location considerations; and (3) enemy capabilities in the field. See also **technical evaluation.**

operational intelligence — Intelligence that is required for planning and conducting campaigns and major operations to accomplish strategic objectives within theaters or operational areas. See also **intelligence; strategic intelligence; tactical intelligence.** (JP 2-0)

operational level of war — The level of war at which campaigns and major operations are planned, conducted, and sustained to accomplish strategic objectives within theaters or other operational areas. Activities at this level link tactics and strategy by establishing operational objectives needed to accomplish the strategic objectives, sequencing events to achieve the operational objectives, initiating actions, and applying resources to bring about and sustain these events. These activities imply a broader dimension of time or space than do tactics; they ensure the logistic and administrative support of tactical forces, and provide the means by which tactical successes are exploited to achieve strategic objectives. See also

strategic level of war; tactical level of war. (JP 3-0)

operationally ready — 1. A unit, ship, or weapon system capable of performing the missions or functions for which organized or designed. Incorporates both equipment readiness and personnel readiness. 2. Personnel available and qualified to perform assigned missions or functions.

operational necessity — A mission associated with war or peacetime operations in which the consequences of an action justify the risk of loss of aircraft and crew. See also **mission.** (JP 3-04.1)

operational procedures — (*) The detailed methods by which headquarters and units carry out their operational tasks.

operational reach — The distance and duration across which a unit can successfully employ military capabilities. (JP 3-0)

operational readiness — (*) The capability of a unit/formation, ship, weapon system, or equipment to perform the missions or functions for which it is organized or designed. May be used in a general sense or to express a level or degree of readiness. Also called **OR.** See also **combat readiness.**

operational readiness evaluation — (*) An evaluation of the operational capability and effectiveness of a unit or any portion thereof.

operational requirement — See **military requirement.**

Operational Requirements Document — A formatted statement containing performance and related operational parameters for the proposed concept or system. Prepared by the user or user's representative at each milestone beginning with Milestone I, Concept Demonstration Approval of the Requirements Generation Process. Also called **ORD.**

operational reserve — An emergency reserve of men and/or materiel established for the support of a specific operation. See also **reserve supplies.**

operational route — (*) Land route allocated to a command for the conduct of a specific operation; derived from the corresponding basic military route network.

operational support airlift — Operational support airlift (OSA) missions are movements of high-priority passengers and cargo with time, place, or mission-sensitive requirements. OSA aircraft are those fixed-wing aircraft acquired and/or retained exclusively for OSA missions, as well as any other Department of Defense-owned or controlled aircraft, fixed- or rotary-wing, used for OSA purposes. Also called **OSA.** See also **aircraft.** (JP 4-01)

operational testing — A continuing process of evaluation that may be applied to either operational personnel or situations to determine their validity or reliability.

operational training — (*) Training that develops, maintains, or improves the operational readiness of individuals or units.

operation and maintenance — Maintenance and repair of real property, operation of utilities, and provision of other services such as refuse collection and disposal, entomology, snow removal, and ice alleviation. Also called **O&M.** (JP 4-04)

operation annexes — Those amplifying instructions that are of such a nature, or are so voluminous or technical, as to make their inclusion in the body of the plan or order undesirable.

operation exposure guide — The maximum amount of nuclear radiation that the commander considers a unit may be permitted to receive while performing a particular mission or missions. Also called **OEG.** See also **radiation exposure status.**

operation map — A map showing the location and strength of friendly forces involved in an operation. It may indicate predicted movement and location of enemy forces. See also **map.**

operation order — A directive issued by a commander to subordinate commanders for the purpose of effecting the coordinated execution of an operation. Also called **OPORD.**

operation plan — Any plan, except for the Single Integrated Operational Plan, for the conduct of military operations. Plans are prepared by combatant commanders in response to requirements established by the Chairman of the Joint Chiefs of Staff and by commanders of subordinate commands in response to requirements tasked by the establishing unified commander. Operation plans are prepared in either a complete format (OPLAN) or as a concept plan (CONPLAN). The CONPLAN can be published with or without a time-phased force and deployment data (TPFDD) file. a. **OPLAN** — An operation plan for the conduct of joint operations that can be used as a basis for development of an operation order (OPORD). An OPLAN identifies the forces and supplies required to execute the CINC's Strategic Concept and a movement schedule of these resources to the theater of operations. The forces and supplies are identified in TPFDD files. OPLANs will include all phases of the tasked operation. The plan is prepared with the appropriate annexes, appendixes, and TPFDD files as described in the Joint Operation Planning and Execution System manuals containing planning policies, procedures, and formats. Also called **OPLAN.** b. **CONPLAN** — An operation plan in an abbreviated format that would require considerable expansion or alteration to convert it into an OPLAN or OPORD. A CONPLAN contains the CINC's Strategic Concept and those annexes and appendixes deemed necessary by the combatant commander to complete planning. Generally, detailed support requirements are not calculated and TPFDD files are not prepared. c. **CONPLAN with TPFDD** — A CONPLAN with TPFDD is the same as a CONPLAN except that it requires more detailed planning for phased deployment of forces. Also called **CONPLAN.** See also **operation order; time-phased force and deployment data.** (JP 5-0)

operations center — The facility or location on an installation, base, or facility used by the commander to command, control, and coordinate all crisis activities. See also **base defense operations center; command center.** (JP 3-07.2)

operations research — The analytical study of military problems undertaken to provide responsible commanders and staff agencies with a scientific basis for decision on action to improve military operations. Also called **operational research; operations analysis.**

operations security — A process of identifying critical information and subsequently analyzing friendly actions attendant to military operations and other activities to: a. identify those actions that can be observed by adversary intelligence systems; b. determine indicators that hostile intelligence systems might obtain that could be interpreted or pieced together to derive critical information in time to be useful to adversaries; and c. select and execute measures that eliminate or reduce to an acceptable level the vulnerabilities of friendly actions to adversary exploitation.

Also called **OPSEC.** See also **command and control warfare; operations security indicators; operations security measures; operations security planning guidance; operations security vulnerabilbity.** (JP 3-07.2)

operations security indicators — Friendly detectable actions and open-source information that can be interpreted or pieced together by an adversary to derive critical information.

operations security measures — Methods and means to gain and maintain essential secrecy about critical information. The following categories apply. a. **action control** — The objective is to eliminate indicators or the vulnerability of actions to exploitation by adversary intelligence systems. Personnel will select what actions to undertake; decide whether or not to execute actions; and determine the "who," "when," "where," and "how" for actions necessary to accomplish tasks. b. **countermeasures** — The objective is to disrupt effective adversary information gathering or prevent their recognition of indicators when collected materials are processed. Use diversions, camouflage, concealment, jamming, threats, police powers, and force against adversary information gathering and processing capabilities. c. **counteranalysis** — The objective is to prevent accurate interpretations of indicators during adversary analysis of collected materials. This is done by confusing the adversary analyst through deception techniques such as covers.

operations security planning guidance — Guidance that serves as the blueprint for operations security planning by all functional elements throughout the organization. It defines the critical information that requires protection from adversary appreciations, taking into account friendly and adversary goals, estimated key adversary questions, probable adversary knowledge, desirable and harmful adversary appreciations, and pertinent intelligence system threats. It also should outline provisional operations security measures to ensure the requisite essential secrecy.

operations security vulnerability — A condition in which friendly actions provide operations security indicators that may be obtained and accurately evaluated by an adversary in time to provide a basis for effective adversary decisionmaking.

operations to restore order — Operations intended to halt violence and support, reinstate, or establish civil authorities. They are designed to return an unstable and lawless environment to the point where indigenous police forces can effectively enforce the law and restore civil authority. See also **operation; peace operations.** (JP 3-07.3)

opportune lift — That portion of lift capability available for use after planned requirements have been met.

opportunity target — See **target of opportunity.**

opposite numbers — Officers (including foreign) having corresponding duty assignments within their respective Military Services or establishments.

optical axis — **(*)** In a lens element, the straight line which passes through the centers of curvature of the lens surfaces. In an optical system, the line formed by the coinciding principal axes of the series of optical elements.

optical minehunting — **(*)** The use of an optical system (e.g., television or towed diver) to detect and classify mines or minelike objects on or protruding from the seabed.

optimum height — **(*)** The height of an explosion which will produce the maximum effect against a given target.

optimum height of burst — **(*)** For nuclear weapons and for a particular target (or area), the height at which it is estimated a weapon of a specified energy yield will produce a certain desired effect over the maximum possible area.

orbital injection — The process of providing a space vehicle with sufficient velocity to establish an orbit.

orbit determination — The process of describing the past, present, or predicted position of a satellite in terms of orbital parameters.

orbit point — **(*)** A geographically or electronically defined location used in stationing aircraft in flight during tactical operations when a predetermined pattern is not established. See also **holding point.**

order — **(*)** A communication, written, oral, or by signal, which conveys instructions from a superior to a subordinate. **(DOD only)** In a broad sense, the terms "order" and "command" are synonymous. However, an order implies discretion as to the details of execution whereas a command does not.

order and shipping time — The time elapsing between the initiation of stock replenishment action for a specific activity and the receipt by that activity of the materiel resulting from such action. Order and shipping time is applicable only to materiel within the supply system, and it is composed of the distinct elements, order time, and shipping time. See also **level of supply.**

order of battle — **(*)** The identification, strength, command structure, and disposition of the personnel, units, and equipment of any military force. Also called **OB; OOB.** (JP 2-01.3)

order time — 1. The time elapsing between the initiation of stock replenishment action and submittal of requisition or order. 2. The time elapsing between the submittal of requisition or order and shipment of materiel by the supplying activity. See also **order and shipping time.**

ordinary transport — **(*)** In railway terminology, transport of a load whose size, weight, or preparation does not entail special difficulties vis-à-vis the facilities or equipment of the railway systems to be used. See also **exceptional transport.**

ordnance — Explosives, chemicals, pyrotechnics, and similar stores, e.g., bombs, guns and ammunition, flares, smoke, or napalm.

organic — Assigned to and forming an essential part of a military organization. Organic parts of a unit are those listed in its table of organization for the Army, Air Force, and Marine Corps, and are assigned to the administrative organizations of the operating forces for the Navy.

organizational equipment — Referring to method of use: signifies that equipment (other than individual equipment) used in furtherance of the common mission of an organization or unit. See also **equipment.**

organizational maintenance — That maintenance that is the responsibility of and performed by a using organization on its assigned equipment. Its phases normally consist of inspecting, servicing, lubricating, and adjusting, as well as the replacing of parts, minor assemblies, and subassemblies.

organization for combat — In amphibious operations, task organization of landing force units for combat, involving combinations of command, ground and aviation combat, combat support, and combat service support units for accomplishment of missions ashore. See also **amphibious operation; task organization.** (JP 3-02)

organization for embarkation — In amphibious operations, the organization for embarkation consisting of temporary landing force task organizations established by the commander, landing force and a temporary organization of Navy forces established by the commander, amphibious task force for the purpose of simplifying planning and facilitating the execution of embarkation. See also **amphibious operation; embarkation; landing force; task organization.** (JP 3-02)

organization for landing — In amphibious operations, the specific tactical grouping of the landing force for the assault. (JP 3-02)

organization of the ground — **(*)** The development of a defensive position by strengthening the natural defenses of the terrain and by assignment of the occupying troops to specific localities.

Organized Crime Drug Enforcement Task Force — A network of 13 regional organized crime drug enforcement task forces designed to coordinate Federal law enforcement efforts to combat the national and international organizations that cultivate, process, and distribute illicit drugs. Also called **OCDETF.** (JP 3-07.4)

origin — Beginning point of a deployment where unit or non-unit-related cargo or personnel are located.

original destination — **(*)** In naval control of shipping, the original final destination of a convoy or an individual ship (whether in convoy or independent). This is particularly applicable to the original destination of a voyage begun in peacetime.

original negative — See **generation (photography).**

original positive — See **generation (photography).**

originating medical facility — **(*)** A medical facility that initially transfers a patient to another medical facility.

originator — The command by whose authority a message is sent. The responsibility of the originator includes the responsibility for the functions of the drafter and the releasing officer. See also **releasing officer.**

oropesa sweep — **(*)** In naval mine warfare, a form of sweep in which a length of sweep wire is towed by a single ship, lateral displacement being caused by an otter and depth being controlled at the ship end by a kite and at the other end by a float and float wire.

orthomorphic projection — **(*)** A projection in which the scale, although varying throughout the map, is the same in all directions at any point, so that very small areas are represented by correct shape and bearings are correct.

oscillating mine — **(*)** A mine, hydrostatically controlled, which maintains a pre-set depth below the surface of the water independently of the rise and fall of the tide. See also **mine.**

other detainee — Person in the custody of the US Armed Forces who has not been classified as an enemy prisoner of war (article 4, Geneva Convention of 1949 Relative to the Treatment of Prisoners of

War (GPW)), retained personnel (article 33, GPW), or civilian internee (article 78, Geneva Convention). Also called **OD.** See also **civilian internee; custody; detainee; prisoner of war; retained personnel.** (JP 1-0)

other war reserve materiel requirement — War reserve materiel requirement less the pre-positioned war reserve materiel requirement.

other war reserve materiel requirement, balance — That portion of the other war reserve materiel requirement that has not been acquired or funded. This level consists of the other war reserve materiel requirement less the other war reserve materiel requirement protectable.

other war reserve materiel requirement, protectable — The portion of the other war reserve materiel requirement that is protected for purposes of procurement, funding, and inventory management.

other war reserve stock — The quantity of an item acquired and placed in stock against the other war reserve materiel requirement.

otter — **(*)** In naval mine warfare, a device which, when towed, displaces itself sideways to a predetermined distance.

outbound traffic — Traffic originating in the continental United States destined for overseas or overseas traffic moving in a general direction away from the continental United States.

outer fix — A fix in the destination terminal area, other than the approach fix, to which aircraft are normally cleared by an air route traffic control center or a terminal area traffic control facility, and from which aircraft are cleared to the approach fix or final approach course.

outer landing ship areas — In amphibious operations, areas to which landing ships proceed initially after their arrival in the objective area. They are usually located on the flanks of the outer transport areas. (JP 3-02)

outer transport area — In amphibious operations, an area inside the antisubmarine screen to which assault transports proceed initially after arrival in the objective area. See also **inner transport area; transport area.**

outline map — **(*)** A map which represents just sufficient geographic information to permit the correlation of additional data placed upon it.

outline plan — **(*)** A preliminary plan which outlines the salient features or principles of a course of action prior to the initiation of detailed planning.

outsized cargo — A single item of cargo, too large for palletization or containerization, that exceeds 1090 inches long by 111 inches wide by 105 inches high. Requires transport by sea or use of a C-5 or C-17 aircraft for transport by air. See also **oversized cargo.** (JP 4-01.6)

overhaul — The restoration of an item to a completely serviceable condition as prescribed by maintenance serviceability standards. See also **rebuild; repair.**

overhead clearance — The vertical distance between the route surface and any obstruction above it.

overlap — 1. In photography, the amount by which one photograph includes the same area covered by another, customarily expressed as a percentage. The overlap between successive air photographs on a flight line is called "forward overlap." The overlap between photographs in adjacent parallel flight lines is called "side overlap." 2. In cartography, that portion of a map or chart that overlaps the area covered by another of the same series. 3. In naval mine warfare, the width of that part of the swept path of a ship or formation that is also swept by an adjacent sweeper or formation or is reswept on the next adjacent lap.

overlay — A printing or drawing on a transparent or semi-transparent medium at the same scale as a map, chart, etc., to show details not appearing or requiring special emphasis on the original.

overpressure — **(*)** The pressure resulting from the blast wave of an explosion. It is referred to as "positive" when it exceeds atmospheric pressure and "negative" during the passage of the wave when resulting pressures are less than atmospheric pressure.

overprint — **(*)** Information printed or stamped upon a map or chart, in addition to that originally printed, to show data of importance or special use.

overseas — All locations, including Alaska and Hawaii, outside the continental United States.

Overseas Environmental Baseline Guidance Document — A set of objective criteria and management practices developed by the Department of Defense to protect human health and the environment. Also called **OEBGD.** (JP 4-04)

overseas search and rescue region — Overseas unified command areas (or portions thereof not included within the inland region or the maritime region). See also **search and rescue region.**

oversized cargo — Large items of specific equipment such as a barge, side loadable warping tug, causeway section, powered, or causeway section, nonpowered. Requires transport by sea. See also **outsized cargo.** (JP 4-01.6)

over the beach operations — See **logistics over-the-shore operations.**

over-the-horizon amphibious operations — An operational initiative launched from beyond visual and radar range of the shoreline. (JP 3-02)

over-the-horizon radar — A radar system that makes use of the atmospheric reflection and refraction phenomena to extend its range of detection beyond line of sight. Over-the-horizon radars may be either forward scatter or back scatter systems.

overt operation — An operation conducted openly, without concealment. See also **clandestine operation; covert operation.** (JP 3-05.3)

overt peacetime psychological operations programs — Those programs developed by combatant commands, in coordination with the chiefs of US diplomatic missions, that plan, support, and provide for the conduct of psychological operations, during military operations other than war, in support of US regional objectives, policies, interests, and theater military missions. Also called **OP3.** See also **consolidation psychological operations; psychological operations.** (JP 3-53)

P

pace — (*) For ground forces, the speed of a column or element regulated to maintain a prescribed average speed.

pace setter — (*) An individual, selected by the column commander, who travels in the lead vehicle or element to regulate the column speed and establish the pace necessary to meet the required movement order.

packaged forces — Forces of varying size and composition preselected for specific missions in order to facilitate planning and training.

packaged petroleum product — A petroleum product (generally a lubricant, oil, grease, or specialty item) normally packaged by a manufacturer and procured, stored, transported, and issued in containers having a fill capacity of 55 United States gallons (or 45 Imperial gallons, or 205 liters) or less.

packup kit — Service-provided maintenance gear including spare parts and consumables most commonly needed by the deployed helicopter detachment. Supplies are sufficient for a short-term deployment but do not include all material needed for every maintenance task. Also called **PUK.** (JP 3-04.1)

padding — Extraneous text added to a message for the purpose of concealing its beginning, ending, or length.

pallet — (*) 1. A flat base for combining stores or carrying a single item to form a unit load for handling, transportation, and storage by materials handling equipment. 2. **(DOD only)** 463L pallet – An 88" x 108" aluminum flat base used to facilitate the upload and download of aircraft.

palletized load system — A truck with hydraulic load handling mechanism, trailer, and flatrack system capable of self-loading and -unloading. Truck and companion trailer each have a 16.5 ton payload capacity. Also called **PLS.** See also **flatrack.** (JP 4-01.7)

palletized load system flatrack — Topless, sideless container component of palletized load system, some of which conform to International Organization for Standardization specifications. See also **palletized load system.** (JP 4-01.7)

palletized unit load — (*) Quantity of any item, packaged or unpackaged, which is arranged on a pallet in a specified manner and securely strapped or fastened thereto so that the whole is handled as a unit.

panel code — (*) A prearranged code designed for visual communications, usually between friendly units, by making use of marking panels. See also **marking panel.**

panoramic camera — (*) 1. In aerial photography, a camera which, through a system of moving optics or mirrors, scans a wide area of the terrain, usually from horizon to horizon. The camera may be mounted vertically or obliquely within the aircraft, to scan across or along the line of flight. 2. In ground photography, a camera which photographs a wide expanse of terrain by rotating horizontally about the vertical axis through the center of the camera lens.

parachute deployment height — (*) The height above the intended impact point at which the parachute or parachutes are fully deployed.

paradrop — **(*)** Delivery by parachute of personnel or cargo from an aircraft in flight.

parallel chains of command — In amphibious operations, a parallel system of command, responding to the interrelationship of Navy, landing force, Air Force, and other major forces assigned, wherein corresponding commanders are established at each subordinate level of all components to facilitate coordinated planning for, and execution of, the amphibious operation. (JP 3-02.2)

parallel sheaf — In artillery and naval gunfire support, a sheaf in which the planes (lines) of fire of all pieces are parallel. See also **converged sheaf.**

parallel staff — **(*)** A staff in which one officer from each nation, or Service, working in parallel is appointed to each post. See also **multinational staff; integrated staff; joint staff.**

paramilitary forces — Forces or groups distinct from the regular armed forces of any country, but resembling them in organization, equipment, training, or mission.

pararescue team — Specially trained personnel qualified to penetrate to the site of an incident by land or parachute, render medical aid, accomplish survival methods, and rescue survivors. Also called **PRT.**

parlimentaire — An agent employed by a commander of belligerent forces in the field to go in person within the enemy lines for the purpose of communicating or negotiating openly and directly with the enemy commander.

parrot — Identification friend or foe transponder equipment.

partial mission-capable — Material condition of an aircraft or training device indicating that it can perform at least one but not all of its missions. Also called **PMC.** See also **full mission-capable; mission-capable; partial mission-capable, maintenance; partial mission-capable, supply.**

partial mission-capable, maintenance — Material condition of an aircraft or training device indicating that it can perform at least one but not all of its missions because of maintenance requirements existing on the inoperable subsystem(s). Also called **PMCM.** See also **full mission-capable; mission-capable; partial mission-capable; partial mission-capable, supply.**

partial mission-capable, supply — Material condition of an aircraft or training device indicating it can perform at least one but not all of its missions because maintenance required to clear the discrepancy cannot continue due to a supply shortage. Also called **PMCS.** See also **full mission-capable; mission-capable; partial mission-capable; partial mission-capable, maintenance.**

partial mobilization — See **mobilization, Part 2.**

partial storage monitoring — A periodic inspection of major assemblies or components for nuclear weapons, consisting mainly of external observation of humidity, temperatures, and visual damage or deterioration during storage. This type of inspection is also conducted prior to and upon completion of a movement.

partisan warfare — Not to be used. See **guerrilla warfare.**

part number — A combination of numbers, letters, and symbols assigned by a designer, a manufacturer, or vendor to identify a specific part or item of materiel.

pass — 1. A short tactical run or dive by an aircraft at a target. 2. A single sweep through or within firing range of an enemy air formation.

passage of lines — An operation in which a force moves forward or rearward through another force's combat positions with the intention of moving into or out of contact with the enemy. A passage may be designated as a forward or rearward passage of lines.

passenger mile — One passenger transported one mile. For air and ocean transport, use nautical miles; for rail, highway, and inland waterway transport in the continental United States, use statute miles.

passive — **(*)** In surveillance, an adjective applied to actions or equipments which emit no energy capable of being detected.

passive air defense — All measures, other than active air defense, taken to minimize the effectiveness of hostile air and missile threats against friendly forces and assets. These measures include camouflage, concealment, deception, dispersion, reconstitution, redundancy, detection and warning systems, and the use of protective construction. See also **air defense; concealment, deception, dispersion.** (JP 3-01)

passive defense — Measures taken to reduce the probability of and to minimize the effects of damage caused by hostile action without the intention of taking the initiative. See also **active defense.**

passive homing guidance — **(*)** A system of homing guidance wherein the receiver in the missile utilizes radiation from the target.

passive mine — **(*)** 1. A mine whose anticountermining device has been operated preventing the firing mechanism from being actuated. The mine will usually remain passive for a comparatively short time. 2. A mine which does not emit a signal to detect the presence of a target. See also **active mine.**

passive or responsive public affairs policy — A responsive posture by which no direct effort is made to initiate, or participate in, the public discussion about an issue or activity. When a passive policy is in effect, authorities must be prepared to respond to news media inquiries about the issue or activity — to make brief statements to avoid confusion, speculation, misunderstanding, or false information that may prevail if news media queries go unanswered. See also **public affairs.** (JP 3-61)

pass time — **(*)** In road transport, the time that elapses between the moment when the leading vehicle of a column passes a given point and the moment when the last vehicle passes the same point.

password — **(*)** A secret word or distinctive sound used to reply to a challenge. See also **challenge; countersign.**

pathfinder drop zone control — The communication and operation center from which pathfinders exercise aircraft guidance.

pathfinder landing zone control — See **pathfinder drop zone control.**

pathfinders — 1. Experienced aircraft crews who lead a formation to the drop zone, release point, or target. 2. Teams dropped or air landed at an objective to establish and operate navigational aids for the purpose of guiding aircraft to drop and landing zones. 3. A radar device used for navigating or homing to an objective when visibility precludes accurate visual navigation. 4. Teams air delivered into enemy territory for the purpose of determining the best approach and withdrawal lanes, landing zones, and sites for helicopterborne forces.

pathogen — A disease-producing microorganism. (JP 3-11)

patient — A sick, injured, wounded, or other person requiring medical and/or dental care or treatment.

patient movement — The act or process of moving a sick, injured, wounded, or other person to obtain medical and/or dental care or treatment. Functions include medical regulating, patient evacuation, and en route medical care. See also **patient; patient movement items; patient movement requirements center.** (JP 4-02)

patient movement items — The medical equipment and supplies required to support patients during aeromedical evacuation. Also called **PMIs.**

patient movement requirements center — A joint activity that coordinates patient movement. It is the functional merging of joint medical regulating processes, Services' medical regulating processes, and coordination with movement components for patient evacuation. This may be joint, reporting to the joint task force surgeon; theater, reporting to the theater surgeon; or global, reporting to the United States Transportation Command surgeon. Also called **PMRC.** See also **patient.** (JP 4-02)

patrol — **(*)** A detachment of ground, sea, or air forces sent out for the purpose of gathering information or carrying out a destructive, harassing, mopping-up, or security mission. See also **combat air patrol.**

pattern bombing — The systematic covering of a target area with bombs uniformly distributed according to a plan.

pattern laying — **(*)** In land mine warfare, the laying of mines in a fixed relationship to each other.

payload — **(*)** 1. The sum of the weight of passengers and cargo that an aircraft can carry. See also **load.** 2. The warhead, its container, and activating devices in a military missile. 3. The satellite or research vehicle of a space probe or research missile. 4. The load (expressed in tons of cargo or equipment, gallons of liquid, or number of passengers) which the vehicle is designed to transport under specified conditions of operation, in addition to its unladen weight.

payload build-up (missile and space) — The process by which the scientific instrumentation (sensors, detectors, etc.) and necessary mechanical and electronic subassemblies are assembled into a complete operational package capable of achieving the scientific objectives of the mission.

payload integration (missile and space) — The compatible installation of a complete payload package into the spacecraft and space vehicle.

payload (missile) — See **payload, Part 2.**

P-day — That point in time at which the rate of production of an item available for military consumption equals the rate at which the item is required by the Armed Forces.

peace building — Post-conflict actions, predominately diplomatic and economic, that strengthen and rebuild governmental infrastructure and institutions in order to avoid a relapse into conflict. See also **peace enforcement; peacekeeping; peacemaking; peace operations.** (JP 3-07)

peace enforcement — Application of military force, or the threat of its use, normally pursuant to international authorization, to compel compliance with resolutions or sanctions designed to maintain or restore peace and order. See also **peace building; peacekeeping; peacemaking; peace operations.** (JP 3-07)

peacekeeping — Military operations undertaken with the consent of all major parties to a dispute, designed to monitor and facilitate implementation of an agreement (ceasefire, truce, or other such agreement) and support diplomatic efforts to reach a long-term political settlement. See also **peace building; peace enforcement; peacemaking; peace operations.** (JP 3-07)

peacemaking — The process of diplomacy, mediation, negotiation, or other forms of peaceful settlements that arranges an end to a dispute and resolves issues that led to it. See also **peace building; peace enforcement; peacekeeping; peace operations.** (JP 3-07)

peace operations — A broad term that encompasses peacekeeping operations and peace enforcement operations conducted in support of diplomatic efforts to establish and maintain peace. Also called **PO.** See also **peace building; peace enforcement; peacekeeping; and peacemaking.** (JP 3-07)

peacetime force materiel assets — That portion of total materiel assets that is designated to meet the peacetime force materiel requirement. See also **war reserves.**

peacetime force materiel requirement — The quantity of an item required to equip, provide a materiel pipeline, and sustain the United States force structure (active and reserve) and those allied forces designated for United States peacetime support in current Secretary of Defense guidance (including approved supply support arrangements with foreign military sales countries) and to support the scheduled establishment through normal appropriation and procurement leadtime periods.

peacetime materiel consumption and losses — The quantity of an item consumed, lost, or worn out beyond economical repair through normal appropriation and procurement leadtime periods.

peak overpressure — (*) The maximum value of overpressure at a given location which is generally experienced at the instant the shock (or blast) wave reaches that location.

pecuniary liability — A personal, joint, or corporate monetary obligation to make good any lost, damaged, or destroyed property resulting from fault or neglect. It may also result under conditions stipulated in a contract or bond.

pencil beam — (*) A searchlight beam reduced to, or set at, its minimum width.

penetration — (*) In land operations, a form of offensive which seeks to break through the enemy's defense and disrupt the defensive system.

penetration aids — Techniques and/or devices employed by offensive aerospace weapon systems to increase the probability of penetration of enemy defenses.

penetration (air traffic control) — That portion of a published high altitude instrument approach procedure that prescribes a descent path from the fix on which the procedure is based to a fix or altitude from which an approach to the airport is made.

penetration (intelligence) — The recruitment of agents within or the infiltration of agents or technical monitoring devices in an organization or group for the purpose of acquiring information or of influencing its activities.

percentage clearance — (*) In mine warfare, the estimated percentage of mines of specified characteristics which have been cleared from an area or channel.

perception management — Actions to convey and/or deny selected information and indicators to foreign audiences to influence their emotions, motives, and objective reasoning as well as to intelligence systems and leaders at all levels to influence official estimates, ultimately resulting in foreign behaviors and official actions favorable to the originator's objectives. In various ways, perception management combines truth projection, operations security, cover and deception, and psychological operations. See also **psychological operations.**

perils of the sea — Accidents and dangers peculiar to maritime activities, such as storms, waves, and wind; collision; grounding; fire, smoke and noxious fumes; flooding, sinking and capsizing; loss of propulsion or steering; and any other hazards resulting from the unique environment of the sea.

perimeter defense — A defense without an exposed flank, consisting of forces deployed along the perimeter of the defended area.

periodic intelligence summary — A report of the intelligence situation in a tactical operation (normally produced at corps level or its equivalent and higher) usually at intervals of 24 hours, or as directed by the commander. Also called **PERINTSUM.**

period — The time it takes for a satellite to complete one orbit around the earth. As a rule of thumb, satellites with periods of 87.5 minutes are on the verge of reentry.

period of interest — A period of time in which a launch of a missile is expected. Also called **POI.**

perishable cargo — Cargo requiring refrigeration, such as meat, fruit, fresh vegetables, and medical department biologicals.

perishable target — A force or activity at a specific location whose value as a target can decrease substantially during a specified time. A significant decrease in value occurs when the target moves or the operational circumstances change to the extent that the target is no longer lucrative. See also **target.** (JP 3-05.3)

permafrost — Permanently frozen subsoil.

permanent echo — Any dense and fixed radar return caused by reflection of energy from the Earth's surface or manmade structure. Distinguished from "ground clutter" by being from definable locations rather than large areas.

permissive action link — A device included in or attached to a nuclear weapon system to preclude arming and/or launching until the insertion of a prescribed discrete code or combination. It may include equipment and cabling external to the weapon or weapon system to activate components within the weapon or weapon system.

permissive environment — See **operational environment.**

persistency — **(*)** In biological or chemical warfare, the characteristic of an agent which pertains to the duration of its effectiveness under determined conditions after its dispersal.

persistent agent — A chemical agent that, when released, remains able to cause casualties for more than 24 hours to several days or weeks. (JP 3-11)

personal effects — All privately owned moveable, personal property of an individual. Also called **PE.** See also **mortuary affairs; personal property.** (JP 4-06)

personal locator beacon — **(*)** An emergency radio locator beacon with a two-way speech facility carried by crew members, either on their person or in their survival equipment, and capable of providing homing signals to assist search and rescue operations. Also called **PLB.** See also **crash locator beacon; emergency locator beacon.**

personal property — Property of any kind or any interest therein, except real property, records of the Federal Government, and naval vessels of the following categories: surface combatants, support ships, and submarines.

person authorized to direct disposition of remains — A person, usually primary next of kin, who is authorized to direct disposition of remains. Also called **PADD.** See also **mortuary affairs.** (JP 4-06)

person eligible to receive effects — The person authorized by law to receive the personal effects of a deceased military member. Receipt of personal effects does not constitute ownership. Also called **PERE.** See also **mortuary affairs; personal effects.** (JP 4-06)

person in custody — Any person under the direct control and protection of US forces.

personnel — Those individuals required in either a military or civilian capacity to accomplish the assigned mission.

personnel increment number — A seven-character, alphanumeric field that uniquely describes a non-unit-related personnel entry (line) in a Joint Operation Planning and Execution System time-phased force and deployment data. Also called **PIN.**

personnel reaction time (nuclear) — **(*)** The time required by personnel to take prescribed protective measures after receipt of a nuclear strike warning.

personnel recovery — The aggregation of military, civil, and political efforts to obtain the release or recovery of personnel from uncertain or hostile environments and denied areas whether they are captured, missing, or isolated. That includes US, allied, coalition, friendly military, or paramilitary, and others as designated by the National Command Authorities. Personnel recovery (PR) is the umbrella term for operations that are focused on the task of recovering captured, missing, or isolated personnel from harm's way. PR includes but is not limited to theater search and rescue; combat search and rescue; search and rescue; survival, evasion, resistance, and escape; evasion and escape; and the coordination of negotiated as well as forcible recovery options. PR can occur through military action, action by nongovernmental organizations, other US Government-approved action, and/or diplomatic initiatives, or through any of these. Also called **PR.** See also **combat search and rescue; evasion; evasion and escape; personnel; recovery; search and rescue.** (JP 3-50.21)

personnel security investigation — An inquiry into the activities of an individual, designed to develop pertinent information pertaining to trustworthiness and suitability for a position of trust as related to loyalty, character, emotional stability, and reliability. Also called **PSI.**

perspective grid — **(*)** A network of lines, drawn or superimposed on a photograph, to represent the perspective of a systematic network of lines on the ground or datum plane.

petroleum intersectional service — **(*)** An intersectional or interzonal service in a theater of operations that operates pipelines and related facilities for the supply of bulk petroleum products to theater Army elements and other forces as directed.

petroleum, oils, and lubricants — **(*)** A broad term which includes all petroleum and associated products used by the Armed Forces. Also called **POL.**

phase line — A line utilized for control and coordination of military operations, usually an easily identified feature in the operational area.

phases of military government — 1. **assault** — That period which commences with the first contact with civilians ashore and extends to the establishment of military government control ashore by the landing force. 2. **consolidation** — That period which commences with the establishment of military government ashore by the landing force and extends to the establishment of control by occupation forces. 3. **occupation** — That period which commences when an area has been occupied in fact, and the military commander within that area is in a position to enforce public safety and order. See also **civil affairs; military occupation.**

phonetic alphabet — A list of standard words used to identify letters in a message transmitted by radio or telephone. The following are the authorized words, listed in order, for each letter in the alphabet: ALFA, BRAVO, CHARLIE, DELTA, ECHO, FOXTROT, GOLF, HOTEL, INDIA, JULIETT, KILO, LIMA, MIKE, NOVEMBER, OSCAR, PAPA, QUEBEC, ROMEO, SIERRA, TANGO, UNIFORM, VICTOR, WHISKEY, X-RAY, YANKEE, and ZULU.

phoney minefield — **(*)** An area free of live mines used to simulate a minefield, or section of a minefield, with the object of deceiving the enemy. See also **gap, minefield.**

photoflash bomb — **(*)** A bomb designed to produce a brief and intense illumination for medium altitude night photography.

photoflash cartridge — **(*)** A pyrotechnic cartridge designed to produce a brief and intense illumination for low altitude night photography.

photogrammetric control — **(*)** Control established by photogrammetric methods as distinguished from control established by ground methods. Also called **minor control.**

photogrammetry — **(*)** The science or art of obtaining reliable measurements from photographic images.

photographic coverage — The extent to which an area is covered by photography from one mission or a series of missions or in a period of time. Coverage, in this sense, conveys the idea of availability of photography and is not a synonym for the word "photography."

photographic intelligence — The collected products of photographic interpretation, classified and evaluated for intelligence use. Also called **PHOTINT.**

photographic interpretation — See **imagery interpretation.**

photographic panorama — A continuous photograph or an assemblage of overlapping oblique or ground photographs that have been matched and joined together to form a continuous photographic representation of the area.

photographic reading — **(*)** The simple recognition of natural or manmade features from photographs not involving imagery interpretation techniques.

photographic scale — **(*)** The ratio of a distance measured on a photograph or mosaic to the corresponding distance on the ground, classified as follows:
a. **very large scale** — 1:4,999 and larger;
b. **large scale** — 1:5,000 to 1:9,999;
c. **medium scale** — 1:10,000 to 1:24,999;
d. **small scale** — 1:25,000 to 1:49,999;
e. **very small scale** — 1:50,000 and smaller.
See also **scale.**

photographic strip — **(*)** Series of successive overlapping photographs taken along a selected course or direction.

photo interpretation key — See **imagery interpretation key.**

photomap — **(*)** A reproduction of a photograph or photomosaic upon which the grid lines, marginal data, contours, place names, boundaries, and other data may be added.

photo nadir — **(*)** The point at which a vertical line through the perspective center of the camera lens intersects the photo plane.

physical characteristics — Those military characteristics of equipment that are primarily physical in nature, such as weight, shape, volume, water-proofing, and sturdiness.

physical damage assessment — The estimate of the quantitative extent of physical damage (through munition blast, fragmentation, and/or fire damage effects) to a target resulting from the application of military force. This assessment is based upon observed or interpreted damage. See also **damage assessment; target.** (JP 3-60)

physical security — **(*)** That part of security concerned with physical measures designed to safeguard personnel; to prevent unauthorized access to equipment, installations, material, and documents; and to safeguard them against espionage, sabotage, damage, and theft. See also **communications security; security.**

pictomap — A topographic map in which the photographic imagery of a standard mosaic has been converted into interpretable colors and symbols by means of a pictomap process.

pictorial symbolization — **(*)** The use of symbols which convey the visual character of the features they represent.

Pierson-Moskowitz scale — A scale that categorizes the force of progressively higher wind speeds. See also **sea state.** (JP 4-01.6)

pillbox — **(*)** A small, low fortification that houses machine guns, antitank weapons, etc. A pillbox is usually made of concrete, steel, or filled sandbags.

pilot's trace — **(*)** A rough overlay to a map made by the pilot of a photographic reconnaissance aircraft during or immediately after a sortie. It shows the location, direction, number, and order of photographic runs made, together with the camera(s) used on each run.

pinpoint — **(*)** 1. A precisely identified point, especially on the ground, that locates a very small target, a reference point for rendezvous or for other purposes; the coordinates that define this point. 2. The ground position of aircraft determined by direct observation of the ground.

pinpoint photograph — **(*)** A single photograph or a stereo pair of a specific object or target.

pinpoint target — **(*)** In artillery and naval gunfire support, a target less than 50 meters in diameter.

pipeline — **(*)** In logistics, the channel of support or a specific portion thereof by means of which materiel or personnel flow from sources of procurement to their point of use.

piracy — An illegal act of violence, depredation (e.g., plundering, robbing, or pillaging), or detention in or over international waters committed for private ends by the crew or passengers of a private ship or aircraft against another ship or aircraft or against persons or property on board such ship or aircraft.

pitch — **(*)** 1. The movement of an aircraft or ship about its transverse axis. 2. In air photography, the camera rotation about the transverse axis of the aircraft. Also called **tip.**

pitch angle — **(*)** The angle between the aircraft's longitudinal axis and the horizontal plane. Also called **inclination angle.**

plan for landing — In amphibious operations, a collective term referring to all individually prepared naval and landing force documents which, taken together, present in detail all instructions for execution of the ship-to-shore movement. (JP 3-02.2)

plan identification number — 1. A command-unique four-digit number followed by a suffix indicating the Joint Strategic Capabilities Plan (JSCP) year for which the plan is written, e.g., "2220-95". 2. In the Joint Operation Planning and Execution System (JOPES) database, a five-digit number representing the command-unique four-digit identifier, followed by a one-character, alphabetic suffix indicating the operation plan option, or a one-digit number numeric value indicating the JSCP year for which the plan is written. Also called **PID.**

planimetric map — A map representing only the horizontal position of features. Sometimes called a line map. See also **map.**

plan information capability — The capability that allows a supported command to enter and update key elements of information in an operation plan stored in the Joint Operation Planning and Execution System.

planned airlift requests — Requests generated to meet airlift requirements that can be forecast or where requirements can be anticipated and published in the air tasking order. See also **air tasking order.** (JP 3-17)

planned target (nuclear) — A nuclear target planned on an area or point in which a need is anticipated. A planned nuclear target may be scheduled or on call. Firing data for a planned nuclear target may or may not be determined in advance. Coordination and warning of friendly troops and aircraft are mandatory.

planned targets — Targets that are known to exist in an operational area, and against which effects are scheduled in advance or are on-call. Examples range from targets on joint target lists in the applicable campaign plans, to targets detected in sufficient time to list in the air tasking order, mission-type orders, or fire support plans. Planned targets have two subcategories: scheduled or on-call. See also **on-call targets; operational area; scheduled targets; target.** (JP 3-60)

planning and direction — See **intelligence cycle.** See also **direction.** (JP 2-0)

planning directive — In amphibious operations, the plan issued by the designated commander, following receipt of the order initiating the amphibious operation, to ensure that the planning process and interdependent plans developed by the amphibious force will be coordinated, completed in the time allowed, and important aspects not overlooked. See also **amphibious force; amphibious operation.** (JP 3-02)

planning factor — **(*)** A multiplier used in planning to estimate the amount and type of effort involved in a contemplated operation. Planning factors are often expressed as rates, ratios, or lengths of time.

planning factors database — Databases created and maintained by the Military Services for the purpose of identifying all geospatial information and services requirements for emerging and existing forces and systems. The database identifies: unit requirements, at the information content level, for geospatial data and services; system requirements for standard Department of Defense geospatial data and services; research, development, test, and evaluation requirements for developmental systems, identified by milestone; and initial operating capability and full operating capability for emerging systems. Also called **PFDB.** See also **data; database; geospatial information and services.** (JP 2-03)

planning order — 1. An order issued by the Chairman of the Joint Chiefs of Staff (CJCS) to initiate execution planning. The planning order will normally follow a commander's estimate and a planning order will normally take the place of the CJCS alert order. National Command Authorities approval of a selected course of action is not required before issuing a CJCS planning order. 2. A planning directive that provides essential planning guidance and directs the initiation of execution planning before the directing authority approves a military course of action. See also **execution planning.** (JP 5-0)

planning phase — In amphibious operations, the phase normally denoted by the period extending from the issuance of the order initiating the amphibious operation up to the embarkation phase. The planning phase may occur during movement or at any other time upon receipt of a new mission or change in the operational situation. See also **amphibious operation.** (JP 3-02)

plan position indicator — **(*)** A cathode ray tube on which radar returns are so displayed as to bear the same relationship to the transmitter as the objects giving rise to them.

plant equipment — Personal property of a capital nature, consisting of equipment, furniture, vehicles, machine tools, test equipment, and accessory and auxiliary items, but excluding special tooling and special test equipment, used or capable of use in the manufacture of supplies or for any administrative or general plant purpose.

plastic zone — **(*)** The region beyond the rupture zone associated with crater formation resulting from an explosion in which there is no visible rupture, but in which the soil is permanently deformed and compressed to a high density. See also **rupture zone.**

plate — (*) 1. In cartography: a. a printing plate of zinc, aluminum, or engraved copper; b. collective term for all "states" of an engraved map reproduced from the same engraved printing plate; c. all detail to appear on a map or chart which will be reproduced from a single printing plate (e.g., the "blue plate" or the "contour plate"). 2. In photography, a transparent medium, usually glass, coated with a photographic emulsion. See also **diapositive.**

platform drop — (*) The airdrop of loaded platforms from rear loading aircraft with roller conveyors. See also **airdrop; airdrop platform.**

plot — (*) 1. Map, chart, or graph representing data of any sort. 2. Representation on a diagram or chart of the position or course of a target in terms of angles and distances from positions; location of a position on a map or a chart. 3. The visual display of a single location of an airborne object at a particular instant of time. 4. A portion of a map or overlay on which are drawn the outlines of the areas covered by one or more photographs. See also **master plot.**

point defense — The defense or protection of special vital elements and installations; e.g., command and control facilities or air bases. (JP 3-52)

point designation grid — (*) A system of lines, having no relation to the actual scale, or orientation, drawn on a map, chart, or air photograph dividing it into squares so that points can be more readily located.

pointee-talkee — A language aid containing selected phrases in English opposite a translation in a foreign language. It is used by pointing to appropriate phrases. See also **evasion aid.** (JP 3-50.3)

point of no return — (*) A point along an aircraft track beyond which its endurance will not permit return to its own or some other associated base on its own fuel supply.

point-to-point sealift — The movement of troops and/or cargo in Military Sealift Command nucleus or commercial shipping between established ports, in administrative landings, or during logistics over-the-shore operations. See also **administrative landing; administrative movement; logistics over-the-shore operations**.

poised mine — (*) A mine in which the ship counter setting has been run down to "one" and which is ready to detonate at the next actuation. See also **mine.**

polar coordinates — (*) 1. Coordinates derived from the distance and angular measurements from a fixed point (pole). 2. In artillery and naval gunfire support, the direction, distance, and vertical correction from the observer/spotter position to the target.

polar orbit — A satellite orbit in which the satellite passes over the North and South Poles on each orbit, and eventually passes over all points on the earth. The angle of inclination between the equator and a polar orbit is 90 degrees.

polar plot — (*) The method of locating a target or point on the map by means of polar coordinates.

political intelligence — Intelligence concerning foreign and domestic policies of governments and the activities of political movements.

political warfare — Aggressive use of political means to achieve national objectives.

politico-military gaming — Simulation of situations involving the interaction of political, military, sociological, psychological, economic, scientific, and other appropriate factors.

pool — 1. Maintenance and control of a supply of resources or personnel upon which other activities may draw. The primary purpose of a pool is to promote maximum efficiency of use of the pooled resources or personnel, e.g., a petroleum pool or a labor and equipment pool. 2. Any combination of resources which serves a common purpose.

port capacity — **(*)** The estimated capacity of a port or an anchorage to clear cargo in 24 hours usually expressed in tons. See also **beach capacity; clearance capacity.**

port complex — **(*)** A port complex comprises one or more port areas of varying importance whose activities are geographically linked either because these areas are dependent on a common inland transport system or because they constitute a common initial destination for convoys.

port designator — **(*)** A group of letters identifying ports in convoy titles or messages.

port evacuation of cargoes — **(*)** The removal of cargoes from a threatened port to alternative storage sites.

port evacuation of shipping — **(*)** The movement of merchant ships from a threatened port for their own protection.

port of debarkation — The geographic point at which cargo or personnel are discharged. This may be a seaport or aerial port of debarkation; for unit requirements; it may or may not coincide with the destination. Also called **POD.** See also **port of embarkation.**

port of embarkation — The geographic point in a routing scheme from which cargo or personnel depart. This may be a seaport or aerial port from which personnel and equipment flow to a port of debarkation; for unit and nonunit requirements, it may or may not coincide with the origin. Also called **POE.** See also **port of debarkation.**

port of support — The geographic point (seaport or airport) in an objective area that is the terminal point for strategic deployment for non-unit-related supplies. Each component designates ports of support for four categories of resupply: general cargo; ammunition; petroleum, oils, and lubricants; and air deliveries. Also called **POS.**

port operations group — A task-organized unit, located at the seaport of embarkation and/or debarkation under the control of the landing force support party and/or combat service support element, that assists and provides support in the loading and/or unloading and staging of personnel, supplies, and equipment from shipping. Also called **POG.** See also **combat service support element; landing force support party; task organization.** (JP 4-01.8)

port security — **(*)** The safeguarding of vessels, harbors, ports, waterfront facilities, and cargo from internal threats such as destruction, loss, or injury from sabotage or other subversive acts; accidents; thefts; or other causes of similar nature. See also **harbor defense; physical security; security.**

port support activity — A tailorable support organization composed of mobilization station assets that ensures the equipment of the deploying units is ready to load. The port support activity (PSA) operates unique equipment in conjunction with ship loading operations. The PSA is operationally controlled by the military port commander or terminal transfer unit commander. Also called **PSA.** See also **support.** (JP 4-01.8)

positional defense — See **position defense.**

position defense — **(*)** The type of defense in which the bulk of the defending force is disposed in selected tactical localities where the decisive battle is to be fought. Principal reliance is placed on the ability of the forces in the defended localities to maintain their positions and to control the terrain between them. The reserve is used to add depth, to block, or restore the battle position by counterattack.

positive control — A method of airspace control that relies on positive identification, tracking, and direction of aircraft within an airspace, conducted with electronic means by an agency having the authority and responsibility therein.

positive identification and radar advisory zone — A specified area established for identification and flight following of aircraft in the vicinity of a fleet-defended area. Also called **PIRAZ.**

positive phase of the shock wave — The period during which the pressure rises very sharply to a value that is higher than ambient and then decreases rapidly to the ambient pressure. See also **negative phase of the shock wave.**

Posse Comitatus Act — Prohibits search, seizure, or arrest powers to US military personnel. Amended in 1981 under Public Law 97-86 to permit increased Department of Defense support of drug interdiction and other law enforcement activities. (Title 18, "Use of Army and Air Force as Posse Comitatus" - United States Code, Section 1385)

postattack period — In nuclear warfare, that period which extends from the termination of the final attack until political authorities agree to terminate hostilities. See also **posthostilities period; transattack period.**

posthostilities period — That period subsequent to the date of ratification by political authorities of agreements to terminate hostilities.

poststrike reconnaissance — Missions undertaken for the purpose of gathering information used to measure results of a strike.

power projection — The ability of a nation to apply all or some of its elements of national power - political, economic, informational, or military - to rapidly and effectively deploy and sustain forces in and from multiple dispersed locations to respond to crises, to contribute to deterrence, and to enhance regional stability. See also **elements of national power.** (JP 3-35)

PPI gauge — See **international loading gauge.**

practice mine — **(*)** 1. In land mine warfare, an inert mine to which is fitted a fuze and a device to indicate, in a non-lethal fashion, that the fuze has been activated. See also **mine.** 2. In naval mine warfare, an inert-filled mine but complete with assembly, suitable for instruction and for practice in preparation. See also **drill mine.**

prearranged fire — **(*)** Fire that is formally planned and executed against targets or target areas of known location. Such fire is usually planned well in advance and is executed at a predetermined time or during a predetermined period of time. See also **fire; on-call; scheduled fire.**

preassault operation — Operations conducted by the amphibious force upon its arrival in the operational area and prior to H-hour and/or L-hour. See also **amphibious force; times.** (JP 3-02)

precautionary launch — The launching of nuclear loaded aircraft under imminent nuclear attack so as to preclude friendly aircraft destruction and loss of weapons on the ground and/or carrier.

precautionary search and rescue and/or combat search and rescue — The planning and pre-positioning of aircraft, ships, or ground forces and facilities before an operation to provide search and rescue (SAR) or combat search and rescue (CSAR) assistance if needed. The planning of precautionary SAR or CSAR is usually done by plans personnel with SAR or CSAR expertise and background on an operations staff, a joint search and rescue center, or a rescue coordination center. Also called **precautionary SAR and/or CSAR.** See also **combat search and rescue; joint combat search and rescue operation; search and rescue.** (JP 3-50.2)

precedence — 1. **communications** — A designation assigned to a message by the originator to indicate to communications personnel the relative order of handling and to the addressee the order in which the message is to be noted. Examples of communication precedence from most immediate to least are flash, immediate, priority, and routine. 2. **reconnaissance** — A letter designation, assigned by a unit requesting several reconnaissance missions, to indicate the relative order of importance (within an established priority) of the mission requested. 3. **evacuation** — The assignment of a priority for medical evacuation that is based on patient condition, advice of the senior medical person at the scene, and the tactical situation. See also **flash message; immediate message; priority message; routine message.**

precession — See **apparent precession.**

precipitation static — Charged precipitation particles that strike antennas and gradually charge the antenna, which ultimately discharges across the insulator, causing a burst of static. Also called **P-STATIC.** (JP 3-51)

precise frequency — A frequency requirement accurate to within one part in 1,000,000,000.

precise time — A time requirement accurate to within 10 milliseconds.

precision approach — An approach in which range, azimuth, and glide slope information are provided to the pilot. See also **final approach; nonprecision approach.** (JP 3-04.1)

precision bombing — Bombing directed at a specific point target.

precision-guided munitions — A weapon that uses a seeker to detect electromagnetic energy reflected from a target or reference point and, through processing, provides guidance commands to a control system that guides the weapon to the target. Also called **PGM.** See also **munitions.** (JP 3-09.1)

precursor — Any chemical reactant which takes place at any stage in the production by whatever method of a toxic chemical. This includes any key component of a binary or multicomponent chemical system. See also **toxic chemical.** (JP 3-11)

precursor chemical — Compounds that are required in the synthetic or extraction processes of drug production, and become incorporated into the drug molecule. Not used in the production of cocaine or heroin. (JP 3-07.4)

precursor front — **(*)** An air pressure wave which moves ahead of the main blast wave for some distance as a result of a nuclear explosion of appropriate yield and low burst height over a heat-absorbing (or dusty) surface. The pressure at the precursor front increases more gradually than in a true (or ideal) shock wave, so that the behavior in the precursor region is said to be non-ideal.

precursor sweeping — **(*)** The sweeping of an area by relatively safe means in order to reduce the risk to mine countermeasures vessels in subsequent operations.

predicted fire — **(*)** Fire that is delivered without adjustment.

predominant height — **(*)** In air reconnaissance, the height of 51 percent or more of the structures within an area of similar surface material.

preemptive attack — An attack initiated on the basis of incontrovertible evidence that an enemy attack is imminent.

preinitiation — The initiation of the fission chain reaction in the active material of a nuclear weapon at any time earlier than that at which either the designed or the maximum compression or degree of assembly is attained.

prelanding operations — In amphibious operations, operations conducted between the commencement of the assault phase and the commencement of the ship-to-shore movement by the main body of the amphibious task force. They encompass similar preparations conducted by the advanced force but focus on the landing area, concentrating specifically on the landing beaches and the helicopter landing zones to be used by the main landing force. Prelanding operations also encompass final preparations for the ship-to-shore movement. (JP 3-02)

pre-launch survivability — The probability that a delivery and/or launch vehicle will survive an enemy attack under an established condition of warning.

preliminary communications search — In search and rescue operations, consists of contacting and checking major facilities within the areas where the craft might be or might have been seen. A preliminary communications search is normally conducted during the uncertainty phase. Also called **PRECOM.** See also **extended communications search; search and rescue incident classification, Subpart a.**

preliminary demolition target — **(*)** A target, other than a reserved demolition target, which is earmarked for demolition and which can be executed immediately after preparation, provided that prior authority has been granted. See also **demolition target; reserved demolition target.**

preliminary movement schedule — A projection of the routing of movement requirements reflected in the time-phased force and deployment data, from origin to destination, including identification of origins, ports of embarkation, ports of debarkation, and en route stops; associated time frames for arrival and departure at each location; type of lift assets required to accomplish the move; and cargo details by carrier. Schedules are sufficiently detailed to support comparative analysis of requirements against capabilities and to develop location workloads for reception and onward movement.

preload loading — **(*)** The loading of selected items aboard ship at one port prior to the main loading of the ship at another. See also **loading.**

premature dud — See **flare dud.**

preparation fire — Fire delivered on a target preparatory to an assault. See also **fire.**

preplanned air support — **(*)** Air support in accordance with a program, planned in advance of operations. See also **air support.**

preplanned mission request — A request for an air strike on a target that can be anticipated sufficiently in advance to permit detailed mission coordination and planning.

preplanned nuclear support — Nuclear support planned in advance of operations. See also **immediate nuclear support; nuclear support.**

pre-position — **(*)** To place military units, equipment, or supplies at or near the point of planned use or at a designated location to reduce reaction time, and to ensure timely support of a specific force during initial phases of an operation.

pre-positioned war reserve materiel requirement, balance — That portion of the pre-positioned war reserve materiel requirement that has not been acquired or funded. This level consists of the pre-positioned war reserve materiel requirement, less the pre-positioned war reserve requirement, protectable.

pre-positioned war reserve materiel requirement, protectable — That portion of the pre-positioned war reserve materiel requirement that is protected for purposes of procurement, funding, and inventory management.

pre-positioned war reserve requirement — That portion of the war reserve materiel requirement that the current Secretary of Defense guidance dictates be reserved and positioned at or near the point of planned use or issue to the user prior to hostilities to reduce reaction time and to assure timely support of a specific force or project until replenishment can be effected.

pre-positioned war reserve stock — The assets that are designated to satisfy the pre-positioned war reserve materiel requirement. Also called **PWRS.**

prescribed nuclear load — **(*)** A specified quantity of nuclear weapons to be carried by a delivery unit. The establishment and replenishment of this load after each expenditure is a command decision and is dependent upon the tactical situation, the nuclear logistical situation, and the capability of the unit to transport and utilize the load. It may vary from day to day and among similar delivery units.

prescribed nuclear stockage — **(*)** A specified quantity of nuclear weapons, components of nuclear weapons, and warhead test equipment to be stocked in special ammunition supply points or other logistical installations. The establishment and replenishment of this stockage is a command decision and is dependent upon the tactical situation, the allocation, the capability of the logistical support unit to store and maintain the nuclear weapons, and the nuclear logistical situation. The prescribed stockage may vary from time to time and among similar logistical support units.

preset guidance — A technique of missile control wherein a predetermined flight path is set into the control mechanism and cannot be adjusted after launching.

Presidential Callup — Procedures by which the President brings all or part of the Army National Guard or Air National Guard to active Federal service under section 12406 and Chapter 15 of title 10 (DOD), US Code. See also **active duty; federal service; Presidential Reserve Callup Authority.** (JP 4-05)

Presidential Reserve Callup Authority — Provision of a public law (US Code, Title 10 (DOD), section 12304) that provides the President a means to activate, without a declaration of national emergency, not more than 200,000 members of the Selected Reserve and the Individual Ready Reserve (of whom not more than 30,000 may be members of the Individual Ready Reserve), for not more than 270 days to meet the support requirements of any operational mission. Members called under this provision may not be used for disaster relief or to suppress insurrection. This authority has particular utility when used in circumstances in which the escalatory national or international signals of partial or full mobilization would be undesirable. Forces available under this authority can provide a tailored, limited-scope, deterrent, or operational response, or may be used as a precursor to any subsequent mobilization. Also called **PRCA.** See also **Individual Ready Reserve; mobilization; Presidential Callup; Selected Reserve.** (JP 3-57)

pressure altitude — **(*)** An atmospheric pressure expressed in terms of altitude which corresponds to that pressure in the standard atmosphere. See also **altitude.**

pressure breathing — **(*)** The technique of breathing which is required when oxygen is supplied direct to an individual at a pressure higher than the ambient barometric pressure.

pressure front — See **shock front.**

pressure mine — **(*)** 1. In land mine warfare, a mine whose fuze responds to the direct pressure of a target. 2. In naval mine warfare, a mine whose circuit responds to the hydrodynamic pressure field of a target. See also **mine.**

pressure mine circuit — See **pressure mine.**

pressurized cabin — The occupied space of an aircraft in which the air pressure has been increased above that of the ambient atmosphere by compression of the ambient atmosphere into the space.

prestrike reconnaissance — Missions undertaken for the purpose of obtaining complete information about known targets for use by the strike force.

prevention — The security procedures undertaken by the public and private sector in order to discourage terrorist acts. See also **antiterrorism.** (JP 3-07.2)

prevention of stripping equipment — See **antirecovery device.**

preventive deployment — The deployment of military forces to deter violence at the interface or zone of potential conflict where tension is rising among parties. Forces may be employed in such a way that they are indistinguishable from a peacekeeping force in terms of equipment, force posture, and activities. See also **peace enforcement; peacekeeping; peace operations.** (JP 3-07)

preventive diplomacy — Diplomatic actions taken in advance of a predictable crisis to prevent or limit violence. (JP 3-07)

preventive maintenance — The care and servicing by personnel for the purpose of maintaining equipment and facilities in satisfactory operating condition by providing for systematic inspection, detection, and correction of incipient failures either before they occur or before they develop into major defects.

preventive medicine — The anticipation, communication, prediction, identification, prevention, education, risk assessment, and control of communicable diseases, illnesses and exposure to endemic, occupational, and environmental threats. These threats include nonbattle injuries, combat stress responses, weapons of mass destruction, and other threats to the health and readiness of military personnel. Communicable diseases include anthropod-, vector-, food-, waste-, and waterborne diseases. Preventative medicine measures include field sanitation, medical surveillance, pest and vector control, disease risk assessment, environmental and occupational health surveillance, waste (human, hazardous, and medical) disposal, food safety inspection, and potable water surveillance. Also called **PVNTMED.** (JP 4-02)

preventive war — A war initiated in the belief that military conflict, while not imminent, is inevitable, and that to delay would involve greater risk.

prewithdrawal demolition target — A target prepared for demolition preliminary to a withdrawal, the demolition of which can be executed as soon after preparation as convenient on the orders of the officer to whom the responsibility for such demolitions has been delegated. See also **demolition target.**

primary aircraft authorization — The number of aircraft authorized to a unit for performance of its operational mission. The primary authorization forms the basis for the allocation of operating resources to include manpower, support equipment, and flying-hour funds. Also called **PAA.**

primary aircraft inventory — The aircraft assigned to meet the primary aircraft authorization. Also called **PAI.**

primary censorship — Armed forces censorship performed by personnel of a company, battery, squadron, ship, station, base, or similar unit on the personal communications of persons assigned, attached, or otherwise under the jurisdiction of a unit. See also **censorship.**

primary control officer — In amphibious operations, the officer embarked in a primary control ship assigned to control the movement of landing craft, amphibious vehicles, and landing ships to and from a colored beach. Also called **PCO.** (JP 3-02)

primary control ship — In amphibious operations, a ship of the task force designated to provide support for the primary control officer and a combat information center control team for a colored beach. Also called **PCS.** (JP 3-02)

primary flight control — The controlling agency on aviation ships and amphibious aviation assault ships that is responsible for air traffic control of aircraft within 5 nautical miles of the ship. On Coast Guard cutters, primary flight control duties are performed by a combat information center, and the term "PRIFLY" is not used. Also called **PRIFLY.** See also **amphibious aviation assault ship; aviation ship.** (JP 3-04.1)

primary imagery dissemination — See **electronic imagery dissemination.**

primary imagery dissemination system — See **electronic imagery dissemination.**

primary interest — Principal, although not exclusive, interest and responsibility for accomplishment of a given mission, including responsibility for reconciling the activities of other agencies that possess collateral interest in the program.

primary operating stocks — Logistic resources on hand or on order necessary to support day-to-day operational requirements, and which, in part, can also be used to offset sustaining combat requirements. Also called **POS.**

primary review authority — The organization assigned by the lead agent to perform the actions and coordination necessary to develop and maintain the assigned joint publication under the cognizance of the lead agent. Also called **PRA.** See also **joint publication; lead agent.** (JP 1-01)

primary target — An object of high publicity value to terrorists. See also **antiterrorism; secondary targets.** (JP 3-07.2)

primed charge — **(*)** A charge ready in all aspects for ignition.

prime mover — A vehicle, including heavy construction equipment, possessing military characteristics, designed primarily for towing heavy, wheeled weapons and frequently providing facilities for the transportation of the crew of, and ammunition for, the weapon.

prime vendor — A contracting process that provides commercial products to regionally grouped military and federal customers from commercial distributors using electronic commerce. Customers typically receive materiel delivery through the vendor's commercial distribution system. Also called **PV.** See also **distribution system.** (JP 4-09)

principal building — A building aboard a diplomatic or consular compound where classified information may be handled, stored, discussed, or processed, but that does not house the offices of the chief of mission or principal officer.

principal items — End items and replacement assemblies of such importance that management techniques require centralized individual item management throughout the supply system, to include depot level, base level, and items in the hands of using units. These specifically include the items where, in the judgment of the Services, there is a need for central inventory control, including centralized computation of requirements, central procurement, central direction of distribution, and central knowledge and control of all assets owned by the Services.

principal officer — The officer in charge of a diplomatic mission, consular office, or other Foreign Service post, such as a United States Liaison Office.

principal operational interest — When used in connection with an established facility operated by one Service for joint use by two or more Services, "principal operational interest" indicates a requirement for the greatest use of, or the greatest need for, the services of that facility. The term may be applied to a Service, but is more applicable to a command.

principal parallel — **(*)** On an oblique photograph, a line parallel to the true horizon and passing through the principal point.

principal plane — **(*)** A vertical plane which contains the principal point of an oblique photograph, the perspective center of the lens, and the ground nadir.

principal scale — **(*)** In cartography, the scale of a reduced or generating globe representing the sphere or spheroid, defined by the fractional relation of their respective radii. Also called **nominal scale.** See also **scale.**

principal vertical — (*) On an oblique photograph, a line perpendicular to the true horizon and passing through the principal point.

printing size of a map or chart — (*) The dimensions of the smallest rectangle which will contain a map or chart, including all the printed material in its margin.

print reference — (*) A reference to an individual print in an air photographic sortie.

priority designator — A two-digit issue and priority code (01 through 15) placed in military standard requisitioning and issue procedure requisitions. It is based upon a combination of factors that relate the mission of the requisitioner and the urgency of need or the end use and is used to provide a means of assigning relative rankings to competing demands placed on the Department of Defense supply system.

priority intelligence requirements — Those intelligence requirements for which a commander has an anticipated and stated priority in the task of planning and decisionmaking. Also called **PIRs.** See also **information requirements; intelligence; intelligence cycle; intelligence requirement.** (JP 2-0)

priority message — A category of precedence reserved for messages that require expeditious action by the addressee(s) and/or furnish essential information for the conduct of operations in progress when routine precedence will not suffice. See also **precedence.**

priority national intelligence objectives — A guide for the coordination of intelligence collection and production in response to requirements relating to the formulation and execution of national security policy. They are compiled annually by the Washington Intelligence Community and flow directly from the intelligence mission as set forth by the National Security Council. They are specific enough to provide a basis for planning the allocation of collection and research resources, but not so specific as to constitute in themselves research and collection requirements.

priority of immediate mission requests — See **emergency priority; urgent priority.**

priority system for mission requests for tactical reconnaissance — A system that assigns each tactical reconnaissance request the appropriate priority as follows. **Priority I** — Takes precedence over all other requests except those previously assigned priority I. The results of these requests are of paramount importance to the immediate battle situation or objective. **Priority II** — The results of these requirements are in support of the general battle situation and will be accomplished as soon as possible after priority I requests. These are requests to gain current battle information. **Priority III** — The results of these requests update the intelligence database but do not affect the immediate battle situation. **Priority IV** — The results of these requests are of a routine nature. These results will be fulfilled when the reconnaissance effort permits. See also **precedence.**

prior permission — (*) Permission granted by the appropriate authority prior to the commencement of a flight or a series of flights landing in or flying over the territory of the nation concerned.

prisoner of war — A detained person as defined in Articles 4 and 5 of the Geneva Convention Relative to the Treatment of Prisoners of War of August 12, 1949. In particular, one who, while engaged in combat under orders of his or her government, is captured by the armed forces of the enemy. As such, he or she is entitled to the combatant's privilege of immunity from the municipal law of the capturing state for warlike acts which do not amount to breaches of the law of armed conflict. For example, a prisoner of war may be, but is not limited to, any person belonging to one of the following categories who has fallen into the power of the enemy: a member of the armed forces, organized militia or volunteer corps; a person who accompanies the armed forces without actually being a member thereof; a member of a merchant marine or civilian aircraft crew not qualifying for more favorable treatment; or individuals who, on the approach of the enemy, spontaneously take up arms to resist the invading forces. Also called **POW or PW.**

prisoner of war branch camp — (*) A subsidiary camp under the supervision and administration of a prisoner of war camp.

prisoner of war camp — An installation established for the internment and administration of prisoners of war.

prisoner of war censorship — The censorship of the communications to and from enemy prisoners of war and civilian internees held by the United States Armed Forces. See also **censorship.**

prisoner of war compound — (*) A subdivision of a prisoner of war enclosure.

prisoner of war enclosure — (*) A subdivision of a prisoner of war camp.

prisoner of war personnel record — (*) A form for recording the photograph, fingerprints, and other pertinent personal data concerning the prisoner of war, including that required by the Geneva Convention.

proactive measures — In antiterrorism, measures taken in the preventive stage of antiterrorism designed to harden targets and detect actions before they occur. (JP 3-07.2)

proactive mine countermeasures — Measures intended to prevent the enemy from successfully laying mines. See also **mine countermeasures.** (JP 3-15)

probability of damage — (*) The probability that damage will occur to a target expressed as a percentage or as a decimal. Also called **PD.**

probable error — See **horizontal error.**

probable error deflection — Error in deflection that is exceeded as often as not.

probable error height of burst — Error in height of burst that projectile and/or missile fuzes may be expected to exceed as often as not.

probable error range — Error in range that is exceeded as often as not.

probably destroyed — (*) In air operations, a damage assessment on an enemy aircraft seen to break off combat in circumstances which lead to the conclusion that it must be a loss although it is not actually seen to crash.

probe — In information operations, any attempt to gather information about an automated information system or its on-line users. See also **information; information operations; information system.** (JP 3-13)

procedural control — (*) A method of airspace control which relies on a combination of previously agreed and promulgated orders and procedures.

procedure — A procedure begins with a specific, documentable event that causes an activity to occur. The activity must produce a product that normally affects another external organization. Frequently, that product will be the event that causes another procedure to occur. It is important to recognize that a procedure determines "what" an organization must do at critical periods but does not direct "how" it will be done.

procedure turn — **(*)** An aircraft maneuver in which a turn is made away from a designated track followed by a turn in the opposite direction, both turns being executed at a constant rate so as to permit the aircraft to intercept and proceed along the reciprocal of the designated track.

procedure word — A word or phrase limited to radio telephone procedure used to facilitate communication by conveying information in a condensed standard form. Also called **proword.**

processing — **(*)** 1. In photography, the operations necessary to produce negatives, diapositives, or prints from exposed films, plates, or paper. 2. **(DOD only)** A system of operations designed to convert raw data into useful information. (JP 2-0)

proclamation — A document published to the inhabitants of an area that sets forth the basis of authority and scope of activities of a commander in a given area and which defines the obligations, liabilities, duties, and rights of the population affected.

procurement lead time — The interval in months between the initiation of procurement action and receipt into the supply system of the production model (excludes prototypes) purchased as the result of such actions. It is composed of two elements, production lead time and administrative lead time. See also **administrative lead time; initiation of procurement action; level of supply; production lead time; receipt into the supply system.**

producer countries — In counterdrug operations, countries where naturally occurring plants such as coca, cannabis, or poppies are cultivated for later refinement into illicit drugs. See also **counterdrug operations.** (JP 3-07.4)

production base — The total national industrial production capacity available for the manufacture of items to meet materiel requirements.

production lead time — The time interval between the placement of a contract and receipt into the supply system of materiel purchased. Two entries are provided: a. **initial** — The time interval if the item is not under production as of the date of contract placement; and b. **reorder** — The time interval if the item is under production as of the date of contract placement. See also **procurement lead time.**

production logistics — That part of logistics concerning research, design, development, manufacture, and acceptance of materiel. In consequence, production logistics includes: standardization and interoperability, contracting, quality assurance, initial provisioning, transportability, reliability and defect analysis, safety standards, specifications and production processes, trials and testing (including provision of necessary facilities), equipment documentation, configuration control, and modifications.

production loss appraisal — An estimate of damage inflicted on an industry in terms of quantities of finished products denied the enemy from the moment of attack through the period of reconstruction to the point when full production is resumed.

proficiency training aircraft — Aircraft required to maintain the proficiency of pilots and other aircrew members who are assigned to nonflying duties.

profile — See **flight profile.**

program aircraft — The total of the active and reserve aircraft. See also **aircraft.**

Programmed Forces — The forces that exist for each year of the Future Years Defense Program. They contain the major combat and tactical support forces that are expected to execute the national strategy within manpower, fiscal, and other constraints. See also **current force; force; Intermediate Force Planning Level.**

program of nuclear cooperation — **(*)** Presidentially approved bilateral proposals for the United States to provide nuclear weapons and specified support to user nations who desire to commit delivery units to NATO in nuclear only or dual capable roles. After presidential approval in principle, negotiations will be initiated with the user nation to develop detailed support arrangements.

progress payment — Payment made as work progresses under a contract, upon the basis of costs incurred, of percentage of completion accomplished, or of a particular stage of completion. The term does not include payments for partial deliveries accepted by the Government under a contract or partial payments on contract termination claims.

prohibited area — A specified area within the land areas of a state or its internal waters, archipelagic waters, or territorial sea adjacent thereto over which the flight of aircraft is prohibited. May also refer to land or sea areas to which access is prohibited. See also **closed area; danger area; restricted area.**

projected map display — **(*)** The displayed image of a map or chart projected through an optical or electro-optical system onto a viewing surface.

projection print — An enlarged or reduced photographic print made by projection of the image of a negative or a transparency onto a sensitized surface.

proliferation (nuclear weapons) — The process by which one nation after another comes into possession of, or into the right to determine the use of, nuclear weapons; each nation becomes potentially able to launch a nuclear attack upon another nation.

prompt radiation — The gamma rays produced in fission and as a result of other neutron reactions and nuclear excitation of the weapon materials appearing within a second or less after a nuclear explosion. The radiations from these sources are known either as prompt or instantaneous gamma rays. See also **induced radiation; initial radiation; residual radiation.**

proofing — The verification that a breached lane is free of live mines by passing a mine roller or other mine-resistant vehicle through as the lead vehicle.

propaganda — Any form of communication in support of national objectives designed to influence the opinions, emotions, attitudes, or behavior of any group in order to benefit the sponsor, either directly or indirectly. See also **black propaganda; grey propaganda; white propaganda.**

propelled mine — See **mobile mine.**

proper authority — An authority authorized to call an opposing force hostile; may be either the National Command Authorities (NCA), the affected CINC, and/or any commander so delegated by either the NCA or the CINC.

proper clearance — A clearance for entry of units into specified defense areas by civil or military authorities having responsibility for granting such clearance.

property — 1. Anything that may be owned. 2. As used in the military establishment, this term is usually confined to tangible property, including real estate and materiel. For special purposes and as used in certain statutes, this term may exclude such items as the public domain, certain lands, certain categories of naval vessels, and records of the Federal Government.

property account — A formal record of property and property transactions in terms of quantity and/or cost, generally by item. An official record of Government property required to be maintained.

proportional navigation — A method of homing navigation in which the missile turn rate is directly proportional to the turn rate in space of the line of sight.

protected emblems — The red cross, red crescent, and other symbols that designate that persons, places, or equipment so marked have a protected status under the law of war.

protected frequencies — Those friendly frequencies used for a particular operation, identified and protected to prevent them from being inadvertently jammed by friendly forces while active electronic warfare operations are directed against hostile forces. These frequencies are of such critical importance that jamming should be restricted unless absolutely necessary or until coordination with the using unit is made. They are generally time-oriented, may change with the tactical situation, and must be updated periodically. See also **electronic warfare.** (JP 3-51)

protected persons/places — Persons (such as enemy prisoners of war) and places (such as hospitals) that enjoy special protections under the law of war. They may or may not be marked with protected emblems.

protected site — **(*)** A facility which is protected by the use of camouflage or concealment, selective siting, construction of facilities designed to prevent damage from fragments caused by conventional weapons, or a combination of such measures.

protection — Measures that are taken to keep nuclear, biological, and chemical hazards from having an adverse effect on personnel, equipment, or critical assets and facilities. Protection consists of five groups of activities: hardening of positions; protecting personnel; assuming mission-oriented protective posture; using physical defense measures; and reacting to attack. See also **mission-oriented protective posture.** (JP 3-11)

protection of shipping — The use of proportionate force by US warships, military aircraft, and other forces, when necessary for the protection of US flag vessels and aircraft, US citizens (whether embarked in US or foreign vessels), and their property against unlawful violence. This protection may be extended (consistent with international law) to foreign flag vessels, aircraft, and persons. (JP 3-07)

protective clothing — **(*)** Clothing especially designed, fabricated, or treated to protect personnel against hazards caused by extreme changes in physical environment, dangerous working conditions, or enemy action.

protective mask — A protective ensemble designed to protect the wearer's face and eyes and prevent the breathing of air contaminated with chemical and/or biological agents. See also **mission-oriented protective posture.** (JP 3-11)

protective minefield — **(*)** 1. In land mine warfare, a minefield employed to assist a unit in its local, close-in protection. 2. In naval mine warfare, a minefield laid in friendly territorial waters to protect ports, harbors, anchorages, coasts, and coastal routes. See also **minefield.**

prototype — A model suitable for evaluation of design, performance, and production potential.

provisioning — See **initial provisioning.**

proword — See **procedure word.**

proximity fuze — **(*)** A fuze wherein primary initiation occurs by remotely sensing the presence, distance, and/or direction of a target or its associated environment by means of a signal generated by the fuze or emitted by the target, or by detecting a disturbance of a natural field surrounding the target.

pseudopursuit navigation — A method of homing navigation in which the missile is directed toward the instantaneous target position in azimuth, while pursuit navigation in elevation is delayed until more favorable angle of attack on the target is achieved.

psychological consolidation activities — Planned psychological activities across the range of military operations directed at the civilian population located in areas under friendly control in order to achieve a desired behavior that supports the military objectives and the operational freedom of the supported commanders.

psychological operations — Planned operations to convey selected information and indicators to foreign audiences to influence their emotions, motives, objective reasoning, and ultimately the behavior of foreign governments, organizations, groups, and individuals. The purpose of psychological operations is to induce or reinforce foreign attitudes and behavior favorable to the originator's objectives. Also called **PSYOP.** See also **consolidation psychological operations; overt peacetime psychological operations programs; perception management.**

psychological warfare — The planned use of propaganda and other psychological actions having the primary purpose of influencing the opinions, emotions, attitudes, and behavior of hostile foreign groups in such a way as to support the achievement of national objectives. Also called **PSYWAR.**

public affairs — Those public information, command information, and community relations activities directed toward both the external and internal publics with interest in the Department of Defense. Also called **PA.** See also **command information; community relations; public information.**

public affairs assessment — An analysis of the news media and public environments to evaluate the degree of understanding about strategic and operational objectives and military activities and to identify levels of public support. Includes judgments about the public affairs impact of pending decisions and recommendations about the structure of public affairs support for the assigned mission. See also **assessment; public affairs.** (JP 3-61)

public affairs ground rules — Conditions established by a military command to govern the conduct of news gathering and the release and/or use of specified information during an operation or during a specific period of time. See also **public affairs.** (JP 3-61)

public affairs guidance — Normally, a package of information to support the public discussion of defense issues and operations. Such guidance can range from a telephonic response to a specific question to a more comprehensive package. Included could be an approved public affairs policy, news statements, answers to anticipated media questions, and community relations guidance. The public affairs guidance also addresses the method(s), timing, location, and other details governing the release of information to the public. Public affairs guidance is approved by the Assistant to the Secretary of Defense (Public Affairs). Also called **PAG.** See also **community relations; public affairs.** (JP 3-61)

public information — Information of a military nature, the dissemination of which through public news media is not inconsistent with security, and the release of which is considered desirable or nonobjectionable to the responsible releasing agency.

public key infrastructure — An enterprise-wide service (i.e. data integrity, user identification and authentication, user non-repudiation, data confidentiality, encryption, and digital signature) that supports digital signatures and other public key-based security mechanisms for Department of Defense functional enterprise programs, including generation, production, distribution, control, and accounting of public key certificates. A public key infrastructure provides the means to bind public keys to their owners and helps in the distribution of reliable public keys in large heterogeneous networks. Public keys are bound to their owners by public key certificates. These certificates contain information such as the owner's name and the associated public key and are issued by a reliable certification authority. Also called **PKI.**

pull-up point — **(*)** The point at which an aircraft must start to climb from a low-level approach in order to gain sufficient height from which to execute the attack or retirement. See also **contact point.**

pulse code — A system of using selected pulse-repetition frequencies to allow a specific laser seeker to acquire a target illuminated by a specific laser designator. See also **laser; laser designator; laser seeker.** (JP 3-09.1)

pulse duration — In radar, measurement of pulse transmission time in microseconds; that is, the time the radar's transmitter is energized during each cycle. Also called **pulse length and pulse width.**

pulsejet — **(*)** A jet-propulsion engine containing neither compressor nor turbine. Equipped with valves in the front which open and shut, it takes in air to create thrust in rapid periodic bursts rather than continuously.

pulse repetition frequency — 1. In lasers, the number of pulses that occur each second. 2. In radar, the number of pulses that occur each second. Pulse repetition frequency should not be confused with transmission frequency, which is determined by the rate at which cycles are repeated within the transmitted pulse. Also called **PRF.** See also **laser.** (JP 3-09.1)

pulsing — **(*)** In naval mine warfare, a method of operating magnetic and acoustic sweeps in which the sweep is energized by current which varies or is intermittent in accordance with a predetermined schedule.

purchase description — A statement outlining the essential characteristics and functions of an item, service, or materiel required to meet the minimum needs of the Government. It is used when a specification is not available or when specific procurement specifications are not required by the individual Military Departments or the Department of Defense.

purchase notice agreements — Agreements concerning the purchase of brand-name items for resale purposes established by each Military Service under the control of the Defense Logistics Agency.

purchasing office — Any installation or activity, or any division, office, branch, section, unit, or other organizational element of an installation or activity charged with the functions of procuring supplies or services.

pursuit — **(*)** An offensive operation designed to catch or cut off a hostile force attempting to escape, with the aim of destroying it.

pyrotechnic — A mixture of chemicals which, when ignited, is capable of reacting exothermically to produce light, heat, smoke, sound or gas.

pyrotechnic delay — **(*)** A pyrotechnic device added to a firing system which transmits the ignition flame after a predetermined delay.

Q

q-message — **(*)** A classified message relating to navigational dangers, navigational aids, mined areas, and searched or swept channels.

Q-route — A system of preplanned shipping lanes in mined or potentially mined waters used to minimize the area the mine countermeasures commander has to keep clear of mines in order to provide safe passage for friendly shipping. (JP 3-15)

quadrant elevation — **(*)** The angle between the horizontal plane and the axis of the bore when the weapon is laid. **(DOD only)** It is the algebraic sum of the elevation, angle of site, and complementary angle of site.

quadruple container — A quadruple container box 57.5" x 96" x 96" with a metal frame, pallet base, and International Organization for Standardization (ISO) corner fittings. Four of these boxes can be lashed together to form a 20- foot American National Standards Institute and/or ISO intermodal container. Also called **QUADCON.**

quay — A structure of solid construction along a shore or bank that provides berthing and generally provides cargo-handling facilities. A similar facility of open construction is called a wharf. See also **wharf.** (JP 4-01.5)

Intentionally Blank

R

radar — A radio detection device that provides information on range, azimuth, and/or elevation of objects.

radar advisory — The term used to indicate that the provision of advice and information is based on radar observation.

radar altimetry area — **(*)** A large and comparatively level terrain area with a defined elevation which can be used in determining the altitude of airborne equipment by the use of radar.

radar beacon — A receiver-transmitter combination that sends out a coded signal when triggered by the proper type of pulse, enabling determination of range and bearing information by the interrogating station or aircraft. Also called **RB.**

radar camouflage — **(*)** The use of radar absorbent or reflecting materials to change the radar echoing properties of a surface of an object.

radar clutter — **(*)** Unwanted signals, echoes, or images on the face of the display tube which interfere with observation of desired signals.

radar countermeasures — See **electronic warfare; chaff.**

radar coverage — **(*)** The limits within which objects can be detected by one or more radar stations.

radar danning — **(*)** In naval mine warfare, a method of navigating by using radar to keep the required distance from a line of dan buoys.

radar deception — See **electromagnetic deception.**

radar exploitation report — A formatted statement of the results of a tactical radar imagery reconnaissance mission. The report includes the interpretation of the sensor imagery. Also called **RADAREXREP.**

radar fire — **(*)** Gunfire aimed at a target which is tracked by radar. See also **fire.**

radar guardship — **(*)** Any ship which has been assigned the task by the officer in tactical command of maintaining the radar watch.

radar horizon — **(*)** The locus of points at which the rays from a radar antenna become tangential to the Earth's surface. On the open sea this locus is horizontal, but on land it varies according to the topographical features of the terrain.

radar imagery — Imagery produced by recording radar waves reflected from a given target surface.

radar intelligence — Intelligence derived from data collected by radar. Also called **RADINT.** See also **intelligence.** (JP 2-0)

radar netting — **(*)** The linking of several radars to a single center to provide integrated target information.

radar netting station — **(*)** A center which can receive data from radar tracking stations and exchange this data among other radar tracking stations, thus forming a radar netting system. See also **radar netting unit; radar tracking station.**

radar netting unit — Optional electronic equipment that converts the operations central of certain air defense fire distribution systems to a radar netting station. See also **radar netting station.**

radar picket — **(*)** Any ship, aircraft, or vehicle, stationed at a distance from the force protected, for the purpose of increasing the radar detection range.

radar reconnaissance — Reconnaissance by means of radar to obtain information on enemy activity and to determine the nature of terrain.

radarscope overlay — **(*)** A transparent overlay for placing on the radarscope for comparison and identification of radar returns.

radarscope photography — **(*)** A film record of the returns shown by a radar screen.

radar signal film — The film on which is recorded all the reflected signals acquired by a coherent radar, and that must be viewed or processed through an optical correlator to permit interpretation.

radar silence — **(*)** An imposed discipline prohibiting the transmission by radar of electromagnetic signals on some or all frequencies.

radar spoking — Periodic flashes of the rotating time base on a radial display. Sometimes caused by mutual interference.

radar tracking station — A radar facility that has the capability of tracking moving targets.

radial — A magnetic bearing extending from a very high frequency omni-range and/or tactical air navigation station.

radial displacement — **(*)** On vertical photographs, the apparent "leaning out," or the apparent displacement of the top of any object having height in relation to its base. The direction of displacement is radial from the principal point on a true vertical, or from the isocentre on a vertical photograph distorted by tip or tilt.

radiant exposure — See **thermal exposure.**

radiation dose — **(*)** The total amount of ionizing radiation absorbed by material or tissues, expressed in centigrays. **(DOD only)** The term radiation dose is often used in the sense of the exposure dose expressed in roentgens, which is a measure of the total amount of ionization that the quantity of radiation could produce in air. This could be distinguished from the absorbed dose, also given in rads, which represents the energy absorbed from the radiation per gram of specified body tissue. Further, the biological dose, in rems, is a measure of the biological effectiveness of the radiation exposure. See also **absorbed dose; exposure dose.**

radiation dose rate — **(*)** The radiation dose (dosage) absorbed per unit of time. **(DOD only)** A radiation dose rate can be set at some particular unit of time (e.g., H + 1 hour) and would be called H + 1 radiation dose rate.

radiation exposure state — **(*)** The condition of a unit, or exceptionally an individual, deduced from the cumulative whole body radiation dose(s) received. It is expressed as a symbol which indicates the potential for future operations and the degree of risk if exposed to additional nuclear radiation.

radiation exposure status — Criteria to assist the commander in measuring unit exposure to radiation based on total past cumulative dose in centigray (cGy). Categories are as follows: (a) **radiation exposure status-0** — No previous exposure history. Also called **RES-0**; (b) **radiation exposure status-1** — Negligible radiation exposure history (greater than 0, but less than 70 cGy). Also called **RES-1**; (c) **radiation**

exposure status-2 — Significant but not a dangerous dose of radiation (greater than 70, but less than 150 cGy). Also called **RES-2**; (d) **radiation exposure status-3** — Unit has already received a dose of radiation which makes further exposure dangerous (greater then 150 cGy). Also called **RES-3.**

radiation intelligence — Intelligence derived from the collection and analysis of non-information-bearing elements extracted from the electromagnetic energy unintentionally emanated by foreign devices, equipments, and systems, excluding those generated by the detonation of atomic or nuclear weapons.

radiation intensity — (*) The radiation dose rate at a given time and place. It may be used, coupled with a figure, to denote the radiation intensity at a given number of hours after a nuclear burst, e.g., RI-3 is the radiation intensity 3 hours after the time of burst. Also called **RI.**

radiation scattering — (*) The diversion of radiation (thermal, electromagnetic, or nuclear) from its original path as a result of interaction or collisions with atoms, molecules, or larger particles in the atmosphere or other media between the source of the radiation (e.g., a nuclear explosion) and a point at some distance away. As a result of scattering, radiation (especially gamma rays and neutrons) will be received at such a point from many directions instead of only from the direction of the source.

radiation sickness — (*) An illness resulting from excessive exposure to ionizing radiation. The earliest symptoms are nausea, vomiting, and diarrhea, which may be followed by loss of hair, hemorrhage, inflammation of the mouth and throat, and general loss of energy.

radioactive decay — (*) The decrease in the radiation intensity of any radioactive material with respect to time.

radioactive decay curve — (*) A graph line representing the decrease of radioactivity with the passage of time.

radioactive decay rate — The time rate of the disintegration of radioactive material generally accompanied by the emission of particles and/or gamma radiation.

radioactivity — The spontaneous emission of radiation, generally alpha or beta particles, often accompanied by gamma rays, from the nuclei of an unstable isotope.

radioactivity concentration guide — (*) The amount of any specified radioisotope that is acceptable in air and water for continuous consumption.

radio and wire integration — The combining of wire circuits with radio facilities. Also called **RWI.**

radio approach aids — (*) Equipment making use of radio to determine the position of an aircraft with considerable accuracy from the time it is in the vicinity of an airfield or carrier until it reaches a position from which landing can be carried out.

radio beacon — (*) A radio transmitter which emits a distinctive or characteristic signal used for the determination of bearings, courses, or location.

radio countermeasures — See **electronic warfare.**

radio deception — The employment of radio to deceive the enemy. Radio deception includes sending false dispatches, using deceptive headings, employing enemy call signs, etc. See also **electronic warfare.**

radio detection — **(*)** The detection of the presence of an object by radio-location without precise determination of its position.

radio direction finding — **(*)** Radio-location in which only the direction of a station is determined by means of its emissions.

radio direction finding database — The aggregate of information, acquired by both airborne and surface means, necessary to provide support to radio direction-finding operations to produce fixes on target transmitters and/or emitters. The resultant bearings and fixes serve as a basis for tactical decisions concerning military operations, including exercises, planned or underway.

radio fix — The location of a ship or aircraft by determining the direction of radio signals coming to the ship or aircraft from two or more sending stations, the locations of which are known.

radio guard — A ship, aircraft, or radio station designated to listen for and record transmissions and to handle traffic on a designated frequency for a certain unit or units.

radiological defense — **(*)** Defensive measures taken against the radiation hazards resulting from the employment of nuclear and radiological weapons.

radiological environment — **(*)** Conditions found in an area resulting from the presence of a radiological hazard.

radiological monitoring — See **monitoring.**

radiological operation — **(*)** The employment of radioactive materials or radiation producing devices to cause casualties or restrict the use of terrain. It includes the intentional employment of fallout from nuclear weapons.

radiological survey — **(*)** The directed effort to determine the distribution and dose rates of radiation in an area.

radiological survey flight altitude — The altitude at which an aircraft is flown during an aerial radiological survey.

radio magnetic indicator — **(*)** An instrument which displays aircraft heading and bearing to selected radio navigation aids.

radio navigation — **(*)** Radio-location intended for the determination of position or direction or for obstruction warning in navigation.

radio range finding — **(*)** Radio-location in which the distance of an object is determined by means of its radio emissions, whether independent, reflected, or retransmitted on the same or other wavelength.

radio range station — **(*)** A radio navigation land station in the aeronautical radio navigation service providing radio equi-signal zones. (In certain instances a radio range station may be placed on board a ship.)

radio silence — **(*)** A condition in which all or certain radio equipment capable of radiation is kept inoperative. **(DOD only)** (Note: In combined or United States Joint or intra-Service communications the frequency bands and/or types of equipment affected will be specified.)

radio telegraphy — The transmission of telegraphic codes by means of radio.

radio telephony — **(*)** The transmission of speech by means of modulated radio waves.

radius of action — **(*)** The maximum distance a ship, aircraft, or vehicle can travel away from its base along a given course with normal combat load and return without refueling, allowing for all safety and operating factors.

radius of damage — The distance from ground zero at which there is a 0.50 probability of achieving the desired damage.

radius of integration — The distance from ground zero that indicates the area within which the effects of both the nuclear detonation and conventional weapons are to be integrated.

radius of safety — **(*)** The horizontal distance from ground zero beyond which the weapon effects on friendly troops are acceptable.

raid — **(*)** An operation, usually small scale, involving a swift penetration of hostile territory to secure information, confuse the enemy, or to destroy installations. It ends with a planned withdrawal upon completion of the assigned mission.

railhead — **(*)** A point on a railway where loads are transferred between trains and other means of transport. See also **navigation head.**

railway line capacity — **(*)** The maximum number of trains which can be moved in each direction over a specified section of track in a 24 hour period. See also **route capacity.**

railway loading ramp — **(*)** A sloping platform situated at the end or beside a track and rising to the level of the floor of the rail cars or wagons.

rainfall (nuclear) — The water that is precipitated from the base surge clouds after an underwater burst of a nuclear weapon. This rain is radioactive and presents an important secondary effect of such a burst.

rainout — **(*)** Radioactive material in the atmosphere brought down by precipitation.

ramjet — **(*)** A jet-propulsion engine containing neither compressor nor turbine which depends for its operation on the air compression accomplished by the forward motion of the engine. See also **pulsejet.**

random minelaying — **(*)** In land mine warfare, the laying of mines without regard to pattern.

range — 1. The distance between any given point and an object or target. 2. Extent or distance limiting the operation or action of something, such as the range of an aircraft, ship, or gun. 3. The distance that can be covered over a hard surface by a ground vehicle, with its rated payload, using the fuel in its tank and its cans normally carried as part of the ground vehicle equipment. 4. Area equipped for practice in shooting at targets. In this meaning, also called **target range.**

range marker — **(*)** A single calibration blip fed onto the time base of a radial display. The rotation of the time base shows the single blips as a circle on the plan position indicator scope. It may be used to measure range.

range markers — Two upright markers that may be lighted at night and placed so that, when aligned, the direction indicated assists in piloting. They may be used in amphibious operations to aid in beaching landing ships or craft.

Rangers — Rapidly deployable airborne light infantry organized and trained to conduct highly complex joint direct action operations in coordination with or in support of other special operations units of all Services. Rangers also can execute direct action operations in support of conventional nonspecial operations missions conducted by a combatant commander and can operate as conventional light infantry when properly augmented with other elements of combined arms.

range spread — The technique used to place the mean point of impact of two or more units 100 meters apart on the gun-target line.

ranging — **(*)** The process of establishing target distance. Types of ranging include echo, intermittent, manual, navigational, explosive echo, optical, radar, etc. See also **spot.**

rated load — **(*)** The designed safe operating load for the equipment under prescribed conditions.

rate of fire — **(*)** The number of rounds fired per weapon per minute.

rate of march — **(*)** The average number of miles or kilometers to be traveled in a given period of time, including all ordered halts. It is expressed in miles or kilometers in the hour. See also **pace.**

ratification — The declaration by which a nation formally accepts, with or without reservation, the content of a standardization agreement. See also **implementation; reservation; subscription.**

rationalization — Any action that increases the effectiveness of allied forces through more efficient or effective use of defense resources committed to the alliance. Rationalization includes consolidation, reassignment of national priorities to higher alliance needs, standardization, specialization, mutual support or improved interoperability, and greater cooperation. Rationalization applies to both weapons and/or materiel resources and non-weapons military matters.

ration dense — Foods that, through processing, have been reduced in volume and quantity to a small compact package without appreciable loss of food value, quality, or acceptance, with a high yield in relation to space occupied, such as dehydrates and concentrates.

ratio print — A print the scale of which has been changed from that of the negative by photographic enlargement or reduction.

ratline — An organized effort for moving personnel and/or material by clandestine means across a denied area or border.

R-day — See **times.**

reaction time — 1. The elapsed time between the initiation of an action and the required response. 2. The time required between the receipt of an order directing an operation and the arrival of the initial element of the force concerned in the designated area.

readiness — The ability of US military forces to fight and meet the demands of the national military strategy. Readiness is the synthesis of two distinct but interrelated levels. a. **unit readiness** — The ability to provide capabilities required by the combatant commanders to execute their assigned missions. This is derived from the ability of each unit to deliver the outputs for which it was designed. b. **joint readiness** — The combatant commander's ability to integrate and synchronize ready combat and support forces to execute his or her assigned missions. See also **military capability; national military strategy.**

readiness condition — See **operational readiness.**

readiness planning — Operation planning required for peacetime operations. Its objective is the maintenance of high states of readiness and the deterrence of potential enemies. It includes planning activities that influence day-to-day operations and the peacetime posture of forces. As such, its focus is on general capabilities and readiness rather than the specifics of a particular crisis, either actual or potential. The assignment of geographic responsibilities to combatant commanders, establishment of readiness standards and levels, development of peacetime deployment patterns, coordination of reconnaissance and surveillance assets and capabilities, and planning of joint exercises are examples of readiness planning. No formal joint planning system exists for readiness planning such as exists for joint operation and execution planning.

ready position — **(*)** In helicopter operations, a designated place where a helicopter load of troops and/or equipment waits for pick-up.

Ready Reserve — The Selected Reserve, Individual Ready Reserve, and Inactive National Guard liable for active duty as prescribed by law (US Code, title 10 (DOD), sections 10142, 12301, and 12302). See also **active duty; Inactive National Guard; Individual Ready Reserve; Selected Reserve.** (JP 4-05)

Ready Reserve Force — A force composed of ships acquired by the Maritime Administration (MARAD) with Navy funding and newer ships acquired by the MARAD for the National Defense Reserve Fleet (NDRF). Although part of the NDRF, ships of the Ready Reserve Force are maintained in a higher state of readiness and can be made available without mobilization or congressionally declared state of emergency. Also called **RRF.** See also **National Defense Reserve Fleet.** (JP 3-02.2)

ready-to-load date — The date when a unit will be ready to move from the origin, i.e., mobilization station. Also called **RLD.**

reallocation authority — **(*)** The authority given to NATO commanders and normally negotiated in peacetime, to reallocate in an "emergency in war" national logistic resources controlled by the combat forces under their command, and made available by nations, in order to influence the battle logistically.

real property — Lands, buildings, structures, utilities systems, improvements, and appurtenances thereto. Includes equipment attached to and made part of buildings and structures (such as heating systems) but not movable equipment (such as plant equipment).

real time — Pertaining to the timeliness of data or information which has been delayed only by the time required for electronic communication. This implies that there are no noticeable delays. See also **near real time.**

rear area — For any particular command, the area extending forward from its rear boundary to the rear of the area assigned to the next lower level of command. This area is provided primarily for the performance of support functions. See also **Army service area.** (JP 3-10)

rear area operations center/rear tactical operations center — A command and control facility that serves as an area and/or subarea commander's planning, coordinating, monitoring, advising, and directing agency for area security operations. (JP 3-10.1)

rear echelon — **(*)** Elements of a force which are not required in the objective area.

rear guard — 1. The rearmost elements of an advancing or a withdrawing force. It has the following functions: to protect the rear of a column from hostile forces; during the withdrawal, to delay the enemy; during the advance, to keep supply routes open. 2. Security detachment that a moving ground force details to the rear to keep it informed and covered. See also **guard.**

rearming — 1. An operation that replenishes the prescribed stores of ammunition, bombs, and other armament items for an aircraft, naval ship, tank, or armored vehicle (including replacement of defective ordnance equipment) in order to make it ready for combat service. 2. Resetting the fuze on a bomb or on an artillery, mortar, or rocket projectile so that it will detonate at the desired time.

reattack recommendation — An assessment, derived from the results of battle damage assessment and munitions effectiveness assessment, providing the commander systematic advice on reattack of targets and further target selection to achieve objectives. The reattack recommendation considers objective achievement, target, and aimpoint selection, attack timing, tactics, and weapon system and munitions selection. The reattack recommendation is a combined operations and intelligence function. Also called **RR.** See also **assessment; battle damage assessment; munitions effectiveness assessment; target.** (JP 3-60)

rebuild — The restoration of an item to a standard as nearly as possible to its original condition in appearance, performance, and life expectancy. See also **overhaul; repair.**

receipt — A transmission made by a receiving station to indicate that a message has been satisfactorily received.

receipt into the supply system — That point in time when the first item or first quantity of the item of the contract has been received at or is en route to point of first delivery after inspection and acceptance. See also **procurement lead time.**

receiving ship — The ship in a replenishment unit that receives the rig(s).

reception — 1. All ground arrangements connected with the delivery and disposition of air or sea drops. Includes selection and preparation of site, signals for warning and approach, facilitation of secure departure of agents, speedy collection of delivered articles, and their prompt removal to storage places having maximum security. When a group is involved, it may be called a reception committee. 2. Arrangements to welcome and provide secure quarters or transportation for defectors, escapees, evaders, or incoming agents. 3. The process of receiving, offloading, marshalling, and transporting of personnel, equipment, and materiel from the strategic and/or intratheater deployment phase to a sea, air, or surface transportation point of debarkation to the marshalling area. (JP 4-01.8)

receptivity — **(*)** The vulnerability of a target audience to particular psychological operations media.

reclama — A request to duly constituted authority to reconsider its decision or its proposed action.

recognition — 1. The determination by any means of the individuality of persons, or of objects such as aircraft, ships, or tanks, or of phenomena such as communications-electronics patterns. 2. In ground combat operations, the determination that an object is similar within a category of something already known; e.g., tank, truck, man.

recognition signal — Any prearranged signal by which individuals or units may identify each other.

recompression chamber — See **hyperbaric chamber.**

reconnaissance — **(*)** A mission undertaken to obtain, by visual observation or other detection methods, information about the activities and resources of an enemy or potential enemy, or to secure data concerning the meteorological, hydrographic, or geographic characteristics of a particular area. Also called **RECON.**

reconnaissance by fire — **(*)** A method of reconnaissance in which fire is placed on a suspected enemy position to cause the enemy to disclose a presence by movement or return of fire.

reconnaissance exploitation report — **(*)** A standard message format used to report the results of a tactical air reconnaissance mission. Whenever possible the report should include the interpretation of sensor imagery. Also called **RECCEXREP.**

reconnaissance in force — **(*)** An offensive operation designed to discover and/or test the enemy's strength or to obtain other information.

reconnaissance patrol — See **patrol.**

reconnaissance photography — Photography taken to obtain information on the results of bombing, or on enemy movements, concentrations, activities, and forces. The primary purposes do not include making maps, charts, or mosaics.

reconstitution site — A location selected by the surviving command authority as the site at which a damaged or destroyed headquarters can be reformed from survivors of the attack and/or personnel from other sources, predesignated as **replacements.**

record information — All forms (e.g., narrative, graphic, data, computer memory) of information registered in either temporary or permanent form so that it can be retrieved, reproduced, or preserved.

recoverable item — An item that normally is not consumed in use and is subject to return for repair or disposal. See also **reparable item.**

recovery — 1. In air (aviation) operations, that phase of a mission which involves the return of an aircraft to a land base or platform afloat. 2. The retrieval of a mine from the location where emplaced. 3. Actions taken to rescue or extract personnel for return to friendly control. 4. Actions taken to extricate damaged or disabled equipment for return to friendly control or repair at another location. See also **evader; evasion; evasion and recovery; recovery; recovery force.**

recovery activation signal — In evasion and recovery operations, a precoordinated signal from an evader that indicates his or her presence in an area to a receiving or observing source that indicates "I am here, start the recovery planning." Also called **RAS**. See also **evader; evasion; evasion and recovery; recovery operations; signal.** (JP 3-50.3)

recovery airfield — Any airfield, military or civil, at which aircraft might land post-H-hour. It is not expected that combat missions would be conducted from a recovery airfield. See also **airfield.**

recovery and reconstitution — 1. Those actions taken by one nation prior to, during, and following an attack by an enemy nation to minimize the effects of the attack, rehabilitate the national economy, provide for the welfare of the populace, and maximize the combat potential of remaining forces and supporting activities. 2. Those actions taken by a military force during or after operational employment to restore its combat capability to full operational readiness. See also **recovery.** (JP 3-35)

recovery force — In evasion and recovery operations, an organization consisting of personnel and equipment with a mission of seeking out evaders, contacting them, and returning them to friendly control. See also **evader; evasion; evasion and recovery; recovery operations.** (JP 3-50.3)

recovery operations — Operations conducted to search for, locate, identify, rescue, and return personnel, sensitive equipment, or items critical to national security. (JP 3-07)

recovery procedures — See **explosive ordnance disposal procedures.**

recovery site — In evasion and escape usage, an area from which an evader or an escapee can be evacuated. See also **escapee; evader; evasion; evasion and escape.** (JP 3-50.3)

recovery vehicle — In combat search and rescue, the vehicle (aircraft, maritime, or land), on which isolated personnel are boarded and transported from the pickup site.

recovery zone — A designated geographic area from which special operations forces can be extracted by air, boat, or other means. Also called **RZ.** (JP 3-05.5)

rectification — **(*)** In photogrammetry, the process of projecting a tilted or oblique photograph on to a horizontal reference plane.

recuperation — Not to be used. See **recovery and reconstitution.**

recurring demand — A request by an authorized requisitioner to satisfy a materiel requirement for consumption or stock replenishment that is anticipated to recur periodically. Demands for which the probability of future occurrence is unknown will be considered as recurring. Recurring demands will be considered by the supporting supply system in order to procure, store, and distribute materiel to meet similar demands in the future.

redeployment — The transfer of forces and materiel to support another joint force commander's operational requirements, or to return personnel, equipment, and materiel to the home and/or demobilization stations for reintegration and/or out-processing. See also **deployment.** (JP 3-35)

redeployment airfield — **(*)** An airfield not occupied in its entirety in peacetime, but available immediately upon outbreak of war for use and occupation by units redeployed from their peacetime locations. It must have substantially the same standard of operational facilities as the main airfield. See also **airfield; departure airfield; diversion airfield; main airfield.**

RED HORSE — Air Force units wartime-structured to provide a heavy engineer capability. They have a responsibility across the operational area, are not tied to a specific base, and are not responsible for base operation and maintenance. These units are mobile, rapidly deployable, and largely self-sufficient for limited periods of time. (JP 3-34)

redistribution — The act of effecting transfer in control, utilization, or location of materiel between units or activities within or among the Military Services or between the Military Services and other Federal agencies.

reduced charge — 1. The smaller of the two propelling charges available for naval guns. 2. Charge employing a reduced amount of propellant to fire a gun at short ranges as compared to a normal charge. See also **normal charge.**

reduced lighting — **(*)** The reduction in brightness of ground vehicle lights by either reducing power or by screening in such a way that any visible light is limited in output. See also **normal lighting.**

reduced operational status — Applies to the Military Sealift Command ships withdrawn from full operational status (FOS) because of decreased operational requirements. A ship in reduced operational status is crewed in accordance with shipboard maintenance and possible future operational requirements, with crew size predetermined contractually. The condition of readiness in terms of calendar days required to attain FOS is designated by the numeral following the acronym ROS (i.e., ROS-5). Also called **ROS.** See also **Military Sealift Command.** (JP 3-02.2)

reduction — The creation of lanes through a minefield or obstacle to allow passage of the attacking ground force. (JP 3-15)

reduction (photographic) — The production of a negative, diapositive, or print at a scale smaller than the original.

reefer — 1. A refrigerator. 2. A motor vehicle, railroad freight car, ship, aircraft, or other conveyance, so constructed and insulated as to protect commodities from either heat or cold.

reentry phase — That portion of the trajectory of a ballistic missile or space vehicle where there is a significant interaction of the vehicle and the Earth's atmosphere. See also **boost phase; midcourse phase; terminal phase.**

reentry vehicle — **(*)** That part of a space vehicle designed to re-enter the Earth's atmosphere in the terminal portion of its trajectory. Also called **RV.** See also **maneuverable reentry vehicle; multiple reentry vehicle.**

reference datum — As used in the loading of aircraft, an imaginary vertical plane at or near the nose of the aircraft from which all horizontal distances are measured for balance purposes. Diagrams of each aircraft show this reference datum as "balance station zero."

reference diversion point — **(*)** One of a number of positions selected by the routing authority on both sides of the route of a convoy or independent to facilitate diversion at sea.

reference point — **(*)** A prominent, easily located point in the terrain.

reflected shock wave — When a shock wave traveling in a medium strikes the interface between this medium and a denser medium, part of the energy of the shock wave induces a shock wave in the denser medium and the remainder of the energy results in the formation of a reflected shock wave that travels back through the less dense medium.

reflex sight — **(*)** An optical or computing sight that reflects a reticle image (or images) onto a combining glass for superimposition on the target.

refraction — The process by which the direction of a wave is changed when moving into shallow water at an angle to the bathymetric contours. The crest of the wave advancing in shallower water moves more slowly than the crest still advancing in deeper water, causing the wave crest to bend toward alignment with the underwater contours. (JP 4-01.6)

refuge area — **(*)** A coastal area considered safe from enemy attack to which merchant ships may be ordered to proceed when the shipping movement policy is implemented. See also **safe anchorage.**

refugee — A person who, by reason of real or imagined danger, has left their home country or country of their nationality and is unwilling or unable to return. See also **dislocated civilian; displaced person; evacuee; expellee; stateless person.** (JP 3-07.6)

regimental landing team — A task organization for landing comprised of an infantry regiment reinforced by those elements that are required for initiation of its combat function ashore.

regional satellite communications support center — United States Space Command operational element responsible for providing the operational communications planners with a single all-spectrum (extremely high frequency, super-high frequency, ultrahigh frequency, Ku, and Ka) point of contact for accessing and managing satellite communications (SATCOM) resources. Specific tasks include: supporting combatant commanders' (CINCs') deliberate and crisis planning, assisting CINCs in day-to-day management of apportioned resources and allocating non-apportioned resources, assisting theater spectrum managers, and facilitating SATCOM interface to the defense information infrastructure. Also called **RSSC.**

register — **(*)** In cartography, the correct position of one component of a composite map image in relation to the other components, at each stage of production.

registration — The adjustment of fire to determine firing data corrections.

registration fire — **(*)** Fire delivered to obtain accurate data for subsequent effective engagement of targets. See also **fire.**

registration point — **(*)** Terrain feature or other designated point on which fire is adjusted for the purpose of obtaining corrections to firing data.

regrade — To determine that certain classified information requires, in the interests of national defense, a higher or a lower degree of protection against unauthorized disclosure than currently provided, coupled with a changing of the classification designation to reflect such higher or lower degree.

regular drill — See **unit training assembly.**

regulated item — **(*)** Any item whose issue to a user is subject to control by an appropriate authority for reasons that may include cost, scarcity, technical or hazardous nature, or operational significance. Also called **controlled item.** See also **critical supplies and materiel.**

regulating point — An anchorage, port, or ocean area to which assault and assault follow-on echelons and follow-up shipping proceed on a schedule, and at which they are normally controlled by the commander, amphibious task force, until needed in the transport area for unloading. See also **assault; commander, amphibious task force.** (JP 3-02)

regulating station — A command agency established to control all movements of personnel and supplies into or out of a given area.

rehabilitation — **(*)** 1. The processing, usually in a relatively quiet area, of units or individuals recently withdrawn from combat or arduous duty, during which units recondition equipment and are rested, furnished special facilities, filled up with replacements, issued replacement supplies and equipment, given training, and generally made ready for employment in future operations. 2. The action performed in restoring an installation to authorized design standards.

rehearsal phase — In amphibious operations, the period during which the prospective operation is practiced for the purpose of: (1) testing adequacy of plans, the timing of detailed operations, and the combat readiness of participating forces; (2) ensuring that all echelons are familiar with plans; and (3) testing communications-information systems. See also **amphibious operation.** (JP 3-02)

reinforcement training unit — See **voluntary training unit.**

reinforcing — A support mission in which the supporting unit assists the supported unit to accomplish the supported unit's mission. Only like units (e.g., artillery to artillery, intelligence to intelligence, armor to armor, etc) can be given a reinforcing/reinforced mission.

reinforcing obstacles — Those obstacles specifically constructed, emplaced, or detonated through military effort and designed to strengthen existing terrain to disrupt, fix, turn, or block enemy movement. See also **obstacle.** (JP 3-15)

relateral tell — **(*)** The relay of information between facilities through the use of a third facility. This type of telling is appropriate between automated facilities in a degraded communications environment. See also **track telling.**

relative altitude — See **vertical separation.**

relative bearing — **(*)** The direction expressed as a horizontal angle normally measured clockwise from the forward point of the longitudinal axis of a vehicle, aircraft, or ship to an object or body. See also **bearing; grid bearing.**

relative biological effectiveness — The ratio of the number of rads of gamma (or X) radiation of a certain energy that will produce a specified biological effect to the number of rads of another radiation required to produce the same effect measures the "relative biological effectiveness" of the latter radiation.

release — **(*)** In air armament, the intentional separation of a free-fall aircraft store, from its suspension equipment, for purposes of employment of the store.

release altitude — Altitude of an aircraft above the ground at the time of release of bombs, rockets, missiles, tow targets, etc.

release point (road) — A well-defined point on a route at which the elements composing a column return under the authority of their respective commanders, each one of these elements continuing its movement towards its own appropriate destination.

releasing commander (nuclear weapons) — A commander who has been delegated authority to approve the use of nuclear weapons within prescribed limits. See also **executing commander (nuclear weapons).**

releasing officer — A properly designated individual who may authorize the sending of a message for and in the name of the originator. See also **originator.**

reliability diagram — **(*)** In cartography, a diagram showing the dates and quality of the source material from which a map or chart has been compiled. See also **information box.**

reliability of source — See **evaluation.**

relief — **(*)** Inequalities of evaluation and the configuration of land features on the surface of the Earth which may be represented on maps or charts by contours, hypsometric tints, shading, or spot elevations.

relief in place — **(*)** An operation in which, by direction of higher authority, all or part of a unit is replaced in an area by the incoming unit. The responsibilities of the replaced elements for the mission and the assigned zone of operations are transferred to the incoming unit. The incoming unit continues the operation as ordered.

religious ministry support — The entire spectrum of professional duties, to include providing for or facilitating essential religious needs and practices; pastoral care; family support programs; religious education; volunteer and community activities; and programs performed to enhance morale and moral, ethical, and personal well being. Enlisted religious support personnel assist the chaplain in providing religious ministry support. See also **command chaplain; command chaplain of the combatant command; lay leader or lay reader; religious ministry support plan; religious ministry support team; Service component command chaplain.** (JP 1-05)

religious ministry support plan — A plan that describes the way in which religious support personnel will provide religious support to all members of a joint force. When approved by the commander, it may be included as an annex to operation plans. See also **command chaplain; command chaplain of the combatant command; lay leader or lay reader; religious ministry support; religious ministry support team; Service component command chaplain.** (JP 1-05)

religious ministry support team — A team that is composed of a chaplain and an Army chaplain assistant, Navy religious program specialist, Air Force chaplain service support personnel, and/or Coast Guard yeoman. The team works together in designing, implementing, and executing the command religious program. See also **command chaplain; command chaplain of the combatant command; lay leader or lay reader; religious ministry support; religious ministry support plan; Service component command chaplain.** (JP 1-05)

relocatable building — A building designed to be readily moved, erected, disassembled, stored, and reused. All types of buildings or building forms designed to provide relocatable capabilities are included in this definition. In classifying buildings as relocatable, the estimated funded and unfunded costs for average building disassembly, repackaging (including normal repair and refurbishment of components), and nonrecoverable building components, including typical foundations, may not exceed 20 percent of the building acquisition cost. Excluded from this definition are building types and forms that are provided as an integral part of a mobile equipment item and that are incidental portions of such equipment components, such as communications vans or trailers. (JP 4-04)

remain-behind equipment — Unit equipment left by deploying forces at their bases when they deploy. (JP 3-02.2)

remaining forces — The total surviving United States forces at any given stage of combat operations.

remote delivery — **(*)** In mine warfare, the delivery of mines to a target area by any means other than direct emplacement. The exact position of mines so laid may not be known.

remotely piloted vehicle — **(*)** An unmanned vehicle capable of being controlled from a distant location through a communication link. It is normally designed to be recoverable. See also **drone.**

render safe procedures — See **explosive ordnance disposal procedures.**

rendezvous area — In an amphibious operation, the area in which the landing craft and amphibious vehicles rendezvous to form waves after being loaded, and prior to movement to the line of departure.

reorder cycle — The interval between successive reorder (procurement) actions.

reorder point — 1. That point at which time a stock replenishment requisition would be submitted to maintain the predetermined or calculated stockage objective. 2. The sum of the safety level of supply plus the level for order and shipping time equals the reorder point. See also **level of supply.**

repair — The restoration of an item to serviceable condition through correction of a specific failure or unserviceable condition. See also **overhaul; rebuild.**

repair and restoration — Repair, beyond emergency repair, of war-damaged facilities to restore operational capability in accordance with combatant command standards of construction, including repair and restoration of pavement surfaces. Normally, repairs to facilities will be made using materials similar to those of the original construction. For severely damaged facilities (i.e., essentially destroyed), restoration may require reconstruction. (JP 4-04)

repair cycle — The stages through which a reparable item passes from the time of its removal or replacement until it is reinstalled or placed in stock in a serviceable condition.

repair cycle aircraft — Aircraft in the active inventory that are in or awaiting depot maintenance, including those in transit to or from depot maintenance.

reparable item — An item that can be reconditioned or economically repaired for reuse when it becomes unserviceable. See also **recoverable item.**

repatriate — A person who returns to his or her country or citizenship, having left said native country either against his or her will, or as one of a group who left for reason of politics, religion, or other pertinent reasons.

repatriation — 1. The procedure whereby American citizens and their families are officially processed back into the United States subsequent to an evacuation. See also **evacuation.** 2. The release and return of enemy prisoners of war to their own country in accordance with the 1949 Geneva Convention Relative to the Treatment of Prisoners of War. (JP 1-0)

repeater-jammer — **(*)** A receiver transmitter device which amplifies, multiplies, and retransmits the signals received, for purposes of deception or jamming.

replacement demand — A demand representing replacement of items consumed or worn out.

replacement factor — **(*)** The estimated percentage of equipment or repair parts in use that will require replacement during a given period due to wearing out beyond repair, enemy action, abandonment, pilferage, and other causes except catastrophes.

replacements — Personnel required to take the place of others who depart a unit.

replenishment at sea — **(*)** Those operations required to make a transfer of personnel and/or supplies when at sea.

reported unit — A unit designation that has been mentioned in an agent report, captured document, or interrogation report, but for which available information is insufficient to include the unit in accepted order of battle holdings.

reporting post — **(*)** An element of the control and reporting system used to extend the radar coverage of the control and reporting center. It does not undertake the control of aircraft.

reporting time interval — 1. In surveillance, the time interval between the detection of an event and the receipt of a report by the user. 2. In communications, the time for transmission of data or a report from the originating terminal to the end receiver. See also **near real time.**

representative downwind direction — **(*)** During the forecast period, the mean surface downwind direction in the hazard area towards which the cloud travels.

representative downwind speed — **(*)** The mean surface downwind speed in the hazard area during the forecast period.

representative fraction — The scale of a map, chart, or photograph expressed as a fraction or ratio. See also **scale.**

request for information — 1. Any specific time-sensitive ad hoc requirement for intelligence information or products to support an ongoing crisis or operation not necessarily related to standing requirements or scheduled intelligence production. A request for information can be initiated to respond to operational requirements and will be validated in accordance with the theater command's procedures. 2. The National Security Agency/Central Security Service uses this term to state ad hoc signals intelligence requirements. Also called **RFI.** See also **information; intelligence.** (JP 2-01)

request modify — **(*)** In artillery and naval gunfire support, a request by any person, other than the person authorized to make modifications to a fire plan, for a modification.

required delivery date — The date that a force must arrive at the destination and complete unloading. Also called **RDD.**

required supply rate (ammunition) — The amount of ammunition expressed in terms of rounds per weapon per day for ammunition items fired by weapons (and in terms of other units of measure per day for bulk allotment and other items) estimated to be required to sustain operations of any designated force without restriction for a specified period. Tactical commanders use this rate to state their requirements for ammunition to support planned tactical operations at specified intervals. The required supply rate is submitted through command channels. It is consolidated at each echelon and is considered by each commander in subsequently determining the controlled supply rate within the command. Also called **RSR.** See also **ammunition controlled supply rate.**

requirements — See **military requirement.**

requirements capability — This capability provides a Joint Operation Planning and Execution System user with the ability to identify, update, review, and delete data on forces and sustainment required to support an operation plan or course of action.

requirements management system — A system for the management of theater and national imagery collection requirements that provides automated tools for users in support of submission, review, and validation of imagery nominations as requirements to be tasked on national or Department of Defense imagery collection, production, and exploitation resources. Also called **RMS.** See also **imagery.** (JP 2-01)

requisition — **(*)** 1. An authoritative demand or request especially for personnel, supplies, or services authorized but not made available without specific request. 2. **(DOD only)** To demand or require services from an invaded or conquered nation.

requisitioning objective — The maximum quantities of materiel to be maintained on hand and on order to sustain current operations. It will consist of the sum of stocks represented by the operating level, safety level, and the order and shipping time or procurement lead time, as appropriate. See also **level of supply.**

rescue combat air patrol — An aircraft patrol provided over a combat search and rescue objective area for the purpose of intercepting and destroying hostile aircraft. Its primary mission is to protect the search and rescue task forces during recovery operations. Also called **RESCAP.** See also **combat air patrol.**

rescue coordination center — A primary search and rescue facility suitably staffed by supervisory personnel and equipped for coordinating and controlling search and rescue and/or combat search and rescue operations. The facility is operated unilaterally by personnel of a single Service or component. For Navy component operations, this facility may be called a rescue coordination team. Also called **RCC (or RCT for Navy component).** See also **combat search and rescue; joint search and rescue center; search and rescue.** (JP 3-50.2)

rescue ship — **(*)** In shipping control, a ship of a convoy stationed at the rear of a convoy column to rescue survivors.

research — All effort directed toward increased knowledge of natural phenomena and environment and toward the solution of problems in all fields of science. This includes basic and applied research.

reseau — **(*)** A grid system of a standard size in the image plane of a photographic system used for mensuration purposes.

reservation — The stated qualification by a nation that describes the part of a standardization agreement that it will not implement or will implement only with limitations. See also **implementation; ratification; subscription.**

reserve — 1. Portion of a body of troops that is kept to the rear, or withheld from action at the beginning of an engagement, in order to be available for a decisive movement. 2. Members of the Military Services who are not in active service but who are subject to call to active duty. 3. Portion of an appropriation or contract authorization held or set aside for future operations or contingencies and, in respect to which, administrative authorization to incur commitments or obligations has been withheld. See also **operational reserve; reserve supplies.**

reserve aircraft — Those aircraft that have been accumulated in excess of immediate needs for active aircraft and are retained in the inventory against possible future needs. See also **aircraft.**

reserve component category — The category that identifies an individual's status in a reserve component. The three reserve component categories are Ready Reserve, Standby Reserve, and Retired Reserve. Each reservist is identified by a specific reserve component category designation.

Reserve Components — Reserve Components of the Armed Forces of the United States are: a. the Army National Guard of the United States; b. the Army Reserve; c. the Naval Reserve; d. the Marine Corps Reserve; e. the Air National Guard of the United States; f. the Air Force Reserve; and g. the Coast Guard Reserve. Also called **RC.** See also **component; reserve.** (JP 4-05.1)

reserved demolition target — A target for demolition, the destruction of which must be controlled at a specific level of command because it plays a vital part in the tactical, operational, or strategic plan, or because of the importance of the structure itself, or because the demolition may be executed in the face of the enemy. See also **demolition target.**

reserved obstacles — Those demolition obstacles that are deemed critical to the plan for which the authority to detonate is reserved by the designating commander. See also **obstacle.** (JP 3-15)

reserved route — **(*)** In road traffic, a specific route allocated exclusively to an authority or formation. See also **route.**

reserve supplies — Supplies accumulated in excess of immediate needs for the purpose of ensuring continuity of an adequate supply. Also called **reserves.** See also **battle reserves; beach reserves; contingency retention stock; economic retention stock; individual reserves; initial reserves; unit reserves.**

residual capabilities assessment — Provides an automated or manual crisis action capability to assess the effects of weapons of mass destruction events for operations planning. Residual capabilities assessment tasks include, but are not limited to, assessment of infrastructure and facility damage, fallout prediction, weapons effect analysis, population impact assessment, and tracking strategic assets.

residual contamination — **(*)** Contamination which remains after steps have been taken to remove it. These steps may consist of nothing more than allowing the contamination to decay normally.

residual forces — Unexpended portions of the remaining United States forces that have an immediate combat potential for continued military operations, and that have been deliberately withheld from utilization.

residual radiation — (*) Nuclear radiation caused by fallout, artificial dispersion of radioactive material, or irradiation which results from a nuclear explosion and persists longer than one minute after burst. See also **contamination; induced radiation; initial radiation.**

residual radioactivity — Nuclear radiation that results from radioactive sources and persists for longer than one minute. Sources of residual radioactivity created by nuclear explosions include fission fragments and radioactive matter created primarily by neutron activation, but may also be created by gamma and other radiation activation. Other possible sources of residual radioactivity include radioactive material created and dispersed by means other than nuclear explosion. See also **contamination; induced radiation; initial radiation.**

resistance movement — An organized effort by some portion of the civil population of a country to resist the legally established government or an occupying power and to disrupt civil order and stability.

resolution — 1. A measurement of the smallest detail that can be distinguished by a sensor system under specific conditions. 2. A formal expression of an official body such as Congress, the United Nations Security Council, or North Atlantic Treaty Organization North Atlantic Committee that may provide the basis for or set limits on a military operation.

resource management operations — The execution of the resource management mission that includes providing advice and guidance to the commander, developing command resource requirements, identifying sources of funding, determining cost, acquiring funds, distributing and controlling funds, tracking costs and obligations, cost capturing and reimbursement procedures, and establishing a management control process. See also **financial management.** (JP 1-06)

resources — The forces, materiel, and other assets or capabilities apportioned or allocated to the commander of a unified or specified command.

response force — A mobile force with appropriate fire support designated, usually by the area commander, to deal with Level II threats in the rear area. Also called **RF.** (JP 3-10.1)

responsibility — 1. The obligation to carry forward an assigned task to a successful conclusion. With responsibility goes authority to direct and take the necessary action to ensure success. 2. The obligation for the proper custody, care, and safekeeping of property or funds entrusted to the possession or supervision of an individual. See also **accountability.**

responsor — (*) An electronic device used to receive an electronic challenge and display a reply thereto.

rest and recuperation — The withdrawal of individuals from combat or duty in a combat area for short periods of rest and recuperation. Also called **R&R.** See also **rehabilitation.**

restitution — (*) The process of determining the true planimetric position of objects whose images appear on photographs.

restitution factor — See **correlation factor.**

restorative and rehabilitative care — A period of minimal care and increasing physical activity necessary to restore patients to functional health and allow their return to duty or useful and productive life. Restorative and rehabilitative treatment may be available in theater on a limited basis due to the theater evacuation policy. This treatment is normally provided in the continental United States. See also **evacuation policy; patient; theater.** (JP 4-02)

restraint of loads — The process of binding, lashing, and wedging items into one unit or into its transporter in a manner that will ensure immobility during transit.

restricted area — 1. An area (land, sea, or air) in which there are special restrictive measures employed to prevent or minimize interference between friendly forces. 2. An area under military jurisdiction in which special security measures are employed to prevent unauthorized entry. See also **air surface zone; controlled firing area; restricted areas (air).**

restricted areas (air) — Designated areas established by appropriate authority over which flight of aircraft is restricted. They are shown on aeronautical charts, published in notices to airmen, and provided in publications of aids to air navigation. See also **restricted area.**

restricted dangerous air cargo — **(*)** Cargo which does not belong to the highly dangerous category but which is hazardous and requires, for transport by cargo or passenger aircraft, extra precautions in packing and handling.

restricted data — All data (information) concerning: a. design, manufacture, or use of atomic weapons; b. the production of special nuclear material; or c. the use of special nuclear material in the production of energy, but shall not include data declassified or removed from the restricted data category pursuant to Section 142 of the Atomic Energy Act. (Section 11w, Atomic Energy Act of 1954, as amended.) See also **formerly restricted data.**

restricted operations area — **(*)** Airspace of defined dimensions, designated by the airspace control authority, in response to specific operational situations/requirements within which the operation of one or more airspace users is restricted. Also called **ROA.**

restricted target — A target that has specific restrictions imposed upon it. Actions that exceed specified restrictions are prohibited until coordinated and approved by the establishing headquarters. See also **target.** (JP 3-60)

restricted target list — A list of restricted targets nominated by elements of the joint force and approved by the joint force commander. This list also includes restricted targets directed by higher authorities. Also called **RTL.** See also **restricted target; target.** (JP 3-60)

restrictive fire area — An area in which specific restrictions are imposed and into which fires that exceed those restrictions will not be delivered without coordination with the establishing headquarters. Also called **RFA.** See also **fires.** (JP 3-09)

restrictive fire line — A line established between converging friendly surface forces that prohibits fires or their effects across that line. Also called **RFL.** See also **fires.** (JP 3-09)

restrictive fire plan — **(*)** A safety measure for friendly aircraft which establishes airspace that is reasonably safe from friendly surface delivered non-nuclear fires.

resupply — **(*)** The act of replenishing stocks in order to maintain required levels of supply.

resuscitative care — The aggressive management of life- and limb-threatening injuries. Interventions include emergency medical treatment, advanced trauma management, and lifesaving surgery to enable the patient to tolerate evacuation to the next level of care. See also **evacuation; patient.** (JP 4-02)

retain — 1. When used in the context of deliberate planning, the directed command will keep the referenced operation plan or operation plan in concept format, and any associated joint operation planning system or Joint Operation Planning and Execution System automated data processing files in an inactive library or status. The plan and its associated files will not be maintained unless directed by follow-on guidance. See also **archive; maintain.** 2. A tactical task to occupy and hold a terrain feature to ensure that it is free of enemy occupation or use.

retained personnel — Enemy personnel who come within any of the categories below are eligible to be certified as retained personnel. a. Medical personnel exclusively engaged in the: (1) Search for collection, transport, or treatment of the wounded or sick; (2) Prevention of disease; and/or (3) Staff administration of medical units and establishments exclusively. b. Chaplains attached to enemy armed forces. c. Staff of national Red Cross societies and other voluntary aid societies duly recognized and authorized by their governments. The staffs of such societies must be subject to military laws and regulations. Also called **RP.** See also **personnel.** (JP 1-0)

reticle — **(*)** A mark such as a cross or a system of lines lying in the image plane of a viewing apparatus. It may be used singly as a reference mark on certain types of monocular instruments or as one of a pair to form a floating mark as in certain types of stereoscopes. See also **graticule.**

Retired Reserve — All Reserve members who receive retirement pay on the basis of their active duty and/or Reserve service; those members who are otherwise eligible for retirement pay but have not reached age 60 and who have not elected discharge and are not voluntary members of the Ready or Standby Reserve. See also **active duty; Ready Reserve; Standby Reserve.** (JP 4-05)

retirement — **(*)** An operation in which a force out of contact moves away from the enemy.

retrofit action — Action taken to modify inservice equipment.

retrograde cargo — Cargo evacuated from a theater.

retrograde movement — Any movement of a command to the rear, or away from the enemy. It may be forced by the enemy or may be made voluntarily. Such movements may be classified as withdrawal, retirement, or delaying action.

retrograde operation — See **retrograde movement.**

retrograde personnel — Personnel evacuated from a theater who may include medical patients, noncombatants, and civilians.

returned to military control — The status of a person whose casualty status of "duty status - whereabouts unknown" or "missing" has been changed due to the person's return or recovery by US military authority. Also called **RMC.** See also **casualty status; duty status - whereabouts unknown; missing.**

return load — **(*)** Personnel and/or cargo to be transported by a returning carrier.

return to base — An order to proceed to the point indicated by the displayed information or by verbal communication. This point is being used to return the aircraft to a place at which the aircraft can land. Command heading, speed, and altitude may be used, if desired. Also called **RTB.**

revolutionary — An individual attempting to effect a social or political change through the use of extreme measures. See also **antiterrorism.** (JP 3-07.2)

revolving fund — A fund established to finance a cycle of operations to which reimbursements and collections are returned for reuse in a manner such as will maintain the principal of the fund, e.g., working capital funds, industrial funds, and loan funds.

riding off — Maneuvering one's unit in order to cause an opposing force to turn away. Also known as **blocking; defensive blocking.**

right (left) bank — See **left (right) bank.**

right (or left) — See **left (or right).**

riot control agent — Any chemical, that is not listed in the Chemical Weapons Convention, which can produce rapidly in humans sensory irritate or disabling physical effects which disappear within a short time following termination of exposure. See also **chemical warfare.** (JP 3-11)

riot control operations — The employment of riot control agents and/or special tactics, formations, and equipment in the control of violent disorders.

rising mine — **(*)** In naval mine warfare, a mine having positive buoyancy which is released from a sinker by a ship influence or by a timing device. The mine may fire by contact, hydrostatic pressure, or other means.

risk — 1. Probability and severity of loss linked to hazards. 2. See degree of risk. See also **hazard; risk management.** (JP 5-00.2)

risk assessment — The identification and assessment of hazards (first two steps of risk management process).

risk management — The process of identifying, assessing, and controlling risks arising from operational factors and making decisions that balance risk cost with mission benefits. Also called **RM.** See also **risk.** (JP 3-0)

riverine area — An inland or coastal area comprising both land and water, characterized by limited land lines of communication, with extensive water surface and/or inland waterways that provide natural routes for surface transportation and communications.

riverine operations — Operations conducted by forces organized to cope with and exploit the unique characteristics of a riverine area, to locate and destroy hostile forces, and/or to achieve or maintain control of the riverine area. Joint riverine operations combine land, naval, and air operations, as appropriate, and are suited to the nature of the specific riverine area in which operations are to be conducted.

road block — **(*)** A barrier or obstacle (usually covered by fire) used to block or limit the movement of hostile vehicles along a route.

road capacity — The maximum traffic flow obtainable on a given roadway using all available lanes; usually expressed in vehicles per hour or vehicles per day.

road clearance time — **(*)** The total time a column requires to travel over and clear a section of the road.

road hazard sign — **(*)** A sign used to indicate traffic hazards. Military hazard signs should be used in a communications zone area only in accordance with existing agreements with national authorities.

road net — The system of roads available within a particular locality or area.

road space — **(*)** The length of roadway allocated to and/or actually occupied by a column on a route, expressed in miles or kilometers.

rocket propulsion — Reaction propulsion wherein both the fuel and the oxidizer, generating the hot gases expended through a nozzle, are carried as part of the rocket engine. Specifically, rocket propulsion differs from jet propulsion in that jet propulsion utilizes atmospheric air as an oxidizer, whereas rocket propulsion utilizes nitric acid or a similar compound as an oxidizer. See also **jet propulsion.**

roentgen — **(*)** A unit of exposure dose of gamma (or X-) radiation. In field dosimetry, one roentgen is essentially equal to one rad.

roentgen equivalent mammal — One roentgen equivalent mammal is the quantity of ionizing radiation of any type which, when absorbed by man or other mammal, produces a physiological effect equivalent to that produced by the absorption of 1 roentgen of X-ray or gamma radiation. Also called **REM.**

role number — **(*)** In the medical field, the classification of treatment facilities according to their different capabilities.

roll back — The process of progressive destruction and/or neutralization of the opposing defenses, starting at the periphery and working inward, to permit deeper penetration of succeeding defense positions.

roll-in-point — The point at which aircraft enter the final leg of the attack, e.g., dive, glide.

roll-on/roll-off discharge facility — Provides a means of disembarking vehicles from a roll-on and roll-off ship to lighterage. The roll-on/roll-off discharge facility consists of six causeway sections, nonpowered assembled into a platform that is two sections long and three sections wide. When use of landing craft, utility, as lighters is being considered, a seventh "sea end" causeway section, on-powered, fitted with a rhino horn is required. The roll-on/roll-off discharge facility assembly includes fendering, lighting, and a ramp for vehicle movement from ship to the platform. Also called **RRDF.** See also **facility; lighterage.** (JP 4-01.6)

roll-up — The process for orderly dismantling of facilities no longer required in support of operations and available for transfer to other areas.

romper — **(*)** A ship which has moved more than 10 nautical miles ahead of its convoy, and is unable to rejoin it. See also **straggler.**

rope — **(*)** An element of chaff consisting of a long roll of metallic foil or wire which is designed for broad, low-frequency responses. See also **chaff.**

rough terrain container handler — A piece of materials handling equipment used to pick up and move containers. Also called **RTCH.** See also **materials handling equipment.** (JP 4-01.6)

route — **(*)** The prescribed course to be traveled from a specific point of origin to a specific destination. See also **axial route;**

controlled route; dispatch route; lateral route; reserved route; signed route; supervised route.

route capacity — (*) 1. The maximum traffic flow of vehicles in one direction at the most restricted point on the route. 2. The maximum number of metric tons which can be moved in one direction over a particular route in one hour. It is the product of the maximum traffic flow and the average payload of the vehicles using the route. See also **railway line capacity.**

route classification — (*) Classification assigned to a route using factors of minimum width, worst route type, least bridge, raft, or culvert military load classification, and obstructions to traffic flow. See also **military load classification.**

route lanes — (*) A series of parallel tracks for the routing of independently sailed ships.

routine message — A category of precedence to be used for all types of messages that justify transmission by rapid means unless of sufficient urgency to require a higher precedence. See also **precedence.**

routine supplies — Those items delivered as a result of normal requisitioning procedures to replace expended supplies or to build up reserve stocks. See also **follow-up supplies; supplies.** (JP 3-17)

routing indicator — A group of letters assigned to indicate: a. the geographic location of a station; b. a fixed headquarters of a command, activity, or unit at a geographic location; and c. the general location of a tape relay or tributary station to facilitate the routing of traffic over the tape relay networks.

row marker — (*) In land mine warfare, a natural, artificial, or specially installed marker, located at the start or finish of a mine row where mines are laid by individual rows. See also **marker.**

rules of engagement — Directives issued by competent military authority that delineate the circumstances and limitations under which United States forces will initiate and/or continue combat engagement with other forces encountered. Also called **ROE.** See also **law of war.**

run — 1. That part of a flight of one photographic reconnaissance aircraft during which photographs are taken. 2. The transit of a sweeper-sweep combination or of a mine-hunter operating its equipment through a lap. This term may also be applied to a transit of any formation of sweepers.

runway visual range — (*) The maximum distance in the direction of takeoff or landing at which the runway, or specified lights or markers delineating it, can be seen from a position above a specified point on its center line at a height corresponding to the average eye level of pilots at touch-down.

rupture zone — (*) The region immediately adjacent to the crater boundary in which the stresses produced by the explosion have exceeded the ultimate strength of the medium. It is characterized by the appearance of numerous radial cracks of various sizes. See also **plastic zone.**

ruse — In military deception, a trick of war designed to deceive the adversary, usually involving the deliberate exposure of false information to the adversary's intelligence collection system. (JP 3-58)

S

sabotage — An act or acts with intent to injure, interfere with, or obstruct the national defense of a country by willfully injuring or destroying, or attempting to injure or destroy, any national defense or war materiel, premises, or utilities, to include human and natural resources.

sabotage alert team — See **security alert team.**

saboteur — One who commits sabotage. See also **antiterrorism; countersabotage; sabotage.** (JP 3-07.2)

safe anchorage — **(*)** An anchorage considered safe from enemy attack to which merchant ships may be ordered to proceed when the shipping movement policy is implemented. See also **refuge area.**

safe area — A designated area in hostile territory that offers the evader or escapee a reasonable chance of avoiding capture and of surviving until he or she can be evacuated.

safe burst height — **(*)** The height of burst at or above which the level of fallout or damage to ground installations is at a predetermined level acceptable to the military commander. See also **types of burst.**

safe current — **(*)** In naval mine warfare, the maximum current that can be supplied to a sweep in a given waveform and pulse cycle which does not produce a danger area with respect to the mines being swept for.

safe depth — **(*)** In naval mine warfare, the shallowest depth of water in which a ship will not actuate a bottom mine of the type under consideration. Safe depth is usually quoted for conditions of ship upright, calm sea, and a given speed.

safe distance — **(*)** In naval mine warfare, the horizontal range from the edge of the explosion damage area to the center of the sweeper.

safe haven — 1. Designated area(s) to which noncombatants of the United States Government's responsibility and commercial vehicles and materiel may be evacuated during a domestic or other valid emergency. 2. Temporary storage provided to Department of Energy classified shipment transporters at Department of Defense facilities in order to assure safety and security of nuclear material and/or nonnuclear classified material. Also includes parking for commercial vehicles containing Class A or Class B explosives. 3. A protected body of water or the well deck of an amphibious ship used by small craft operating offshore for refuge from storms or heavy seas. (JP 4-01.6)

safe house — An innocent-appearing house or premises established by an organization for the purpose of conducting clandestine or covert activity in relative security.

safe separation distance — **(*)** The minimum distance between the delivery system and the weapon beyond which the hazards associated with functioning (detonation) are acceptable.

safety and arming mechanism — **(*)** A dual function device which prevents the unintended activation of a main charge or propulsion unit prior to arming, but allows activation thereafter upon receipt of the appropriate stimuli.

safety device — **(*)** A device which prevents unintentional functioning.

safety distance — **(*)** In road transport, the distance between vehicles traveling in column specified by the command in light of safety requirements.

safety fuze — A pyrotechnic contained in a flexible and weather-proof sheath burning at a timed and constant rate; used to transmit a flame to the detonator.

safety height — See **altitude; minimum safe altitude.**

safety lane — **(*)** Specified sea lane designated for use in transit by submarine and surface ships to prevent attack by friendly forces.

safety level of supply — The quantity of materiel, in addition to the operating level of supply, required to be on hand to permit continuous operations in the event of minor interruption of normal replenishment or unpredictable fluctuations in demand. See also **level of supply.**

safety line — **(*)** In land mine warfare, demarcation line for trip wire or wire-actuated mines in a minefield. It serves to protect the laying personnel. After the minefield is laid, this line is neither marked on the ground nor plotted on the minefield record.

safety wire — **(*)** A cable, wire, or lanyard attached to the aircraft and routed to an expendable aircraft store to prevent arming initiation prior to store release. See also **arming wire.**

safety zone — **(*)** An area (land, sea, or air) reserved for noncombat operations of friendly aircraft, surface ships, submarines, or ground forces. (Note: DOD does not use the word "submarines".)

safing — As applied to weapons and ammunition, the changing from a state of readiness for initiation to a safe condition. Also called **de-arming.**

safing and arming mechanism — A mechanism whose primary purpose is to prevent an unintended functioning of the main charge of the ammunition prior to completion of the arming delay and, in turn, allow the explosive train of the ammunition to function after arming.

salted weapon — **(*)** A nuclear weapon which has, in addition to its normal components, certain elements or isotopes which capture neutrons at the time of the explosion and produce radioactive products over and above the usual radioactive weapon debris. See also **minimum residual radioactivity weapon.**

salvage — 1. Property that has some value in excess of its basic material content but is in such condition that it has no reasonable prospect of use for any purpose as a unit and its repair or rehabilitation for use as a unit is clearly impractical. 2. The saving or rescuing of condemned, discarded, or abandoned property, and of materials contained therein for reuse, refabrication, or scrapping.

salvage group — In an amphibious operation, a naval task organization designated and equipped to rescue personnel and to salvage equipment and material.

salvage operation — 1. The recovery, evacuation, and reclamation of damaged, discarded, condemned, or abandoned allied or enemy materiel, ships, craft, and floating equipment for reuse, repair, refabrication, or scrapping. 2. Naval salvage operations include harbor and channel clearance, diving, hazardous towing and rescue tug services, and the recovery of materiel, ships, craft, and floating equipment sunk offshore or elsewhere stranded.

salvo — 1. In naval gunfire support, a method of fire in which a number of weapons are fired at the same target simultaneously.

2. In close air support or air interdiction operations, a method of delivery in which the release mechanisms are operated to release or fire all ordnance of a specific type simultaneously.

sanction enforcement and maritime intercept operations — Operations that employ coercive measures to interdict the movement of certain types of designated items into or out of a nation or specified area. (JP 3-07)

sanctuary — A nation or area near or contiguous to the combat area that, by tacit agreement between the warring powers, is exempt from attack and therefore serves as a refuge for staging, logistic, or other activities of the combatant powers.

sanitize — To revise a report or other document in such a fashion as to prevent identification of sources, or of the actual persons and places with which it is concerned, or of the means by which it was acquired. Usually involves deletion or substitution of names and other key details.

satellite and missile surveillance — The systematic observation of aerospace for the purpose of detecting, tracking, and characterizing objects, events, and phenomena associated with satellites and inflight missiles, friendly and enemy. See also **surveillance.**

S-bend distortion — See **S-curve distortion.**

scale — **(*)** The ratio or fraction between the distance on a map, chart, or photograph and the corresponding distance on the surface of the Earth. See also **conversion scale; graphic scale; photographic scale; principal scale.**

scale (photographic) — See **photographic scale.**

scaling law — **(*)** A mathematical relationship which permits the effects of a nuclear explosion of given energy yield to be determined as a function of distance from the explosion (or from ground zero) provided the corresponding effect is known as a function of distance for a reference explosion, e.g., of 1-kiloton energy yield.

scan — 1. The path periodically followed by a radiation beam. 2. In electronic intelligence, the motion of an electronic beam through space looking for a target. Scanning is produced by the motion of the antenna or by lobe switching. See also **electronic intelligence.**

scan line — **(*)** The line produced on a recording medium frame by a single sweep of a scanner.

scan period — The period taken by a radar, sonar, etc., to complete a scan pattern and return to a starting point.

scan rate — **(*)** The rate at which individual scans are recorded.

scan type — The path made in space by a point on the radar beam; for example, circular, helical, conical, spiral, or sector.

scatterable mine — **(*)** In land mine warfare, a mine laid without regard to classical pattern and which is designed to be delivered by aircraft, artillery, missile, ground dispenser, or by hand. Once laid, it normally has a limited life. See also **mine.**

scene of action commander — In antisubmarine warfare, the commander at the scene of contact. The commander is usually in a ship, or may be in a fixed-wing aircraft, helicopter, or submarine.

scheduled arrival date — The projected arrival date of a specified movement requirement at a specified location.

scheduled fire — **(*)** A type of prearranged fire executed at a predetermined time.

scheduled maintenance — Periodic prescribed inspection and/or servicing of equipment accomplished on a calendar, mileage, or hours of operation basis. See also **organizational maintenance.**

scheduled service (air transport) — A routine air transport service operated in accordance with a timetable.

scheduled speed — **(*)** The planned sustained speed of a convoy through the water which determines the speed classification of that convoy. See also **convoy speed; critical speed; declared speed.**

scheduled targets — Planned targets upon which fires will be delivered at a specific time. See also **planned targets; target.** (JP 3-60)

scheduled target (nuclear) — A planned target on which a nuclear weapon is to be delivered at a specific time during the operation of the supported force. The time is specified in terms of minutes before or after a designated time or in terms of the accomplishment of a predetermined movement or task. Coordination and warning of friendly troops and aircraft are mandatory.

scheduled wave — See **wave.**

schedule of fire — Groups of fires or series of fires fired in a definite sequence according to a definite program. The time of starting the schedule may be ON CALL. For identification purposes, schedules may be referred to by a code name or other designation.

schedule of targets — In artillery, mortar, and naval gunfire support, individual targets, groups, or series of targets to be fired on, in a definite sequence according to a definite program.

scheduling and movement — Joint Operation Planning and Execution System application software providing the capability to create, update, allocate, manifest, and review organic carrier information before and during deployment. It provides the ability to review, analyze, and generate several predefined reports on an extensive variety of scheduling and movement information. Also called **S&M.**

scheduling and movement capability — The capability required by Joint Operation Planning and Execution System planners and operators to allow for review and update of scheduling and movement data before and during implementation of a deployment operation.

scheme of maneuver — Description of how arrayed forces will accomplish the commander's intent. It is the central expression of the commander's concept for operations and governs the design of supporting plans or annexes.

scientific and technical intelligence — The product resulting from the collection, evaluation, analysis, and interpretation of foreign scientific and technical information that covers: a. foreign developments in basic and applied research and in applied engineering techniques; and b. scientific and technical characteristics, capabilities, and limitations of all foreign military systems, weapons, weapon systems, and materiel; the research and development related thereto; and the production methods employed for their manufacture. Also called **S&TI.** See also **intelligence; research; scientific intelligence; technical intelligence.** (JP 2-02)

scientific intelligence — See **scientific and technical intelligence.**

screen — **(*)** 1. An arrangement of ships, aircraft and/or submarines to protect a main body or convoy. 2. In cartography, a sheet of transparent film, glass, or plastic carrying a "ruling" or other regularly repeated pattern which may be used in conjunction with a mask, either photographically or photomechanically, to produce areas of the pattern. 3. In surveillance, camouflage and concealment, any natural or artificial material, opaque to surveillance sensor(s), interposed between the sensor(s) and the object to be camouflaged or concealed. See also **concealment.** 4. A security element whose primary task is to observe, identify, and report information, and which only fights in self-protection. See also **flank guard**; **guard.** 5. **(DOD only)** A task to maintain surveillance; provide early warning to the main body; or impede, destroy, and harass enemy reconnaissance within its capability without becoming decisively engaged. See also **security operations.**

screening group — In amphibious operations, a task organization of ships that furnishes protection to the task force en route to the objective area and during operations in the objective area. (JP 3-02)

scribing — **(*)** In cartography, a method of preparing a map or chart by cutting the lines into a prepared coating.

S-curve distortion — **(*)** The distortion in the image produced by a scanning sensor which results from the forward displacement of the sensor during the time of lateral scan.

S-Day — See **times.**

sea-air-land team — A naval force specially organized, trained, and equipped to conduct special operations in maritime, littoral, and riverine environments. Also called **SEAL team.** (JP 3-05.3)

sea areas — Areas in the amphibious objective area designated for the stationing of amphibious task force ships. Sea areas include inner transport area, sea echelon area, fire support area, etc. See also **amphibious objective area; fire support area; inner transport area; sea echelon area.** (JP 3-02)

sea barge — A type of barge-ship that can carry up to 38 loaded barges. It may also carry tugs, stacked causeway sections, various watercraft, or heavy lift equipment to better support joint logistics over-the-shore operations.

seabasing — In amphibious operations, a technique of basing certain landing force support functions aboard ship which decreases shore-based presence. See also **amphibious operation.** (JP 3-02)

seaborne forces — US or foreign combatants or auxiliary ships, including aircraft and ground forces assigned to or emanating from such vessels and other military forces operating in support of such forces and operating in, on, or over the sea.

sea control operations — The employment of naval forces, supported by land and air forces as appropriate, in order to achieve military objectives in vital sea areas. Such operations include destruction of enemy naval forces, suppression of enemy sea commerce, protection of vital sea lanes, and establishment of local military superiority in areas of naval operations. See also **land control operations.**

sea echelon — **(*)** A portion of the assault shipping which withdraws from or remains out of the transport area during an amphibious landing and operates in designated areas to seaward in an on-call or unscheduled status.

sea echelon area — In amphibious operations, an area to seaward of a transport area from which assault shipping is phased into the transport area, and to which assault shipping withdraws from the transport area. (JP 3-02)

sea echelon plan — In amphibious operations, the distribution plan for amphibious shipping in the transport area to minimize losses due to enemy attack by weapons of mass destruction and to reduce the area to be swept of mines. See also **amphibious operation.** (JP 3-02)

sea frontier — The naval command of a coastal frontier, including the coastal zone in addition to the land area of the coastal frontier and the adjacent sea areas.

sealed cabin — **(*)** The occupied space of an aircraft characterized by walls which do not allow any gaseous exchange between the ambient atmosphere and the inside atmosphere and containing its own ways of regenerating the inside atmosphere.

sealift enhancement program — Special equipment and modifications that adapt merchant-type dry cargo ships and tankers to specific military missions. They are typically installed on Ready Reserve Force ships or ships under Military Sealift Command control. Sealift enhancements fall into three categories: productivity, survivability, and operational enhancements. Also called **SEP.** See also **Military Sealift Command; Ready Reserve; Ready Reserve Force.** (JP 4-01.7)

Sealift Readiness Program — A standby contractual agreement between Military Sealift Command and US ship operators for voluntary provision of private ships for defense use. Call-up of ships may be authorized by joint approval of the Secretary of Defense and the Secretary of Transportation. Also called **SRP.** See also **Military Sealift Command.**

sea port — A land facility designated for reception of personnel or materiel moved by sea, and that serves as an authorized port of entrance into or departure from the country in which located. See also **port of debarkation; port of embarkation.**

search — 1. An operation to locate an enemy force known or believed to be at sea. 2. A systematic reconnaissance of a defined area, so that all parts of the area have passed within visibility. 3. To distribute gunfire over an area in depth by successive changes in gun elevation.

search and rescue — The use of aircraft, surface craft (land or water), submarines, specialized rescue teams, and equipment to search for and rescue personnel in distress on land or at sea. Also called **SAR**. See also **combat search and rescue; combat search and rescue mission coordinator; component search and rescue controller; isolated personnel; joint combat search and rescue operation; joint search and rescue center; joint search and rescue center director; rescue coordination center; search and rescue mission coordinator.** (JP 3-50.2)

search and rescue alert notice — An alerting message used for United States domestic flights. It corresponds to the declaration of the alert phase. Also called **ALNOT.** See also **search and rescue incident classification, subpart b.**

search and rescue incident classification — Three emergency phases into which an incident may be classified or progress, according to the seriousness of the incident and its requirement for rescue service. a. **uncertainty phase** — Doubt exists as to the safety of a craft or person because of knowledge of possible difficulties or because of lack of information concerning progress or position. b. **alert phase** — Apprehension exists for the safety of a craft or person because of definite information that serious difficulties exist that do not amount to a distress or because of a continued lack of information concerning progress or position. c. **distress phase** — Immediate assistance is required by a craft or person because of being threatened by grave or imminent danger or because of continued lack of information concerning progress or position after procedures for the alert phase have been executed.

search and rescue mission coordinator — The designated person or organization selected to direct and coordinate support for a specific search and rescue mission. Also called **SAR mission coordinator.** See also **combat search and rescue; combat search and rescue mission coordinator; component search and rescue controller; search and rescue.** (JP 3-50.2)

search and rescue region — See **inland search and rescue region; maritime search and rescue region; overseas search and rescue region.**

search attack unit — The designation given to one or more ships and/or aircraft separately organized or detached from a formation as a tactical unit to search for and destroy submarines. Also called **SAU.**

searched channel — **(*)** In naval mine warfare, the whole or part of a route or a path which has been searched, swept, or hunted, the width of the channel being specified.

searching fire — **(*)** Fire distributed in depth by successive changes in the elevation of a gun. See also **fire.**

search jammer — See **automatic search jammer.**

search mission — **(*)** In air operations, an air reconnaissance by one or more aircraft dispatched to locate an object or objects known or suspected to be in a specific area.

search radius — In search and rescue operations, a radius centered on a datum point having a length equal to the total probable error plus an additional safety length to ensure a greater than 50 percent probability that the target is in the search area.

search sweeping — **(*)** In naval mine warfare, the operation of sweeping a sample of route or area to determine whether poised mines are present.

SEASHED — A temporary deck in container ships for transport of large military vehicles and outsized breakbulk cargo that will not fit into containers. See also **outsized cargo.** (JP 4-01.6)

sea state — A scale that categorizes the force of progressively higher seas by wave height. This scale is mathematically co-related to the Pierson-Moskowitz scale and the relationship of wind to waves. See also **Pierson-Moskowitz scale.** (JP 4-01.6)

sea surveillance — **(*)** The systematic observation of surface and subsurface sea areas by all available and practicable means primarily for the purpose of locating, identifying and determining the movements of ships, submarines, and other vehicles, friendly and enemy, proceeding on or under the surface of the world's seas and oceans. See also **surveillance.**

sea surveillance system — **(*)** A system for collecting, reporting, correlating, and presenting information supporting and derived from the task of sea surveillance.

seaward launch point — A designated point off the coast from which special operations forces will launch to proceed to the beach to conduct operations. Also called **SLP.** See also **seaward recovery point.** (JP 3-05.5)

seaward recovery point — A designated point off the coast to which special operations forces will proceed for recovery by submarine or other means of recovery. Also called **SRP.** See also **seaward launch point.** (JP 3-05.5)

secondary censorship — Armed forces censorship performed on the personal communications of officers, civilian employees, and accompanying civilians of the Armed Forces of the United States, and on those personal communications of enlisted personnel of the Armed Forces not subject to Armed Forces primary censorship or those requiring reexamination. See also **censorship.**

secondary imagery dissemination — See **electronic imagery dissemination.**

secondary imagery dissemination system — See **electronic imagery dissemination.**

secondary loads — Unit equipment, supplies, and major end items that are transported in the beds of organic vehicles.

secondary port — **(*)** A port with one or more berths, normally at quays, which can accommodate ocean-going ships for discharge.

secondary rescue facilities — Local airbase-ready aircraft, crash boats, and other air, surface, subsurface, and ground elements suitable for rescue missions, including government and privately operated units and facilities.

secondary road — A road supplementing a main road, usually wide enough and suitable for two-way, all-weather traffic at moderate or slow speeds.

secondary targets — Alternative targets of lower publicity value that are attacked when the primary target is unattainable. See also **antiterrorism; primary target.** (JP 3-07.2)

secondary wave breaker system — A series of waves superimposed on another series and differing in height, period, or angle of approach to the beach. (JP 4-01.6)

second strike — The first counterblow of a war. (Generally associated with nuclear operations.)

secret — See **security classification.**

Secretary of a Military Department — The Secretary of the Air Force, Army, or Navy.

SECRET Internet Protocol Router Network — Worldwide SECRET level packet switch network that uses high-speed internet protocol routers and high-capacity Defense Information Systems Network circuitry. Also called **SIPRNET.** See also **Defense Information Systems Network.** (JP 2-01)

section — 1. As applied to ships or naval aircraft, a tactical subdivision of a division. It is normally one-half of a division in the case of ships, and two aircraft in the case of aircraft. 2. A subdivision of an office, installation, territory, works, or organization; especially a major subdivision of a staff. 3. A tactical unit of the Army and Marine Corps. A section is smaller than a platoon and larger than a squad. In some organizations the section, rather than the squad, is the basic tactical unit. 4. An area in a warehouse extending from one wall to the next; usually the largest subdivision of one floor.

sector — **(*)** 1. An area designated by boundaries within which a unit operates, and for which it is responsible. 2. One of the subdivisions of a coastal frontier. See also **area of influence; zone of action.**

sector of fire — **(*)** A defined area which is required to be covered by the fire of individual or crew served weapons or the weapons of a unit.

sector scan — **(*)** Scan in which the antenna oscillates through a selected angle.

secure — **(*)** In an operational context, to gain possession of a position or terrain feature, with or without force, and to make such disposition as will prevent, as far as possible, its destruction or loss by enemy action. See also **denial measure.**

security — 1. Measures taken by a military unit, activity, or installation to protect itself against all acts designed to, or which may, impair its effectiveness. 2. A condition that results from the establishment and maintenance of protective measures that ensure a state of inviolability from hostile acts or influences. 3. With respect to classified matter, the condition that prevents unauthorized persons from having access to official information that is safeguarded in the interests of national security. See also **national security.**

security alert team — Two or more security force members who form the initial reinforcing element responding to security alarms, emergencies, or irregularities. Also called **SAT.**

security assistance — Group of programs authorized by the Foreign Assistance Act of 1961, as amended, and the Arms Export Control Act of 1976, as amended, or other related statutes by which the United States provides defense articles, military training, and other defense-related services by grant, loan, credit, or cash sales in furtherance of national policies and objectives. Also called **SA.**

security assistance organization — All Department of Defense elements located in a foreign country with assigned responsibilities for carrying out security assistance management functions. It includes military assistance advisory groups, military missions and groups, offices of defense and military cooperation, liaison groups, and defense attaché personnel designated to perform security assistance functions. Also called **SAO.** See also **security assistance.** (JP 3-07.1)

security certification — A certification issued by competent authority to indicate that a person has been investigated and is eligible for access to classified matter to the extent stated in the certification.

security classification — A category to which national security information and material is assigned to denote the degree of damage that unauthorized disclosure would cause to national defense or foreign relations of the United States and to denote the degree of protection required. There are three such categories. a. **top secret** — National security information or material that requires the highest degree of protection and the unauthorized disclosure of which could reasonably be expected to cause exceptionally grave damage to the national security. Examples of "exceptionally grave damage" include armed hostilities against the United States or its allies; disruption of foreign relations vitally affecting the national security; the compromise of vital national defense plans or complex cryptologic and communications intelligence systems; the revelation of sensitive intelligence operations; and the disclosure of scientific or technological developments vital to national security. b. **secret** — National security information or material that requires a substantial degree of protection and the unauthorized disclosure of which could reasonably be expected to cause serious damage to the national security. Examples of "serious damage" include disruption of foreign relations significantly affecting the national security; significant impairment of a program or policy directly related to the national security; revelation of significant military plans or intelligence operations; and compromise of significant scientific or technological developments relating to national security. c. **confidential** — National security information or material that requires protection and the unauthorized disclosure of which could reasonably be expected to cause damage to the national security. See also **classification; security.**

security clearance — An administrative determination by competent authority that an individual is eligible, from a security stand-point, for access to classified information.

security countermeasures — Those protective activities required to prevent espionage, sabotage, theft, or unauthorized use of classified or controlled information, systems, or material of the Department of Defense. See also **counterintelligence.** (JP 2-01.2)

security intelligence — **(*)** Intelligence on the identity, capabilities, and intentions of hostile organizations or individuals who are or may be engaged in espionage, sabotage, subversion, or terrorism. See also **counterintelligence; intelligence; security.**

security review — The process of reviewing news media products at some point, usually before transmission, to ensure that no oral, written, or visual information is filed for publication or broadcast that would divulge national security information or would jeopardize ongoing or future operations or that would threaten the safety of the members of the force. See also **security.** (JP 3-61)

sedition — Willfully advocating or teaching the duty or necessity of overthrowing the US government or any political subdivision by force or violence. See also **counterintelligence.** (JP 2-01.2)

seize — To employ combat forces to occupy physically and to control a designated area. See also **combat forces.** (JP 3-18)

seizures — In counterdrug operations, includes drugs and conveyances seized by law enforcement authorities and drug-related assets (monetary instruments, etc.) confiscated based on evidence that they have been derived from or used in illegal narcotics activities. See also **counterdrug operations; law enforcement agency.** (JP 3-07.4)

selected area for evasion — A designated area in hostile territory that offers evaders or escapees a reasonable chance of avoiding capture and of surviving until they can be evacuated. Also called **SAFE.** See also **escapee; evader; hostile.** (JP 3-50.3)

selected area for evasion area intelligence description — In evasion and recovery operations, an in-depth, all-source evasion study designed to assist the recovery of military personnel from a selected area for evasion under hostile conditions. Also called **SAID.** See also **evasion; evasion and recovery; hostile; recovery operations; selected area for evasion.** (JP 3-50.3)

Selected Reserve — Those units and individuals within the Ready Reserve designated by their respective Services and approved by the Joint Chiefs of Staff as so essential to initial wartime missions that they have priority over all other Reserves. All Selected Reservists are in an active status. The Selected Reserve also includes persons performing initial active duty for training. See also **Ready Reserve.**

Selected Reserve strength — The total number of guardsmen and reservists in the Selected Reserve who are subject to the 200K Presidential recall or mobilization under declaration of war or national emergency.

selective identification feature — A capability that, when added to the basic identification friend or foe system, provides the means to transmit, receive, and display selected coded replies.

selective jamming — See **spot jamming.**

selective loading — **(*)** The arrangement and stowage of equipment and supplies aboard ship in a manner designed to facilitate issues to units. See also **loading.**

selective mobilization — See **mobilization, Part 2.**

selective release process — The process involving requesting, analyzing, and obtaining approval for release of weapons to obtain specific, limited damage on selected targets.

selective unloading — In an amphibious operation, the controlled unloading from assault shipping, and movement ashore, of specific items of cargo at the request of the landing force commander. Normally, selective unloading parallels the landing of nonscheduled units during the initial unloading period of the ship-to-shore movement. (JP 3-02.2)

selenodesy — That branch of applied mathematics that determines, by observation and measurement, the exact positions of points and the figures and areas of large portions of the moon's surface, or the shape and size of the moon.

self-defense — A commander has the authority and obligation to use all necessary means available and to take all appropriate action to defend that commander's unit and other US forces in the vicinity from a hostile act or hostile intent. Force used should not exceed that which is necessary to decisively counter the hostile act or intent and ensure the continued safety of US forces or other persons and property they are ordered to protect. US forces may employ such force in self-defense only so long as the hostile force continues to present an imminent threat.

self-destroying fuze — **(*)** A fuze designed to burst a projectile before the end of its flight.

self-protection depth — **(*)** The depth of water where the aggregate danger width relative to mines affected by a minesweeping technique is zero. Safe depth is a particular self-protection depth.

self-sustaining containership — A containership with shipboard-installed cranes capable of loading and off-loading containers without assistance of port crane service. See also **containership.** (JP 4-01.7)

semi-active homing guidance — **(*)** A system of homing guidance wherein the receiver in the missile utilizes radiations from the target which has been illuminated by an outside source.

semi-controlled mosaic — **(*)** A mosaic composed of corrected or uncorrected prints laid so that major ground features match their geographical coordinates. See also **mosaic.**

semi-permanent joint task force — A joint task force that has been assigned an expanded or follow-on mission and will continue to conduct these operations in a specified area for an undetermined period of time. See also **joint task force; mission; operation.** (JP 5-00.2)

senior meteorological and oceanographic officer — Meteorological and oceanographic officer responsible for assisting the combatant commander and staff in developing and executing operational meteorological and oceanographic service concepts in support of a designated joint force. Also called **SMO.** See also **meteorological and oceanographic.** (JP 3-59)

senior officer present afloat — The senior line officer of the Navy, on active service, eligible for command at sea, who is present and in command of any unit of the operating forces afloat in the locality or within an area prescribed by competent authority. This officer is responsible for the administration of matters which collectively affect naval units of the operating forces afloat in the locality prescribed. Also called **SOPA.**

sensitive — Requiring special protection from disclosure that could cause embarrassment, compromise, or threat to the security of the sponsoring power. May be applied to an agency, installation, person, position, document, material, or activity.

sensitive compartmented information — All information and materials bearing special community controls indicating restricted handling within present and future community intelligence collection programs and their end products for which community systems of compartmentation have been or will be formally established. (These controls are over and above the provisions of DOD 5200.1-R, Information Security Program Regulation.) Also called **SCI.**

sensitive compartmented information facility — An accredited area, room, group of rooms, or installation where sensitive compartmented information (SCI) may be stored, used, discussed, and/or electronically processed. Sensitive compartmented information facility (SCIF) procedural and physical measures prevent the free access of persons unless they have been formally indoctrinated for the particular SCI authorized for use or storage within the SCIF. Also called **SCIF.** See also **sensitive compartmented information.** (JP 2-01)

separation zone — **(*)** An area between two adjacent horizontal or vertical areas into which units are not to proceed unless certain safety measures can be fulfilled.

sequel — A major operation that follows the current major operation. Plans for a sequel are based on the possible outcomes (success, stalemate, or defeat) associated with the current operation. See also **branch.** (JP 3-0)

sequence circuit — **(*)** In mine warfare, a circuit which requires actuation by a predetermined sequence of influences of predetermined magnitudes.

sequenced ejection system — See **ejection systems.**

serial — **(*)** 1. An element or a group of elements within a series which is given a numerical or alphabetical designation for convenience in planning, scheduling, and control. 2. **(DOD only)** A serial can be a group of people, vehicles, equipment, or supplies and is used in airborne, air assault, amphibious operations, and convoys.

serial assignment table — A table that is used in amphibious operations and shows the serial number, the title of the unit, the approximate number of personnel; the material, vehicles, or equipment in the serial; the number and type of landing craft and/or amphibious vehicles required to boat the serial; and the ship on which the serial is embarked.

seriously ill or injured — The casualty status of a person whose illness or injury is classified by medical authority to be of such severity that there is cause for immediate concern, but there is not imminent danger to life. Also called **SII.** See also **casualty status.**

seriously wounded — A casualty whose injuries or illness are of such severity that the patient is rendered unable to walk or sit, thereby requiring a litter for movement and evacuation. See also **evacuation; litter; patient.** (JP 4-02)

service ammunition — Ammunition intended for combat rather than for training purposes.

Service component command — A command consisting of the Service component commander and all those Service forces, such as individuals, units, detachments, organizations, and installations under that command, including the support forces that have been assigned to a combatant command or further assigned to a subordinate unified command or joint task force. See also **component; functional component command.** (JP 0-2)

Service component command chaplain — The senior chaplain assigned to the staff of, or designated by, the Service component commander. The component command chaplain is responsible for supervising and coordinating religious ministries within the purview of the component commander and may be supported by a staff of chaplains and enlisted religious support personnel. See also **command chaplain; command chaplain of the combatant command; lay leader or lay reader; religious ministry support; religious ministry support plan; religious ministry support team.** (JP 1-05)

service environment — **(*)** All external conditions, whether natural or induced, to which items of materiel are likely to be subjected throughout their life cycle.

Service force module — A hypothetical force module built per Service doctrine composed of combat, combat support, and combat service support forces and sustainment for an estimated period, e.g., 30 days.

service group — A major naval administration and/or tactical organization, consisting of the commander and the staff, designed to exercise operational and administrative control of assigned squadrons and units in executing their tasks of providing logistic support of fleet operations.

service mine — **(*)** A mine capable of a destructive explosion.

service squadron — An administrative and/or tactical subdivision of a naval service force or service group, consisting of the commander and the staff and organized to exercise operational and administrative control of assigned units in providing logistic support of fleet units as directed.

service test — A test of an item, system of materiel, or technique conducted under simulated or actual operational conditions to determine whether the specified military requirements or characteristics are satisfied. See also **troop test.**

service troops — Those units designed to render supply, maintenance, transportation, evacuation, hospitalization, and other services required by air and ground combat units to carry out effectively their mission in combat. See also **combat service support elements; troops.**

Service-unique container — Any 20- or 40-foot International Organization for Standardization container procured or leased by a Service to meet Service-unique requirements. Also called **component-owned container.** See also **common-use container; component-owned container.** (JP 4-01.7)

Service-unique transportation assets — Transportation assets that are: a. Assigned to a Military Department for functions of the Secretaries of the Military Departments set forth in Sections 3013(b), 5013(b), and 8013(b) of Title 10 of the United States Code, including administrative functions (such as motor pools), intelligence functions, training functions, and maintenance functions; b. Assigned to the Department of the Army for the execution of the missions of the Army Corps of Engineers; c. Assigned to the Department of the Navy as the special mission support force of missile range instrumentation ships, ocean survey ships, cable ships, oceanographic research ships, acoustic research ships, and naval test support ships; the naval fleet auxiliary force of fleet ammunition ships, fleet stores ships, fleet ocean tugs, and fleet oilers; hospital ships; Marine Corps intermediate maintenance activity ships, Marine Corps helicopter support to senior Federal officials; and, prior to the complete discharge of cargo, maritime pre-positioning ships; d. Assigned to the Department of the Air Force for search and rescue, weather reconnaissance, audiovisual services, aeromedical evacuation functions, and transportation of senior Federal officials. (JP 4-01)

servicing — See **common servicing; cross-servicing; joint servicing.** See also **inter-Service support.**

severe damage — See **nuclear damage, Part 3.**

shaded relief — (*) A cartographic technique that provides an apparent three-dimensional configuration of the terrain on maps and charts by the use of graded shadows that would be cast by high ground if light were shining from the northwest. Shaded relief is usually used in combination with contours. See also **hill shading.**

shadowing — To observe and maintain contact (not necessarily continuously) with a unit or force.

shallow fording — The ability of a self-propelled gun or ground vehicle equipped with built-in waterproofing, with its wheels or tracks in contact with the ground, to negotiate a water obstacle without the use of a special waterproofing kit. See also **flotation.**

shaped charge — (*) A charge shaped so as to concentrate its explosive force in a particular direction.

shared data environment — Automation services that support the implementation and maintenance of data resources that are used by two or more combat support applications. Services provided include: identification of common data, physical data modeling, database segmentation, development of data access and maintenance routines, and database reengineering to use the common data environment. See also **data.** (JP 4-0)

shear link assembly — (*) A device designed to break at a specified mechanical load.

sheet explosive — **(*)** Plastic explosive provided in a sheet form.

sheetlines — Those lines defining the geographic limits of the map or chart detail.

shelf life — **(*)** The length of time during which an item of supply, subject to deterioration or having a limited life which cannot be renewed, is considered serviceable while stored. See also **storage life.**

shell (specify) — **(*)** A command or request indicating the type of projectile to be used.

shelter — An International Organization for Standardization container outfitted with live- or work-in capability. See also **International Organization for Standardization.** (JP 4-01.7)

shielding — **(*)** 1. Material of suitable thickness and physical characteristics used to protect personnel from radiation during the manufacture, handling, and transportation of fissionable and radioactive materials. 2. Obstructions which tend to protect personnel or materials from the effects of a nuclear explosion.

shifting fire — Fire delivered at constant range at varying deflections; used to cover the width of a target that is too great to be covered by an open sheaf.

ship counter — **(*)** In naval mine warfare, a device in a mine which prevents the mine from detonating until a preset number of actuations has taken place.

ship haven — See **moving havens.**

ship influence — **(*)** In naval mine warfare, the magnetic, acoustic, and pressure effects of a ship, or a minesweep simulating a ship, which is detectable by a mine or other sensing devices.

shipping configuration — The manner in which an item is prepared for shipment.

shipping control — See **naval control of shipping.**

shipping designator — A code word assigned to a particular overseas base, port, or area for specific use as an address on shipments to the overseas location concerned. The code word is usually four letters and may be followed by a number to indicate a particular addressee.

shipping lane — **(*)** A term used to indicate the general flow of merchant shipping between two departure/terminal areas.

shipping time — The time elapsing between the shipment of materiel by the supplying activity and receipt of materiel by the requiring activity. See also **order and shipping time.**

ship-to-shore movement — **(*)** That portion of the assault phase of an amphibious operation which includes the deployment of the landing force from the assault shipping to designated landing areas.

shoal — A sandbank or bar that makes water shoal; i.e., a sand-bank that is not rocky and on which there is a water depth of 6 fathoms or less. (JP 4-01.6)

shock front — **(*)** The boundary between the pressure disturbance created by an explosion (in air, water, or earth) and the ambient atmosphere, water, or earth.

shore fire control party — A specially trained unit for control of naval gunfire in support of troops ashore. It consists of a spotting team to adjust fire and a naval gunfire liaison team to perform liaison functions for the supported battalion commander. Also called **SFCP.**

shoreline effect — See **coastal refraction.**

shore party — **(*)** A task organization of the landing force, formed for the purpose of facilitating the landing and movement off the beaches of troops, equipment, and supplies; for the evacuation from the beaches of casualties and enemy prisoners of war; and for facilitating the beaching, retraction, and salvaging of landing ships and craft. It comprises elements of both the naval and landing forces. Also called **beach group.** See also **beachmaster unit; beach party; naval beach group.**

shore-to-shore movement — The assault movement of personnel and materiel directly from a shore staging area to the objective, involving no further transfers between types of craft or ships incident to the assault movement.

shortfall — The lack of forces, equipment, personnel, materiel, or capability, reflected as the difference between the resources identified as a plan requirement and those apportioned to a combatant commander for planning, that would adversely affect the command's ability to accomplish its mission.

short-range air defense engagement zone — See **weapon engagement zone.** (JP 3-52)

short-range ballistic missile — A ballistic missile with a range capability up to about 600 nautical miles. Also called **SRBM.**

short-range transport aircraft — See **transport aircraft.**

short scope buoy — **(*)** A buoy used as a navigational reference which remains nearly vertical over its sinker.

short supply — An item is in short supply when the total of stock on hand and anticipated receipts during a given period are less than the total estimated demand during that period.

short takeoff and landing — **(*)** The ability of an aircraft to clear a 50-foot (15 meters) obstacle within 1,500 feet (450 meters) of commencing takeoff or in landing, to stop within 1,500 feet (450 meters) after passing over a 50-foot (15 meters) obstacle. Also called **STOL.**

short takeoff and vertical landing aircraft — **(*)** Fixed-wing aircraft capable of clearing a 15-meter (50-foot) obstacle within 450 meters (1,500 feet) of commencing takeoff run, and capable of landing vertically. Also called **STOVL.** See also **short takeoff and landing.**

short title — **(*)** A short, identifying combination of letters, and/or numbers assigned to a document or device for purposes of brevity and/or security.

short ton — 2,000 pounds. Also called **S/T or STON.** (JP 4-01.7)

shouldering — Maneuvering a vessel in contact with an opposing vessel to cause the opposing vessel to turn away. Shouldering is undertaken with the intent of minimizing damage to the opposing vessel.

show of force — An operation designed to demonstrate US resolve that involves increased visibility of US deployed forces in an attempt to defuse a specific situation that, if allowed to continue, may be detrimental to US interests or national objectives. (JP 3-07)

shuttered fuze — **(*)** A fuze in which inadvertent initiation of the detonator will not initiate either the booster or the burst charge.

side-looking airborne radar — (*) An airborne radar, viewing at right angles to the axis of the vehicle, which produces a presentation of terrain or moving targets. Also called **SLAR.**

side oblique air photograph — An oblique photograph taken with the camera axis at right angles to the longitudinal axis of the aircraft.

side overlap — See **overlap.**

sighting — Actual visual contact. Does not include other contacts, which must be reported by type, e.g., radar and sonar contacts. See also **contact report.**

SIGINT direct service — A reporting procedure to provide signals intelligence (SIGINT) to a military commander or other authorized recipient in response to SIGINT requirements. The product may vary from recurring, serialized reports produced by the National Security Agency/Central Security Service to instantaneous aperiodic reports provided to the command or other recipient, usually from a fixed SIGINT activity engaged in collection and processing. See also **signals intelligence.**

SIGINT direct service activity — A signals intelligence (SIGINT) activity composed of collection and associated resources that normally performs in a direct service role under the SIGINT operational control of the Director, National Security Agency/ Chief, Central Security Service. See also **signals intelligence.**

SIGINT direct support — The provision of signals intelligence (SIGINT) information to a military commander by a SIGINT direct support unit in response to SIGINT operational tasking levied by that commander. See also **signals intelligence.**

SIGINT direct support unit — A signals intelligence (SIGINT) unit, usually mobile, designed to perform a SIGINT direct support role for a military commander under delegated authority from the Director, National Security Agency/Chief, Central Security Service. See also **signals intelligence.**

SIGINT operational control — The authoritative direction of signals intelligence (SIGINT) activities, including tasking and allocation of effort, and the authoritative prescription of those uniform techniques and standards by which SIGINT information is collected, processed, and reported. See also **signals intelligence.**

SIGINT operational tasking — The authoritative operational direction of and direct levying of signals intelligence (SIGINT) information needs by a military commander on designated SIGINT resources. These requirements are directive, irrespective of other priorities, and are conditioned only by the capability of those resources to produce such information. Operational tasking includes authority to deploy all or part of the SIGINT resources for which SIGINT operational tasking authority has been delegated. See also **signals intelligence.**

SIGINT operational tasking authority — A military commander's authority to operationally direct and levy signals intelligence (SIGINT) requirements on designated SIGINT resources; includes authority to deploy and redeploy all or part of the SIGINT resources for which SIGINT operational tasking authority has been delegated. Also called **SOTA.** See also **signals intelligence.**

SIGINT resources — Personnel and equipment of any unit, activity, or organizational element engaged in signals intelligence activities. See also **signals intelligence.**

SIGINT support plans — Plans prepared by the National Security Agency/Central Security Service, in coordination with concerned elements of the United States SIGINT system, which specify how the resources of the system will be aligned in crisis or war to support military operations covered by certain Joint Chiefs of Staff and unified and specified command operation plans. See also **signals intelligence.**

signal — **(*)** 1. As applied to electronics, any transmitted electrical impulse. 2. Operationally, a type of message, the text of which consists of one or more letters, words, characters, signal flags, visual displays, or special sounds with prearranged meaning, and which is conveyed or transmitted by visual, acoustical, or electrical means.

signal center — A combination of signal communication facilities operated by the Army in the field and consisting of a communications center, telephone switching central and appropriate means of signal communications. See also **communications center.**

signal letters — See **international call sign.**

signal operation instructions — A series of orders issued for technical control and coordination of the signal communication activities of a command. In Marine Corps usage, these instructions are designated communication operation instructions.

signal security — A generic term that includes both communications security and electronics security. See also **security.**

signals intelligence — 1. A category of intelligence comprising either individually or in combination all communications intelligence, electronic intelligence, and foreign instrumentation signals intelligence, however transmitted. 2. Intelligence derived from communications, electronic, and foreign instrumentation signals. Also called **SIGINT.** See also **communications intelligence; electronic intelligence; foreign instrumentation signals intelligence; intelligence.** (JP 2-0)

signal-to-noise ratio — The ratio of the amplitude of the desired signal to the amplitude of noise signals at a given point in time.

signature equipment — **(*)** Any item of equipment which reveals the type and nature of the unit or formation to which it belongs.

signed route — A route along which a unit has placed directional signs bearing its unit identification symbol. The signs are for the unit's use only and must comply with movement regulations.

significant wave height — The average height of the third of waves observed during a given period of time. Significant wave height is used for evaluating the impact of waves and breakers on watercraft in the open sea and surf zones. See also **surf zone.** (JP 4-01.6)

Silver Triangle — The South American region consisting of Peru, Bolivia, and Colombia that is historically known to be a major illegal drug production area. (JP 3-07.4)

simulative electromagnetic deception — See **electromagnetic deception.**

simultaneous engagement — The concurrent engagement of hostile targets by combination of interceptor aircraft and surface-to-air missiles.

single-anchor leg mooring — A mooring facility dedicated to the offshore petroleum discharge system. Once installed, it permits a tanker to remain on station and pump in much higher sea states than is possible with a spread moor. Also called **SALM.** See also **offshore petroleum discharge system.** (JP 4-01.6)

single department purchase — A method of purchase whereby one Military Department buys commodities for another Military Department or Departments.

single flow route — (*) A route at least one-and-a-half lanes wide allowing the passage of a column of vehicles, and permitting isolated vehicles to pass or travel in the opposite direction at predetermined points. See also **double flow route.**

single integrated theater logistic manager — Service component or agency, usually in a mature theater, that is designated by the combatant commander or subunified commander as the single in-theater manager for planning and execution of a specific common-user logistic (CUL) item or related items. Single integrated logistic managers are normally long-term in nature with responsibilities that include planning, coordination, control, and execution of a specific CUL function (or similar CUL functions) at the theater level, in both peacetime and during actual operations, within the parameters of combatant commander's directives. Also called **SITLM.** See also **agency.** (JP 4-07)

single manager for transportation — The United States Transportation Command is the Department of Defense single manager for transportation, other than Service-unique or theater-assigned transportation assets. See also **Service-unique transportation assets; theater-assigned transportation assets; United States Transportation Command.** (JP 4-01)

single port manager — Through its transportation component commands, US Transportation Command is the Department of Defense-designated single port manager for all common-user aerial and sea ports worldwide. The single port manager performs those functions necessary to support the strategic flow of the deploying forces' equipment and sustainment from the aerial and sea port of embarkation and hand-off to the combatant commander in the aerial and sea port of debarkation (APOD and SPOD). The single port manager is responsible for providing strategic deployment status information to the combatant commander and to manage workload of the APOD and SPOD operator based on the commander's priorities and guidance. The single port manager is responsible through all phases of the theater aerial and sea port operations continuum, from an unimproved airfield and bare beach deployment to a commercial contract supported deployment. Also called **SPM.** See also **Military Traffic Management Command; transportation component command; United States Transportation Command.** (JP 4-0)

single-service manager — A component commander, designated by the combatant commander, who has been assigned responsibility and delegated the authority to coordinate specific theater personnel support activities such as theater postal operations. See also **component.** (JP 1-0)

single-spot ship — Those ships certified to have less than three adjacent landing areas. See also **spot.** (JP 3-04.1)

sinker — (*) In naval mine warfare, a heavy weight to which a buoyant mine is moored. The sinker generally houses the mooring rope drum and depth-setting mechanism and for mines laid by ships, it also serves as a launching trolley.

situation map — **(*)** A map showing the tactical or the administrative situation at a particular time. See also **map.**

situation report — **(*)** A report giving the situation in the area of a reporting unit or formation. Also called **SITREP.**

situation template — A depiction of assumed adversary dispositions, based on adversary doctrine and the effects of the battlespace if the adversary should adopt a particular course of action. In effect, situation templates are the doctrinal templates depicting a particular operation modified to account for the effects of the battlespace environment and the adversary's current situation (training and experience levels, logistic status, losses, dispositions). Normally, the situation template depicts adversary units two levels of command below the friendly force, as well as the expected locations of high-value targets. Situation templates use time-phase lines to indicate movement of forces and the expected flow of the operation. Usually, the situation template depicts a critical point in the course of action. Situation templates are one part of a adversary course of action model. Models may contain more than one situation template. See also **course of action; doctrinal template**. (JP 2-01.3)

skim sweeping — **(*)** In naval mine warfare, the technique of wire sweeping to a fixed depth over deep-laid moored mines to cut any shallow enough to endanger surface shipping.

skin paint — A radar indication caused by the reflected radar signal from an object.

skin tracking — The tracking of an object by means of a skin paint.

slant range — **(*)** The line of sight distance between two points, not at the same level relative to a specific datum.

slated items — Bulk petroleum and packaged bulk petroleum items that are requisitioned for overseas use by means of a consolidated requirement document, prepared and submitted through joint petroleum office channels. Packaged petroleum items are requisitioned in accordance with normal requisitioning procedures.

slice — An average logistic planning factor used to obtain estimates of requirements for personnel and materiel. (e.g., a personnel slice generally consists of the total strength of the stated basic combatant elements, plus its proportionate share of all supporting and higher headquarters personnel.)

slightly wounded — A casualty whose injuries or illness are relatively minor, permitting the patient to walk and/or sit. See also **patient; walking patient.** (JP 4-02)

small arms — Man portable, individual, and crew-served weapon systems used mainly against personnel and lightly armored or unarmored equipment.

small arms ammunition — Ammunition for small arms, i.e., all ammunition up to and including 20 millimeters (.787 inches).

small austere airfield — Unsophisticated airfield, usually with a short runway, that is limited in one or a combination of the following: taxiway systems, ramp space, security, materials handling equipment, aircraft servicing, maintenance, navigation aids, weather observing sensors, and communications. Also called **SAAF.** See also **airfield.** (JP 4-01.1)

small-lot storage — Generally considered to be a quantity of less than one pallet stack, stacked to maximum storage height. Thus, the term refers to a lot consisting of from one container to two or more pallet loads, but is not of sufficient quantity to form a complete pallet column. See also **storage.**

small-scale map — A map having a scale smaller than 1:600,000. See also **map.**

smoke screen — A cloud of smoke used to conceal ground maneuver, obstacle breaching, recovery operations, and amphibious assault operations as well as key assembly areas, supply routes, and logistic facilities.

snagline mine — (*) A contact mine with a buoyant line attached to one of the horns or switches which may be caught up and pulled by the hull or propellers of a ship.

soft missile base — (*) A launching base that is not protected against a nuclear explosion.

soil shear strength — The maximum resistance of a soil to shearing stresses.

solatium payments — Monetary compensation given to alleviate grief, suffering, and anxiety resulting from injuries, and property or personal loss. (JP 1-06)

solenoid sweep — (*) In naval mine warfare, a magnetic sweep consisting of a horizontal axis coil wound on a floating iron tube.

sonar — A sonic device used primarily for the detection and location of underwater objects. (This term is derived from the words "sound navigation and ranging.")

sonic — Of or pertaining to sound or the speed of sound. See also **speed of sound.**

sonobuoy — A sonar device used to detect submerged submarines that, when activated, relays information by radio. It may be active directional or nondirectional, or it may be passive directional or nondirectional.

sortie — (*) In air operations, an operational flight by one aircraft.

sortie allotment message — The means by which the joint force commander allots excess sorties to meet requirements of subordinate commanders that are expressed in their air employment and/or allocation plan. Also called **SORTIEALOT.** (JP 3-56.1)

sortie number — (*) A reference used to identify the images taken by all the sensors during one air reconnaissance sortie.

sortie plot — An overlay representing the area on a map covered by imagery taken during one sortie.

sortie reference — See **sortie number.**

sorting — In counterdrug operations, the process involved in differentiating traffic that could be involved in drug trafficking from legitimate air traffic. Initial sorting criteria are established jointly by the US Coast Guard and US Customs Service, coordinated with Department of Defense counterparts, and disseminated as required. See also **counterdrug operations.** (JP 3-07.4)

source — 1. A person, thing, or activity from which information is obtained. 2. In clandestine activities, a person (agent), normally a foreign national, in the employ of an intelligence activity for intelligence purposes. 3. In interrogation activities, any person who furnishes information, either with or without the knowledge that the information is being used for intelligence purposes. In this context, a controlled source is in the employment or under the control of the intelligence activity and knows that the information is to be used for intelligence purposes. An uncontrolled source is a voluntary contributor of information and may or may not know that the information is to be used for intelligence purposes. See also **agent; collection agency.**

space assignment — An assignment to the individual Departments/Services by the appropriate transportation operating agency of movement capability which completely or partially satisfies the stated requirements of the Departments/Services for the operating month and that has been accepted by them without the necessity for referral to the Joint Transportation Board for allocation.

space available mail — A transportation category for military mail transported to and from overseas bases by air on a space-available basis. Also called **SAM.**

space control operations — Operations that provide freedom of action in space for friendly forces while, when directed, denying it to an enemy, and include the broad aspects of protection of US and US allied space systems and negation of enemy space systems. Space control operations encompass all elements of the space defense mission.

space defense — All defensive measures designed to destroy attacking enemy vehicles (including missiles) while in space, or to nullify or reduce the effectiveness of such attack. See also **aerospace defense.**

space environment — The region beginning at the lower boundary of the Earth's ionosphere (approximately 50 km) and extending outward that contains solid particles (asteroids and meteoroids), energetic charged particles (ions, protons, electrons, etc.), and electromagnetic and ionizing radiation (x-rays, extreme ultraviolet, gamma rays, etc.). See also **ionosphere.** (JP 3-59)

space support operations — Operations required to ensure that space control and support of terrestrial forces are maintained. They include activities such as launching and deploying space vehicles, maintaining and sustaining space vehicles while on orbit, and recovering space vehicles if required.

space systems — All of the devices and organizations forming the space network. The network includes spacecraft, ground control stations, and associated terminals.

space weather — A term used to describe the environment and other natural phenomena occurring above 50 kilometers altitude that can degrade Department of Defense communications (satellite communications and skywave), global positioning system, radar, and satellite operations.

span of detonation (atomic demolition munition employment) — That total period of time, resulting from a timer error, between the earliest and the latest possible detonation time. 1. **early time** — The earliest possible time that an atomic demolition munition can detonate; 2. **fire time** — That time the atomic demolition munition will detonate should the timers function precisely without error; 3. **late time** — The latest possible time that an atomic demolition munition can detonate.

special access program — A sensitive program, approved in writing by a head of agency with original top secret classification authority, that imposes need-to-know and access controls beyond those normally provided for access to confidential, secret, or top secret information. The level of controls is based on the criticality of the program and the assessed hostile intelligence threat. The program may be an acquisition program, an intelligence program, or an operations and support program. Also called **SAP.** (JP 3-05.3)

special activities — Activities conducted in support of national foreign policy objectives that are planned and executed so that the role of the US Government is not apparent or acknowledged publicly. They are also functions in support of such activities but are not intended to influence US political processes, public opinion, policies, or media and do not include diplomatic activities or the collection and production of intelligence or related support functions. (JP 3-05)

special agent — A person, either United States military or civilian, who is a specialist in military security or the collection of intelligence or counterintelligence information.

special air operation — An air operation conducted in support of special operations and other clandestine, covert, and psychological activities. (JP 3-05.3)

special ammunition supply point — A mobile supply point where special ammunition is stored and issued to delivery units.

special assignment airlift requirements — Airlift requirements, including Chairman of the Joint Chiefs of Staff -directed or -coordinated exercises, that require special consideration due to the number of passengers involved, weight or size of cargo, urgency of movement, sensitivity, or other valid factors that preclude the use of channel airlift. See also **airlift requirement; channel airlift.**

special boat squadron — A permanent Navy echelon III major command to which two or more special boat units are assigned for some operational and all administrative purposes. The squadron is tasked with the training and deployment of these special boat units and may augment naval special warfare task groups and task units. Also called **SBS.** (JP 3-05.3)

special boat unit — Those US Navy forces organized, trained, and equipped to conduct or support naval special warfare, riverine warfare, coastal patrol and interdiction, and joint special operations with patrol boats or other combatant craft designed primarily for special operations support. Also called **SBU.** (JP 3-05)

special cargo — Cargo that requires special handling or protection, such as pyrotechnics, detonators, watches, and precision instruments.

special-equipment vehicle — A vehicle consisting of a general-purpose chassis with special-purpose body and/or mounted equipments designed to meet a specialized requirement.

special forces — US Army forces organized, trained, and equipped specifically to conduct special operations. Special forces have five primary missions: unconventional warfare, foreign internal defense, direct action, special reconnaissance, and counterterrorism. Counterterrorism is a special mission for specially organized, trained, and equipped special forces units designated in theater contingency plans. Also called **SF.** (JP 3-05)

special forces group — A combat arms organization capable of planning, conducting, and supporting special operations activities in all operational environments in peace, conflict, and war. It consists of a group headquarters and headquarters company, a support company, and special forces battalions. The group can operate as a single unit, but normally the battalions plan and conduct operations from widely separated locations. The group provides general operational direction and synchronizes the activities of subordinate battalions. Although principally structured for unconventional warfare, special forces group units are capable of task-organizing to meet specific requirements. Also called **SFG.** (JP 3-05)

special forces operations base — A command, control, and support base established and operated by a special forces group or battalion from organic and attached resources. The base commander and his staff coordinate and synchronize the activities of subordinate and forward-deployed forces. A special forces operations base is normally established for an extended period of time to support a series of operations. Also called **SFOB.** (JP 3-05)

special hazard — **(*)** In aircraft crash rescue and fire-fighting activities: fuels, materials, components, or situations that could increase the risks normally associated with military aircraft accidents and could require special procedures, equipment, or extinguishing agents.

special information operations — Information operations that by their sensitive nature and due to their potential effect or impact, security requirements, or risk to the national security of the United States, require a special review and approval process. Also called **SIO.** See also **information; information operations; operation.** (JP 3-13)

special interest target — In counterdrug operations, a contact that may be outside initial sorting criteria but still requires special handling, such as controlled deliveries or other unusual situations. Also called **SIT.** See also **suspect; track of interest.**

specialist intelligence report — A category of specialized, technical reports used in the dissemination of intelligence. Also called **SPIREP.** See also **intelligence reporting.**

specialization — An arrangement within an alliance wherein a member or group of members most suited by virtue of technical skills, location, or other qualifications assume(s) greater responsibility for a specific task or significant portion thereof for one or more other members.

special mission unit — A generic term to represent a group of operations and support personnel from designated organizations that is task-organized to perform highly classified activities. Also called **SMU.** (JP 3-05.3)

special operations — Operations conducted by specially organized, trained, and equipped military and paramilitary forces to achieve military, political, economic, or informational objectives by unconventional military means in hostile, denied, or politically sensitive areas. These operations are conducted across the full range of military operations, independently or in coordination with operations of conventional, non-special operations forces. Political-military considerations frequently shape special operations, requiring clandestine, covert, or low visibility techniques and oversight at the national level. Special operations differ from conventional operations in degree of physical and political risk, operational techniques, mode of employment, independence from friendly support, and dependence on detailed operational intelligence and indigenous assets. Also called **SO.** (JP 3-05)

special operations combat control team — A team of Air Force personnel organized, trained, and equipped to conduct and support special operations. Under clandestine, covert, or low-visibility conditions, these teams establish and control air assault zones; assist aircraft by verbal control, positioning, and operating navigation aids; conduct limited offensive direct action and special reconnaissance operations; and assist in the insertion and extraction of special operations forces. Also called **SOCCT.** See also **combat control team.** (JP 3-05.3)

special operations command — A subordinate unified or other joint command established by a joint force commander to plan, coordinate, conduct, and support joint special operations within the joint force commander's assigned operational area. Also called **SOC.** See also **special operations.** (JP 3-05)

special operations command and control element — A special operations command and control element (SOCCE) that is the focal point for the synchronization of special operations forces activities with conventional forces. It performs command and control functions according to mission requirements. It normally collocates with the command post of the supported force. The SOCCE can also receive special operations forces operational, intelligence, and target acquisition reports directly from deployed special operations elements and provide them to the supported component headquarters. The SOCCE remains under the operational control of the joint force special operations component commander or commander, joint special operations task force. Also called **SOCCE.** See also **command and control; joint force special operations component commander; special operations; special operations forces.** (JP 3-05.1)

special operations forces — Those Active and Reserve Component forces of the Military Services designated by the Secretary of Defense and specifically organized, trained, and equipped to conduct and support special operations. Also called **SOF.** See also **Air Force special operations forces; Army special operations forces; naval special warfare forces.** (JP 3-05.3)

special operations liaison element — A special operations liaison team provided by the joint force special operations component commander to the joint force air component commander (if designated) to coordinate, deconflict, and integrate special operations air and surface operations with conventional air operations. Also called **SOLE.** See also **joint force air component commander; joint force special operations component commander; special operations.** (JP 3-05)

special operations mission planning folder — The package that contains the materials required to execute a given special operations mission. It will include the mission tasking letter, mission tasking package, original feasibility assessment (as desired), initial assessment (as desired), target intelligence package, plan of execution, infiltration and exfiltration plan of execution, and other documentation as required or desired. Also called **SOMPF.** (JP 3-05.5)

special operations naval mobile environment team — A team of Navy personnel organized, trained, and equipped to support naval special warfare forces by providing weather, oceanographic, mapping, charting, and geodesy support. Also called **SONMET.** (JP 3-05.3)

special operations-peculiar — Equipment, material, supplies, and services required for special operations mission support for which there is no broad conventional force requirement. This includes standard items used by other Department of Defense (DOD) forces but modified for special operations forces (SOF); items initially designed for, or used by, SOF until adapted for use as Service-common by other DOD forces; and items approved by the Commander in Chief, US Special Operations Command (USCINCSOC) as critically urgent for the immediate accomplishment of a special operations mission but not normally procured by USCINCSOC. Also called **SO-peculiar.** See also **special operations.** (JP 3-05)

special operations terminal attack controller — United States Air Force combat control personnel certified to perform the terminal attack control function in support of special operations forces missions. Special operations terminal attack controller operations emphasize the employment of night infrared, laser, and beacon tactics and equipment. Also called **SOTAC.** See also **special operations; special tactics team; terminal.** (JP 3-09.1)

special operations weather team/tactical element — A task-organized team of Air Force personnel organized, trained, and equipped to collect critical weather observations from data-sparse areas. These teams are trained to operate independently in permissive or uncertain environments, or as augmentation to other special operations elements in hostile environments, in direct support of special operations. Also called **SOWT/TE.** (JP 3-05)

special operations wing — An Air Force special operations wing. Also called **SOW.** (JP 3-05.5)

special (or project) equipment — Equipment not authorized in standard equipment publications but determined as essential in connection with a contemplated operation, function, or mission. See also **equipment.**

special purpose Marine air-ground task force — A Marine air-ground task force organized, trained, and equipped with narrowly focused capabilities. It is designed to accomplish a specific mission, often of limited scope and duration. It may be any size, but normally it is a relatively small force — the size of a Marine expeditionary unit or smaller. Also called **SPMAGTF.** See also **aviation combat element; combat service support element; command element; ground combat element; Marine air-ground task force; Marine expeditionary force; Marine expeditionary force (forward); Marine expeditionary unit; task force.**

special-purpose vehicle — A vehicle incorporating a special chassis and designed to meet a specialized requirement.

special reconnaissance — Reconnaissance and surveillance actions conducted by special operations forces to obtain or verify, by visual observation or other collection methods, information concerning the capabilities, intentions, and activities of an actual or potential enemy or to secure data concerning the meteorological, hydrographic, or geographic characteristics of a particular area. It includes target acquisition, area assessment, and post-strike reconnaissance. Also called **SR.** (JP 3-05.5)

special staff — All staff officers having duties at a headquarters and not included in the general (coordinating) staff group or in the personal staff group. The special staff includes certain technical specialists and heads of services, e.g., quartermaster officer, antiaircraft officer, transportation officer, etc. See also **staff.**

special tactics team — US Air Force special operations forces consisting of combat control, pararescue, and combat weather personnel who are organized, trained, and equipped to establish and control the air-ground interface at an airhead in the objective area. Functions include assault zone reconnaissance and surveillance, establishment, and terminal control; combat search and rescue; combat casualty care and evacuation staging; special operations terminal attack control; and tactical weather observations and forecasting. Also called **STT.** See also **air traffic controller; combat search and rescue; special operations; special operations forces; special operations terminal attack controller.** (JP 3-05.1)

special unloading berth — Berths established in the vicinity of the approach lanes into which transports may move for unloading, thus reducing the running time for landing craft and assisting in the dispersion of transports. (JP 3-02.2)

special weapons — A term sometimes used to indicate weapons grouped for special procedures, for security, or other reasons. Specific terminology, e.g., "nuclear weapons" or "guided missiles," is preferable.

specific intelligence collection requirement — An identified gap in intelligence holdings that may be satisfied only by collection action, and that has been validated by the appropriate requirements control authority. Also called **SICR.**

specific search — Reconnaissance of a limited number of points for specific information.

specified combatant command — See **specified command.** (JP 0-2)

specified command — A command that has a broad, continuing mission, normally functional, and is established and so designated by the President through the Secretary of Defense with the advice and assistance of the Chairman of the Joint Chiefs of Staff. It normally is composed of forces from a single Military Department. Also called **specified combatant command.** (JP 5-0)

spectrozonal photography — **(*)** A photographic technique whereby the natural spectral emissions of all objects are selectively filtered in order to image only those objects within a particular spectral band or zone and eliminate the unwanted background.

spectrum management — Planning, coordinating, and managing joint use of the electromagnetic spectrum through operational, engineering, and administrative procedures. The objective of spectrum management is to enable electronic systems to perform their functions in the intended environment without causing or suffering unacceptable interference. See also **electromagnetic spectrum.** (JP 3-51)

speed of advance — **(*)** In naval usage, the speed expected to be made good over the ground. Also called **SOA.** See also **pace; rate of march.**

speed of sound — **(*)** The speed at which sound travels in a given medium under specified conditions. The speed of sound at sea level in the International Standard Atmosphere is 1108 ft/second, 658 knots, 1215 km/hour. See also **hypersonic; sonic; subsonic; supersonic; transonic.**

spillover — The part of the laser spot that is not on the target because of beam divergence or standoff range, improper boresighting of laser designator, or poor operator illuminating procedures. See also **laser spot.** (JP 3-09.1)

spin stabilization — Directional stability of a projectile obtained by the action of gyroscopic forces that result from spinning of the body about its axis of symmetry.

split cameras — **(*)** An assembly of two cameras disposed at a fixed overlapping angle relative to each other.

split pair — See **split vertical photography.**

split-up — See **break-up.**

split vertical photography — **(*)** Photographs taken simultaneously by two cameras mounted at an angle from the vertical, one tilted to the left and one to the right, to obtain a small side overlap.

spoiling attack — A tactical maneuver employed to seriously impair a hostile attack while the enemy is in the process of forming or assembling for an attack. Usually employed by armored units in defense by an attack on enemy assembly positions in front of a main line of resistance or battle position.

sponsor — Military member or civilian employee with dependents.

spoke — The portion of the hub and spoke distribution system that refers to transportation mode operators responsible for scheduled delivery to a customer of the "hub". See also **distribution; distribution system; hub; hub and spoke distribution.** (JP 4-01.4)

spot — **(*)** 1. To determine by observation, deviations of ordnance from the target for the purpose of supplying necessary information for the adjustment of fire. 2. To place in a proper location. 3. **(DOD only)** An approved shipboard helicopter landing site. See also **ordnance.** (JP 3-04.1)

spot elevation — **(*)** A point on a map or chart whose elevation is noted.

spot jamming — **(*)** The jamming of a specific channel or frequency. See also **barrage jamming; electronic warfare; jamming.**

spot net — Radio communication net used by a spotter in calling fire.

spot report — A concise narrative report of essential information covering events or conditions that may have an immediate and significant effect on current planning and operations that is afforded the most expeditious means of transmission consistent with requisite security. Also called **SPOTREP.** (Note: In reconnaissance and surveillance usage, spot report is not to be used.) See **Joint Tactical Air Reconnaissance/Surveillance Mission Report.**

spot size — **(*)** The size of the electron spot on the face of the cathode ray tube.

spotter — An observer stationed for the purpose of observing and reporting results of naval gunfire to the firing agency and who also may be employed in designating targets. See also **field artillery observer; naval gunfire spotting team.**

spotting line — **(*)** Any straight line to which the fall of shot of projectiles is related or fire is adjusted by an observer or a spotter. See also **gun-target line; observer-target line.**

spray dome — **(*)** The mound of water spray thrown up into the air when the shock wave from an underwater detonation of a nuclear weapon reaches the surface.

spreader bar — A device specially designed to permit the lifting and handling of containers or vehicles and breakbulk cargo. (JP 4-01.6)

spreading fire — A notification by the spotter or the naval gunfire ship, depending on who is controlling the fire, to indicate that fire is about to be distributed over an area.

sprocket — **(*)** In naval mine warfare, an anti-sweep device included in a mine mooring to allow a sweep wire to pass through the mooring without parting the mine from its sinker.

squadron — 1. An organization consisting of two or more divisions of ships, or two or more divisions (Navy) or flights of aircraft. It is normally but not necessarily composed of ships or aircraft of the same type. 2. The basic administrative aviation unit of the Army, Navy, Marine Corps, and Air Force. 3. Battalion-sized ground or aviation units in US Army cavalry regiments.

squib — A small pyrotechnic device that may be used to fire the igniter in a rocket or for some similar purpose. Not to be confused with a detonator that explodes.

squirt — **(*)** In air-to-air refuelling, a means of providing visual detection of a nearby aircraft. In practice this is achieved by the donor aircraft dumping fuel and/or the receiver aircraft selecting afterburners, if so equipped.

staballoy — Metal alloys made from high-density depleted uranium mixed with other metals for use in kinetic energy penetrators for armor-piercing munitions. Several different metals, such as titanium or molybdenum, can be used for the purpose. The various staballoy metals have low radioactivity that is not considered to be a significant health hazard.

stabilized glide slope indicator — An electrohydraulic optical landing aid for use on air-capable ships. With it, a pilot can visually establish and maintain the proper glide slope for a safe approach and landing. The visual acquisition range is approximately 3 miles at night under optimal conditions. Also called **SGSI.** See also **air-capable ship.** (JP 3-04.1)

stabilized patient — A patient whose airway is secured, hemorrhage is controlled, shock treated, and fractures are immobilized. See also **patient.** (JP 4-02.2)

stable base film — **(*)** A particular type of film having a high stability in regard to shrinkage and stretching.

stable patient — A patient for whom no inflight medical intervention is expected but the potential for medical intervention exists. See also **patient.** (JP 4-02.2)

staff — See **multinational staff; general staff; integrated staff; joint staff; parallel staff; special staff.**

staff estimates — Assessments of courses of action by the various staff elements of a command that serve as the foundation of the commander's estimate.

staff supervision — The process of advising other staff officers and individuals subordinate to the commander of the commander's plans and policies, interpreting those plans and policies, assisting such subordinates in carrying them out, determining the extent to which they are being followed, and advising the commander thereof.

stage — **(*)** 1. An element of the missile or propulsion system that generally separates from the missile at burnout or cut-off. Stages are numbered chronologically in order of burning. 2. To process, in a specified area, troops which are in transit from one locality to another. See also **marshalling; staging area.**

staged crews — Aircrews specifically positioned at intermediate airfields to take over aircraft operating on air routes, thus relieving complementary crews of flying fatigue and speeding up the flow rate of the aircraft concerned.

staging — Assembling, holding, and organizing arriving personnel, equipment, and sustaining materiel in preparation for onward movement. The organizing and preparation for movement of personnel, equipment, and materiel at designated areas to incrementally build forces capable of meeting the operational commander's requirements. See also **staging area.** (JP 3-35)

staging area — 1. **Amphibious or airborne** — A general locality between the mounting area and the objective of an amphibious or airborne expedition, through which the expedition or parts thereof pass after mounting, for refueling, regrouping of ships, and/or exercise, inspection, and redistribution of troops. 2. **Other movements** — A general locality established for the concentration of troop units and transient personnel between movements over the lines of communications. Also called **SA.** See also **airborne; marshalling; stage; staging.** (JP 3-35)

staging base — 1. An advanced naval base for the anchoring, fueling, and refitting of transports and cargo ships as well as replenishment of mobile service squadrons. 2. A landing and takeoff area with minimum servicing, supply, and shelter provided for the temporary occupancy of military aircraft during the course of movement from one location to another.

standard advanced base units — Personnel and materiel organized to function as advanced base units, including the functional components that are employed in the establishment of naval advanced bases. Such advanced base units may establish repair bases, supply bases, supply depots, airfields, air bases, or other naval shore establishments at overseas locations.

standardization — The process by which the Department of Defense achieves the closest practicable cooperation among the Services and Defense agencies for the most efficient use of research, development, and production resources, and agrees to adopt on the broadest possible basis the use of: a. common or compatible operational, administrative, and logistic procedures; b. common or compatible technical procedures and criteria; c. common, compatible, or interchangeable supplies, components, weapons, or equipment; and d. common or compatible tactical doctrine with corresponding organizational compatibility.

standard operating procedure — See **standing operating procedure.**

standard parallel — **(*)** A parallel on a map or chart along which the scale is as stated for that map or chart.

standard pattern — **(*)** In land mine warfare, the agreed pattern to which mines are normally laid.

standard positioning system — One of two levels of service provided by the global positioning system, the standard positioning system normally offers users a horizontal accuracy of 100 meters or better with a 95% probability. Also called **SPS.**

standard route — In naval control of shipping, a pre-planned single track that is assigned a code name and connects positions within the main shipping lanes.

standard unit — A type unit whose unit-type code and movement characteristics are described in the type unit characteristics file.

standard use Army aircraft flight route — Routes established below the coordinating altitude to facilitate the movement of Army aviation assets. Routes are normally located in the corps through brigade rear areas of operation and do not require approval by the airspace control authority. Also called **SAAFR.** (JP 3-52)

Standby Reserve — Those units and members of the Reserve Components (other than those in the Ready Reserve or Retired Reserve) who are liable for active duty only, as provided in the US Code, title 10 (DOD), sections 10151, 12301, and 12306. See also **active duty; Ready Reserve; Reserve Components; Retired Reserve.** (JP 4-05)

stand fast — **(*)** In artillery, the order at which all action on the position ceases immediately.

standing operating procedure — **(*)** A set of instructions covering those features of operations which lend themselves to a definite or standardized procedure without loss of effectiveness. The procedure is applicable unless ordered otherwise. Also called **SOP.**

standing order — **(*)** A promulgated order which remains in force until amended or cancelled.

state and regional defense airlift — The program for use during an emergency of civil aircraft other than air carrier aircraft.

stateless person — Civilian who has been denationalized or whose country of origin cannot be determined or who cannot establish a right to the nationality claimed. See also **dislocated civilian; displaced person; evacuee; expellee; refugee.** (JP 3-07.6)

state of readiness — See **defense readiness condition; weapons readiness state.**

state of readiness--state 1--safe — The state of a demolition target upon or within which the demolition charge has been placed and secured. The firing or initiating circuits have been installed, but not connected to the demolition charge. Detonators or initiators have not been connected nor installed. See also **state of readiness--state 2--armed.**

state of readiness--state 2--armed — The state of a demolition target in which the demolition charges are in place, the firing and priming circuits are installed and complete, and the charge is ready for immediate firing. See also **state of readiness--state 1--safe.**

static air temperature — **(*)** The temperature at a point at rest relative to the ambient air.

static line (air transport) — A line attached to a parachute pack and to a strop or anchor cable in an aircraft so that, when the load is dropped, the parachute is deployed automatically.

static line cable — See **anchor cable.**

static marking — **(*)** Marks on photographic negatives and other imagery caused by unwanted discharges of static electricity.

station — 1. A general term meaning any military or naval activity at a fixed land location. 2. A particular kind of activity to which other activities or individuals may come for a specific service, often of a technical nature, e.g., aid station. 3. An assigned or prescribed position in a naval formation or cruising disposition; or an assigned area in an approach, contact, or battle disposition. 4. Any place of duty or post or position in the field to which an individual, group of individuals, or a unit may be assigned. 5. One or more transmitters or receivers or a combination of transmitters and receivers, including the accessory equipment necessary at one location, for carrying on radio communication service. Each station will be classified by the service in which it operates permanently or temporarily.

station authentication — A security measure designed to establish the authenticity of a transmitting or receiving station.

station time — **(*)** In air transport operations, the time at which crews, passengers, and cargo are to be on board and ready for the flight.

status-of-forces agreement — An agreement that defines the legal position of a visiting military force deployed in the territory of a friendly state. Agreements delineating the status of visiting military forces may be bilateral or multilateral. Provisions pertaining to the status of visiting forces may be set forth in a separate agreement, or they may form a part of a more comprehensive agreement. These provisions describe how the authorities of a visiting force may control members of that force and the amenability of the force or its members to the local law or to the authority of local officials. To the extent that agreements delineate matters affecting the relations between a military force and civilian authorities and population, they may be considered as civil affairs agreements. Also called **SOFA.** See also **civil affairs agreement.** (JP 3-07.2)

stay behind — Agent or agent organization established in a given country to be activated in the event of hostile overrun or other circumstances under which normal access would be denied.

stay behind force — **(*)** A force which is left in position to conduct a specified mission when the remainder of the force withdraws or retires from the area.

stellar guidance — A system wherein a guided missile may follow a predetermined course with reference primarily to the relative position of the missile and certain preselected celestial bodies.

stepped-up separation — **(*)** The vertical separation in a formation of aircraft measured from an aircraft ahead upward to the next aircraft behind or in echelon.

stereographic coverage — Photographic coverage with overlapping air photographs to provide a three-dimensional presentation of the picture; 60 percent overlap is considered normal and 53 percent is generally regarded as the minimum.

sterilize — **(*)** 1. In naval mine warfare, to permanently render a mine incapable of firing by means of a device (e.g., sterilizer) within the mine. 2. **(DOD only)** To remove from material to be used in covert and clandestine operations, marks or devices which can identify it as emanating from the sponsoring nation or organization.

sterilizer — **(*)** In mine warfare, a device included in mines to render the mine permanently inoperative on expiration of a pre-determined time after laying.

stick (air transport) — A number of paratroopers who jump from one aperture or door of an aircraft during one run over a drop zone.

stick commander (air transport) — A designated individual who controls paratroops from the time they enter the aircraft until their exit. See also **jumpmaster.**

stimulants — Controlled drugs that make the user feel stronger, more decisive, and self-possessed; includes cocaine and amphetamines. (JP 3-07.4)

stockage objective — The maximum quantities of materiel to be maintained on hand to sustain current operations. It will consist of the sum of stocks represented by the operating level and the safety level. See also **level of supply.**

stock control — **(*)** Process of maintaining inventory data on the quantity, location, and condition of supplies and equipment due-in, on-hand, and due-out, to determine quantities of material and equipment available and/or required for issue and to facilitate distribution and management of materiel. See also **inventory control.**

stock coordination — A supply management function exercised usually at department level that controls the assignment of material cognizance for items or categories of material to inventory managers.

stock level — See **level of supply.**

Stock Number — See **National Stock Number.**

stockpile to target sequence — 1. The order of events involved in removing a nuclear weapon from storage and assembling, testing, transporting, and delivering it on the target. 2. A document that defines the logistic and employment concepts and related physical environments involved in the delivery of a nuclear weapon from the stockpile to the target. It may also define the logistic flow involved in moving nuclear weapons to and from the stockpile for quality assurance testing, modification and retrofit, and the recycling of limited life components.

stock record account — A basic record showing by item the receipt and issuance of property, the balances on hand, and such other identifying or stock control data as may be required by proper authority.

stop-loss — Presidential authority under Title 10 US Code 12305 to suspend laws relating to promotion, retirement, or separation of any member of the Armed Forces determined essential to the national security of the United States ("laws relating to promotion" broadly includes, among others, grade tables, current general or flag officer authorizations, and E8 and 9 limits). This authority may be exercised by the President only if Reservists are serving on active duty under Title 10 authorities for Presidential Selected Reserve Callup, partial mobilization, or full mobilization. See also **mobilization; partial mobilization; Presidential Selected Reserve Callup Authority.** (JP 1-0)

storage — 1. The retention of data in any form, usually for the purpose of orderly retrieval and documentation. 2. A device consisting of electronic, electrostatic, electrical, hardware, or other elements into which data may be entered, and from which data may be obtained as desired. See also **ammunition and toxic material open space; bin storage; bulk storage; igloo space; large-lot storage; medium-lot storage; open improved storage space; open unimproved wet space; small-lot storage.**

storage life — **(*)** The length of time for which an item of supply, including explosives, given specific storage conditions, may be expected to remain serviceable and, if relevant, safe. See also **shelf life.**

storage or stowage — Storage is the act of placing material or ammunition and other supplies onboard the vessel. Stowage relates to the act of securing those items stored in such a manner that they do not shift or move during at-sea periods, using methods and equipment as approved by higher authority. See also **storage; stowage.** (JP 3-04.1)

stores — See **naval stores; supplies.**

stowage — The method of placing cargo into a single hold or compartment of a ship to prevent damage, shifting, etc. (JP 3-02)

stowage diagram — **(*)** A scaled drawing included in the loading plan of a vessel for each deck or platform showing the exact location of all cargo. See also **stowage plan.**

stowage factor — The number that expresses the space, in cubic feet, occupied by a long ton of any commodity as prepared for shipment, including all crating or packaging.

stowage plan — A completed stowage diagram showing what materiel has been loaded and its stowage location in each hold, between-deck compartment, or other space in a ship, including deck space. Each port of discharge is indicated by colors or other appropriate means. Deck and between-deck cargo normally is shown in perspective, while cargo stowed in the lower hold is shown in profile, except that vehicles usually are shown in perspective regardless of stowage. See also **stowage diagram.**

strafing — The delivery of automatic weapons fire by aircraft on ground targets.

straggler — **(*)** 1. Any personnel, vehicles, ships, or aircraft which, without apparent purpose or assigned mission, become separated from their unit, column, or formation. 2. A ship separated from its convoy by more than 5 nautical miles, through inability to keep up, and unable to rejoin before dark, or over 10 nautical miles from its convoy whether or not it can rejoin before dark. See also **romper.**

strapping — 1. An operation by which supply containers, such as cartons or boxes, are reinforced by bands, metal straps, or wire, placed at specified intervals around them, drawn taut, and then sealed or clamped by a machine. 2. Measurement of storage tanks and calculation of volume to provide tables for conversion of depth of product in linear units of measurement to volume of contents.

strategic advantage — The overall relative power relationship of opponents that enables one nation or group of nations effectively to control the course of a military or political situation.

strategic airlift — The common-user airlift linking theaters to the continental United States (CONUS) and to other theaters as well as the airlift within CONUS. These airlift assets are assigned to the Commander in Chief, United States Transportation Command. Due to the intertheater ranges usually involved, strategic airlift is normally comprised of the heavy, longer range, intercontinental airlift assets, but may be augmented with shorter range aircraft when required. Also called **intertheater airlift.** See also **theater airlift.** (JP 4-01.1)

strategic air transport — The movement of personnel and materiel by air in accordance with a strategic plan.

strategic air transport operations — **(*)** The carriage of passengers and cargo between theaters by means of: a. scheduled service; b. special flight; c. air logistic support; d. aeromedical evacuation.

strategic air warfare — Air combat and supporting operations designed to effect, through the systematic application of force to a selected series of vital targets, the progressive destruction and disintegration of the enemy's war-making capacity to a point where the enemy no longer retains the ability or the will to wage war. Vital targets may include key manufacturing systems, sources of raw material, critical material, stockpiles, power systems, transportation systems, communication facilities, concentration of uncommitted elements of enemy armed forces, key agricultural areas, and other such target systems.

strategic concentration — (*) The assembly of designated forces in areas from which it is intended that operations of the assembled force shall begin so that they are best disposed to initiate the plan of campaign.

strategic concept — The course of action accepted as the result of the estimate of the strategic situation. It is a statement of what is to be done in broad terms sufficiently flexible to permit its use in framing the military, diplomatic, economic, informational, and other measures which stem from it. See also **basic undertakings.**

strategic estimate — The estimate of the broad strategic factors that influence the determination of missions, objectives, and courses of action. The estimate is continuous and includes the strategic direction received from the National Command Authorities or the authoritative body of an alliance or coalition. See also **commander's estimate of the situation; estimate; logistic estimate of the situation; national intelligence estimate.** (JP 3-0)

strategic intelligence — Intelligence that is required for the formulation of strategy, policy, and military plans and operations at national and theater levels. See also **intelligence; operational intelligence; tactical intelligence.** (JP 2-0)

strategic level of war — The level of war at which a nation, often as a member of a group of nations, determines national or multinational (alliance or coalition) security objectives and guidance, and develops and uses national resources to accomplish these objectives. Activities at this level establish national and multinational military objectives; sequence initiatives; define limits and assess risks for the use of military and other instruments of national power; develop global plans or theater war plans to achieve these objectives; and provide military forces and other capabilities in accordance with strategic plans. See also **operational level of war; tactical level of war.** (JP 3-0)

strategic map — A map of medium scale or smaller used for planning of operations, including the movement, concentration, and supply of troops. See also **map.**

strategic material (critical) — Material required for essential uses in a war emergency, the procurement of which in adequate quantity, quality, or time, is sufficiently uncertain, for any reason, to require prior provision of the supply thereof.

strategic mining — A long-term mining operation designed to deny the enemy the use of specific sea routes or sea areas. (JP 3-15)

strategic mission — A mission directed against one or more of a selected series of enemy targets with the purpose of progressive destruction and disintegration of the enemy's warmaking capacity and will to make war. Targets include key manufacturing systems, sources of raw material, critical material, stockpiles, power systems, transportation systems, communication facilities, and other such target systems. As opposed to tactical operations, strategic operations are designed to have a long-range rather than immediate effect on the enemy and its military forces.

strategic mobility — The capability to deploy and sustain military forces worldwide in support of national strategy. See also **mobility.**

strategic plan — A plan for the overall conduct of a war.

strategic psychological activities — (*) Planned psychological activities in peace, crisis, and war which pursue objectives to gain the support and cooperation of friendly and neutral countries and to reduce the will and the capacity of hostile or potentially hostile countries to wage war.

strategic sealift — The afloat pre-positioning and ocean movement of military materiel in support of US and multinational forces. Sealift forces include organic and commercially acquired shipping and shipping services, including chartered foreign-flag vessels and associated shipping services. (JP 4-01.5)

strategic sealift forces — Sealift forces composed of ships, cargo handling and delivery systems, and the necessary operating personnel. They include US Navy, US Marine Corps, and US Army elements with Active and Reserve components. Merchant marine vessels manned by civilian mariners may constitute part of this force. See also **force.** (JP 4-01.6)

strategic sealift shipping — Common-user ships of the Military Sealift Command (MSC) force, including pre-positioned ships after their pre-positioning mission has been completed and they have been returned to the operational control of MSC. See also **Military Sealift Command; Military Sealift Command force.** (JP 4-01.6)

strategic transport aircraft — (*) Aircraft designed primarily for the carriage of personnel and/or cargo over long distances.

strategic vulnerability — The susceptibility of vital instruments of national power to being seriously decreased or adversely changed by the application of actions within the capability of another nation to impose. Strategic vulnerability may pertain to political, geographic, economic, informational, scientific, sociological, or military factors.

strategic warning — A warning prior to the initiation of a threatening act. See also **strategic warning lead time; strategic warning post-decision time; strategic warning pre-decision time; tactical warning; warning; warning of war.**

strategic warning lead time — That time between the receipt of strategic warning and the beginning of hostilities. This time may include two action periods: strategic warning pre-decision time and strategic warning post-decision time. See also **commander's estimate of the situation; strategic concept; strategic warning.**

strategic warning post-decision time — That time beginning after the decision, made at the highest levels of government(s) in response to strategic warning, is ordered executed and ending with the start of hostilities or termination of the threat. It is that part of strategic warning lead time available for executing pre-hostility actions to strengthen the national strategic posture; however, some preparatory actions may be initiated in the predecision period. See also **strategic warning; strategic warning lead time.**

strategic warning pre-decision time — That time which begins upon receipt of strategic warning and ends when a decision is ordered executed. It is that part of strategic warning lead time available to the highest levels of government(s) to determine that strategic course of action to be executed. See also **strategic warning; strategic warning lead time.**

strategy — The art and science of developing and employing instruments of national power in a synchronized and integrated fashion to achieve theater, national, and/or multinational objectives. See also **military strategy; national strategy.** (JP 3-0)

strategy determination — The Joint Operation Planning and Execution System function in which analysis of changing events in the international environment and the development of national strategy to respond to those events is conducted. In joint operation planning, the responsibility for recommending military strategy to the National Command Authorities lies with the Chairman of the Joint Chiefs of Staff, in consultation with the other members of the Joint Chiefs of Staff and in concert with supported commanders. In the deliberate planning process, the Joint Strategic Capabilities Plan is produced as a result of this process. In the crisis assessment phase of the crisis action planning process, crisis action planning procedures are used to formulate decisions for direct development of possible military courses of action.

stretcher — See **litter.**

stretch out — A reduction in the delivery rate specified for a program without a reduction in the total quantity to be delivered.

strike — **(*)** An attack which is intended to inflict damage on, seize, or destroy an objective.

strikedown — A term used to describe the movement of aircraft from the flight deck to the hangar deck level. See also **aircraft; flight deck.** (JP 3-04.1)

strike photography — **(*)** Air photographs taken during an air strike.

strip marker — **(*)** In land mine warfare, a marker, natural, artificial, or specially installed, located at the start and finish of a mine strip. See also **marker.**

strip plot — **(*)** A portion of a map or overlay on which a number of photographs taken along a flight line is delineated without defining the outlines of individual prints.

strong point — **(*)** A key point in a defensive position, usually strongly fortified and heavily armed with automatic weapons, around which other positions are grouped for its protection.

structured message text — **(*)** A message text composed of paragraphs ordered in a specified sequence, each paragraph characterized by an identifier and containing information in free form. It is designed to facilitate manual handling and processing. See also **formatted message text; free form message text.**

stuffing — Packing of cargo into a container. See also **unstuffing.** (JP 4-01.7)

subassembly — **(*)** In logistics, a portion of an assembly, consisting of two or more parts, that can be provisioned and replaced as an entity. See also **assembly; component.**

subgravity — A condition in which the resultant ambient acceleration is between 0 and 1 G.

subkiloton weapon — **(*)** A nuclear weapon producing a yield below one kiloton. See also **kiloton weapon; megaton weapon; nominal weapon.**

submarine operating authority — (*) The naval commander exercising operational control of submarines. Also called **SUBOPAUTH.**

submarine operations area — A geographic area defined for submarine operations for peacetime or warfare activities.

submarine patrol area — A restricted area established to allow submarine operations: a. unimpeded by the operation of, or possible attack from, friendly forces in wartime; b. without submerged mutual interference in peacetime.

submunition — (*) Any munition that, to perform its task, separates from a parent munition.

subordinate command — A command consisting of the commander and all those individuals, units, detachments, organizations, or installations that have been placed under the command by the authority establishing the subordinate command.

subordinate unified command — A command established by commanders of unified commands, when so authorized through the Chairman of the Joint Chiefs of Staff, to conduct operations on a continuing basis in accordance with the criteria set forth for unified commands. A subordinate unified command may be established on an area or functional basis. Commanders of subordinate unified commands have functions and responsibilities similar to those of the commanders of unified commands and exercise operational control of assigned commands and forces within the assigned operational area. Also called **subunified command.** See also **area command; functional component command; operational control; subordinate command; unified command.** (JP 0-2)

subscription — An agreement by a nation's Military Services to agree to accept and abide by, with or without reservation, the details of a standardization agreement. See also **implementation; ratification; reservation.**

subsidiary landing — (*) In an amphibious operation, a landing usually made outside the designated landing area, the purpose of which is to support the main landing.

subsonic — Of or pertaining to speeds less than the speed of sound. See also **speed of sound.**

substitute transport-type vehicle — A wheeled vehicle designed to perform, within certain limitations, the same military function as military transport vehicles, but not requiring all the special characteristics thereof. They are developed from civilian designs by addition of certain features, or from military designs by deletion of certain features.

subunified command — See **subordinate unified command.** (JP 0-2)

subversion — Action designed to undermine the military, economic, psychological, or political strength or morale of a regime. See also **unconventional warfare.**

subversion of Department of Defense personnel — Actions designed to undermine the loyalty, morale, or discipline of Department of Defense military and civilian personnel.

subversive activity — Anyone lending aid, comfort, and moral support to individuals, groups, or organizations that advocate the overthrow of incumbent governments by force and violence is subversive and is engaged in subversive activity. All willful acts that are intended to be detrimental to the best interests of the government and that do not fall into the categories of treason, sedition, sabotage, or espionage will be placed in the category of subversive activity.

subversive political action — A planned series of activities designed to accomplish political objectives by influencing, dominating, or displacing individuals or groups who are so placed as to affect the decisions and actions of another government.

summit — The highest altitude above mean sea level that a projectile reaches in its flight from the gun to the target; the algebraic sum of the maximum ordinate and the altitude of the gun.

sun synchronous orbit — An orbit where the satellite's orbital plane is at a fixed orientation to the sun, i.e., the orbit precesses about the earth at the same rate that the earth orbits the sun. It has the characteristics of maintaining similar sun angles along its ground trace for all orbits, and typically has an inclination from 96 to 98 degrees, depending on orbit altitude and orbit shape (eccentricity).

supercargo — Personnel that accompany cargo on board a ship for the purpose of accomplishing en route maintenance and security.

supersonic — Of or pertaining to speed in excess of the speed of sound. See also **speed of sound.**

supervised route — **(*)** In road traffic, a roadway over which limited control is exercised by means of traffic control posts, traffic patrols, or both. Movement credit is required for its use by a column of vehicles or a vehicle of exceptional size or weight. See also **route.**

supplementary facilities — **(*)** Facilities required at a particular location to provide a specified minimum of support for reinforcing forces, which exceed the facilities required to support in-place forces.

supplies — In logistics, all materiel and items used in the equipment, support, and maintenance of military forces. See also **assembly; component; equipment; subassembly.**

supply — The procurement, distribution, maintenance while in storage, and salvage of supplies, including the determination of kind and quantity of supplies. a. **producer phase** — That phase of military supply that extends from determination of procurement schedules to acceptance of finished supplies by the Military Services. b. **consumer phase** — That phase of military supply which extends from receipt of finished supplies by the Military Services through issue for use or consumption.

supply by air — See **airdrop; air movement.**

supply chain — The linked activities associated with providing materiel from a raw materiel stage to an end user as a finished product. See also **supply; supply chain management.** (JP 4-09)

supply chain management — A cross-functional approach to procuring, producing, and delivering products and services to customers. The broad management scope includes sub-suppliers, suppliers, internal information, and funds flow. See also **supply; supply chain.** (JP 4-09)

supply control — The process by which an item of supply is controlled within the supply system, including requisitioning, receipt, storage, stock control, shipment, disposition, identification, and accounting.

supplying ship — **(*)** The ship in a replenishment unit that provides the personnel and/or supplies to be transferred.

supply management — See **inventory control.**

supply point — A location where supplies, services, and materiels are located and issued. These locations are temporary and mobile, normally being occupied for up to 72 hours.

supply support activity — Activities assigned a Department of Defense activity address code and that have a supply support mission, i.e., direct support supply units, missile support elements, and maintenance support units. Also called **SSA.** (JP 4-01.7)

supply transaction reporting — Reporting on individual transactions affecting the stock status of materiel to the appropriate supply accounting activity as they occur.

support — 1. The action of a force that aids, protects, complements, or sustains another force in accordance with a directive requiring such action. 2. A unit that helps another unit in battle. 3. An element of a command that assists, protects, or supplies other forces in combat. See also **close support; direct support; general support; interdepartmental or agency support; international logistic support; inter-Service support; mutual support.** (JP 0-2)

supported commander — 1. The commander having primary responsibility for all aspects of a task assigned by the Joint Strategic Capabilities Plan or other joint operation planning authority. In the context of joint operation planning, this term refers to the commander who prepares operation plans or operation orders in response to requirements of the Chairman of the Joint Chiefs of Staff. 2. In the context of a support command relationship, the commander who receives assistance from another commander's force or capabilities, and who is responsible for ensuring that the supporting commander understands the assistance required. See also **support; supporting commander.** (JP 3-0)

supporting aircraft — All active aircraft other than unit aircraft. See also **aircraft.**

supporting arms — Weapons and weapons systems of all types employed to support forces by indirect or direct fire.

supporting arms coordination center — A single location on board an amphibious command ship in which all communication facilities incident to the coordination of fire support of the artillery, air, and naval gunfire are centralized. This is the naval counterpart to the fire support coordination center utilized by the landing force. Also called **SACC.** See also **fire support coordination center.**

supporting artillery — Artillery that executes fire missions in support of a specific unit, usually infantry, but remains under the command of the next higher artillery commander.

supporting attack — **(*)** An offensive operation carried out in conjunction with a main attack and designed to achieve one or more of the following: a. deceive the enemy; b. destroy or pin down enemy forces which could interfere with the main attack; c. control ground whose occupation by the enemy will hinder the main attack; or d. force the enemy to commit reserves prematurely or in an indecisive area.

supporting commander — 1. A commander who provides augmentation forces or other support to a supported commander or who develops a supporting plan. Includes the designated combatant commands and Defense agencies as appropriate. 2. In the context of a support command relationship, the commander who aids, protects, complements, or sustains another commander's force, and who is responsible for providing the assistance required by the supported commander. See also **support; supported commander.** (JP 3-0)

supporting fire — **(*)** Fire delivered by supporting units to assist or protect a unit in combat. See also **direct supporting fire.**

supporting forces — Forces stationed in or to be deployed to an operational area to provide support for the execution of an operation order. Combatant command (command authority) of supporting forces is not passed to the supported commander.

supporting operations — In amphibious operations, those operations conducted by forces other than those conducted by the amphibious force. See also **amphibious force; amphibious operation.** (JP 3-02)

supporting plan — An operation plan prepared by a supporting commander or a subordinate commander to satisfy the requests or requirements of the supported commander's plan. See also **supported commander; supporting commander.**

support items — Items subordinate to or associated with an end item (i.e., spares, repair parts, tools, test equipment, and sundry materiel) and required to operate, service, repair, or overhaul an end item.

support to counterinsurgency — Support provided to a government in the military, paramilitary, political, economic, psychological, and civic actions it undertakes to defeat insurgency. See also **support to insurgency.** (JP 3-07)

support to insurgency — Support provided to an organized movement aimed at the overthrow of a constituted government through use of subversion and armed conflict. See also **support to counterinsurgency.** (JP 3-07)

suppression — Temporary or transient degradation by an opposing force of the performance of a weapons system below the level needed to fulfill its mission objectives.

suppression mission — A mission to suppress an actual or suspected weapons system for the purpose of degrading its performance below the level needed to fulfill its mission objectives at a specific time for a specified duration.

suppression of enemy air defenses — That activity that neutralizes, destroys, or temporarily degrades surface-based enemy air defenses by destructive and/or disruptive means. Also called **SEAD.** See also **electromagnetic spectrum; electronic warfare.**

suppressive fire — Fires on or about a weapons system to degrade its performance below the level needed to fulfill its mission objectives, during the conduct of the fire mission. See also **fire.**

surface action group — A temporary or standing organization of combatant ships, other than carriers, tailored for a specific tactical mission. Also called **SAG.** See **group; mission.** (JP 3-33)

surface code — See **panel code.**

surface combatant — A ship constructed and armed for combat use with the capability to conduct operations in multiple maritime roles against air, surface and subsurface threats, and land targets.

surface smuggling event — In counterdrug operations, the sighting of a suspected drug smuggling vessel or arrival of a suspected drug smuggling vessel. See also **arrival zone; counterdrug operations; transit zone.** (JP 3-07.4)

surface-to-air guided missile — **(*)** A surface-launched guided missile for use against air targets.

surface-to-air missile envelope — That air space within the kill capabilities of a specific surface-to-air missile system.

surface-to-air missile installation — A surface-to-air missile site with the surface-to-air missile system hardware installed.

surface-to-air missile site — A plot of ground prepared in such a manner that it will readily accept the hardware used in surface-to-air missile system.

surface-to-air weapon — A surface-launched weapon for use against airborne targets. Future developments in air defense systems may lead to the employment of weapons other than missiles. Examples include rockets, directed-energy weapons, and air defense guns. (JP 3-09.3)

surface-to-surface guided missile — **(*)** A surface-launched guided missile for use against surface targets.

surface warfare — That portion of maritime warfare in which operations are conducted to destroy or neutralize enemy naval surface forces and merchant vessels. Also called **SUW.** (JP 3-33)

surface zero — See **ground zero.**

surf line — The point offshore where waves and swells are affected by the underwater surface and become breakers. See also **breaker.** (JP 4-01.6)

surf zone — The area of water from the surf line to the beach. See also **surf line.** (JP 4-01.6)

surplus property — Any excess property not required for the needs and for the discharge of the responsibilities of all federal agencies, including the Department of Defense, as determined by the General Services Administration.

surprise dosage attack — **(*)** A chemical operation which establishes on target a dosage sufficient to produce the desired casualties before the troops can mask or otherwise protect themselves.

surveillance — **(*)** The systematic observation of aerospace, surface, or subsurface areas, places, persons, or things, by visual, aural, electronic, photographic, or other means. See also **air surveillance; satellite and missile surveillance; sea surveillance.**

surveillance approach — An instrument approach conducted in accordance with directions issued by a controller referring to the surveillance radar display.

survey — The directed effort to determine the location and the nature of a chemical, biological, and radiological hazard in an area. (JP 3-11)

survey control point — A survey station used to coordinate survey control.

survey information center — A place where survey data are collected, correlated, and made available to subordinate units.

survey, liaison, and reconnaissance party — A task organization formed from the Marine air-ground task force and Navy support element, which is introduced into the objective area prior to arrival of the fly-in echelon (FIE). The survey, liaison, and reconnaissance party conducts initial reconnaissance, establishes liaison with in theater authorities and initiates preparations for arrival of the main body of the FIE and the maritime pre-positioning ships squadron. Also called **SLRP.**

survey photography — See **air cartographic photography.**

survivability — Concept which includes all aspects of protecting personnel, weapons, and supplies while simultaneously deceiving the enemy. Survivability tactics include building a good defense; employing frequent movement; using concealment, deception, and camouflage; and constructing fighting and protective positions for both individuals and equipment. (JP 3-34)

susceptibility — **(*)** The vulnerability of a target audience to particular forms of psychological operations approach.

suspect — 1. In counterdrug operations, a track of interest where correlating information actually ties the track of interest to alleged illegal drug operations. See also **counterdrug operations; special interest target; track of interest.** 2. An identity applied to a track that is potentially hostile because of its characteristics, behavior, origin, or nationality. See also **assumed friend; hostile; neutral; unknown**.

suspension equipment — **(*)** All aircraft devices such as racks, adapters, missile launchers, and pylons used for carriage, employment, and jettison of aircraft stores.

suspension strop — **(*)** A length of webbing or wire rope between the helicopter and cargo sling.

sustainability — See **military capability.**

sustained attrition minefield — **(*)** In naval mine warfare, a minefield which is replenished to maintain its danger to the enemy in the face of countermeasures.

sustained rate of fire — **(*)** Actual rate of fire that a weapon can continue to deliver for an indefinite length of time without seriously overheating.

sustaining stocks — **(*)** Stocks to support the execution of approved operation plans beyond the initial predetermined period covered by basic stocks until resupply is available for support of continued operations.

sustainment — The provision of personnel, logistic, and other support required to maintain and prolong operations or combat until successful accomplishment or revision of the mission or of the national objective.

swell — Ocean waves that have traveled out of their fetch. Swell characteristically exhibits a more regular and longer period and has flatter crests than waves within their fetch. (JP 4-01.6)

sweep — To employ technical means to uncover planted microphones or other surveillance devices. See also **technical survey.**

sweeper track — See **hunter track.**

sweep jamming — **(*)** A narrow band of jamming that is swept back and forth over a relatively wide operating band of frequencies.

swept path — **(*)** In naval mine warfare, the width of the lane swept by the mechanical sweep at all depths less than the sweep depth.

switch horn — **(*)** In naval mine warfare, a switch in a mine operated by a projecting spike. See also **horn.**

sympathetic detonation — **(*)** Detonation of a charge by exploding another charge adjacent to it.

synchronization — 1. The arrangement of military actions in time, space, and purpose to produce maximum relative combat power at a decisive place and time. 2. In the intelligence context, application of intelligence sources and methods in concert with the operation plan. (JP 2-0)

synchronized clock — A technique of timing the delivery of fires by placing all units on a common time. The synchronized clock uses a specific hour and minute based on either local or universal time. Local time is established using the local time zone. (JP 3-09.3)

synthesis — In intelligence usage, the examining and combining of processed information with other information and intelligence for final interpretation.

synthetic exercise — **(*)** An exercise in which enemy and/or friendly forces are generated, displayed, and moved by electronic or other means on simulators, radar scopes, or other training devices.

systems architecture — Descriptions, including graphics, of systems and interconnections providing for or supporting warfighting functions.

systems design — The preparation of an assembly of methods, procedures, or techniques united by regulated interaction to form an organized whole.

systems support contractors — Contract personnel, normally with high levels of technical expertise, hired to support specific military systems. See also **external support contractors; theater support contractors.** (JP 4-07)

T

table of allowance — An equipment allowance document that prescribes basic allowances of organizational equipment, and provides the control to develop, revise, or change equipment authorization inventory data. Also called **TOA.**

TABOO frequencies — Any friendly frequency of such importance that it must never be deliberately jammed or interfered with by friendly forces. Normally, these frequencies include international distress, CEASE BUZZER, safety, and controller frequencies. These frequencies are generally long standing. However, they may be time-oriented in that, as the combat or exercise situation changes, the restrictions may be removed. See also **CEASE BUZZER; electronic warfare.** (JP 3-51)

TACAN — **(*)** An ultrahigh frequency electronic air navigation system, able to provide continuous bearing and slant range to a selected station. The term is derived from tactical air navigation.

tacit arms control agreement — An arms control course of action in which two or more nations participate without any formal agreement having been made.

tactical aeromedical evacuation — **(*)** That phase of evacuation which provides airlift for patients from the combat zone to points outside the combat zone, and between points within the communications zone.

tactical air commander (ashore) — The officer (aviator) responsible to the landing force commander for control and coordination of air operations within the landing force commander's area of operations when control of these operations is passed ashore.

tactical air control center — The principal air operations installation (ship-based) from which all aircraft and air warning functions of tactical air operations are controlled. Also called **Navy TACC.** (JP 3-09.3)

tactical air coordinator (airborne) — An officer who coordinates, from an aircraft, the actions of other aircraft engaged in air support of ground or sea forces. Also called **TAC(A).** See also **forward observer.**

tactical air direction center — An air operations installation under the overall control of the tactical air control center (afloat) or tactical air command center, from which aircraft and air warning service functions of tactical air operations in an area of concern are directed. Also called **TADC.**

tactical airfield fuel dispensing system — A tactical aircraft refueling system deployed by a Marine air-ground task force in support of air operations at an expeditionary airfield or a forward arming and refueling point. Also called **TAFDS.**

tactical air groups (shore-based) — Task organizations of tactical air units assigned to the amphibious task force that are to be land-based within, or sufficiently close to, the objective area to provide tactical air support to the amphibious task force. (JP 3-02)

tactical air officer (afloat) — The officer (aviator) under the amphibious task force commander who coordinates planning of all phases of air participation of the amphibious operation and air operations of supporting forces en route to and in the objective area. Until control is passed ashore, this officer exercises control over all operations of the tactical air control center (afloat) and is charged with the following: a. control of all aircraft in the objective area assigned for tactical air operations, including offensive and defensive air; b. control of all other aircraft entering or passing through the objective area; and c. control of all air warning facilities in the objective area.

tactical air operation — An air operation involving the employment of air power in coordination with ground or naval forces to: a. gain and maintain air superiority; b. prevent movement of enemy forces into and within the objective area and to seek out and destroy these forces and their supporting installations; c. join with ground or naval forces in operations within the objective area, in order to assist directly in attainment of their immediate objective.

tactical air operations center — The principal air control agency of the US Marine air command and control system responsible for airspace control and management. It provides real-time surveillance, direction, positive control, and navigational assistance for friendly aircraft. It performs real-time direction and control of all antiair warfare operations, to include manned interceptors and surface-to-air weapons. It is subordinate to the tactical air command center. Also called **TAOC.** (JP 3-09.3)

tactical air reconnaissance — The use of air vehicles to obtain information concerning terrain, weather, and the disposition, composition, movement, installations, lines of communications, electronic and communication emissions of enemy forces. Also included are artillery and naval gunfire adjustment, and systematic and random observation of ground battle areas, targets, and/or sectors of airspace.

tactical air support — **(*)** Air operations carried out in coordination with surface forces and which directly assist land or maritime operations. See also **air support.**

tactical air support element — An element of a US Army division, corps, or field army tactical operations center consisting of Army component intelligence staff officer and Army component operations staff officer air personnel who coordinate and integrate tactical air support with current tactical ground operations.

tactical air transport operations — **(*)** The carriage of passengers and cargo within a theater by means of: a. airborne operations: (1) parachute assault, (2) helicopterborne assault, (3) air landing; b. air logistic support; c. special missions; d. aeromedical evacuation missions.

tactical assembly area — An area that is generally out of the reach of light artillery and the location where units make final preparations (pre-combat checks and inspections) and rest, prior to moving to the line of departure. See also **assembly area; line of departure.** (JP 3-35)

tactical call sign — **(*)** A call sign which identifies a tactical command or tactical communication facility. See also **call sign.**

tactical combat force — A combat unit, with appropriate combat support and combat service support assets, that is assigned the mission of defeating Level III threats. Also called **TCF.** (JP 3-10.1)

tactical command, control, communications, and computer system(s) — The facilities, equipment, communications, procedures, and personnel essential to theater-level and below-theater-level commanders for planning, directing, and controlling operations of assigned and attached forces pursuant to the mission assigned and that provide for the conveyance and/or exchange of data and information from one person or force to another. See also **command, control, and computer systems.** (JP 6-0)

tactical concept — (*) A statement, in broad outline, which provides a common basis for future development of tactical doctrine. See also **tactical sub-concept.**

tactical control — Command authority over assigned or attached forces or commands, or military capability or forces made available for tasking, that is limited to the detailed direction and control of movements or maneuvers within the operational area necessary to accomplish missions or tasks assigned. Tactical control is inherent in operational control. Tactical control may be delegated to, and exercised at any level at or below the level of combatant command. When forces are transferred between combatant commands, the command relationship the gaining commander will exercise (and the losing commander will relinquish) over these forces must be specified by the Secretary of Defense. Tactical control provides sufficient authority for controlling and directing the application of force or tactical use of combat support assets within the assigned mission or task. Also called **TACON.** See also **combatant command; combatant command (command authority); operational control.** (JP 0-2)

tactical deception group — A task organization that conducts deception operations against the enemy, including electronic, communication, visual, and other methods designed to misinform and confuse the enemy. (JP 3-02)

tactical digital information link — A Joint Staff-approved, standardized communication link suitable for transmission of digital information. Tactical digital information links interface two or more command and control or weapons systems via a single or multiple network architecture and multiple communication media for exchange of tactical information. Also called **TADIL.**

tactical diversion — See **diversion.**

tactical event system — Current architecture for reporting theater ballistic missile events. The tactical event system is composed of three independent processing and reporting elements: the joint tactical ground stations, attack launch early warning, and tactical detection and reporting. Also called **TES.**

tactical exploitation of national capabilities — Congressionally mandated program to improve the combat effectiveness of the Services through more effective military use of national programs. Also called **TENCAP.**

tactical information processing and interpretation system — A tactical, mobile, land-based, automated information-handling system designed to store and retrieve intelligence information and to process and interpret imagery or nonimagery data. Also called **TIPI.**

tactical intelligence — Intelligence that is required for planning and conducting tactical operations. Also called **TACINTEL.** See also **intelligence.** (JP 2-0)

tactical intelligence and related activities — Those activities outside the National Foreign Intelligence Program that accomplish the following: a. respond to operational commanders' tasking for time-sensitive information on foreign entities; b. respond to national intelligence community tasking of systems whose primary mission is support to operating forces; c. train personnel for intelligence duties; d. provide an intelligence reserve; or e. are devoted to research and development of intelligence or related capabilities. Specifically excluded are programs that are so closely integrated with a weapon system that their primary function is to provide immediate-use targeting data. Also called **TIARA.**

tactical level of war — The level of war at which battles and engagements are planned and executed to accomplish military objectives assigned to tactical units or task forces. Activities at this level focus on the ordered arrangement and maneuver of combat elements in relation to each other and to the enemy to achieve combat objectives. See also **operational level of war; strategic level of war.**

tactical loading — See **combat loading; unit loading.**

tactical locality — **(*)** An area of terrain which, because of its location or features, possesses a tactical significance in the particular circumstances existing at a particular time.

tactical-logistical group — Representatives designated by troop commanders to assist Navy control officers aboard control ships in the ship-to-shore movement of troops, equipment, and supplies. Also called **TACLOG group.**

tactical map — A large-scale map used for tactical and administrative purposes. See also **map.**

tactical minefield — A minefield that is employed to directly attack enemy maneuver as part of a formation obstacle plan and is laid to delay, channel, or break up an enemy advance, giving the defending element a positional advantage over the attacker.

tactical mining — **(*)** In naval mine warfare, mining designed to influence a specific operation or to counter a known or presumed tactical aim of the enemy. Implicit in tactical mining is a limited period of effectiveness of the minefield.

tactical nuclear weapon employment — The use of nuclear weapons by land, sea, or air forces against opposing forces, supporting installations or facilities, in support of operations that contribute to the accomplishment of a military mission of limited scope, or in support of the military commander's scheme of maneuver, usually limited to the area of military operations.

tactical obstacles — Those obstacles employed to disrupt enemy formations, to turn them into a desired area, to fix them in position under direct and indirect fires, and to block enemy penetrations. (JP 3-15)

tactical operations center — A physical groupment of those elements of a general and special staff concerned with the current tactical operations and the tactical support thereof. Also called **TOC.** See also **command post.**

tactical range — **(*)** A range in which realistic targets are in use and a certain freedom of maneuver is allowed.

tactical reserve — A part of a force held under the control of the commander as a maneuvering force to influence future action.

tactical security — **(*)** In operations, the measures necessary to deny information to the enemy and to ensure that a force retains its freedom of action and is warned or protected against an unexpected encounter with the enemy or an attack. See also **physical security; security.** (JP 3-07.2)

tactical sub-concept — **(*)** A statement, in broad outline, for a specific field of military capability within a tactical concept which provides a common basis both for equipment and weapon system development and for future development of tactical doctrine. See also **tactical concept.**

tactical transport aircraft — (*) Aircraft designed primarily for the carriage of personnel and/or cargo over short or medium distances.

tactical troops — Combat troops, together with any service troops required for their direct support, who are organized under one commander to operate as a unit and engage the enemy in combat. See also **troops.**

tactical unit — An organization of troops, aircraft, or ships that is intended to serve as a single unit in combat. It may include service units required for its direct support.

tactical vehicle — See **military designed vehicle.**

tactical warning — 1. A warning after initiation of a threatening or hostile act based on an evaluation of information from all available sources. 2. In satellite and missile surveillance, a notification to operational command centers that a specific threat event is occurring. The component elements that describe threat events are as follows: a. **country of origin** — Country or countries initiating hostilities; b. **event type and size** — Identification of the type of event and determination of the size or number of weapons; c. **country under attack** — Determined by observing trajectory of an object and predicting its impact point; and d. **event time** — Time the hostile event occurred. Also called **integrated tactical warning.** See also **attack assessment; strategic warning.**

tactical warning and assessment — A composite term. See separate definitions for tactical warning and for attack assessment.

tactical warning and attack assessment — A composite term. See separate definitions for tactical warning and for attack assessment. Also called **TW/AA.**

tactics — 1. The employment of units in combat. 2. The ordered arrangement and maneuver of units in relation to each other and/or to the enemy in order to use their full potentialities.

tagline — A line attached to a draft of cargo or container to provide control and minimize pendulation of cargo during lifting operations. See also **container; draft.** (JP 4-01.6)

tanker airlift control center — The Air Mobility Command direct reporting unit responsible for tasking and controlling operational missions for all activities involving forces supporting US Transportation Command's global air mobility mission. The tanker airlift control center is comprised of the following functions: current operations, command and control, logistic operations, aerial port operations, aeromedical evacuation, flight planning, diplomatic clearances, weather, and intelligence. Also called **TACC.** See also **tanker airlift control element.** (JP 3-17)

tanker airlift control element — A mobile command and control organization deployed to support strategic and theater air mobility operations at fixed, en route, and deployed locations where air mobility operational support is nonexistent or insufficient. The tanker airlift control element (TALCE) provides on-site management of air mobility airfield operations to include command and control, communications, aerial port services, maintenance, security, transportation, weather, intelligence, and other support functions, as necessary. The TALCE is composed of mission support elements from various units and deploys in support of peacetime, contingency, and emergency relief operations on both planned and "no notice" basis. Also called **TALCE.** See also **tanker airlift control center.** (JP 3-17)

tare weight — The weight of a container deducted from gross weight to obtain net weight or the weight of an empty container. (JP 4-01.7)

target — 1. An area, complex, installation, force, equipment, capability, function, or behavior identified for possible action to support the commander's objectives, guidance, and intent. Targets fall into two general categories: planned and immediate. 2. In intelligence usage, a country, area, installation, agency, or person against which intelligence operations are directed. 3. An area designated and numbered for future firing. 4. In gunfire support usage, an impact burst that hits the target. Also called **TGT.** See also **objective area.** (JP 3-60)

target acquisition — **(*)** The detection, identification, and location of a target in sufficient detail to permit the effective employment of weapons. Also called **TA.** See also **target analysis.**

target analysis — **(*)** An examination of potential targets to determine military importance, priority of attack, and weapons required to obtain a desired level of damage or casualties. See also **target acquisition.**

target approach point — **(*)** In air transport operations, a navigational check point over which the final turn into the drop zone/landing zone is made. See also **initial point.**

target area of interest — The geographical area where high-value targets can be acquired and engaged by friendly forces. Not all target areas of interest will form part of the friendly course of action; only target areas of interest associated with high priority targets are of interest to the staff. These are identified during staff planning and wargaming. Target areas of interest differ from engagement areas in degree. Engagement areas plan for the use of all available weapons; target areas of interest might be engaged by a single weapon. Also called **TAI.** See also **area of interest; high-value target; target.** (JP 2-01.3)

target area survey base — **(*)** A base line used for the locating of targets or other points by the intersection of observations from two stations located at opposite ends on the line.

target array — A graphic representation of enemy forces, personnel, and facilities in a specific situation, accompanied by a target analysis.

target audience — **(*)** An individual or group selected for influence or attack by means of psychological operations.

target base line — A line connecting prime targets along the periphery of a geographic area.

target bearing — 1. **true** — The true compass bearing of a target from a firing ship. 2. **relative** — The bearing of a target measured in the horizontal from the bow of one's own ship clockwise from 0 degrees to 360 degrees, or from the nose of one's own aircraft in hours of the clock.

target classification — A grouping of targets in accordance with their threat to the amphibious task force and its component elements: targets not to be fired upon prior to D-day and targets not to be destroyed except on direct orders.

target complex — **(*)** A geographically integrated series of target concentrations. See also **target.**

target component — A set of targets within a target system performing a similar function. See also **target; target critical damage point.** (JP 3-60)

target concentration — **(*)** A grouping of geographically proximate targets. See also **target; target complex.**

target critical damage point — The part of a target component that is most vital. Also called **critical node.** See also **target; target component.** (JP 3-05.5)

target data inventory — A basic targeting program that provides a standardized target data in support of the requirements of the Joint Chiefs of Staff, Military Departments, and unified and specified commands for target planning coordination and weapons application. Also called **TDI.**

target date — **(*)** The date on which it is desired that an action be accomplished or initiated.

target discrimination — **(*)** The ability of a surveillance or guidance system to identify or engage any one target when multiple targets are present.

target dossier — **(*)** A file of assembled target intelligence about a specific geographic area.

target folder — **(*)** A folder containing target intelligence and related materials prepared for planning and executing action against a specific target.

target information center — The agency or activity responsible for collecting, displaying, evaluating, and disseminating information pertaining to potential targets. See also **target.** (JP 3-02)

targeting — The process of selecting and prioritizing targets and matching the appropriate response to them, taking account of operational requirements and capabilities. See also **joint targeting coordination board; target.** (JP 3-60)

targeting effects — The cumulative results of actions taken to attack targets and target systems by lethal and nonlethal means. See also **target; targeting.** (JP 3-60)

target intelligence — **(*)** Intelligence that portrays and locates the components of a target or target complex and indicates its vulnerability and relative importance. See also **target; target complex.**

target list — The listing of targets maintained and promulgated by the senior echelon of command; it contains those targets that are to be engaged by supporting arms, as distinguished from a "list of targets" that may be maintained by any echelon as confirmed, suspected, or possible targets for informational and planning purposes. See also **joint target list; list of targets.**

target materials — Graphic, textual, tabular, digital, video, or other presentations of target intelligence, primarily designed to support operations against designated targets by one or more weapon(s) systems. Target materials are suitable for training, planning, executing, and evaluating military operations. See also **Air Target Materials Program.** (JP 2-0)

target nomination list — A list of targets nominated by component commanders, national agencies, or the joint force commander staff for potential inclusion on the joint integrated prioritized target list to support joint force commander objectives and priorities. Also called **TNL.** See also **joint integrated prioritized target list; target.** (JP 3-60)

target of opportunity — A target visible to a surface or air sensor or observer, which is within range of available weapons and against which fire has not been scheduled or requested. See also **target.** (JP 3-60)

target overlay — (*) A transparent sheet which, when superimposed on a particular chart, map, drawing, tracing or other representation, depicts target locations and designations. The target overlay may also show boundaries between maneuver elements, objectives and friendly forward dispositions.

target pattern — The flight path of aircraft during the attack phase. Also called **attack pattern.**

target priority — A grouping of targets with the indicated sequence of attack.

target range — See **range.**

target response (nuclear) — The effect on men, material, and equipment of blast, heat, light, and nuclear radiation resulting from the explosion of a nuclear weapon.

target signature — (*) 1. The characteristic pattern of a target displayed by detection and identification equipment. 2. In naval mine warfare, the variation in the influence field produced by the passage of a ship or sweep.

target stress point — The weakest point (most vulnerable to damage) on the critical damage point. Also called **vulnerable node.** See also **target critical damage point.** (JP 3-05.5)

target system — (*) 1. All the targets situated in a particular geographic area and functionally related. 2. **(DOD only)** A group of targets that are so related that their destruction will produce some particular effect desired by the attacker. See also **target; target complex.**

target system component — A set of targets belonging to one or more groups of industries and basic utilities required to produce component parts of an end product such as periscopes, or one type of a series of interrelated commodities, such as aviation gasoline.

task component — A subdivision of a fleet, task force, task group, or task unit, organized by the respective commander or by higher authority for the accomplishment of specific tasks.

task element — A component of a naval task unit organized by the commander of a task unit or higher authority.

task force — (*) 1. A temporary grouping of units, under one commander, formed for the purpose of carrying out a specific operation or mission. 2. A semi-permanent organization of units, under one commander, formed for the purpose of carrying out a continuing specific task. 3. A component of a fleet organized by the commander of a task fleet or higher authority for the accomplishment of a specific task or tasks. Also called **TF.** See also **force.**

task group — A component of a naval task force organized by the commander of a task force or higher authority. Also called **TG.**

tasking order — A method used to task and to disseminate to components, subordinate units, and command and control agencies projected targets and specific missions. In addition, the tasking order provides specific instructions concerning the mission planning agent, targets, and other control agencies, as well as general instructions for accomplishment of the mission. Also called **TASKORD.** See also **mission; order; target.** (JP 3-05.1)

task organization — 1. In the Navy, an organization which assigns to responsible commanders the means with which to accomplish their assigned tasks in any planned action. 2. An organization table pertaining to a specific naval directive.

task-organizing — The act of designing an operating force, support staff, or logistic package of specific size and composition to meet a unique task or mission. Characteristics to examine when task-organizing the force include, but are not limited to: training, experience, equipage, sustainability, operating environment, enemy threat, and mobility. (JP 3-05)

task unit — A component of a naval task group organized by the commander of a task group or higher authority.

taxiway — **(*)** A specially prepared or designated path on an airfield for the use of taxiing aircraft.

T-day — See **times.**

tear line — A physical line on an intelligence message or document separating categories of information that have been approved for foreign disclosure and release. Normally, the intelligence below the tear line is that which has been previously cleared for disclosure or release. (JP 2-0)

technical analysis — **(*)** In imagery interpretation, the precise description of details appearing on imagery.

technical architecture — A minimal set of rules governing the arrangement, interaction, and interdependence of the parts or elements whose purpose is to ensure that a conformant system satisfies a specified set of requirements.

technical assistance — The providing of advice, assistance, and training pertaining to the installation, operation, and maintenance of equipment.

technical characteristics — Those characteristics of equipment that pertain primarily to the engineering principles involved in producing equipment possessing desired military characteristics; e.g., for electronic equipment, technical characteristics include such items as circuitry as well as types and arrangement of components.

technical documentation — Visual information documentation (with or without sound as an integral documentation component) of an actual event made for purposes of evaluation. Typically, technical documentation contributes to the study of human or mechanical factors, procedures, and processes in the fields of medicine, science, logistics, research, development, test and evaluation, intelligence, investigations, and armament delivery. Also called **TECDOC.** See also **visual information documentation.**

technical escort — An individual technically qualified and properly equipped to accompany designated material requiring a high degree of safety or security during shipment.

technical evaluation — The study and investigations by a developing agency to determine the technical suitability of material, equipment, or a system for use in the Military Services. See also **operational evaluation.**

technical information — Information, including scientific information, that relates to research, development, engineering, test, evaluation, production, operation, use, and maintenance of munitions and other military supplies and equipment.

technical intelligence — Intelligence derived from exploitation of foreign material, produced for strategic, operational, and tactical level commanders. Technical intelligence begins when an individual service member finds something new on the battlefield and takes the proper steps to report it. The item is then exploited at succeedingly higher levels until a countermeasure is produced to neutralize the adversary's technological advantage. Also called **TECHINT.** See also **exploitation; intelligence.** (JP 2-01.3)

technical operational intelligence — A Defense Intelligence Agency initiative to provide enhanced scientific and technical intelligence to the commanders of unified commands and their subordinates through a closed loop system involving all Service and Defense Intelligence Agency scientific and technical intelligence centers. Through a system manager in the National Military Joint Intelligence Center, the technical operational intelligence program provides timely collection, analysis, and dissemination of area of responsibility-specific scientific and technical intelligence to combatant commanders and their subordinates for planning, training, and executing joint operations. Also called **TOPINT.** (JP 2-0)

technical review authority — The organization tasked to provide specialized technical or administrative expertise to the primary review authority or coordinating review authority for joint publications. Also called **TRA.** See also **coordinating review authority; joint publication; primary review authority.** (JP 1-01)

technical specification — A detailed description of technical requirements, usually with specific acceptance criteria, stated in terms suitable to form the basis for the actual design development and production processes of an item having the qualities specified in the operational characteristics. See also **operational characteristics.**

technical supply operations — Operations performed by supply units or technical supply elements of supply and maintenance units in acquiring, accounting for, storing, and issuing Class II and IV items needed by supported units and maintenance activities.

technical surveillance countermeasures — Techniques and measures to detect and neutralize a wide variety of hostile penetration technologies that are used to obtain unauthorized access to classified and sensitive information. Technical penetrations include the employment of optical, electro-optical, electromagnetic, fluidic, and acoustic means as the sensor and transmission medium, or the use of various types of stimulation or modification to equipment or building components for the direct or indirect transmission of information meant to be protected. Also called **TSCM.** See also **counterintelligence.** (JP 2-01.2)

technical survey — A complete electronic and physical inspection to ascertain that offices, conference rooms, war rooms, and other similar locations where classified information is discussed are free of monitoring systems. See also **sweep.**

telecommunication — **(*)** Any transmission, emission, or reception of signs, signals, writings, images, sounds, or information of any nature by wire, radio, visual, or other electromagnetic systems. (JP 6-0)

telecommunications center — A facility, normally serving more than one organization or terminal, responsible for transmission, receipt, acceptance, processing, and distribution of incoming and outgoing messages.

teleconference — (*) A conference between persons remote from one another but linked by a telecommunications system.

teleprocessing — The combining of telecommunications and computer operations interacting in the automatic processing, reception, and transmission of data and/or information.

television imagery — Imagery acquired by a television camera and recorded or transmitted electronically.

telling — See **track telling.**

temperature gradient — At sea, a temperature gradient is the change of temperature with depth; a positive gradient is a temperature increase with an increase in depth, and a negative gradient is a temperature decrease with an increase in depth.

tempest — An unclassified term referring to technical investigations for compromising emanations from electrically operated information processing equipment; these investigations are conducted in support of emanations and emissions security. See also **counterintelligence.** (JP 2-01.2)

temporary interment — A site for the purpose of: a. the interment of the remains if the circumstances permit; or b. the reburial of remains exhumed from an emergency interment. See also **emergency interment; group interment; mortuary affairs; trench interment.** (JP 4-06)

terminal — A facility designed to transfer cargo from one means of conveyance to another. (Conveyance is the piece of equipment used to transport cargo; i.e., railcar to truck or truck to truck. This is as opposed to mode, which is the type of equipment; i.e., ship to rail, rail to truck.) See also **facility.** (JP 4-01.6)

terminal clearance capacity — The amount of cargo or personnel that can be moved through and out of a terminal on a daily basis.

terminal control — 1. The authority to direct the maneuver of aircraft which are delivering ordnance, passengers, or cargo to a specific location or target. Terminal control is a type of air control. 2. Any electronic, mechanical, or visual control given to aircraft to facilitate target acquisition and resolution. (JP 3-09.3)

terminal control area — A control area or portion thereof normally situated at the confluence of air traffic service routes in the vicinity of one or more major airfields. See also **airway; control area; controlled airspace; control zone.**

terminal guidance — 1. The guidance applied to a guided missile between midcourse guidance and arrival in the vicinity of the target. 2. Electronic, mechanical, visual, or other assistance given an aircraft pilot to facilitate arrival at, operation within or over, landing upon, or departure from an air landing or airdrop facility.

terminal operations — The reception, processing, and staging of passengers; the receipt, transit, storage, and marshalling of cargo; the loading and unloading of ships or aircraft; and the manifesting and forwarding of cargo and passengers to destination.

terminal phase — That portion of the trajectory of a ballistic missile between reentry into the atmosphere or the end of the mid-course phase and impact or arrival in the vicinity of the target. See also **boost phase; midcourse phase; reentry phase.**

terminal velocity — **(*)** 1. Hypothetical maximum speed a body could attain along a specified flight path under given conditions of weight and thrust if diving through an unlimited distance in air of specified uniform density. 2. Remaining speed of a projectile at the point in its downward path where it is level with the muzzle of the weapon.

terms of reference — Terms of reference allude to a mutual agreement under which a command, element, or unit exercises authority or undertakes specific missions or tasks relative to another command, element, or unit. Also called **TORs.** (JP 3-0)

terrain analysis — **(*)** The collection, analysis, evaluation, and interpretation of geographic information on the natural and manmade features of the terrain, combined with other relevant factors, to predict the effect of the terrain on military operations.

terrain avoidance system — **(*)** A system which provides the pilot or navigator of an aircraft with a situation display of the ground or obstacles which project above either a horizontal plane through the aircraft or a plane parallel to it, so that the pilot can maneuver the aircraft to avoid the obstruction.

terrain clearance system — **(*)** A system which provides the pilot, or autopilot, of an aircraft with climb or dive signals such that the aircraft will maintain a selected height over flat ground and clear the peaks of undulating ground within the selected height in a vertical plane through the flight vector. This system differs from terrain following in that the aircraft need not descend into a valley to follow the ground contour.

terrain exercise — An exercise in which a stated military situation is solved on the ground, the troops being imaginary and the solution usually being in writing.

terrain flight — **(*)** Flight close to the Earth's surface during which airspeed, height, and/or altitude are adapted to the contours and cover of the ground in order to avoid enemy detection and fire. Also called **TERF.**

terrain following system — **(*)** A system which provides the pilot or autopilot of an aircraft with climb or dive signals such that the aircraft will maintain as closely as possible a selected height above a ground contour in a vertical plane through the flight vector.

terrain intelligence — Intelligence on the military significance of natural and manmade characteristics of an area.

terrain study — An analysis and interpretation of natural and manmade features of an area, their effects on military operations, and the effect of weather and climate on these features.

terrestrial environment — The Earth's land area, including its manmade and natural surface and sub-surface features, and its interfaces and interactions with the atmosphere and the oceans.

terrestrial reference guidance — The technique of providing intelligence to a missile from certain characteristics of the surface over which the missile is flown, thereby achieving flight along a predetermined path.

territorial airspace — Airspace above land territory, internal waters, archipelagic waters, and territorial seas.

territorial sea — A belt of ocean space adjacent to and measured from the coastal state's baseline to a maximum width of 12 nm. Throughout the vertical and horizontal planes of the territorial sea, the coastal state exercises sovereign jurisdiction, subject to the right of innocent passage of vessels on the surface and the right of transit passage in, under, and over international straits. Territorial sea areas that are a continuation of sea lanes through archipelagoes are subject to archipelagic sealane passage, with the same transit rights as those that apply to international straits.

terrorism — The calculated use of unlawful violence or threat of unlawful violence to inculcate fear; intended to coerce or to intimidate governments or societies in the pursuit of goals that are generally political, religious, or ideological. See also **antiterrorism; combatting terrorism; counterterrorism; terrorist; terrorist groups; terrorist threat conditions.** (JP 3-07.2)

terrorist — An individual who uses violence, terror, and intimidation to achieve a result. See also **terrorism.** (JP 3-07.2)

terrorist groups — Any element, regardless of size or espoused cause, that commits acts of violence or threatens violence in pursuit of its political, religious, or ideological objectives. See also **terrorism.** (JP 3-07.2)

terrorist threat condition — A Chairman of the Joint Chiefs of Staff-approved program standardizing the Military Services' identification of and recommended responses to terrorist threats against US personnel and facilities. This program facilitates inter-Service coordination and support for antiterrorism activities. Also called **THREATCON.** There are four THREATCONs above normal. a. **THREATCON ALPHA** — This condition applies when there is a general threat of possible terrorist activity against personnel and facilities, the nature and extent of which are unpredictable, and circumstances do not justify full implementation of THREATCON BRAVO measures. However, it may be necessary to implement certain measures from higher THREATCONs resulting from intelligence received or as a deterrent. The measures in this THREATCON must be capable of being maintained indefinitely. b. **THREATCON BRAVO** — This condition applies when an increased and more predictable threat of terrorist activity exists. The measures in this THREATCON must be capable of being maintained for weeks without causing undue hardship, affecting operational capability, and aggravating relations with local authorities. c. **THREATCON CHARLIE** — This condition applies when an incident occurs or intelligence is received indicating some form of terrorist action against personnel and facilities is imminent. Implementation of measures in this THREATCON for more than a short period probably will create hardship and affect the peacetime activities of the unit and its personnel. d. **THREATCON DELTA** — This condition applies in the immediate area where a terrorist attack has occurred or when intelligence has been received that terrorist action against a specific location or person is likely. Normally, this THREATCON is declared as a localized condition. See also **antiterrorism; force protection.** (JP 3-07.2)

terrorist threat level — An intelligence threat assessment of the level of terrorist threat faced by US personnel and interests in a foreign country. The assessment is based on a continuous intelligence analysis of a minimum of five elements: terrorist group existence, capability, history, trends, and targeting. There are five threat levels: NEGLIGIBLE, LOW, MEDIUM, HIGH, and CRITICAL. Threat levels should not be confused with terrorist threat conditions (THREATCON). Threat level assessments are provided to senior leaders to assist them in determining the appropriate local THREATCON. (Department of State also makes threat assessments, which may differ from those determined by Department of Defense.)

test depth — **(*)** The depth to which the submarine is tested by actual or simulated submergence. See also **maximum operating depth.**

tests — See **service test; troop test.**

theater — The geographical area outside the continental United States for which a commander of a combatant command has been assigned responsibility.

theater airlift — That airlift assigned or attached to a combatant commander other than Commander in Chief, US Transportation Command, that provides air movement and delivery of personnel and equipment directly into objective areas through air landing, airdrop, extraction, or other delivery techniques; and the air logistic support of all theater forces, including those engaged in combat operations, to meet specific theater objectives and requirements. Also called **intratheater airlift.** See also **strategic airlift.** (JP 4-01.1)

theater airlift liaison officer — An officer specially trained to implement the theater air control system and to control tactical airlift assets. Theater airlift liaison officers are highly qualified, rated airlift officers with tactical (airdrop) airlift experience and assigned duties, supporting US Army units. Also called **TALO.** (JP 3-17)

theater-assigned transportation assets — Transportation assets that are assigned under the combatant command (command authority) of a geographic combatant commander. See also **combatant command (command authority); single manager for transportation.** (JP 4-01)

theater distribution — The flow of personnel, equipment, and materiel within theater to meet the geographic combatant commander's missions. See also **distribution; theater; theater distribution system.** (JP 4-01.4)

theater distribution management — The function of optimizing the distribution networks to achieve the effective and efficient flow of personnel, equipment, and materiel to meet the combatant commander's requirements. See also **distribution; theater; theater distribution.** (JP 4-01.4)

theater distribution system — A distribution system comprised of four independent and mutually supported networks within theater to meet the geographic combatant commander's requirements; the physical network; the financial network; the information network; and the communications network. See also **distribution; distribution plan; distribution system; theater; theater distribution.** (JP 4-01.4)

theater missile — A missile, which may be a ballistic missile, a cruise missile, or an air-to-surface missile (not including short-range, non-nuclear, direct fire missiles, bombs, or rockets such as Maverick or wire-guided missiles), whose target is within a given theater of operation. Also called **TM.** See also **joint theater missile defense.** (JP 3-01)

theater of operations — A subarea within a theater of war defined by the geographic combatant commander required to conduct or support specific combat operations. Different theaters of operations within the same theater of war will normally be geographically separate and focused on different enemy forces. Theaters of operations are usually of significant size, allowing for operations over extended periods of time. Also called **TO.** See also **theater of war.** (JP 5-0)

theater of war — Defined by the National Command Authorities or the geographic combatant commander, the area of air, land, and water that is, or may become, directly involved in the conduct of the war. A theater of war does not normally encompass the geographic combatant commander's entire area of responsibility and may contain more than one theater of operations. See also **area of responsibility; theater of operations.** (JP 5-0)

theater strategic environment — A composite of the conditions, circumstances, and influences in the theater that describes the diplomatic-military situation, affect the employment of military forces, and affect the decisions of the operational chain of command. See also **theater.** (JP 5-00.1)

theater strategy — The art and science of developing integrated strategic concepts and courses of action directed toward securing the objectives of national and alliance or coalition security policy and strategy by the use of force, threatened use of force, or operations not involving the use of force within a theater. See also **military strategy; national military strategy; national security strategy; strategy.** (JP 3-0)

theater support contractors — Contract personnel hired in, and operating in, a specific operational area. See also **external support contractors; systems support contractors.** (JP 4-07)

thermal crossover — The natural phenomenon that normally occurs twice daily when temperature conditions are such that there is a loss of contrast between two adjacent objects on infrared imagery.

thermal energy — The energy emitted from the fireball as thermal radiation. The total amount of thermal energy received per unit area at a specified distance from a nuclear explosion is generally expressed in terms of calories per square centimeter.

thermal exposure — The total normal component of thermal radiation striking a given surface throughout the course of a detonation; expressed in calories per square centimeter or megajoules per square meter.

thermal imagery — **(*)** Imagery produced by sensing and recording the thermal energy emitted or reflected from the objects which are imaged.

thermal pulse — The radiant power versus time pulse from a nuclear weapon detonation.

thermal radiation — **(*)** 1. The heat and light produced by a nuclear explosion. 2. **(DOD only)** Electromagnetic radiations emitted from a heat or light source as a consequence of its temperature; it consists essentially of ultraviolet, visible, and infrared radiations.

thermal shadow — **(*)** The tone contrast difference of infrared linescan imagery which is caused by a thermal gradient which persists as a result of a shadow of an object which has been moved.

thermal X-rays — **(*)** The electromagnetic radiation, mainly in the soft (low-energy) X-ray region, emitted by the debris of a nuclear weapon by virtue of its extremely high temperature.

thermonuclear — An adjective referring to the process (or processes) in which very high temperatures are used to bring about the fusion of light nuclei with the accompanying release of energy.

thermonuclear weapon — **(*)** A weapon in which very high temperatures are used to bring about the fusion of light nuclei such as those of hydrogen isotopes (e.g., deuterium and tritium) with the accompanying release of energy. The high temperatures required are obtained by means of fission.

thorough decontamination — Decontamination carried out by a unit, with or without external support, to reduce contamination on personnel, equipment, materiel, and/or working areas equal to natural background or to the lowest possible levels, to permit the partial or total removal of individual protective equipment and to maintain operations with minimum degradation. This may include terrain decontamination beyond the scope of operational decontamination. See also **immediate decontamination; operational decontamination.**

threat analysis — In antiterrorism, a continual process of compiling and examining all available information concerning potential terrorist activities by terrorist groups which could target a facility. A threat analysis will review the factors of a terrorist group's existence, capability, intentions, history, and targeting, as well as the security environment within which friendly forces operate. Threat analysis is an essential step in identifying probability of terrorist attack and results in a threat assessment. See also **antiterrorism.** (JP 3-07.2)

threat and vulnerability assessment — In antiterrorism, the pairing of a facility's threat analysis and vulnerability analysis. See also **antiterrorism.** (JP 3-07.2)

THREATCON ALPHA — See **terrorist threat condition.**

THREATCON BRAVO — See **terrorist threat condition.**

THREATCON CHARLIE — See **terrorist threat condition.**

THREATCON DELTA — See **terrorist threat condition.**

threat identification and assessment — The Joint Operation Planning and Execution System function that provides: timely warning of potential threats to US interests; intelligence collection requirements; the effects of environmental, physical, and health hazards, and cultural factors on friendly and enemy operations; and determines the enemy military posture and possible intentions.

threat-oriented munitions — **(*)** In stockpile planning, munitions intended to neutralize a finite assessed threat and for which the total requirement is determined by an agreed mathematical model. See also **level-of-effort munitions.**

threshold — **(*)** The beginning of that portion of the runway usable for landing.

throughput — The average quantity of cargo and passengers that can pass through a port on a daily basis from arrival at the port to loading onto a ship or plane, or from the discharge from a ship or plane to the exit (clearance) from the port complex. Throughput is usually expressed in measurement tons, short tons, or passengers. Reception and storage limitation may affect final throughput.

tie down diagram — **(*)** A drawing indicating the prescribed method of securing a particular item of cargo within a specific type of vehicle.

tie down point — **(*)** An attachment point provided on or within a vehicle for securing cargo.

tie down point pattern — **(*)** The pattern of tie down points within a vehicle.

tilt angle — **(*)** The angle between the optical axis of an air camera and the vertical at the time of exposure.

time and frequency standard — A reference value of time and time interval. Standards of time and frequency are determined by astronomical observations and by the operation of atomic clocks and other advanced timekeeping instruments. They are disseminated by transport of clocks, radio transmissions, satellite relay, and other means.

time-definite delivery — The delivery of requested logistics support at a time and destination specified by the receiving activity. See also **logistic support.** (JP 4-0)

time fuze — **(*)** A fuze which contains a graduated time element to regulate the time interval after which the fuze will function.

time interval — Duration of a segment of time without reference to when the time interval begins or ends. Time intervals may be given in seconds of time or fractions thereof.

time of attack — The hour at which the attack is to be launched. If a line of departure is prescribed, it is the hour at which the line is to be crossed by the leading elements of the attack.

time of delivery — The time at which the addressee or responsible relay agency receipts for a message.

time of flight — In artillery, mortar, and naval gunfire support, the time in seconds from the instant a weapon is fired, launched, or released from the delivery vehicle or weapons system to the instant it strikes or detonates.

time of origin — The time at which a message is released for transmission.

time of receipt — The time at which a receiving station completes reception of a message.

time on target — 1. Time at which aircraft are scheduled to attack/photograph the target. 2. The actual time at which aircraft attack/photograph the target. 3. The time at which a nuclear detonation is planned at a specified desired ground zero.

time over target conflict — A situation wherein two or more delivery vehicles are scheduled such that their proximity violates the established separation criteria for yield, time, distance, or all three.

time over target (nuclear) — See **time on target — Part 3.**

time-phased force and deployment data — The Joint Operation Planning and Execution System database portion of an operation plan; it contains time-phased force data, non-unit-related cargo and personnel data, and movement data for the operation plan, including the following: a. In-place units; b. Units to be deployed to support the operation plan with a priority indicating the desired sequence for their arrival at the port of debarkation; c. Routing of forces to be deployed; d. Movement data associated with deploying forces; e. Estimates of non-unit-related cargo and personnel movements to be conducted concurrently with the deployment of forces; and f. Estimate of transportation requirements that must be fulfilled by common-user lift resources as well as those requirements that can be fulfilled by assigned or attached transportation resources. Also called **TPFDD.** See also **time-phased force and deployment data maintenance; time-phased force and deployment data refinement; time-phased force and deployment list.** (JP 5-0)

time-phased force and deployment data maintenance — The deliberate planning process that requires a supported commander to incorporate changes to time-phased force and deployment data (TPFDD) that occur after the TPFDD becomes effective for execution. TPFDD maintenance is conducted by the supported combatant commander in coordination with the supporting combatant commanders, Service components, US Transportation Command, and other agencies as required. At designated intervals, changes to data in the TPFDD, including force structure, standard reference files, and Services' type unit characteristics files, are updated in Joint Operation Planning and Execution System (JOPES) to ensure currency of deployment data. TPFDD maintenance may also be used to update the TPFDD for Chairman of the Joint Chiefs of Staff or Joint Strategic Capabilities Plan submission in lieu of refinement during the JOPES plan development phase. Also called **TPFDD maintenance.** See also **time-phased force and deployment data; time-phased force and deployment data refinement; time-phased force and deployment list.**

time-phased force and deployment data refinement — For both global and regional operation plan development, the process consists of several discrete phases time-phased force and deployment data (TPFDD) that may be conducted sequentially or concurrently, in whole or in part. These phases are concept, plan development, and review. The plan development phase consists of several subphases: forces, logistics, and transportation, with shortfall identification associated with each phase. The plan development phases are collectively referred to as TPFDD refinement. The normal TPFDD refinement process consists of sequentially refining force, logistic (non-unit-related personnel and sustainment), and transportation data to develop a TPFDD file that supports a feasible and adequate overlapping of several refinement phases. The decision is made by the supported commander, unless otherwise directed by the Chairman of the Joint Chiefs of Staff. For global planning, refinement conferences are conducted by the Joint Staff in conjunction with US Transportation Command. TPFDD refinement is conducted in coordination with supported and supporting commanders, Services, the Joint Staff, and other supporting agencies. Commander in Chief, US Transportation Command, will normally host refinement conferences at the request of the Joint Staff or the supported commander. Also called **TPFDD refinement.** See also **time-phased force and deployment data; time-phased force and deployment data maintenance; time-phased force and deployment list.**

time-phased force and deployment list — A Joint Operation Planning and Execution System database located at Appendix 1 to Annex A of deliberate plans. It identifies types and/or actual units required to support the operation plan and indicates origin and ports of debarkation or ocean area. This listing is to include both a. In-place units; and b. Units to be deployed to support the deliberate plan. Also called **TPFDL.** See also **Joint Operation Planning and Execution System; time-phased force and deployment data; time-phased force and deployment data maintenance; time-phased force and deployment data refinement.** (JP 3-35)

times — (C-, D-, M-days end at 2400 hours Universal Time (Zulu time) and are assumed to be 24 hours long for planning.) The Chairman of the Joint Chiefs of Staff normally coordinates the proposed date with the commanders of the appropriate unified and specified commands, as well as any recommended changes to C-day. L-hour will be established per plan, crisis, or theater of operations and will apply to both air and surface movements. Normally, L-hour will be established to allow C-day to be a 24-hour day. a. **C-day.** The unnamed day on which a deployment operation commences or is to commence. The deployment may be movement of troops, cargo, weapon systems, or a combination of these elements using any or all types of transport. The letter "C" will be the only one used to denote the above. The highest command or headquarters responsible for coordinating the planning will specify the exact meaning of C-day within the aforementioned definition. The command or headquarters directly responsible for the execution of the operation, if other than the one coordinating the planning, will do so in light of the meaning specified by the highest command or headquarters coordinating the planning. b. **D-day.** The unnamed day on which a particular operation commences or is to commence. c. **F-hour.** The effective time of announcement by the Secretary of Defense to the Military Departments of a decision to mobilize Reserve units. d. **H-hour.** The specific hour on D-day at which a particular operation commences. e. **H-hour (amphibious operations).** For amphibious operations, the time the first assault elements are scheduled to touch down on the beach, or a landing zone, and in some cases the commencement of countermine breaching operations. f. **L-hour.** The specific hour on C-day at which a deployment operation commences or is to commence. g. **L-hour (amphibious operations).** In amphibious operations, the time at which the first helicopter of the helicopter-borne assault wave touches down in the landing zone. h. **M-day.** The term used to designate the unnamed day on which full mobilization commences or is due to commence. i. **N-day.** The unnamed day an active duty unit is notified for deployment or redeployment. j. **R-day.** Redeployment day. The day on which redeployment of major combat, combat support, and combat service support forces begins in an operation. k. S-day. The day the President authorizes Selective Reserve callup (not more than 200,000). l. **T-day.** The effective day coincident with Presidential declaration of national emergency and authorization of partial mobilization (not more than 1,000,000 personnel exclusive of the 200,000 callup). m. **W-day.** Declared by the National Command Authorities, W-day is associated with an adversary decision to prepare for war (unambiguous strategic warning). (JP 3-02)

time-sensitive targets — Those targets requiring immediate response because they pose (or will soon pose) a danger to friendly forces or are highly lucrative, fleeting targets of opportunity. Also called **TSTs.** (JP 3-60)

time slot — **(*)** Period of time during which certain activities are governed by specific regulations.

time to target — The number of minutes and seconds to elapse before aircraft ordnance impacts on target. Also called **TTT.** (JP 3-09.3)

tip — See **pitch.**

tips — External fuel tanks.

title block — See **information box.**

TNT equivalent — **(*)** A measure of the energy released from the detonation of a nuclear weapon, or from the explosion of a given quantity of fissionable material, in terms of the amount of TNT (trinitrotoluene) which could release the same amount of energy when exploded.

tolerance dose — The amount of radiation that may be received by an individual within a specified period with negligible results.

tone down — See **attenuation.**

tophandler — A device specially designed to permit the lifting and handling of containers from the top with rough terrain container handlers. See also **container.** (JP 4-01.6)

topographic base — See **chart base.**

topographic engineering — Those engineering tasks that provide geospatial information and services to commanders and staffs across the range of military operations. These tasks include terrain analyses, terrain visualization, digitized terrain products, nonstandard map products, and baseline survey data. See also **geospatial information and services.** (JP 3-34)

topographic map — A map that presents the vertical position of features in measurable form as well as their horizontal positions. See also **map.**

topography — The configuration of the ground to include its relief and all features. Topography addresses both dry land and the sea floor (underwater topography). (JP 4-01.6)

top secret — See **security classification.**

torpedo defense net — **(*)** A net employed to close an inner harbor to torpedoes fired from seaward or to protect an individual ship at anchor or underway.

toss bombing — A method of bombing where an aircraft flies on a line towards the target, pulls up in a vertical plane, releasing the bomb at an angle that will compensate for the effect of gravity drop on the bomb. Similar to loft bombing; unrestricted as to altitude. See also **loft bombing.**

total active aircraft authorization — The sum of the primary and backup aircraft authorizations.

total active aircraft inventory — The sum of the primary and backup aircraft assigned to meet the total active aircraft authorization.

total asset visibility — The capability to provide users with timely and accurate information on the location, movement, status, and identity of units, personnel, equipment, materiel, and supplies. It also includes the capability to act upon that information to improve overall performance of the Department of Defense's logistic practices. Also called **TAV.** See also **automated identification technology; in-transit visibility; joint total asset visibility.** (JP 4-01.8)

total dosage attack — **(*)** A chemical operation which does not involve a time limit within which to produce the required toxic level.

total materiel assets — The total quantity of an item available in the military system worldwide and all funded procurement of the item with adjustments to provide for transfers out of or into the inventory through the appropriation and procurement lead-time periods. It includes peacetime force materiel assets and war reserve stock.

total materiel requirement — The sum of the peacetime force material requirement and the war reserve material requirement.

total mobilization — See **mobilization.**

total overall aircraft inventory — The sum of the total active aircraft inventory and the inactive aircraft inventory. Also called **TOAI.**

total pressure — **(*)** The sum of dynamic and static pressures.

touchdown zone — **(*)** 1. For fixed wing aircraft — The first 3,000 feet or 1,000 meters of runway beginning at the threshold. 2. For rotary wings and vectored thrust aircraft — That portion of the helicopter landing area or runway used for landing.

toxic chemical — Any chemical which, through its chemical action on life processes, can cause death, temporary incapacitation, or permanent harm to humans or animals. This includes all such chemicals, regardless of their origin or of their method of production, and regardless of whether they are produced in facilities, in munitions or elsewhere. (JP 3-11)

toxic chemical, biological, or radiological attack — An attack directed at personnel, animals, or crops, using injurious agents of chemical, biological, or radiological origin.

toxin — See **toxin agent.** (JP 3-11)

toxin agent — A poison formed as a specific secretion product in the metabolism of a vegetable or animal organism, as distinguished from inorganic poisons. Such poisons can also be manufactured by synthetic processes.

track — 1. A series of related contacts displayed on a data display console or other display device. 2. To display or record the successive positions of a moving object. 3. To lock onto a point of radiation and obtain guidance therefrom. 4. To keep a gun properly aimed, or to point continuously a target-locating instrument at a moving target. 5. The actual path of an aircraft above or a ship on the surface of the Earth. The course is the path that is planned; the track is the path that is actually taken. 6. One of the two endless belts on which a full-track or half-track vehicle runs. 7. A metal part forming a path for a moving object; e.g., the track around the inside of a vehicle for moving a mounted machine gun.

track correlation — Correlating track information for identification purposes using all available data.

track management — Defined set of procedures whereby the commander ensures accurate friendly and enemy unit and/or platform locations, and a dissemination procedure for filtering, combining, and passing that information to higher, adjacent, and subordinate commanders.

track of interest — In counterdrug operations, contacts that meet the initial sorting criteria applicable in the area where the contacts are detected. Also called **TOI.** See also **special interest target; suspect.**

track production area — **(*)** An area in which tracks are produced by one radar station.

track symbology — **(*)** Symbols used to display tracks on a data display console or other display device.

track telling — The process of communicating air surveillance and tactical data information between command and control systems or between facilities within the systems. Telling may be classified into the following types: back tell; cross tell; forward tell; lateral tell; overlap tell; and relateral tell.

trafficability — Capability of terrain to bear traffic. It refers to the extent to which the terrain will permit continued movement of any or all types of traffic.

traffic circulation map — A map showing traffic routes and the measures for traffic regulation. It indicates the roads for use of certain classes of traffic, the location of traffic control stations, and the directions in which traffic may move. Also called **circulation map.** See also **map.**

traffic control police — Any persons ordered by a military commander and/or by national authorities to facilitate the movement of traffic and to prevent and/or report any breach of road traffic regulations.

traffic density — (*) The average number of vehicles that occupy one mile or one kilometer of road space, expressed in vehicles per mile or per kilometer.

traffic flow — (*) The total number of vehicles passing a given point in a given time. Traffic flow is expressed as vehicles per hour.

traffic flow security — The protection resulting from features, inherent in some cryptoequipment, that conceal the presence of valid messages on a communications circuit, normally achieved by causing the circuit to appear busy at all times.

traffic information (radar) — Information issued to alert an aircraft to any radar targets observed on the radar display that may be in such proximity to its position or intended route of flight to warrant its attention.

traffic management — The direction, control, and supervision of all functions incident to the procurement and use of freight and passenger transportation services.

traffic pattern — The traffic flow that is prescribed for aircraft landing at, taxiing on, and taking off from an airport. The usual components of a traffic pattern are upwind leg, crosswind leg, downwind leg, base leg, and final approach.

train — 1. A service force or group of service elements that provides logistic support, e.g., an organization of naval auxiliary ships or merchant ships or merchant ships attached to a fleet for this purpose; similarly, the vehicles and operating personnel that furnish supply, evacuation, and maintenance services to a land unit. 2. Bombs dropped in short intervals or sequence.

trained strength in units — Those reservists assigned to units who have completed initial active duty for training of 12 weeks or its equivalent and are eligible for deployment overseas on land when mobilized under proper authority. Excludes personnel in non-deployable accounts or a training pipeline.

train headway — The interval of time between two trains boarded by the same unit at the same point.

training aids — Any item developed or procured with the primary intent that it shall assist in training and the process of learning.

training and readiness oversight — The authority that combatant commanders may exercise over assigned Reserve Component (RC) forces when not on active duty or when on active duty for training. As a matter of Department of Defense policy, this authority includes: a. Providing guidance to Service component commanders on operational requirements and priorities to be addressed in Military Department training and readiness programs; b. Commenting on Service component program recommendations and budget requests; c. Coordinating and approving participation by assigned RC forces in joint exercises and other joint training when on active duty for training or performing inactive duty for training; d. Obtaining and reviewing readiness and inspection reports on assigned RC forces; and e. Coordinating and reviewing mobilization plans (including post-mobilization training activities and deployability validation procedures) developed for assigned RC forces. Also called **TRO.** See also **combatant commander.** (JP 0-2)

training and retirement category — The category identifying (by specific training and retirement category designator) a reservist's training or retirement status in a reserve component category and Reserve Component.

training-pay category — A designation identifying the number of days of training and pay required for members of Reserve Components.

training period — An authorized and scheduled regular inactive duty training period. A training period must be at least two hours for retirement point credit and four hours for pay. Previously used interchangeably with other common terms such as drills, drill period, assemblies, periods of instruction, etc.

training pipeline — A Reserve Component category designation that identifies untrained officer and enlisted personnel who have not completed initial active duty for training of 12 weeks or its equivalent. See also **nondeployable account.**

training unit — A unit established to provide military training to individual reservists or to Reserve Component units.

train path — **(*)** In railway terminology, the timing of a possible movement of a train along a given route. All the train paths on a given route constitute a timetable.

trajectory — See **ballistic trajectory.**

transattack period — 1. In nuclear warfare, the period from the initiation of the attack to its termination. 2. As applied to the Single Integrated Operational Plan, the period that extends from execution (or enemy attack, whichever is sooner) to termination of the Single Integrated Operational Plan. See also **postattack period.**

transfer loader — **(*)** A wheeled or tracked vehicle with a platform capable of vertical and horizontal adjustment used in the loading and unloading of aircraft, ships, or other vehicles.

transient — 1. Personnel, ships, or craft stopping temporarily at a post, station, or port to which they are not assigned or attached, and having destination elsewhere. 2. An independent merchant ship calling at a port and sailing within 12 hours, and for which routing instructions to a further port have been promulgated. 3. An individual awaiting orders, transport, etc., at a post or station to which he or she is not attached or assigned.

transient forces — Forces that pass or stage through, or base temporarily within, the operational area of another command but are not under its operational control. See also **force; transient.** (JP 0-2)

transit area — See **staging area.**

transit bearing — **(*)** A bearing determined by noting the time at which two features on the Earth's surface have the same relative bearing.

transition altitude — The altitude at or below which the vertical position of an aircraft is controlled by reference to true altitude.

transition layer — **(*)** The airspace between the transition altitude and the transition level.

transition level — **(*)** The lowest flight level available for use above the transition altitude. See also **altitude; transition altitude.**

transit passage — The nonsuspendable right of continuous and expeditious navigation and/or overflight in the normal mode through an international strait linking one part of the high seas (or exclusive economic zone) with another.

transit route — **(*)** A sea route which crosses open waters normally joining two coastal routes.

transit zone — The path taken by either airborne or seaborne smugglers. Zone can include transfer operations to another carrier (airdrop, at-sea transfer, etc.). See also **arrival zone.** (JP 3-07.4)

transmission factor (nuclear) — The ratio of the dose inside the shielding material to the outside (ambient) dose. Transmission factor is used to calculate the dose received through the shielding material. See also **half thickness; shielding.**

transmission security — See **communications security.**

transonic — **(*)** Of or pertaining to the speed of a body in a surrounding fluid when the relative speed of the fluid is subsonic in some places and supersonic in others. This is encountered when passing from subsonic to supersonic speed and vice versa. See also **speed of sound.**

transponder — **(*)** A receiver-transmitter which will generate a reply signal, upon proper interrogation. See also **responsor.**

transportability — The capability of material to be moved by towing, self-propulsion, or carrier via any means, such as railways, highways, waterways, pipelines, oceans, and airways.

transport aircraft — **(*)** Aircraft designed primarily for the carriage of personnel and/or cargo. Transport aircraft may be classed according to range, as follows: a. **Short-range** — Not to exceed 1200 nautical miles at normal cruising conditions (2222 km). b. **Medium-range** — Between 1200 and 3500 nautical miles at normal cruising conditions (2222 and 6482 km). c. **Long-range** — Exceeds 3500 nautical miles at normal cruising conditions (6482 km). See also **strategic transport aircraft; tactical transport aircraft.**

transport area — In amphibious operations, an area assigned to a transport organization for the purpose of debarking troops and equipment. See also **inner transport area; outer transport area.**

transportation closure — The actual arrival date of a specified movement requirement at port of debarkation.

transportation component command — The three component commands of United States Transportation Command: Air Force Air Mobility Command; Navy Military Sealift Command; and Army Military Traffic Management Command. Each transportation component command remains a major command of its parent Service and continues to organize, train, and equip its forces as specified by law. Each transportation component command also continues to perform Service-unique missions. Also called **TCC.** See also

United States Transportation Command. (JP 4-0)

transportation emergency — A situation created by a shortage of normal transportation capability and of a magnitude sufficient to frustrate military movement requirements, and which requires extraordinary action by the President or other designated authority to ensure continued movement of essential Department of Defense traffic.

transportation feasibility — Operation plans and operation plans in concept format are considered transportation feasible when the capability to move forces, equipment, and supplies exists from the point of origin to the final destination according to the plan. Transportation feasibility determination will require concurrent analysis and assessment of available strategic and theater lift assets, transportation infrastructure, and competing demands and restrictions. a. The supported commander of a combatant command (CINC) will analyze deployment, joint reception, staging, onward movement, and integration (JRSOI), and theater distribution of forces, equipment, and supplies to final destination. b. Supporting CINCs will provide an assessment on movement of forces from point of origin to aerial port of embarkation and/or seaport of embarkation. c. The Commander in Chief, United States Transportation Command will assess the strategic leg of the time-phased force and deployment data for transportation feasibility, indicating to the Chairman of the Joint Chiefs of Staff and supported CINC that movements arrive at the port of debarkation consistent with the supported CINC's assessment of JRSOI and theater distribution. d. Following analysis of all inputs, the supported CINC is responsible for declaring a plan end-to-end executable. See also **operation plan.** (JP 3-35)

transportation movement requirement — The need for transport of units, personnel, or materiel from a specified origin to a specified destination within a specified timeframe. (JP 4-01)

transportation operating agencies — Those Federal agencies having responsibilities under national emergency conditions for the operational direction of one or more forms of transportation. Also called **federal modal agencies; federal transport agencies.**

transportation priorities — Indicators assigned to eligible traffic that establish its movement precedence. Appropriate priority systems apply to the movement of traffic by sea and air. In times of emergency, priorities may be applicable to continental United States movements by land, water, or air.

transportation system — All the land, water, and air routes and transportation assets engaged in the movement of US forces and their supplies across the range of military operations, involving both mature and immature theaters and at the strategic, operational, and tactical levels of war. (JP 4-0)

transport control center (air transport) — The operations center through which the air transport force commander exercises control over the air transport system.

transport group — An element that directly deploys and supports the landing of the landing force (LF), and is functionally designated as **a transport group in the amphibious task force organization**. A transport group provides for the embarkation, movement to the objective, landing, and logistic support of the LF. Transport groups comprise all sealift and airlift in which the LF is embarked. They are categorized as follows: a. airlifted groups; b. Navy amphibious ship transport groups; and c. strategic sealift shipping groups. (JP 3-02.2)

transporting (ordnance) — The movement or repositioning of ordnance or explosive devices along established explosive routes (does not apply to the aircraft flight line). See also **ordnance.** (JP 3-04.1)

transshipment point — **(*)** A location where material is transferred between vehicles.

traverse — **(*)** 1. To turn a weapon to the right or left on its mount. 2. A method of surveying in which lengths and directions of lines between points on the earth are obtained by or from field measurements, and used in determining positions of the points

traverse level — **(*)** That vertical displacement above low-level air defense systems, expressed both as a height and altitude, at which aircraft can cross the area.

traverse racking test load value — Externally applied force in pounds or kilograms at the top-corner fitting that will strain or stretch end structures of the container sideways. (JP 4-01.7)

treason — Violation of the allegiance owed to one's sovereign or state; betrayal of one's country.

trench interment — A method of interment in which remains are placed head-to-toe. Used only for temporary multiple burials. See also **emergency interment; group interment; mortuary affairs, temporary interment.** (JP 4-06)

trend — The straying of the fall of shot, such as might be caused by incorrect speed settings of the fire support ship.

triangulation station — **(*)** A point on the Earth, the position of which is determined by triangulation. Also called **trig point.**

tri-camera photography — **(*)** Photography obtained by simultaneous exposure of three cameras systematically disposed in the air vehicle at fixed overlapping angles relative to each other in order to cover a wide field. See also **fan camera photography.**

trig list — A list published by certain Army units that includes essential information of accurately located survey points.

trim — The difference in draft at the bow and stern of a vessel or the manner in which a vessel floats in the water based on the distribution of cargo, stores and ballast aboard the vessel. See also **draft; watercraft.** (JP 4-01.6)

triple point — The intersection of the incident, reflected, and fused (or Mach) shock fronts accompanying an air burst. The height of the triple point above the surface, i.e., the height of the Mach stem, increases with increasing distance from a given explosion.

troop basis — An approved list of those military units and individuals (including civilians) required for the performance of a particular mission by numbers, organization and equipment and, in the case of larger commands, by deployment.

troops — A collective term for uniformed military personnel (usually not applicable to naval personnel afloat). See also **airborne troops; combat service support elements; combat support troops; service troops; tactical troops.**

troop safety (nuclear) — An element that defines a distance from the proposed burst location beyond which personnel meeting the criteria described under degree of risk will be safe to the degree prescribed. It is expressed as a combination of a degree of risk and vulnerability category. See also **emergency risk (nuclear); negligible risk (nuclear); unwarned exposed; warned protected.**

troop space cargo — Cargo such as sea or barracks bags, bedding rolls or hammocks, locker trunks, and office equipment, normally stowed in an accessible place. This cargo will also include normal hand-carried combat equipment and weapons to be carried ashore by the assault troops.

troop test — A test conducted in the field for the purpose of evaluating operational or organizational concepts, doctrine, tactics, and techniques, or to gain further information on material. See also **service test.**

tropical storm — A tropical cyclone in which the surface wind speed is at least 34, but not more than 63 knots.

tropopause — **(*)** The transition zone between the stratosphere and the troposphere. The tropopause normally occurs at an altitude of about 25,000 to 45,000 feet (8 to 15 kilometers) in polar and temperate zones, and at 55,000 feet (20 kilometers) in the tropics.

true airspeed indicator — An instrument which displays the speed of the aircraft relative to the ambient air.

true altitude — The height of an aircraft as measured from mean sea level.

true bearing — The direction to an object from a point; expressed as a horizontal angle measured clockwise from true north.

true convergence — The angle at which one meridian is inclined to another on the surface of the Earth. See also **convergence.**

true horizon — **(*)** 1. The boundary of a horizontal plane passing through a point of vision. 2. In photogrammetry, the boundary of a horizontal plane passing through the perspective center of a lens system.

true north — **(*)** The direction from an observer's position to the geographic North Pole. The north direction of any geographic meridian.

turbojet — A jet engine whose air is supplied by a turbine-driven compressor, the turbine being activated by exhaust gases.

turnaround — **(*)** The length of time between arriving at a point and being ready to depart from that point. It is used in this sense for the loading, unloading, re-fueling, and re-arming, where appropriate, of vehicles, aircraft, and ships. See also **turnaround cycle.**

turnaround cycle — **(*)** A term used in conjunction with vehicles, ships, and aircraft, and comprising the following: loading time at departure point; time to and from destination; unloading and loading time at destination; unloading time at returning point; planned maintenance time; and, where applicable, time awaiting facilities. See also **turnaround.**

turning movement — **(*)** A variation of the envelopment in which the attacking force passes around or over the enemy's principal defensive positions to secure objectives deep in the enemy's rear to force the enemy to abandon his position or divert major forces to meet the threat.

turning point — **(*)** In land mine warfare, a point on the centerline of a mine strip or row where it changes direction.

turn-off guidance — Information which enables the pilot of a landing aircraft to select and follow the correct taxiway from the time the aircraft leaves the runway until it may safely be brought to a halt clear of the active runway.

two-person control — The continuous surveillance and control of positive control material at all times by a minimum of two authorized individuals, each capable of detecting incorrect or unauthorized procedures with respect to the task being performed and each familiar with established security requirements. Also called **TPC.**

two-person rule — A system designed to prohibit access by an individual to nuclear weapons and certain designated components by requiring the presence at all times of at least two authorized persons, each capable of detecting incorrect or unauthorized procedures with respect to the task to be performed.

types of burst — See **airburst; fallout safe height of burst; height of burst; high airburst; high altitude burst; low airburst; nuclear airburst; nuclear exoatmospheric burst; nuclear contact-surface burst; nuclear proximity-surface burst; nuclear underground burst; nuclear underwater burst; optimum height of burst; safe burst height.**

type unit — A type of organizational or functional entity established within the Armed Forces and uniquely identified by a five-character, alphanumeric code called a unit type code.

type unit data file — A file that provides standard planning data and movement characteristics for personnel, cargo, and accompanying supplies associated with type units.

U

ultraviolet imagery — That imagery produced as a result of sensing ultraviolet radiations reflected from a given target surface.

unaccounted for — An inclusive term (not a casualty status) applicable to personnel whose person or remains are not recovered or otherwise accounted for following hostile action. Commonly used when referring to personnel who are killed in action and whose bodies are not recovered. See also **casualty; casualty category; casualty status; casualty type.**

unanticipated immediate targets — Those immediate targets that are unknown or not expected to exist in an operational area. See also **operational area; target.** (JP 3-60)

uncertain environment — See **operational environment.** (JP 3-05.3)

uncharged demolition target — **(*)** A demolition target for which charges have been calculated, prepared, and stored in a safe place, and for which execution procedures have been established. See also **demolition target.**

unclassified matter — **(*)** Official matter which does not require the application of security safeguards, but the disclosure of which may be subject to control for other reasons. See also **classified matter.**

unconventional assisted recovery — Evader recovery conducted by directed unconventional warfare forces, dedicated extraction teams, and/or unconventional assisted recovery mechanisms operated by guerrilla groups or other clandestine organizations to seek out, contact, authenticate, support, and return evaders to friendly control. Also called **UAR.** See also **assisted recovery; authenticate; evader; recovery.** (JP 3-50.3)

unconventional assisted recovery coordination center — A compartmented special operations forces (SOF) facility suitably staffed by supervisory personnel and tactical planners to coordinate, synchronize and de-conflict non-conventional assisted recovery (NAR) operations on a 24-hour basis within the geographical area assigned to the joint force commander. The unconventional assisted recovery coordination center (UARCC) is an integral part of the joint force commander's (JFC's) comprehensive personnel recovery architecture and the functional equivalent of a component rescue coordination center. When directed by the JFC, through the joint force special operations component commander, the special operations command Operations Directorate establishes the UARCC (normally within the Joint Operations Center (JOC)) to serve as the focal point for all NAR operations. The UARCC interfaces and coordinates with the JOC, joint search and rescue center, component rescue coordination centers (RCCs) (including the SOF RCC) and the special activities cell. Also called **UARCC.** See also **joint operations center; joint search and rescue center; special operations forces; unconventional assisted recovery.** (JP 3-05.1)

unconventional assisted recovery mechanism — That entity, group of entities, or organizations within enemy-held or hostile areas that operates to receive, support, move, and exfiltrate military personnel or selected individuals to friendly control. Also called **UARM.** See also **assisted recovery; recovery; unconventional assisted recovery.** (JP 3-50.3)

unconventional recovery operation — Evader recovery operations conducted by unconventional forces. See also **evader; recovery operations.** (JP 3-50.3)

unconventional warfare — A broad spectrum of military and paramilitary operations, normally of long duration, predominantly conducted by indigenous or surrogate forces who are organized, trained, equipped, supported, and directed in varying degrees by an external source. It includes guerrilla warfare and other direct offensive, low visibility, covert, or clandestine operations, as well as the indirect activities of subversion, sabotage, intelligence activities, and evasion and escape. Also called **UW.** (JP 3-05.5)

unconventional warfare forces — US forces having an existing unconventional warfare capability.

undersea warfare — Operations conducted to establish battlespace dominance in the underwater environment, which permits friendly forces to accomplish the full range of potential missions and denies an opposing force the effective use of underwater systems and weapons. It includes offensive and defensive submarine, antisubmarine, and mine warfare operations. Also called **USW.** See also **antisubmarine warfare; mine warfare.**

understowed cargo — See **flatted cargo.**

underwater demolition — (*) The destruction or neutralization of underwater obstacles; this is normally accomplished by underwater demolition teams.

underwater demolition team — A group of officers and enlisted specially trained and equipped for making hydrographic reconnaissance of approaches to prospective landing beaches; for effecting demolition of obstacles and clearing mines in certain areas; locating, improving, and marking of useable channels; channel and harbor clearance; acquisition of pertinent data during pre-assault operations, including military information; observing the hinterland to gain information useful to the landing force; and for performing miscellaneous underwater and surface tasks within their capabilities. Also called **UDT.**

underway replenishment — See **replenishment at sea.**

underway replenishment force — (*) A task force of fleet auxiliaries (consisting of oilers, ammunition ships, stores issue ships, etc.) adequately protected by escorts furnished by the responsible operational commander. The function of this force is to provide underway logistic support for naval forces. See also **force.**

underway replenishment group — A task group configured to provide logistic replenishment of ships underway by transfer-at-sea methods.

unexpended weapons or ordnance — Airborne weapons that have not been subjected to attempts to fire or drop and are presumed to be in normal operating conditions and can be fired or jettisoned if necessary. See also **ordnance.** (JP 3-04.1)

unexploded explosive ordnance — (*) Explosive ordnance which has been primed, fused, armed or otherwise prepared for action, and which has been fired, dropped, launched, projected, or placed in such a manner as to constitute a hazard to operations, installations, personnel, or material and remains unexploded either by malfunction or design or for any other cause. Also called **UXO.** See also **explosive ordnance.** (JP 3-15)

unified action — A broad generic term that describes the wide scope of actions (including the synchronization of activities with governmental and nongovernmental agencies) taking place within unified commands, subordinate unified commands, or joint task forces under the overall direction of the commanders of those commands. See also **joint task force; subordinate unified command; unified command.** (JP 0-2)

Unified Action Armed Forces — A publication setting forth the policies, principles, doctrines, and functions governing the activities and performance of the Armed Forces of the United States when two or more Military Departments or Service elements thereof are acting together. Also called **UNAAF.** (JP 0-2)

unified combatant command — See **unified command.** (JP 0-2)

unified command — A command with a broad continuing mission under a single commander and composed of significant assigned components of two or more Military Departments, that is established and so designated by the President through the Secretary of Defense with the advice and assistance of the Chairman of the Joint Chiefs of Staff. Also called **unified combatant command.** See also **combatant command; subordinate unified command.** (JP 0-2)

Unified Command Plan — The document, approved by the President, that sets forth basic guidance to all unified combatant commanders; establishes their missions, responsibilities, and force structure; delineates the general geographical area of responsibility for geographic combatant commanders; and specifies functional responsibilities for functional combatant commanders. Also called **UCP.** See also **combatant command; combatant commander.** (JP 0-2)

uniformed services — The Army, Navy, Air Force, Marine Corps, Coast Guard, National Oceanic and Atmospheric Administration, and Public Health Services. See also **Military Department; Military Service.**

unilateral arms control measure — An arms control course of action taken by a nation without any compensating concession being required of other nations.

unintentional radiation exploitation — Exploitation for operational purposes of noninformation-bearing elements of electromagnetic energy unintentionally emanated by targets of interest.

unintentional radiation intelligence — Intelligence derived from the collection and analysis of noninformation-bearing elements extracted from the electromagnetic energy unintentionally emanated by foreign devices, equipment, and systems, excluding those generated by the detonation of nuclear weapons. Also called **RINT.** See also **intelligence.** (JP 2-0)

uni-Service command — A command comprised of forces of a single Service.

unit — 1. Any military element whose structure is prescribed by competent authority, such as a table of organization and equipment; specifically, part of an organization. 2. An organization title of a subdivision of a group in a task force. 3. A standard or basic quantity into which an item of supply is divided, issued, or used. In this meaning, also called **unit of issue.** 4. With regard to Reserve Components of the Armed Forces, denotes a Selected Reserve unit organized, equipped, and trained for mobilization to serve on active duty as a unit or to augment or be augmented by another unit. Headquarters and support functions without wartime missions are not considered units.

unit aircraft — Those aircraft provided an aircraft unit for the performance of a flying mission. See also **aircraft.**

unit combat readiness — See **combat readiness.**

unit commitment status — **(*)** The degree of commitment of any unit designated and categorized as a force allocated to NATO.

unit designation list — A list of actual units by unit identification code designated to fulfill requirements of a force list.

United States — Includes the land area, internal waters, territorial sea, and airspace of the United States, including the following: a. US territories, possessions, and commonwealths; and b. Other areas over which the US Government has complete jurisdiction and control or has exclusive authority or defense responsibility.

United States Armed Forces — Used to denote collectively only the regular components of the Army, Navy, Air Force, Marine Corps, and Coast Guard. See also **Armed Forces of the United States.**

United States Civil Authorities — Those elected and appointed public officials and employees who constitute the governments of the 50 States, District of Columbia, Commonwealth of Puerto Rico, US possessions and territories, and political subdivisions thereof.

United States Civilian Internee Information Center — The national center of information in the United States for enemy and US civilian internees.

United States controlled shipping — That shipping under US flag and selected ships under foreign flag considered to be under "effective US control," i.e., that can reasonably be expected to be made available to the United States in time of national emergency. See also **effective US controlled ships.**

United States message text format — A program designed to enhance joint and combined combat effectiveness through standardization of message formats, data elements, and information exchange procedures. Standard message formats with standard information content provides all tactical commanders at the joint interface with a common playing field and a common language. Also called **USMTF.**

United States Military Service-funded foreign training — Training that is provided to foreign nationals in United States Military Service schools and installations under authority other than the Foreign Assistance Act of 1961.

United States Naval Ship — A public vessel of the United States that is in the custody of the Navy and is: a. Operated by the Military Sealift Command and manned by a civil service crew; or b. Operated by a commercial company under contract to the Military Sealift Command and manned by a merchant marine crew. Also called **USNS.** See also **Military Sealift Command.** (JP 3-02.2)

United States Prisoner of War Information Center — The national center of information in the United States for enemy and US prisoners of war.

United States Signals Intelligence System — The unified organization of signals intelligence activities under the direction of the Director, National Security Agency/

Chief, Central Security Service. It consists of the National Security Agency/Central Security Service, the components of the Military Services authorized to conduct signals intelligence, and such other entities (other than the Federal Bureau of Investigation) authorized by the National Security Council or the Secretary of Defense to conduct signals intelligence activities. Also called **USSS.** See also **counterintelligence.** (JP 2-01.2)

United States Transportation Command — The unified command with the mission to provide strategic air, land, and sea transportation for the Department of Defense, across the range of military operations. Also called **USTRANSCOM.** See also **global transportation network; single port manager; transportation component command; unified command.** (JP 4-01)

unit identification code — A six-character, alphanumeric code that uniquely identifies each Active, Reserve, and National Guard unit of the Armed Forces. Also called **UIC.**

unitized load — A single item or a number of items packaged, packed, or arranged in a specified manner and capable of being handled as a unit. Unitization may be accomplished by placing the item or items in a container or by banding them securely together. See also **palletized unit load.**

unit line number — A seven-character alphanumeric code that describes a unique increment of a unit deployment, i.e., advance party, main body, equipment by sea and air, reception team, or trail party, in a Joint Operation Planning and Execution System time-phased force and deployment data. Also called **ULN.**

unit loading — (*) The loading of troop units with their equipment and supplies in the same vessels, aircraft, or land vehicles. See also **loading.**

unit movement control center — A temporary organization activated by major subordinate commands and subordinate units during deployment to control and manage marshalling and movement. Also called **UMCC.** See also **deployment; marshaling; unit.** (JP 4-01.8)

unit of issue — In its special storage meaning, refers to the quantity of an item; as each number, dozen, gallon, pair, pound, ream, set, yard. Usually termed unit of issue to distinguish from "unit price." See also **unit.**

unit movement data — A unit equipment and/or supply listing containing corresponding transportability data. Tailored unit movement data has been modified to reflect a specific movement requirement. Also called **UMD.**

unit personnel and tonnage table — A table included in the loading plan of a combat-loaded ship as a recapitulation of totals of personnel and cargo by type, listing cubic measurements and weight. Also called **UP&TT.**

unit price — The cost or price of an item of supply based on the unit of issue.

unit readiness — See **readiness.**

unit-related equipment and supplies — All equipment and supplies that are assigned to a specific unit or that are designated as accompanying supplies. The logistic dimensions of these items are contained in the type unit characteristics file standard.

unit reserves — Prescribed quantities of supplies carried by a unit as a reserve to cover emergencies. See also **reserve supplies.**

unit training assembly — An authorized and scheduled period of unit inactive duty training of a prescribed length of time.

unit type code — A Joint Chiefs of Staff developed and assigned code, consisting of five characters that uniquely identify a "type unit."

Universal Joint Task List — A menu of capabilities (mission-derived tasks with associated conditions and standards, i.e., the tools) that may be selected by a joint force commander to accomplish the assigned mission. Once identified as essential to mission accomplishment, the tasks are reflected within the command joint mission essential task list. Also called **UJTL.** (JP 3-33)

universal polar stereographic grid — A military grid prescribed for joint use in operations in limited areas and used for operations requiring precise position reporting. It covers areas between the 80 degree parallels and the poles.

Universal Postal Union — A worldwide postal organization to which the United States and most other countries are members. The exchange of mail, except parcel post, between the United States and other nations is governed by the provisions of the Universal Postal Union convention. Also called **UPU.**

Universal Time — A measure of time that conforms, within a close approximation, to the mean diurnal rotation of the Earth and serves as the basis of civil timekeeping. Universal Time (UT1) is determined from observations of the stars, radio sources, and also from ranging observations of the moon and artificial Earth satellites. The scale determined directly from such observations is designated Universal Time Observed (UTO); it is slightly dependent on the place of observation. When UTO is corrected for the shift in longitude of the observing station caused by polar motion, the time scale UT1 is obtained. When an accuracy better than one second is not required, Universal Time can be used to mean Coordinated Universal Time. Also called **ZULU time.** Formerly called Greenwich Mean Time.

universal transverse mercator grid — **(*)** A grid coordinate system based on the transverse mercator projection, applied to maps of the Earth's surface extending to 84 degrees N and 80 degrees S latitudes. Also called **UTM grid.**

unknown — 1. A code meaning "information not available." 2. An unidentified target. An aircraft or ship that has not been determined to be hostile, friendly, or neutral using identification friend or foe and other techniques, but that must be tracked by air defense or naval engagement systems. 3. An identity applied to an evaluated track that has not been identified. See also **assumed friend; friend; hostile; neutral; suspect.**

unlimited war — Not to be used. See **general war.**

unmanned aerial vehicle — A powered, aerial vehicle that does not carry a human operator, uses aerodynamic forces to provide vehicle lift, can fly autonomously or be piloted remotely, can be expendable or recoverable, and can carry a lethal or nonlethal payload. Ballistic or semiballistic vehicles, cruise missiles, and artillery projectiles are not considered unmanned aerial vehicles. Also called **UAV.** (JP 3-55.1)

unplanned immediate targets — Those immediate targets that are known to exist in an operational area but are not detected, located, or selected for action in sufficient time to be included in the normal targeting process. See also **immediate targets; operational area; target.** (JP 3-60)

unpremeditated expansion of a war — Not to be used. See **escalation.**

unscheduled convoy phase — (*) The period in the early days of war when convoys are instituted on an ad hoc basis before the introduction of convoy schedules in the regular convoy phase.

unstuffing — The removal of cargo from a container. Also called **stripping.**

unwanted cargo — (*) A cargo loaded in peacetime which is not required by the consignee country in wartime.

unwarned exposed — (*) The vulnerability of friendly forces to nuclear weapon effects. In this condition, personnel are assumed to be standing in the open at burst time, but have dropped to a prone position by the time the blast wave arrives. They are expected to have areas of bare skin exposed to direct thermal radiation, and some personnel may suffer dazzle. See also **warned exposed; warned protected.**

urgent mining — (*) In naval mine warfare, the laying of mines with correct spacing but not in the ordered or planned positions. The mines may be laid either inside or outside the allowed area in such positions that they will hamper the movements of the enemy more than those of our own forces.

urgent priority — A category of immediate mission request that is lower than emergency priority but takes precedence over ordinary priority; e.g., enemy artillery or mortar fire that is falling on friendly troops and causing casualties or enemy troops or mechanized units moving up in such force as to threaten a breakthrough. See also **immediate mission request; priority of immediate mission requests.**

US commercial assets — US commercial aircraft, spacecraft, flag shipping, offshore, and land-based assets located landward of the outer limit of the continental shelf of the United States, its territories, and possessions, and excluding those privately owned oil rigs operating under foreign license in disputed offshore areas.

use of force policy — Policy guidance issued by the Commandant, US Coast Guard, on the use of force and weapons.

US Defense Representative — A senior US officer in a foreign country representing the Secretary of Defense, the Chairman of the Joint Chiefs of Staff, and the commander of the unified command that coordinates the security matters regarding in-country, non-combat Department of Defense (DOD) elements (i.e., DOD personnel and organizations under the command of a combatant commander but not assigned to, or attached to, the combatant commander). Also called **USDR.**

US forces — All Armed Forces (including the Coast Guard) of the United States, any person in the Armed Forces of the United States, and all equipment of any description that either belongs to the US Armed Forces or is being used (including Type I and II Military Sealift Command vessels), escorted, or conveyed by the US Armed Forces.

US national — US citizen and US permanent and temporary legal resident aliens.

US Transportation Command coordinating instructions — Instructions of the US Transportation Command that establish suspense dates for selected members of the joint planning and execution community to complete updates to the operation plan database. Instructions will ensure that the target date movement requirements will be validated and available for scheduling.

V

validate — Execution procedure used by combatant command components, supporting combatant commanders, and providing organizations to confirm to the supported commander and US Transportation Command that all the information records in a time-phased force and deployment data not only are error-free for automation purposes, but also accurately reflect the current status, attributes, and availability of units and requirements. Unit readiness, movement dates, passengers, and cargo details should be confirmed with the unit before validation occurs.

validation — 1. A process normally associated with the collection of intelligence that provides official status to an identified requirement and confirms that the requirement is appropriate for a given collector and has not been previously satisfied. 2. In computer modeling and simulation, the process of determining the degree to which a model or simulation is an accurate representation of the real world from the perspective of the intended uses of the model or simulation. See also **accreditation; independent review; verification.** (JP 2-0)

valuable cargo — **(*)** Cargo which may be of value during a later stage of the war.

value engineering — An organized effort directed at analyzing the function of Department of Defense systems, equipment, facilities, procedures, and supplies for the purpose of achieving the required function at the lowest total cost of effective ownership, consistent with requirements for performance, reliability, quality, and maintainability.

variability — **(*)** The manner in which the probability of damage to a specific target decreases with the distance from ground zero; or, in damage assessment, a mathematical factor introduced to average the effects of orientation, minor shielding, and uncertainty of target response to the effects considered.

variable safety level — See **safety level of supply.**

variant — 1. One of two or more cipher or code symbols that have the same plain text equivalent. 2. One of several plain text meanings that are represented by a single code group. Also called **alternative.**

variation — The angular difference between true and magnetic north. See also **deviation.**

vectored attack — **(*)** Attack in which a weapon carrier (air, surface, or subsurface) not holding contact on the target is vectored to the weapon delivery point by a unit (air, surface, or subsurface) which holds contact on the target.

vehicle cargo — Wheeled or tracked equipment, including weapons, that require certain deck space, head room, and other definite clearance.

vehicle distance — **(*)** The clearance between vehicles in a column which is measured from the rear of one vehicle to the front of the following vehicle.

vehicle summary and priority table — A table listing all vehicles by priority of debarkation from a combat-loaded ship. It includes the nomenclature, dimensions, square feet, cubic feet, weight, and stowage location of each vehicle; the cargo loaded in each vehicle; and the name of the unit to which the vehicle belongs.

verification — 1. In arms control, any action, including inspection, detection, and identification, taken to ascertain compliance with agreed measures. 2. In computer modeling and simulation, the process of determining that a model or simulation implementation accurately represents the developer's conceptual description and specifications. See also **accreditation; configuration management; independent review; validation.**

verify — **(*)** To ensure that the meaning and phraseology of the transmitted message conveys the exact intention of the originator.

vertex — **(*)** In artillery and naval gunfire support, the highest point in the trajectory of a projectile.

vertex height — See **maximum ordinate.**

vertical air photograph — **(*)** An air photograph taken with the optical axis of the camera perpendicular to the surface of the Earth.

vertical and/or short takeoff and landing — Vertical and/or short takeoff and landing capability for aircraft.

vertical envelopment — A tactical maneuver in which troops, either air-dropped or air-landed, attack the rear and flanks of a force, in effect cutting off or encircling the force.

vertical interval — Difference in altitude between two specified points or locations, e.g., the battery or firing ship and the target; observer location and the target; location of previously fired target and new target; observer and a height of burst; and battery or firing ship and a height of burst, etc.

vertical landing zone — A specified ground area for landing vertical takeoff and landing aircraft to embark or disembark troops and/or cargo. A landing zone may contain one or more landing sites. Also called **VLZ.** See also **landing zone; vertical takeoff and landing aircraft.** (JP 3-02)

vertical loading — **(*)** A type of loading whereby items of like character are vertically tiered throughout the holds of a ship so that selected items are available at any stage of the unloading. See also **loading.**

vertical probable error — The product of the range probable error and the slope of fall.

vertical replenishment — **(*)** The use of a helicopter for the transfer of materiel to or from a ship. Also called **VERTREP.**

vertical separation — **(*)** Separation between aircraft expressed in units of vertical distance.

vertical strip — A single flightline of overlapping photos. Photography of this type is normally taken of long, narrow targets such as beaches or roads.

vertical takeoff and landing aircraft — Fixed-wing aircraft and helicopters capable of taking off or landing vertically. Also called **VTOL aircraft.** See also **vertical landing zone; vertical takeoff and landing aircraft transport area.** (JP 3-02)

vertical takeoff and landing aircraft transport area — Area to the seaward and on the flanks of the outer transport and landing ship areas, but preferably inside the area screen, for launching and/or recovering vertical takeoff and landing aircraft. Also called **VTOL aircraft transport area.** See also **vertical takeoff and landing aircraft.** (JP 3-02)

very seriously ill or injured — The casualty status of a person whose illness or injury is classified by medical authority to be of such severity that life is imminently endangered. Also called **VSII.** See also **casualty status.**

very small aperture terminal — Refers to a fixed satellite terminal whose antenna diameter typically does not exceed two meters. Also called **VSAT.**

vesicant agent — See **blister agent.**

vignetting — (*) A method of producing a band of color or tone on a map or chart, the density of which is reduced uniformly from edge to edge.

visibility range — The horizontal distance (in kilometers or miles) at which a large dark object can just be seen against the horizon sky in daylight.

visual call sign — (*) A call sign provided primarily for visual signaling. See also **call sign.**

visual information — Use of one or more of the various visual media with or without sound. Generally, visual information includes still photography, motion picture photography, video or audio recording, graphic arts, visual aids, models, display, visual presentation services, and the support processes. Also called **VI.**

visual information documentation — Motion media, still photography, and audio recording of technical and nontechnical events while they occur, usually not controlled by the recording crew. Visual information documentation encompasses Combat Camera, operational documentation, and technical documentation. Also called **VIDOC.** See also **combat camera; operational documentation; technical documentation.**

visual meteorological conditions — Weather conditions in which visual flight rules apply; expressed in terms of visibility, ceiling height, and aircraft clearance from clouds along the path of flight. When these criteria do not exist, instrument meteorological conditions prevail and instrument flight rules must be complied with. Also called **VMC.** See also **instrument meteorological conditions.** (JP 3-04.1)

visual mine firing indicator — (*) A device used with exercise mines to indicate that the mine would have detonated had it been poised.

vital area — (*) A designated area or installation to be defended by air defense units.

vital ground — (*) Ground of such importance that it must be retained or controlled for the success of the mission. See also **key terrain.**

voice call sign — (*) A call sign provided primarily for voice communication. See also **call sign.**

Voluntary Intermodal Sealift Agreement — The objective of the Voluntary Intermodal Sealift Agreement (VISA) is to provide the Department of Defense (DOD) with assured access to US flag assets, both vessel capacity and intermodal systems, to meet DOD contingency requirements. VISA is an improvement to the Sealift Readiness Program and both programs will be utilized to provide required lift assets. VISA is modeled after the DOD Civil Reserve Air Fleet program. Carriers contractually commit specified portions of their fleet to meet time-phased DOD contingency requirements. The worldwide intermodal system provided by these carriers provides extensive and flexible capabilities to the Department of Defense. Also called **VISA.** See also **intermodal;**

intermodal systems; Sealift Readiness Program. (JP 4-09)

voluntary tanker agreement — An agreement established by the Maritime Administration to provide for US commercial tanker owners and operators to voluntarily make their vessels available to satisfy Department of Defense needs. It is designed to meet contingency or war requirements for point-to-point petroleum, oils, and lubricants movements, and not to deal with capacity shortages in resupply operations. Also called **VTA.** (JP 4-01.2)

voluntary training — Training in a non-pay status for Individual Ready Reservists and active status Standby Reservists. Participation in voluntary training is for retirement points only and may be achieved by training with Selected Reserve or voluntary training units; by active duty for training; by completion of authorized military correspondence courses; by attendance at designated courses of instruction; by performing equivalent duty; by participation in special military and professional events designated by the Military Departments; or by participation in authorized Civil Defense activities. Retirees may voluntarily train with organizations to which they are properly preassigned by orders for recall to active duty in a national emergency or declaration of war. Such training shall be limited to that training made available within the resources authorized by the Secretary concerned.

voluntary training unit — A unit formed by volunteers to provide Reserve Component training in a non-pay status for Individual Ready Reservists and active status Standby Reservists attached under competent orders and participating in such units for retirement points. Also called **reinforcement training unit.**

VOR — (*) An air navigational radio aid which uses phase comparison of a ground transmitted signal to determine bearing. This term is derived from the words "very high frequency omnidirectional radio range."

vulnerability — 1. The susceptibility of a nation or military force to any action by any means through which its war potential or combat effectiveness may be reduced or its will to fight diminished. 2. The characteristics of a system that cause it to suffer a definite degradation (incapability to perform the designated mission) as a result of having been subjected to a certain level of effects in an unnatural (manmade) hostile environment. 3. In information operations, a weakness in information system security design, procedures, implementation, or internal controls that could be exploited to gain unauthorized access to information or an information system. See also **information; information operations; information system.** (JP 3-13)

vulnerability analysis — In information operations, a systematic examination of an information system or product to determine the adequacy of security measures, identify security deficiencies, provide data from which to predict the effectiveness of proposed security measures, and confirm the adequacy of such measures after implementation. See also **information operations; information system; security; vulnerability.** (JP 3-13)

vulnerability assessment — A Department of Defense, command, or unit-level evaluation (assessment) to determine the vulnerability of a terrorist attack against an installation, unit, exercise, port, ship, residence, facility, or other site. Identifies areas of improvement to withstand, mitigate, or deter acts of violence or terrorism.

vulnerability program — A program to determine the degree of any existing susceptibility of nuclear weapon systems to enemy countermeasures, accidental fire, and accidental shock and to remedy these weaknesses insofar as possible.

vulnerability study — An analysis of the capabilities and limitations of a force in a specific situation to determine vulnerabilities capable of exploitation by an opposing force.

vulnerable area — See **vital area.**

vulnerable node — See **target stress point.**

vulnerable point — See **vital area.**

Intentionally Blank

W

wading crossing — See **deep fording capability; shallow fording.**

walking patient — A patient whose injuries and/or illness are relatively minor, permitting the patient to walk and not require a litter. See also **litter; patient; slightly wounded.** (JP 4-02)

wanted cargo — **(*)** In naval control of shipping, a cargo which is not immediately required by the consignee country but will be needed later.

warble — **(*)** In naval mine warfare, the process of varying the frequency of sound produced by a narrow band noisemaker to ensure that the frequency to which the mine will respond is covered.

warden system — An informal method of communication used to pass information to US citizens during emergencies. See also **noncombatant evacuation operations.** (JP 3-07.5)

war game — A simulation, by whatever means, of a military operation involving two or more opposing forces using rules, data, and procedures designed to depict an actual or assumed real life situation.

warhead — That part of a missile, projectile, torpedo, rocket, or other munition which contains either the nuclear or thermonuclear system, high explosive system, chemical or biological agents, or inert materials intended to inflict damage.

warhead mating — The act of attaching a warhead section to a rocket or missile body, torpedo, airframe, motor, or guidance section.

warhead section — **(*)** A completely assembled warhead, including appropriate skin sections and related components.

war materiel procurement capability — The quantity of an item that can be acquired by orders placed on or after the day an operation commences (D-day) from industry or from any other available source during the period prescribed for war materiel procurement planning purposes.

war materiel requirement — The quantity of an item required to equip and support the approved forces specified in the current Secretary of Defense guidance through the period prescribed for war materiel planning purposes.

warned exposed — **(*)** The vulnerability of friendly forces to nuclear weapon effects. In this condition, personnel are assumed to be prone with all skin covered and with thermal protection at least that provided by a two-layer summer uniform. See also **unwarned exposed; warned protected.**

warned protected — **(*)** The vulnerability of friendly forces to nuclear weapon effects. In this condition, personnel are assumed to have some protection against heat, blast, and radiation such as that afforded in closed armored vehicles or crouched in fox holes with improvised overhead shielding. See also **unwarned exposed; warned exposed.**

warning — 1. A communication and acknowledgment of dangers implicit in a wide spectrum of activities by potential opponents ranging from routine defense measures to substantial increases in readiness and force preparedness and to acts of terrorism or political, economic, or military provocation. 2. Operating procedures, practices, or conditions that may result in injury or death if not carefully observed or followed. (JP 3-04.1)

warning area — See **danger area.**

warning net — A communication system established for the purpose of disseminating warning information of enemy movement or action to all interested commands.

warning of attack — A warning to national policymakers that an adversary is not only preparing its armed forces for war, but intends to launch an attack in the near future. See also **tactical warning; warning; warning of war.**

warning of war — A warning to national policymakers that a state or alliance intends war, or is on a course that substantially increases the risks of war and is taking steps to prepare for war. See also **strategic warning; warning; warning of attack.**

warning order — **(*)** 1. A preliminary notice of an order or action which is to follow. 2. **(DOD only)** A crisis action planning directive issued by the Chairman of the Joint Chiefs of Staff that initiates the development and evaluation of courses of action by a supported commander and requests that a commander's estimate be submitted. 3. **(DOD only)** A planning directive that describes the situation, allocates forces and resources, establishes command relationships, provides other initial planning guidance, and initiates subordinate unit mission planning. (JP 5-0)

warning red — See **air defense warning conditions.**

warning shots — The firing of shots or delivery of ordnance by personnel or weapons systems in the vicinity of a person, vessel, or aircraft as a signal to immediately cease activity. Warning shots are one measure to convince a potentially hostile force to withdraw or cease its threatening actions.

warning white — See **air defense warning conditions.**

warning yellow — See **air defense warning conditions.**

warp — To haul a ship ahead by line or anchor. (JP 4-01.6)

war reserve materiel requirement — That portion of the war materiel requirement required to be on hand on D-day. This level consists of the war materiel requirement less the sum of the peacetime assets assumed to be available on D-day and the war materiel procurement capability.

war reserve materiel requirement, balance — That portion of the war reserve materiel requirement that has not been acquired or funded. This level consists of the war reserve materiel requirement less the war reserve materiel requirement, protectable.

war reserve materiel requirement, protectable — That portion of the war reserve materiel requirement that is either on hand and/or previously funded that shall be protected; if issued for peacetime use, it shall be promptly reconstituted. This level consists of the pre-positioned war reserve materiel requirement, protectable, and the other war reserve materiel requirement, protectable.

war reserve (nuclear) — Nuclear weapons materiel stockpiled in the custody of the Department of Energy or transferred to the custody of the Department of Defense and intended for employment in the event of war.

war reserves — **(*)** Stocks of materiel amassed in peacetime to meet the increase in military requirements consequent upon an outbreak of war. War reserves are intended to provide the interim support essential to sustain operations until resupply can be effected.

war reserve stock — That portion of total materiel assets designated to satisfy the war reserve materiel requirement. Also called **WRS.** See also **reserve; war reserve materiel requirement; war reserves.** (JP 2-03)

war reserve stocks for allies — A Department of Defense program to have the Services procure or retain in their inventories those minimum stockpiles of materiel such as munitions, equipment, and combat-essential consumables to ensure support for selected allied forces in time of war until future in-country production and external resupply can meet the estimated combat consumption.

wartime load — The maximum quantity of supplies of all kinds which a ship can carry. The composition of the load is prescribed by proper authority.

wartime manpower planning system — A standardized Department of Defense (DOD)-wide procedure, structure, and database for computing, compiling, projecting, and portraying the time-phased wartime manpower requirements, demand, and supply of the DOD components. Also called **WARMAPS.** See also **S-day.**

wartime reserve modes — Characteristics and operating procedures of sensor, communications, navigation aids, threat recognition, weapons, and countermeasures systems that will contribute to military effectiveness if unknown to or misunderstood by opposing commanders before they are used, but could be exploited or neutralized if known in advance. Wartime reserve modes are deliberately held in reserve for wartime or emergency use and seldom, if ever, applied or intercepted prior to such use. Also called **WARM.**

watching mine — **(*)** In naval mine warfare, a mine secured to its mooring but showing on the surface, possibly only in certain tidal conditions. See also **floating mine; mine.**

watercraft — Any vessel or craft designed specifically and only for movement on the surface of the water. (JP 4-01.6)

waterspace management — The allocation of waterspace in terms of antisubmarine warfare attack procedures to permit the rapid and effective engagement of hostile submarines while preventing inadvertent attacks on friendly submarines.

water terminal — A facility for berthing ships simultaneously at piers, quays, and/or working anchorages, normally located within sheltered coastal waters adjacent to rail, highway, air, and/or inland water transportation networks. (JP 4-01.5)

wave — **(*)** 1. A formation of forces, landing ships, craft, amphibious vehicles or aircraft, required to beach or land about the same time. Can be classified as to type, function or order as shown: a. assault wave; b. boat wave; c. helicopter wave; d. numbered wave; e. on-call wave; f. scheduled wave. 2. **(DOD only)** An undulation of water caused by the progressive movement of energy from point to point along the surface of the water. (JP 4-01.6)

wave crest — The highest part of a wave. See also **crest; wave.** (JP 4-01.6)

wave height — The vertical distance between trough and crest, usually expressed in feet. See also **wave.** (JP 4-01.6)

wave length — The horizontal distance between successive wave crests measured perpendicular to the crest, usually expressed in feet. See also **crest; wave; wave crest.** (JP 4-01.6)

wave-off — An action to abort a landing, initiated by the bridge, primary flight control, landing safety officer or enlisted man, or pilot at his or her discretion. The response to a wave-off signal is mandatory. See also **abort; primary flight control.** (JP 3-04.1)

wave period — The time it takes for two successive wave crests to pass a given point. See also **wave; wave crest.** (JP 4-01.6)

wave trough — The lowest part of the wave between crests. See also **crest; wave.** (JP 4-01.6)

wave velocity — The speed at which a wave form advances across the sea, usually expressed in knots. See also **wave.** (JP 4-01.6)

way point — 1. In air operations, a point or a series of points in space to which an aircraft, ship, or cruise missile may be vectored. 2. A designated point or series of points loaded and stored in a global positioning system or other electronic navigational aid system to facilitate movement.

W-day — See **times.**

weapon and payload identification — 1. The determination of the type of weapon being used in an attack. 2. The discrimination of a re-entry vehicle from penetration aids being utilized with the re-entry vehicle. See also **attack assessment.**

weapon debris (nuclear) — The residue of a nuclear weapon after it has exploded; that is, materials used for the casing and other components of the weapon, plus unexpended plutonium or uranium, together with fission products.

weaponeering — The process of determining the quantity of a specific type of lethal or nonlethal weapons required to achieve a specific level of damage to a given target, considering target vulnerability, weapons effect, munitions delivery accuracy, damage criteria, probability of kill, and weapon reliability. (JP 2-0)

weapon engagement zone — In air defense, airspace of defined dimensions within which the responsibility for engagement of air threats normally rests with a particular weapon system. Also called **WEZ**. a. **fighter engagement zone.** In air defense, that airspace of defined dimensions within which the responsibility for engagement of air threats normally rests with fighter aircraft. Also called **FEZ**. b. **high-altitude missile engagement zone.** In air defense, that airspace of defined dimensions within which the responsibility for engagement of air threats normally rests with high-altitude surface-to-air missiles. Also called **HIMEZ**. c. **low-altitude missile engagement zone.** In air defense, that airspace of defined dimensions within which the responsibility for engagement of air threats normally rests with low- to medium-altitude surface-to-air missiles. Also called **LOMEZ.** d. **short-range air defense engagement zone.** In air defense, that airspace of defined dimensions within which the responsibility for engagement of air threats normally rests with short-range air defense weapons. It may be established within a low- or high-altitude missile engagement zone. Also called **SHORADEZ**. e. **joint engagement zone.** In air defense, that airspace of defined dimensions within which multiple air defense systems (surface-to-air missiles and aircraft) are simultaneously employed to engage air threats. Also called **JEZ.** (JP 3-52)

weapons assignment — (*) In air defense, the process by which weapons are assigned to individual air weapons controllers for use in accomplishing an assigned mission.

weapons free zone — An air defense zone established for the protection of key assets or facilities, other than air bases, where weapon systems may be fired at any target not positively recognized as friendly. (JP 3-52)

weapons of mass destruction — Weapons that are capable of a high order of destruction and/or of being used in such a manner as to destroy large numbers of people. Weapons of mass destruction can be high explosives or nuclear, biological, chemical, and radiological weapons, but exclude the means of transporting or propelling the weapon where such means is a separable and divisible part of the weapon. Also called **WMD.** See also **destruction; special operations.**

weapons readiness state — The degree of readiness of air defense weapons which can become airborne or be launched to carry out an assigned task. Weapons readiness states are expressed in numbers of weapons and numbers of minutes. Weapon readiness states are defined as follows: a. **2 minutes** — Weapons can be launched within two minutes. b. **5 minutes** — Weapons can be launched within five minutes. c. **15 minutes** — Weapons can be launched within fifteen minutes. d. **30 minutes** — Weapons can be launched within thirty minutes. e. **1 hour** — Weapons can be launched within one hour. f. **3 hours** — Weapons can be launched within three hours. g. **released** — Weapons are released from defense commitment for a specified period of time.

weapons recommendation sheet — (*) A sheet or chart which defines the intention of the attack, and recommends the nature of weapons, and resulting damage expected, tonnage, fuzing, spacing, desired mean points of impact, and intervals of reattack.

weapons state of readiness — See **weapons readiness state.**

weapon(s) system — (*) A combination of one or more weapons with all related equipment, materials, services, personnel, and means of delivery and deployment (if applicable) required for self-sufficiency.

weapon system employment concept — (*) A description in broad terms, based on established outline characteristics, of the application of a particular equipment or weapon system within the framework of tactical concept and future doctrines.

weapon-target line — An imaginary straight line from a weapon to a target.

weather central — An organization that collects, collates, evaluates, and disseminates meteorological information in such manner that it becomes a principal source of such information for a given area.

weather deck — A deck having no overhead protection; uppermost deck. (JP 4-01.6)

weather minimum — The worst weather conditions under which aviation operations may be conducted under either visual or instrument flight rules. Usually prescribed by directives and standing operating procedures in terms of minimum ceiling, visibility, or specific hazards to flight.

weight and balance sheet — (*) A sheet which records the distribution of weight in an aircraft and shows the center of gravity of an aircraft at takeoff and landing.

wellness — Force health protection program that consolidates and incorporates physical and mental fitness, health promotion, and environmental and occupational health. See also **force health protection.** (JP 4-02)

wharf — A structure built of open rather than solid construction along a shore or a bank that provides cargo-handling facilities. A similar facility of solid construction is called a quay. See also **quay.** (JP 4-01.5)

wheel load capacity — The capacity of airfield runways, taxiways, parking areas, or roadways to bear the pressures exerted by aircraft or vehicles in a gross weight static configuration.

white cap — A small wave breaking offshore as a result of the action of strong winds. See also **wave.** (JP 4-01.6)

whiteout — **(*)** Loss of orientation with respect to the horizon caused by sun reflecting on snow and overcast sky.

white propaganda — Propaganda disseminated and acknowledged by the sponsor or by an accredited agency thereof. See also **propaganda.**

Wilson cloud — See **condensation cloud.**

winch — A hoisting machine used for loading and discharging cargo and stores or for hauling in lines. See also **stores.** (JP 4-01.6)

wind shear — A change of wind direction and magnitude.

wind velocity — **(*)** The horizontal direction and speed of air motion.

wing — 1. An Air Force unit composed normally of one primary mission group and the necessary supporting organizations, i.e., organizations designed to render supply, maintenance, hospitalization, and other services required by the primary mission groups. Primary mission groups may be functional, such as combat, training, transport, or service. 2. A fleet air wing is the basic organizational and administrative unit for naval-, land-, and tender-based aviation. Such wings are mobile units to which are assigned aircraft squadrons and tenders for administrative organization control. 3. A balanced Marine Corps task organization of aircraft groups and squadrons, together with appropriate command, air control, administrative, service, and maintenance units. A standard Marine Corps aircraft wing contains the aviation elements normally required for the air support of a Marine division. 4. A flank unit; that part of a military force to the right or left of the main body.

wingman — An aviator subordinate to and in support of the designated section leader; also, the aircraft flown in this role.

withdrawal operation — A planned retrograde operation in which a force in contact disengages from an enemy force and moves in a direction away from the enemy.

withhold (nuclear) — The limiting of authority to employ nuclear weapons by denying their use within specified geographical areas or certain countries.

working anchorage — An anchorage where ships lie to discharge cargoes over-side to coasters or lighters. See also **emergency anchorage.**

working capital fund — A revolving fund established to finance inventories of supplies and other stores, or to provide working capital for industrial-type activities.

work order — A specific or blanket authorization to perform certain work —

usually broader in scope than a job order. It is sometimes used synonymously with job order.

world geographic reference system — See **georef.**

wounded — See **seriously wounded; slightly wounded**.

wounded in action — A casualty category applicable to a hostile casualty, other than the victim of a terrorist activity, who has incurred an injury due to an external agent or cause. The term encompasses all kinds of wounds and other injuries incurred in action, whether there is a piercing of the body, as in a penetration or perforated wound, or none, as in the contused wound. These include fractures, burns, blast concussions, all effects of biological and chemical warfare agents, and the effects of an exposure to ionizing radiation or any other destructive weapon or agent. The hostile casualty's status may be categorized as "very seriously ill or injured," "seriously ill or injured," "incapacitating illness or injury," or "not seriously injured." Also called **WIA.** See also **casualty category.**

wreckage locator chart — A chart indicating the geographic location of all known aircraft wreckage sites and all known vessel wrecks that show above low water or can be seen from the air. It consists of a visual plot of each wreckage, numbered in chronological order, and cross referenced with a wreckage locator file containing all pertinent data concerning the wreckage.

Intentionally Blank

X

Intentionally Blank

Intentionally Blank

Y

yaw — **(*)** 1. The rotation of an aircraft, ship, or missile about its vertical axis so as to cause the longitudinal axis of the aircraft, ship, or missile to deviate from the flight line or heading in its horizontal plane. 2. Angle between the longitudinal axis of a projectile at any moment and the tangent to the trajectory in the corresponding point of flight of the projectile.

yield — See **nuclear yields.**

Intentionally Blank

Z

zero-length launching — **(*)** A technique in which the first motion of the missile or aircraft removes it from the launcher.

zero point — The location of the center of a burst of a nuclear weapon at the instant of detonation. The zero point may be in the air, or on or beneath the surface of land or water, depending upon the type of burst, and it is thus to be distinguished from ground zero.

zone I (nuclear) — A circular area determined by using minimum safe distance I as the radius and the desired ground zero as the center from which all armed forces are evacuated. If evacuation is not possible or if a commander elects a higher degree of risk, maximum protective measures will be required.

zone II (nuclear) — A circular area (less zone I) determined by using minimum safe distance II as the radius and the desired ground zero as the center in which all personnel require maximum protection. Maximum protection denotes that armed forces personnel are in "buttoned up" tanks or crouched in foxholes with improvised overhead shielding.

zone III (nuclear) — A circular area (less zones I and II) determined by using minimum safe distance III as the radius and the desired ground zero as the center in which all personnel require minimum protection. Minimum protection denotes that armed forces personnel are prone on open ground with all skin areas covered and with an overall thermal protection at least equal to that provided by a two-layer uniform.

zone of action — **(*)** A tactical subdivision of a larger area, the responsibility for which is assigned to a tactical unit; generally applied to offensive action. See also **sector.**

zone of fire — An area into which a designated ground unit or fire support ship delivers, or is prepared to deliver, fire support. Fire may or may not be observed.

ZULU time — See **Universal Time.**

Intentionally Blank

APPENDIX A
ABBREVIATIONS AND ACRONYMS

A

A	analog
A&P	administrative and personnel
A2C2	Army airspace command and control
AA	avenue of approach
AAA	antiaircraft artillery; arrival and assembly area; assign alternate area
AAAS	amphibious aviation assault ship
AABB	American Association of Blood Banks
AABFS	amphibious assault bulk fuel system
AABWS	amphibious assault bulk water system
AADC	area air defense commander
AAFES	Army and Air Force Exchange Service
AAFIF	automated air facility information file
AAFSF	amphibious assault fuel supply facility
AAGS	Army air-ground system
AAMDC	area air and missile defense command
AAOE	arrival and assembly operations element
AAOG	arrival and assembly operations group
AAP	Allied administrative publication; assign alternate parent
AAR	after action report; after action review
AAT	automatic analog test
AAU	analog applique unit
AAV	amphibious assault vehicle
AAW	antiair warfare
AB	airbase
ABCA	American, British, Canadian, Australian Armies Standardization Program
ABCCC	airborne battlefield command and control center
ABD	airbase defense
ABFC	advanced base functional component
ABFS	amphibious bulk fuel system
ABGD	air base ground defense
ABL	airborne laser
ABM	antiballistic missile
ABN	airborne
ABNCP	Airborne Command Post
ABO	air base operability; blood typing system
A/C	aircraft
AC	Active Component; aircraft commander; alternating current
AC-130	Hercules
ACA	airlift clearance authority; airspace control authority; airspace coordination area

ACAA	automatic chemical agent alarm
ACAPS	area communications electronics capabilities
ACB	amphibious construction battalion
ACC	Air Combat Command; air component commander; area coordination center
ACCON	acoustic condition
ACCS	air command and control system
ACCSA	Allied Communications and Computer Security Agency
ACDO	assistant command duty officer
ACE	airborne command element (USAF); air combat element (NATO); Allied Command Europe; aviation combat element Marine air-ground task force (MAGTF)
ACF	air contingency force
ACI	assign call inhibit
ACIC	Army Counterintelligence Center
ACINT	acoustic intelligence
ACK	acknowledgement
ACL	access control list; allowable cabin load
ACLANT	Allied Command Atlantic
ACM	advanced cruise missile; advanced conventional munitions; air combat maneuver; air contingency Marine air-ground task force (MAGTF); airspace control measure
ACN	assign commercial network
ACO	airspace control order
ACOC	area communications operations center
ACOCC	air combat operations command center
ACP	airspace control plan; Allied Communications Publication; assign common pool
ACR	armored cavalry regiment (Army); assign channel reassignment
ACS	agile combat support; air-capable ship; airspace control system; auxiliary crane ship
ACSA	acquisition and cross-servicing agreement; Allied Communications Security Agency
AC/S, C4I	Assistant Chief of Staff, Command, Control, Communications, Computers, and Intelligence (USMC)
ACT	activity
ACU	assault craft unit
ACV	air cushion vehicle; armored combat vehicle
ACW	advanced conventional weapons
A/D	analog-to-digital
AD	active duty; advanced deployability; air defense; priority add-on
ADA	air defense artillery
A/DACG	arrival/departure airfield control group
ADAL	authorized dental allowance list
ADANS	Air Mobility Command Deployment Analysis System
ADC	area damage control

ADCAP	advanced capability
A/DCG	arrival/departure control group
ADCON	administrative control
ADD	assign on-line diagnostic
ADDO	Assistant Deputy Director for Operations
ADDO(MS)	Assistant Deputy Director for Operations/Military Support
ADE	air defense emergency; assign digit editing
ADF	automatic direction finding
ADIZ	air defense identification zone
ADKC/RCU	Automatic Key Distribution Center/Rekeying Control Unit
ADL	armistice demarcation line; assign XX (SL) routing
ADMIN	administration
ADN	Allied Command Europe desired ground zero number
ADNET	anti-drug network
ADOC	air defense operations center
ADP	automated data process(ing)
ADPE	automated data processing equipment
ADPS	automatic data processing system
ADR	accident data recorder; aircraft damage repair; armament delivery recording
ADRA	Adventist Development and Relief Agency
ADS	amphibian discharge site
ADSIA	Allied Data Systems Interoperability Agency
ADT	active duty for training; assign digital transmission group; automatic digital tester
ADVON	advanced echelon
ADSW	active duty for special work
AE	aeromedical evacuation; assault echelon; attenuation equalizer
AECA	Arms Export Control Act
AECC	aeromedical evacuation coordination center
AECM	aeromedical evacuation crew member
AEF	aerospace expeditionary force
AEG	aerospace expeditionary group
AELT	aeromedical evacuation liaison team
AEOS	aeromedical evacuation operations squadron
AEOT	aeromedical evacuation operations team
AEPS	aircrew escape propulsion system
AES	aeromedical evacuation squadron; aeromedical evacuation system
AETC	Air Education and Training Command
AETF	air and space expeditionary task force
AEU	assign essential user bypass
AEW	airborne early warning; air expeditionary wing
AEW&C	airborne early warning and control
AF	amphibious force
AFAARS	Air Force After Action Reporting System
AFARN	Air Force air request net

AFATDS	Advanced Field Artillery Tactical Data System
AFB	Air Force Base
AFC	area frequency coordinator; automatic frequency control
AFCA	Air Force Communications Agency
AFCAP	Air Force contract augmentation program; Armed Forces contract augmentation program
AFCC	Air Force Component Commander
AFCENT	Allied Forces Central Europe (NATO)
AFCERT	Air Force computer emergency response team
AFCS	automatic flight control system
AFD	assign fixed directory
AFDC	Air Force Doctrine Center
AFDIGS	Air Force digital graphics system
AF/DP	Deputy Chief of Staff for Personnel, United States Air Force
AFEES	Armed Forces Examining and Entrance Station
AFFIS	Air Facilities File Information System
AFFMA	Air Force Frequency Management Agency
AFFOR	Air Force forces
AFI	Air Force Instruction
AFID	anti-fratricide identification device
AF/IL	Deputy Chief of Staff for Installations and Logistics, USAF
AFIRB	Armed Forces Identification Review Board
AFIWC	Air Force Information Warfare Center
AFJMAN	Air Force Joint Manual
AFLC	Air Force Logistics Command
AFLE	Air Force Liaison Element
AFLNO	Air Force liaison officer
AFMAN	Air Force Manual
AFMC	Air Force Materiel Command
AFME	Armed Forces medical examiner
AFMIC	Armed Forces Medical Intelligence Center
AFMLO	Air Force Medical Logistics Office
AFMS	Air Force Medical Service
AFNORTH	Allied Forces Northern Europe (NATO)
AFNORTHWEST	Allied Forces North West Europe (NATO)
AFNSEP	Air Force National Security and Emergency Preparedness Agency
AFOE	assault follow-on echelon
AFOSI	Air Force Office of Special Investigations
AFPAM	Air Force pamphlet
AFPC	Air Force Personnel Center
AFPD	Air Force Policy Directive
AFPEO	Armed Forces Professional Entertainment Overseas
AFR	Air Force Reserve; assign frequency for networking
AFRC	Air Force Reserve Command; Armed Forces Recreation Center
AFRCC	Air Force rescue coordination center

AFRRI	Armed Forces radiological research center
AFRTS	Armed Forces Radio and Television Service
AFS	aeronautical fixed service
AFSATCOM	Air Force satellite communications (system)
AFSC	Armed Forces Staff College; United States Air Force specialty code
AFSOB	Air Force special operations base
AFSOC	Air Force Special Operations Command; Air Force special operations component
AFSOCC	Air Force special operations control center
AFSOD	Air Force special operations detachment
AFSOE	Air Force special operations element
AFSOF	Air Force special operations forces
AFSOUTH	Allied Forces, South (NATO)
AFSPACE	United States Space Command Air Force
AFSPC	Air Force Space Command
AFSPOC	Air Force Space Operations Center
AFTAC	Air Force Technical Applications Center
AFTH	Air Force Theater Hospital
AFTN	Aeronautical Feed Telecommunications Network
AFTO	Air Force technical order
AFWCF	Air Force working capital fund
AF/XO	Deputy Chief of Staff for Plans and Operations, United States Air Force
AF/XOI	Air Force Director of Intelligence, Surveillance, and Reconnaissance
AF/XOO	Director of Operations, United States Air Force
A/G	air to ground
AG	adjutant general (Army)
AGARD	Advisory Group for Aerospace Research and Development
AGE	aerospace ground equipment
AGIL	airborne general illumination lightself
AGL	above ground level
AGM-28A	Hound Dog
AGM-65	Maverick
AGM-69	short range attack missile
AGR	Active Guard and Reserve
AI	airborne interceptor; air interdiction; area of interest
AIA	Air Intelligence Agency
AIASA	annual integrated assessment for security assistance
AIC	air intercept controller; assign individual compressed dial; Atlantic Intelligence Command
AICF/USA	Action Internationale Contre La Faim (International Action Against Hunger)
AIDS	acquired immune deficiency syndrome
AIF	automated installation intelligence file
AIG	addressee indicator group
AIIRS	automated intelligence information reporting system

AIM	Airman's Information Manual
AIM-7	Sparrow
AIM-9	Sidewinder
AIM-54A	Phoenix
AIMD	aircraft intermediate maintenance department
AIQC	antiterrorism instructor qualification course
AIRCENT	Allied Air Forces Central Europe (NATO)
AIREVACCONFIRM	air evacuation confirmation
AIREVACREQ	air evacuation request
AIREVACRESP	air evacuation response
AIRNORTHWEST	Allied Air Forces North West Europe (NATO)
AIRREQRECON	air request reconnaissance
AIRSOUTH	Allied Air Forces Southern Europe (NATO)
AIRSUPREQ	air support request
AIS	automated information system
AIT	automated identification technology
AIU	Automatic Digital Network Interface Unit
AJ	anti-jam
AJBPO	Area Joint Blood Program Office
AJCC	alternate joint communications center
AJ/CM	anti-jam control modem
AJFP	adaptive joint force packaging
AJMRO	area joint medical regulating office
AJNPE	airborne joint nuclear planning element
AK	commercial cargo ship
AKNLDG	acknowledge message
ALCC	airlift control center; airlift coordination cell
ALCE	airlift control element
ALCG	analog line conditioning group
ALCM	air launched cruise missile
ALCOM	United States Alaskan Command
ALCON	all concerned
ALD	accounting line designator; airborne laser designator; available-to-load date
ALE	airlift liaison element
ALERFA	alert phase (ICAO)
ALERT	attack and launch early reporting to theater
ALLOREQ	allocation request
ALMSNSCD	airlift mission schedule
ALNOT	alert notice; search and rescue alert notice
ALO	air liaison officer
ALOC	air lines of communications
ALORD	alert launch order
ALSA	Air Land Sea Application (Center)
ALSS	advanced logistic support site
ALTTSC	alternate Tomahawk strike coordinator
A/M	approach and moor
AM	amplitude modulation

AMAL	authorized medical allowance list
AMB	air mobility branch; ambassador
AMBUS	ambulance bus
AMC	airborne mission commander; Air Mobility Command; Army Materiel Command: midpoint compromise search area
AMCC	alternate military command center
AMCIT	American citizen
AMCM	airborne mine countermeasures
AMD	air mobility division
AME	air mobility element; antenna mounted electronics
AMEDD	Army Medical Department
AMEDDCS	U.S. Army Medical Department Center and School
AMEMB	American Embassy
AMF(L)	ACE Mobile Force (Land) (NATO)
AMH	automated message handler
AMIO	alien migrant interdiction operations
AMMO	ammunition
AMOCC	air mobility operations control center
AMOG	air mobility operations group
AMOPES	Army Mobilization and Operations Planning and Execution System
AMOPS	Army mobilization and operations planning system; Army mobilization operations system
AMP	amplifier
AMPE	automated message processing exchange
AMPN	amplification
AMPSSO	Automated Message Processing System Security Office (or Officer)
AMRAAM	advanced medium-range air-to-air missile
AMS	Army management structure
AMVER	automated mutual-assistance vessel rescue system
AMW	air mobility wing; amphibious warfare
AMX	air mobility express
AN	alphanumeric; analog nonsecure
ANCA	Allied Naval Communications Agency
ANDVT	advanced narrowband digital voice terminal
ANG	Air National Guard
A/NM	administrative/network management
ANMCC	Alternate National Military Command Center
ANN	assign NNX routing
ANR	Alaskan NORAD Region
ANSI	American National Standards Institute
ANX	assign NNXX routing
ANY	assign NYX routing
ANZUS	Australia-New Zealand-United States Treaty
AO	action officer; administration officer; air officer; area of operations; aviation ordnance person

AO&M	administration, operation, and maintenance
AOA	amphibious objective area
AOB	advanced operations base; aviation operations branch
AOC	air operations center; Army operations center
AOCC	air operations control center
AOC-E	Aviation Operations Center-East (USCS)
AOCU	analog orderwire control unit
AOC-W	Aviation Operations Center-West (USCS)
AOD	on-line diagnostic
AOI	area of interest
AOL	area of limitation
AOP	air operations plan; area of probability
AOR	area of responsibility
AOS	area of separation
AOSS	aviation ordnance safety supervisor
AP	average power
APC	aerial port commander; armored personnel carrier; assign preprogrammed conference list
APCC	aerial port control center; alternate processing and correlation center
APF	afloat pre-positioning force
APIC	allied press information center
APO	afloat pre-positioning operations; Army Post Office
APOD	aerial port of debarkation
APOE	aerial port of embarkation
APORT	aerial port
APORTSREP	air operations bases report
APPS	analytical photogrammetric positioning system
APR	assign primary zone routing
APS	afloat pre-positioning ships
APU	auxiliary power unit
AR	air refueling; Army regulation; Army reserve
ARB	alternate recovery base; assign receive bypass lists
ARBS	angle rate bombing system
ARC	air Reserve Components; American (National) Red Cross
ARCENT	United States Army Central Command
ARCP	air refueling control point
ARCT	air refueling control time
ARDF	automatic radio direction finding
AREC	air resource element coordinator
ARFOR	Army forces
ARG	amphibious ready group
ARGO	automatic ranging grid overlay
ARINC	Aeronautical Radio Incorporated
ARIP	air refueling initiation point
ARL-M	airborne reconnaissance low-multifunction
ARM	antiradiation missiles
ARNG	Army National Guard

ARP	air refueling point
ARPERCEN	United States Army Reserve Personnel Center
ARQ	automatic request-repeat
ARRC	Allied Command Europe Rapid Reaction Corps (NATO)
ARRDATE	arrival date
ARS	air rescue service
ARSOA	Army special operations aviation
ARSOC	Army special operations component
ARSOF	Army special operations forces
ARSOTF	Army special operations task force
ARSPACE	Army Space Command
ARSPOC	Army space operations center
ART	air reserve technician
ARTCC	air route traffic control center
ARTS III	Automated Radar Tracking System
AS	analog secure; aviation ship
ASA	automatic spectrum analyzer
ASARS	Advanced Synthetic Aperture Radar System
ASAS	All Source Analysis System
ASAT	antisatellite weapon
ASBP	Armed Services Blood Program
ASBPO	Armed Services Blood Program Office
ASC	acting Service chief; Aeronautical Systems Center; Air Systems Command; assign switch classmark; Automatic Digital Network switching center
ASCC	Air Standardization Coordinating Committee
ASCIET	all Services combat identification evaluation team
ASCII	American Standard Code for Information Interchange
ASCS	air support control section
ASD(A&L)	Assistant Secretary of Defense (Acquisition and Logistics)
ASD(C)	Assistant Secretary of Defense (Comptroller)
ASD(C3I)	Assistant Secretary of Defense (Command, Control, Communications, and Intelligence)
ASD(FM&P)	Assistant Secretary of Defense (Force Management and Personnel)
ASD(FMP)	Assistant Secretary of Defense (Force Management Policy)
ASD(HA)	Assistant Secretary of Defense (Health Affairs)
ASDI	analog simple data interface
ASD(ISA)	Assistant Secretary of Defense (International Security Affairs)
ASD(ISP)	Assistant Secretary of Defense (International Security Policy)
ASD(LA)	Assistant Secretary of Defense (Legislative Affairs)
ASD(P&L)	Assistant Secretary of Defense (Production and Logistics)
ASD(PA)	Assistant Secretary of Defense (Public Affairs)
ASD(PA&E)	Assistant Secretary of Defense (Program Analysis and Evaluation)
ASD(RA)	Assistant Secretary of Defense (Reserve Affairs)

ASD(RSA)	Assistant Secretary of Defense (Regional Security Affairs)
ASD(S&R)	Assistant Secretary of Defense (Strategy and Requirements)
ASD(SO/LIC)	Assistant Secretary of Defense (Special Operations and Low Intensity Conflict)
ASE	aircraft survivability equipment; automated stabilization equipment
ASF	aeromedical staging facility
ASG	area support group
ASI	assign and display switch initialization
ASIF	Airlift Support Industrial Fund
ASL	allowable supply list; archipelagic sea lane; assign switch locator (SL) routing; authorized stockage list (Army)
ASM	armored scout mission; automated scheduling message
ASMD	antiship missile defense
ASO	air support operations
ASOC	air support operations center
ASOFDTG	as of date/time group
ASPP	acquisition systems protection program
ASPPO	Armed Service Production Planning Office
ASR	available supply rate
ASSETREP	transportation assets report
AST	assign secondary traffic channels
ASTS	aeromedical staging squadron
ASW	antisubmarine warfare; average surface wind
ASWBPL	Armed Services Whole Blood Processing Laboratories
ASWC	antisubmarine warfare commander
AT	annual training; antiterrorism
At	total attainable search area
ATA	Airlift Tanker Association; airport traffic area
ATAC	antiterrorism alert center (Navy)
ATACC	advanced tactical air command center
ATACMS	Army Tactical Missile System
ATACO	air tactical actions control officer
ATACS	Army Tactical Communications System
ATAF	Allied Tactical Air Force (NATO)
ATBM	antitactical ballistic missile
ATC	air target chart; Air Threat Conference; air traffic control; air transportable clinic (USAF)
ATCA	Allied Tactical Communications Agency
ATCAA	air traffic control assigned airspace
ATCALS	air traffic control and landing system
ATCC	air traffic control center; Antiterrorism Coordinating Committee
ATCRBS	Air Traffic Control Radar Beacon System
ATCS	air traffic control section
ATDL1	Army tactical data link 1
ATDLS	Advanced Tactical Data Link System

ATDM	adaptive time division multiplexer
ATDS	airborne tactical data system
ATF	amphibious task force; Bureau of Alcohol, Tobacco and Firearms (TREAS)
AT/FP	antiterrorism/force protection
ATG	assign trunk group cluster
ATGM	antitank guided missile; antitank guided munition
ATH	air transportable hospital; assign thresholds
ATHS	Airborne Target Handover System
ATM	air target material; assign traffic metering
ATMCT	air terminal movement control team
ATMP	Air Target Materials Program
ATN	assign thresholds
ATO	air tasking order
ATOC	air tactical operations center; air terminal operations center
ATOCONF	air tasking order/confirmation
ATP	Allied tactical publication
ATR	attrition reserve
ATS	air traffic service; assign terminal service
ATSD(AE)	Assistant to the Secretary of Defense (Atomic Energy)
ATSD(IO)	Assistant to the Secretary of Defense (Intelligence Oversight)
ATT	assign terminal type
ATTU	air transportable treatment unit
AUEL	automated unit equipment list
AUG	application user group
AUIC	active duty unit identification code
AUTODIN	Automatic Digital Network
AUTOSEVOCOM	Automatic Secure Voice Communications Network
AUX	auxiliary
AV	air vehicle
AV-8	Harrier
AVDTG	analog via digital trunk group
AVGAS	aviation gasoline
AVIM	aviation intermediate maintenance
AVL	assign variable location
AVOU	analog voice orderwire unit
AVOW	analog voice orderwire
AVS	audiovisual squadron
AVUM	aviation unit maintenance
AV/VI	audiovisual/visual information
AW	acoustic warfare; air warfare
AWACS	Airborne Warning and Control System
AWADS	adverse weather aerial delivery system
AWC	air warfare commander
AWCAP	airborne weapons corrective action program
AWCCM	acoustic warfare counter-countermeasures
AWCM	acoustic warfare countermeasures

AWN	Automated Weather Network
AWOL	absent without leave
AWS	Air Weather Service
AWSE	armament weapons support equipment
AWSIM	air warfare simulation model
AWSR	Air Weather Service regulation
AXP	ambulance exchange point
AXX	assign XXX routing
AZR	assign zone restriction lists

B

B	cross-over barrier pattern
B-52	Stratofortress
B&A	boat and aircraft
BAF	backup alert force
BAG	baggage
BAH	basic allowance for housing
BAI	backup aircraft inventory; battlefield air interdiction
BAS	basic allowance for subsistence; battalion aid station
BATF	Bureau of Alcohol, Tobacco, and Firearms
B/B	baseband
BB	breakbulk
BBL	barrel (42 US gallons)
BC	bottom current
BCAT	beddown capability assessment tool
BCD	battlefield coordination detachment
BCI	bit count integrity
BCN	beacon
BCOC	base cluster operations center
BCR	baseline change request
BCTP	battle command training program
BCU	beach clearance unit
BDA	battle damage assessment
BDC	blood donor center
BDE	brigade
BDL	beach discharge lighter
BDOC	base defense operations center
BDR	battle damage repair
BDZ	base defense zone
BE	basic encyclopedia
BEN	base encyclopedia number
BE number	basic encyclopedia number
BER	bit error ratio
BES	budget estimate submission
BGC	boat group commander
BI	battle injury
BIA	Bureau of Indian Affairs

BIAS	Battlefield Illumination Assistance System
BIDDS	Base Information Digital Distribution System
BIDE	basic identity data element
BIFC	Boise Interagency Fire Center
BIH	International Time Bureau (Bureau International d'l'Heure)
BINM	Bureau of International Narcotics Matters
BIO	biological; Bureau of International Organizations
BISS	base installation security system
BIT	built-in test
BITE	built-in test equipment
BIU	beach interface unit
BLCP	beach lighterage control point
BLDREP	blood report
BLDSHIPREP	blood shipment report
BLM	Bureau of Land Management
BLOS	beyond line of sight
BLS	beach landing site
BLT	battalion landing team
BM	battle management; beachmaster
BMC4I	Battle Management Command, Control, Communications, Computers, and Intelligence
BMCT	begin morning civil twilight
BMD	ballistic missile defense
BMDO	Ballistic Missile Defense Organization
BMEWS	ballistic missile early warning system
BMU	beachmaster unit
BN	battalion
BMNT	begin morning nautical twilight
BOC	base operations center
BOCCA	Bureau of Coordination of Civil Aircraft (NATO)
BOG	beach operations group
BOH	bottom of hill
BOS	base operating support
BOSG	base operations support group
BP	battle position; block parity
BPD	blood products depot
BPG	beach party group
BPI	bits per inch
BPO	blood program office
BPPBS	bi-annual planning, programming, and budget system
BPS	basic psychological operations study; bits per second
BPSK	biphase shift keying
BPT	beach party team
BPWRR	bulk petroleum war reserve requirement
BPWRS	bulk petroleum war reserve stocks
BR	budget review
BRAC	base realignment and closure

BRC	base recovery course
BS	battle staff; broadcast source
BSA	beach support area; brigade support area
BSC	black station clock
BSC ro	black station clock receive out
BSP	base support plan
BSSG	brigade service support group
BSU	blood supply unit
BTB	believed-to-be
BTC	blood transshipment center
BTG	basic target graphic
BTOC	battalion tactical operations center
BTU	beach termination unit
BULK	bulk cargo
BVR	beyond visual range
BW	bandwidth; biological warfare; biological weapon
BZ	buffer zone

C

C	centigrade; clock; coverage factor; creeping line pattern
C&A	certification and accreditation
C&LAT	cargo and loading analysis table
C2	command and control
C2-attack	an offensive form of command and control warfare
C2E	command and control element
C2IP	Command and Control Initiatives Program
C2IPS	Command and Control Information Processing System
C2P	command and control protection
C2-protect	a defensive form of command and control warfare
C2S	command and control support
C2W	command and control warfare
C3	command, control, and communications
C3AG	Command, Control, and Communications Advisory Group
C3CM	command, control, and communications countermeasures
C3I	command, control, communications, and intelligence
C3IC	coalition coordination, communications, and integration center
C3SMP	Command, Control, and Communications Systems Master Plan
C4	command, control, communications, and computers
C4CM	command, control, communications, and computer countermeasures
C4I	command, control, communications, computers, and intelligence
C4IFTW	command, control, communications, computers, and intelligence for the Warrior

C4ISR	command, control, communications, computers, intelligence, surveillance, and reconnaissance
C4S	command, control, communications, and computer systems
C4 systems	command, control, communications, and computer systems
C-5	Galaxy
C-17	Globemaster III
C-21	Learjet
C-27	Spartan
C-130	Hercules
C-141	Starlifter
CA	civil administration; civil affairs; combat assessment
CAA	civil air augmentation; Command Arrangement Agreements
CAB	combat aviation brigade
CAC	current actions center
CACTIS	community automated intelligence system
CAD	cartridge actuated device; collective address designator
CADRS	concern and deficiency reporting system
CADS	containerized ammunition distribution system
CAEMS	Computer-aided Embarkation Management System
CAF	Canadian Air Force; Combat Air Forces
CAFMS	computer-assisted force management system
CAG	civil affairs group; collective address group
CAIMS	conventional ammunition integrated management system
CAINS	carrier aircraft inertial navigation system
CAL	caliber
CALCM	conventional air-launched cruise missile
CALMS	computer-aided load manifesting system
CAM	chemical agent monitor; crisis action module
CAMPS	consolidated air mobility planning system
CAMT	countering air and missile threats
CANA	convalescent antidote for nerve agent
CANR	Canadian NORAD region
CANUS	Canada-United States
CAO	chief administrative officer; counterair operation
CAOC	combat air operations center; combined air operations center
CAO SOP	standing operating procedures for coordination of atomic operations
CAP	Civil Air Patrol; combat air patrol; configuration and alarm panel; Consolidated Appeals Process (UN); crisis action planning
CAR	Chief of the Army Reserve
CARDA	continental United States airborne reconnaissance for damage assessment; continental United States area reconnaissance for damage assessment

CARE	Cooperative for Assistance and Relief Everywhere (CAREUSA)
CARIBROC	Caribbean Regional Operations Center
CARP	computed air release point; contingency alternate route plan
CARS	combat arms regimental system
CARVER	criticality, accessibility, recuperability, vulnerability, effect, recognizability
CAS	casualty; civil aviation security; close air support
CASEVAC	casualty evacuation
CASP	computer-aided search planning
CASPER	contact area summary position report
CASREP	casualty report
CASREQ	close air support request
CAT	category; crisis action team
CATCC	carrier air traffic control center
CATF	commander, amphibious task force
CAU	crypto ancillary unit; cryptographic auxiliary unit
CAVU	ceiling and visibility unlimited
CAW	carrier air wing
CAW/ESS	crisis action weather and environmental support system
CAX	computer-assisted exercise
C-B	chemical-biological
CB	chemical-biological; construction battalion (SEABEES)
CBBLS	hundreds of barrels
CBD	chemical, biological defense
CBFS	cesium beam frequency standard
CBLTU	common battery line terminal unit
CBMR	capabilities-based munitions requirements
CBMU	construction battalion maintenance unit
CBPO	Consolidated Base Personnel Office
CBPS	chemical biological protective shelter
CBR	chemical, biological, and radiological
CBRN	Caribbean Basin Radar Network; chemical, biological, radiological, or nuclear
CBS	common battery signaling
CBT	combatting terrorism; common battery terminal
CBTZ	combat zone
CBU	cluster bomb unit; conference bridge unit; construction battalion unit
CBW	chemical and biological warfare
C/C	cabin cruiser; cast off and clear
CC	command center; component command (NATO)
CCA	carrier-controlled approach; circuit card assembly; container control activity; contamination control area; contingency capabilities assessment; contract construction agent (DOD)

CCB	Community Counterterrorism Board; Configuration Control Board
CCC	crisis coordination center; critical control circuit; coalition coordination cell; cross-cultural communications course
CCD	camouflage, concealment, and deception
CCE	container control element; continuing criminal enterprise
CCEB	Combined Communications-Electronics Board
CCF	collection coordination facility
CCG	crisis coordination group
CCGD	commander, Coast Guard district
CCIB	command center integration branch
CCIR	commander's critical information requirements; International Radio Consultative Committee
CCIS	common channel interswitch signaling
CCITT	International Telegraph and Telephone Consultative Committee
CCIU	CEF control interface unit
CCJTF	commander, combined joint task force
CCL	communications/computer link
CCO	central control officer; combat cargo officer
CCP	casualty collection point; consolidated cryptologic program; consolidation and containerization point
CCPDS	command center processing and display system
CCRD	CINCs required delivery date
CCS	central control ship; container control site
CCSA	containership cargo stowage adapter
CCSD	command communications service designator; control communications service designator
CCT	collaborative contingency targeting; combat control team
CCTI	Chairman of the Joint Chiefs of Staff commended training issue
CCTV	closed circuit television
CCW	1980 United Nations Convention on Conventional Weapons; continuous carrier wave
CD	channel designator; counterdrug
C-day	unnamed day on which a deployment operation begins
CDC	Centers for Disease Control and Prevention
CDF	combined distribution frame
CDI	cargo disposition instructions; conditioned diphase
C di	conditioned diphase
CDIP	combined defense improvement project
CDIPO	counterdrug intelligence preparation for operations
CDLMS	common data link management system
CDM	cable driver modem
CDMGB	cable driver modem group buffer
CDN	compressed dial number
CDO	command duty officer
CDOC	counterdrug operations center

CDOPS	counterdrug operations
CDP	landing craft air cushion departure point
CDR	commander; continuous data recording
CDRESC	commander, electronic security command
CDREUDAC	Commander, European Command Defense Analysis Center (ELINT) or European Data Analysis Center
CDRFORSCOM	Commander, Forces Command
CDRG	catastrophic disaster response group (FEMA)
CDRL	contract data requirements list
CDRMTMC	Commander, Military Traffic Management Command
CD-ROM	compact disc read-only memory
CDRUSAINSCOM	Commander, United States Army Intelligence and Security Command
CDRUSELEMNORAD	Commander, United States Element, North American Aerospace Defense Command
CDS	Chief of Defence Staff (Canada); container delivery system
CDSO	counterdrug support office
CDSSC	continuity of operations plan designated successor service chief
CDU	counterdrug update
C-E	communications-electronics
CE	civil engineering; command element (MAGTF); communications-electronics; counterespionage
CEB	combat engineer battalion
CEC	civil engineering corps
CECOM	communications-electronics command
CEDREP	communications-electronics deployment report
CEE	captured enemy equipment
CEF	civil engineering file; common equipment facility
CEG	common equipment group
CEI	critical employment indicator
CEM	combined effects munition
CEMC	communications-electronics management center
CEOI	communications-electronics operating instructions
CEP	cable entrance panel; circular error probable
CEPOD	communications-electronics post-deployment report
CERF	Central Emergency Revolving Fund (UN)
CERT	computer emergency response team
CERTSUB	certain submarine
CES	coast earth station
CESE	civil engineering support equipment; communications equipment support element
CESG	communications equipment support group
CESO	civil engineer support office
CESP	civil engineering support plan
CESPG	civil engineering support plan group; civil engineering support planning generator
CF	causeway ferry; drift error confidence factor

CFA	Committee on Food Aid Policies and Programmes (UN)
CFACC	combined force air component commander
CFC	Combined Forces Command, Korea
CFL	Contingency Planning Facilities List; coordinated fire line
CFM	cubic feet per minute
CFR	Code of Federal Regulations
CFSO	counterintelligence force protection source operations
CG	Chairman's guidance; Coast Guard; comptroller general
CGAS	Coast Guard Air Station
CGAUX	Coast Guard Auxiliary
CGC	Coast Guard Cutter
CGCAP	Coast Guard capabilities plan
CGFMFLANT	Commanding General, Fleet Marine Forces, Atlantic
CGFMFPAC	Commanding General, Fleet Marine Forces, Pacific
CGLSMP	Coast Guard logistic support and mobilization plan
CGS	common ground station; continental United States ground station
CH	channel; contingency hospital
CH-53	Sea Stallion
CHAMPUS	Civilian Health and Medical Program for the Uniformed Services
CHB	cargo handling battalion
CHCSS	Chief, Central Security Service
CHE	cargo-handling equipment; container-handling equipment
CHET	customs high endurance tracker
CHOP	change of operational control
CHPPM	US Army Center for Health Promotion and Preventive Medicine
CHRIS	chemical hazard response information system
CHSTR	characteristics of transportation resources
CHSTREP	characteristics of transportation resources report
CI	civilian internee; counterintelligence
CIA	Central Intelligence Agency
CIAP	central intelligence architecture plan; command, control, communications, computers, intelligence surveillance, reconnaissance (C4ISR) integrated architecture program; command intelligence architecture plan; command intelligence architecture program
CIAS	counterintelligence analysis section
CIAT	counterintelligence analytic team
CIB	combined information bureau; controlled image base
CIC	combat information center; combat intelligence center (Marine Corps); combined intelligence center; communications interface controller; content indicator code; counterintelligence center
CID	combat identification; combat intelligence division; criminal investigation division
CIDB	common intelligence database

CIDC	Criminal Investigation Division Command
CIEG/CIEL	common information exchange glossary and language
CIF	CINC Initiative Fund
CIG	communications interface group
CIHO	counterintelligence/human intelligence officer
CIL	critical item list
CILO	counterintelligence liaison officer
CIMIC	civil-military cooperation
CIN	cargo increment number
CINC	combatant commander; commander in chief; commander of a combatant command
CINCAFLANT	Commander in Chief, Air Forces Atlantic
CINCARLANT	Commander in Chief, Army Forces Atlantic
CINCCENT	Commander in Chief, United States Central Command
CINCCFC	Commander in Chief, Combined Forces Command
CINCJFCOM	Commander in Chief, United States Joint Forces Command
CINCLANTFLT	Commander in Chief, Atlantic Fleet
CINCNET	CINCs' network
CINCNORAD	Commander in Chief, North American Aerospace Defense Command
CINCPACAF	Commander in Chief, Pacific Air Forces
CINCPACFLT	Commander in Chief, Pacific Fleet
CINCUNC	Commander in Chief, United Nations Command
CINCUSAREUR	Commander in Chief, United States Army, Europe
CINCUSNAVEUR	Commander in Chief, United States Naval Forces, Europe
CIP	communications interface processor
CIPSU	communications interface processor pseudo line
CIR	continuing intelligence requirement
CIRM	International Radio-Medical Center
CIRV	common interswitch rekeying variable
CIRVIS	communications instructions for reporting vital intelligence sightings
CIS	common item support; Commonwealth of Independent States; communications interface shelter
CISO	counterintelligence support officer
CIV	civilian
CIWG	communications interoperability working group
CJATF	commander, joint amphibious task force
CJB	Congressional Justification Book
CJCS	Chairman of the Joint Chiefs of Staff
CJCSAN	Chairman of the Joint Chiefs of Staff Alerting Network
CJCSI	Chairman of the Joint Chiefs of Staff Instruction
CJCSM	Chairman of the Joint Chiefs of Staff Manual
CJDA	critical joint duty assignment
CJMAO	Central Joint Mortuary Affairs Office; Chief, joint mortuary affairs office
CJTF	combined joint task force (NATO); commander, joint task force

CKT	circuit
CLA	landing craft air cushion launch area
CLEA	civilian law enforcement agency
C-level	category level
CLF	cantilever lifting frame; combat logistics force; commander, landing force
CLGP	cannon-launched guided projectile
CLIPS	communications link interface planning system
CLPSB	CINC Logistic Procurement Support Board
CLS	contracted logistic support
CLZ	craft landing zone; cushion landing zone
CM	Chairman's memorandum; collection manager; configuration management; consequence management; control modem; countermine
Cm	mean coverage factor
CMA	collection management authority
CMAH	commander of a combatant command's Mobile Alternate Headquarters
CMC	Commandant of the Marine Corps
Cmc	midpoint compromise coverage factor
CMD	command; cruise missile defense
CMO	chief military observer; civil-military operations; collections management office(r); configuration management office
CMOC	Cheyenne Mountain Operations Center; civil-military operations center
CMOS	cargo movement operations system; complementary metal-oxide semiconductor
CMP	communications message processor
CMS	cockpit management system; command management system; community management staff; community security materiel system; contingency mutual support; crisis management system
CMTS	comments
CMTU	cartridge magnetic tape unit
CMX	crisis management exercise
CN	counternarcotic
CNA	computer network attack
CNAC	Customs National Aviation Center (USCS)
CNASP	chairman's net assessment for strategic planning
CNC	Counter-Narcotics Center (CIA); Crime and Narcotics Center
CNCE	communications nodal control element
CND	computer network defense; counternarcotics division
CNGB	Chief, National Guard Bureau
CNO	Chief of Naval Operations
CNOG	Chairman, Nuclear Operations Group
CNRF	Commander, Naval Reserve Forces

CNTY	country
CNWDI	critical nuclear weapons design information
CO	commanding officer
COA	course of action
COB	collocated operating base
COBOL	common business-oriented language
COC	combat operations center
COCOM	combatant command (command authority)
COD	carrier onboard delivery; combat operations division
COE	Army Corps of Engineers; common operating environment
COFC	container on flatcar
COGs	centers of gravity
COGARD	Coast Guard
COIN	counterinsurgency
COLDS	cargo offload and discharge system
COLISEUM	community on-line intelligence system for end-users and managers
COLT	combat observation and lasing team
COM	chief of mission; collection operations management; commander
COMACC	Commander, Air Combat Command
COMAFFOR	commander, Air Force forces
COMAFSOC	Commander, Air Force Special Operations Command
COMALF	commander airlift forces
COMARFOR	commander, Army forces
COMCAM	combat camera
COMCARGRU	commander, carrier group
COMCEN	communications center
COMCRUDESGRU	commander, cruiser destroyer group
COMDCAEUR	Commander, Defense Communications Agency Europe
COMDESRON	commander destroyer squadron
COMDT COGARD	Commandant, United States Coast Guard
COMDTINST	Commandant, United States Coast Guard Instruction
COMICEDEFOR	Commander, United States Forces, Iceland
COMIDEASTFOR	Commander, Middle East Forces
COMINEWARCOM	Commander, Mine Warfare Command
COMINT	communications intelligence
COMJCSE	Commander, Joint Communications Support Element
COMJIC	Commander, Joint Intelligence Center
COMJSOTF	commander, joint special operations task force
COMLANDFOR	commander, land forces
COMLANTAREACOGARD	Commander, Coast Guard Atlantic Area
COMLOGGRU	combat logistics group
COMM	communications
COMMARFOR	commander, Marine Corps forces
COMMDZ	Commander, Maritime Defense Zone
COMMZ	communications zone

COMNAV	Committee for European Airspace Coordination Working Group on Communications and Navigation Aids
COMNAVAIRLANT	Commander, Naval Air Force, Atlantic
COMNAVAIRPAC	Commander, Naval Air Force, Pacific
COMNAVAIRSYSCOM	Commander, Naval Air Systems Command
COMNAVCOMTELCOM	Commander, Naval Computer and Telecommunications Command
COMNAVFOR	commander, naval forces
COMNAVSEASYSCOM	Commander, Naval Sea Systems Command
COMNAVSECGRP	Commander, United States Navy Security Group
COMNAVSURFLANT	Commander, Naval Surface Force, Atlantic
COMNAVSURFPAC	Commander, Naval Surface Force, Pacific
COMP	component
COMPACAREACOGARD	Commander, Coast Guard Pacific Area
COMPASS	common operational modeling, planning, and simulation strategy
COMPES	contingency operations mobility planning and execution system
COMPLAN	communications plan
COMPUSEC	computer security
COMSAT	communications satellite
COMSC	Commander, Military Sealift Command
COMSEC	communications security
COMSOCCENT	Commander, Special Operations Command, United States Central Command
COMSOCEUR	Commander, Special Operations Command, United States European Command
COMSOCLANT	Commander Special Operations Command, United States Atlantic Command
COMSOCPAC	Commander Special Operations Command, United States Pacific Command
COMSOCSOUTH	Commander Special Operations Command, United States Southern Command
COMSOF	commander, special operations forces
COMSTAT	communications status
COMSUBLANT	Commander Submarine Force, United States Atlantic Fleet
COMSUBPAC	Commander Submarine Force, United States Pacific Fleet
COMSUPNAVFOR	commander, supporting naval forces
COMTAC	tactical communications
COMUSAFE	Commander in Chief, United States Air Force in Europe
COMUSARCENT	Commander, United States Army Forces, Central Command
COMUSCENTAF	Commander, United States Air Force, Central Command
COMUSFORAZ	Commander, United States Forces, Azores
COMUSJ	Commander, United States Forces, Japan
COMUSK	Commander, United States Forces, Korea

COMUSMARCENT	Commander, United States Marine Forces, Central Command
COMUSNAVCENT	Commander, United States Navy, Central Command
CONCAP	construction capabilities contract; Construction Capabilities Contract Process; construction capabilities contract program
CONEX	container express
CONEXPLAN	contingency and exercise plan
CONOPS	concept of operations
CONPLAN	concept plan; operation plan in concept format
CONR	continental United States North American Aerospace Defense Command Region
CONTRAIL	condensation trail
CONUS	continental United States
CONUSA	Continental United States Army
COOP	continuity of operations
COP	common operational picture
COPE	custodian of postal effects
COPG	chairman, operations planners group
COPS	communications operational planning system
CORE	contingency response program
COS	chief of staff; chief of station; critical occupational specialty
COSCOM	corps support command
COSMIC	North Atlantic Treaty Organization (NATO) security category
COSPAS	*cosmicheskaya sistyema poiska avariynch sudov* - space system for search of distressed vessels (Russian satellite system)
COT	commanding officer of troops
COTHEN	Customs Over-the Horizon Enforcement Network (USCS)
COTP	captain of the port
COTS	cargo offload and transfer system; commercial off-the-shelf
COU	cable orderwire unit
counter C3	counter command, control, and communications
COVCOM	covert communications
CP	check point; collection point; command post; contact point; counterproliferation
CP&I	coastal patrol and interdiction
CPA	Chairman's program assessment; closest point of approach
CPD	combat plans division
CPE	customer premise equipment
CPFL	contingency planning facilities list
CPG	central processor group; Contingency Planning Guidance
CPI	crash position indicator
CPM	civilian personnel manual
CPO	chief petty officer; complete provisions only
CPR	cardiopulmonary resuscitation; Chairman's program recommendation

CPS	characters per second; collective protective shelter
CPU	central processing unit
CPX	command post exercise
CRA	command relationships agreement; coordinating review authority
CRAF	civil reserve air fleet
CRAM	control random access memory
CRB	configuration review board
CRC	circuit routing chart; control and reporting center, CONUS replacement center; COOP response cell; cyclic redundancy rate
CRD	capstone requirements document; chemical reconnaissance detachment; CINC's required date
CRE	control reporting element
CRF	channel reassignment function
CRI	collective routing indicator
CRIF	cargo routing information file
CRITIC	critical information; critical intelligence communication; critical message (intelligence)
CRITICOMM	critical intelligence communications system
CRM	collection requirements management; crew resource management;
CRP	control and reporting post
CRRC	combat rubber raiding craft
CRS	Catholic Relief Services; Chairman's readiness system; coastal radio station; community relations service; container recovery system
CRSP	centralized receiving and shipping point
CRT	cathode ray tube
CRTS	casualty receiving and treatment ship
CR-UAV	close-range unmanned aerial vehicle
CRYPTO	cryptographic
CS	call sign; Chaplain Service (Air Force); circuit switch; coastal station; combat support; controlled space; creeping line single-unit; critical source
CSA	Chief of Staff, United States Army; container stuffing activity
CSAAS	combat support agency assessment system
CSADR	combat support agency director's report
CSAF	Chief of Staff, United States Air Force
CSAM	computer security for acquisition managers
CSAR	combat search and rescue
CSAR3	combat support agency responsiveness and readiness report
CSARTE	combat search and rescue task element
CSARTF	combat search and rescue task force

CSC	commander in chief's (CINC's) Strategic Concept; community support center; creeping line single-unit coordinated; International Convention for Safe Containers
CSCC	coastal sea control commander
CSE	client server environment
CSEL	circuit switch select line; combat survivor evader locator
CSEP	Chairman of the Joint Chiefs of Staff-sponsored exercise program
CSG	Chairman's Staff Group; coordinating subgroup; Cryptologic Services Group (NSA)
CSGN	coordinating subgroup for narcotics
CSH	combat support hospital
CSI	critical safety item; critical sustainability item
CSIF	communications service industrial fund
CSIPG	circuit switch interface planning guide
CSL	combat stores list
CSNP	causeway section, nonpowered
CSNP(BE)	causeway section, nonpowered (beach end)
CSNP(I)	causeway section, nonpowered (intermediate)
CSNP(SE)	causeway section, nonpowered (sea end)
CSO	communications support organization
CSOA	combined special operations area
CSOB	command systems operations branch
CSOD	command systems operation division
CSP	call service position; causeway section, powered; commence search point; contracting support plan; crisis staffing procedures (JCS); cryptologic support package
CSPAR	commander in chief's (CINC's) preparedness assessment report
CSR	commander in chief's (CINC's) summary report
CSRF	common source route file
CSS	central security service; combat service support; communications subsystem; coordinator surface search
CSSA	combat service support area
CSSC	coded switch set controller
CSSE	combat service support element (MAGTF)
CSSU	combat service support unit
CSW	compartment stowage worksheet
CT	communications terminal; control telemetry; counterterrorism
CTA	common table of allowance
CTAPS	contingency Theater Air Control System automated planning system
CTC	counterterrorist center
CTF	combined task force
CTG	commander, task group
CTID	communications transmission identifier

CTOC	corps tactical operations center
CTRIF	Combatting Terrorism Readiness Initiative Fund
CTSS	central targeting support staff
CU	cubic capacity; common unit
CUL	common-user logistics
CULT	common-user land transportation
CV	aircraft carrier; carrier; curriculum vitae
CVBG	carrier battle group
CVISC	combat visual information support center
CVN	aircraft carrier, nuclear
CVR	cockpit voice recorder
CVSD	continuous variable slope delta
CVW	carrier air wing; cryptovariable weekly (GPS)
CW	carrier wave; chemical warfare; continuous wave
CWC	Chemical Weapons Convention; composite warfare commander
CWDE	chemical warfare defense equipment
CWO	communications watch officer
CWP	causeway pier
CWPD	Conventional War Plans Division, Joint Staff (J-7)
CWR	calm water ramp
CY	calendar year

D

D	total drift, data
d	surface drift
D&M	detection and monitoring
D3A	decide, detect, deliver, and assess
D/A	digital-to-analog
DA	data adapter aerospace drift; data administrator; Department of the Army; direct action; Directorate for Administration (DIA)
Da	aerospace drift
DA&M	Director of Administration and Management
DAA	designated approving authority; display alternate area routing lists
DAAS	defense automatic addressing system
DAASO	defense automatic addressing system office
DAB	Defense Acquisition Board
DAC	Directorate for Administrative Counterintelligence and Security Activity
DACB	data adapter control block
DACG	departure airfield control group
DACM	data adapter control mode
DADCAP	dawn and dusk combat air patrol
DAF	Department of the Air Force
DAICC	domestic air interdiction coordinator center

DAL	defended asset list
DALS	downed aviator locator system
DAMA	demand assigned multiple access
DAMES	defense automatic addressing system (DAAS) automated message exchange system
DAN	Diver's Alert Network
DAO	Defense Attaché Office; defense attaché officer; department/agency/organization
DAR	distortion adaptive receiver
DARO	Defense Airborne Reconnaissance Office
DARPA	Defense Advanced Research Projects Agency
DART	disaster assistance response team; downed aircraft recovery team; dynamic analysis and replanning tool
DAS	deep air support (USMC); direct access subscriber; direct air support
DAS3	decentralized automated service support system
DASA	Department of the Army (DA) staff agencies
DASC	direct air support center
DASC-A	direct air support center (airborne)
DASD	Deputy Assistant Secretary of Defense
DASD(H&RA)	Deputy Assistant Secretary of Defense (Humanitarian & Refugee Affairs)
DASD(I)	Deputy Assistant Secretary of Defense (Intelligence)
DASD(P&HA)	Deputy Assistant Secretary of Defense (Peacekeeping and Humanitarian Affairs)
DASSS	decentralized automated service support system
DAT	deployment action team
DATT	defense attaché
DATU	data adapter termination unit
dB	decibel
DBA	database administrator
DBG	database generation
DBI	defense budget issue
DBMS	database management system
DBOF	Defense Business Operations Fund
D/C	downconverter
DC	Deputies Committee; direct current; dislocated civilian
DCA	Defense Communications Agency; defensive counterair; dual-capable aircraft
DCAA	Defense Contract Audit Agency
DCC	damage control center; deployment control center
DCCC	defense collection coordination center
DCCEP	developing country combined exercise program
DCD	data collection device
DCI	Director of Central Intelligence; dual channel interchange
D/CI&SP	Director, Counterintelligence and Security Programs
DCID	Director of Central Intelligence Directive
DCJTF	deputy commander, joint task force

DCM	data channel multiplexer; deputy chief of mission
DCMC	Office of Deputy Chairman, Military Committee
DCMO	deputy chief military observer
DCN	data link coordination net
DCNO	Deputy Chief of Naval Operations
DCO	defense coordinating officer (DOD); dial central office
DCPA	Defense Civil Preparedness Agency
DCPG	digital clock pulse generator
DCS	Defense Communications System; Defense Courier Service; deputy chief of staff; digital computer system
DCSCU	dual capability servo control unit
DC/S for RA	Deputy Chief of Staff for Reserve Affairs
DCSINT	Deputy Chief of Staff for Intelligence
DCSLOG	Deputy Chief of Staff for Logistics, US Army
DCSOPS	Deputy Chief of Staff for Operations and Plans, United States Army
DCSPER	Deputy Chief of Staff for Personnel, United States Army
DD	destroyer (Navy ship)
DDA	Deputy Director for Administration (CIA); designated development activity
D-day	unnamed day on which operations commence or are scheduled to commence
DDC	data distribution center
DDCI	Deputy Director of Central Intelligence (CIA)
DDG	guided missile destroyer
DDI	Deputy Director of Intelligence (CIA)
DDL	digital data link
DDM	digital data modem
DDMS	deputy director for military support
DDN	Defense Data Network
DDO	Deputy Director of Operations (CIA)
DDR&E	director of defense research and engineering
DDS	defense dissemination system; dry deck shelter
DDS&T	Deputy Director for Science & Technology (CIA)
DDWSO	Deputy Director for Wargaming, Simulation, and Operations
DE	damage expectancy; delay equalizer; directed energy
De	total drift error
de	individual drift error
DEA	Drug Enforcement Administration
dea	aerospace drift error
DEACN	Drug Enforcement Administration Communications Network
DEAR	disease and environmental alert report
DEARAS	Department of Defense (DOD) Emergency Authorities Retrieval and Analysis System
DeCA	Defense Commissary Agency
DECL	declassify

DEFCON defense readiness condition
DEFSMAC Defense Special Missile and Astronautics Center
DEL deployable equipment list
DEMARC demarcation
de max maximum drift error
de min minimum drift error
de minimax minimax drift error
DeMS deployment management system
DEMUX demultiplex
DEP Delayed Entry Program; deployed
DEP&S Drug Enforcement Plans and Support
DEPCJTF deputy commander, joint task force
DEPID deployment indicator code
DEPMEDS deployable medical systems
DepOpsDeps Service deputy operations deputies
DESCOM Depot System Command (Army)
DESIGAREA designated area message
DEST destination
DET detachment; detainee
DETRESFA distress phase (ICAO)
DEW directed-energy warfare
DF direction finding; dispersion factor; disposition form
DFAS Defense Finance and Accounting Service
DFC deputy force commander
DFE division force equivalent
DFM deterrent force module
DFO disaster field office (FEMA)
DFR Defense Fuel Region
DFR/E Defense Fuel Region, Europe
DFR/ME Defense Fuel Region, Middle East
DFSC Defense Fuel Supply Center
DFSP Defense Fuel Support Point
DFT deployment for training
DG defense guidance
DGIAP Defense General Intelligence and Applications Program
DGM digital group multiplex
DGZ desired ground zero
DH death due to hostilities
DHHS Department of Health and Human Services
DHS Defense Human Intelligence (HUMINT) Service; Director of Health Services
DI DIA Directorate for Intelligence Production; discrete identifier; dynamic interface
DIA Defense Intelligence Agency
DIAC Defense Intelligence Analysis Center
DIAM Defense Intelligence Agency manual; Defense Intelligence Agency memorandum
DIAR Defense Intelligence Agency (DIA) regulation

DIBITS	digital in-band interswitch trunk signaling
DIBRS	defense incident-based reporting system
DIBTS	digital in-band trunk signaling
DICO	Data Information Coordination Office
DIDHS	Deployable Intelligence Data Handling System
DIDS	Defense Intelligence Dissemination System
DIEB	Defense Intelligence Executive Board
DIEPS	Digital Imagery Exploitation Production System
DIG	digital
DII	defense information infrastructure
DIILS	Defense Institute of International Legal Studies
DILPA	diphase loop modem-A
DIMA	drilling individual mobilization augmentee
DIN	defense intelligence notice
DINET	Defense Industrial Net
DIOC	drug interdiction operations center
DIPC	defense industrial plant equipment center
DIPGM	diphase supergroup modem
DIRINT	Director of Intelligence
DIRJIATF	director, joint inter-agency task force
DIRLAUTH	direct liaison authorized
DIRM	Directorate for Information and Resource Management
DIRMOBFOR	director of mobility forces
DIRNSA	Director, National Security Agency
DIS	daily intelligence summary; defense information system; Defense Investigative Service; distributed interactive simulation
DISA	Defense Information Systems Agency
DISANMOC	Defense Information Systems Agency Network Management and Operations Center
DISCOM	division support command (Army)
DISGM	diphase supergroup
DISN	Defense Information Systems Network
DISN-E	Defense Information Systems Network – Europe
DISO	Defense Intelligence Support Office
DISP	drug investigation support program (FAA)
DISUM	daily intelligence summary
DITDS	defense information threat data system; defense intelligence threat data system
DITSUM	defense intelligence terrorist summary
DJS	Director, Joint Staff
DJSM	Director, Joint Staff memorandum
DJTFAC	deployable joint task force augmentation cell
DLA	Defense Logistics Agency
DLAM	Defense Logistics Agency manual
DLAR	Defense Logistics Agency regulation
DLEA	drug law enforcement agency
DLED	dedicated loop encryption device

DLP	data link processor
DLPMA	diphase loop modem A
DLQ	deck landing qualification
DLR	depot-level repairable
DLSA	Defense Legal Services Agency
DLTM	digital line termination module
DLTU	digital line termination unit
DM	detection and monitoring
dmax	maximum drift distance
DMB	datum marker buoy
DMC	data mode control
DMD	digital message device
DMDC	defense management data center; defense manpower data center
DME	distance measuring equipment
DMI	director military intelligence
dmin	minimum drift distance
DML	data manipulation language
DMO	directory maintenance official
DMOS	duty military occupational specialty
DMPI	designated mean point of impact
DMRD	defense management resource decision
DMRIS	defense medical regulating information system
DMS	defense message system; defense meteorological system; director of military support
DMSB	Defense Medical Standardization Board
DMSO	Defense Modeling and Simulation Office; director of major staff office; Division Medical Supply Office
DMSP	Defense Meteorological Satellite Program
DMSSC	defense medical systems support center
DMT	disaster management team (UN)
DMU	disk memory unit
DMZ	demilitarized zone
DN	digital nonsecure
DNA	Defense Nuclear Agency; deoxyribonucleic acid
DNBI	disease and nonbattle injury
DNBI casualty	disease and nonbattle injury casualty
DNC	digital nautical chart
DNI	Director of Naval Intelligence
DNIF	duty not involving flying
DNMSP	driftnet monitoring support program
DNSO	Defense Network Systems Organization
DNVT	digital nonsecure voice terminal
DNY	display area code (NYX) routing
DO	Directorate for Intelligence Operations (DIA)
DOA	dead on arrival; director of administration
DOB	date of birth; dispersal operating base
DOC	designed operational capability

DOCC	deep operations coordination cell
DOCDIV	documents division
DOCEX	document exploitation
DOD	Department of Defense
DODAAC	Department of Defense activity address code
DODAC DOD	ammunition code
DODD	Department of Defense directive
DODDS	Department of Defense Dependent Schools
DODEX	Department of Defense intelligence system information system extension
DODI	Department of Defense instruction
DODIC	Department of Defense identification code
DODID	Department of Defense Intelligence Digest
DODIIS	Department of Defense Intelligence Information System
DODIPC	Department of Defense intelligence production community
DODIPP	Department of Defense Intelligence Production Program
DOD-JIC	Department of Defense Joint Intelligence Center
DODM	data orderwire diphase modem; Department of Defense manual
DODR	Department of Defense regulation
DOE	Department of Energy
DOF	degree of freedom
DOI	Defense Special Security Communications System (DSSCS) Operating Instructions; Department of Interior
DOJ	Department of Justice
DOL	Department of Labor
DOM	day of month
DOMS	director of military support
DON	Department of the Navy
DOPMA	Defense Officer Personnel Management Act
DOR	date of rank
DOS	date of separation; day of supply; denial of service; Department of State; disk operating system
DOT	Department of Transportation
DOTEO	Department of Transportation emergency organization
DOW	data orderwire; died of wounds
DOX-T	direct operational exchange - tactical
DOY	day of year
DP	Air Force component plans officer (staff); Directorate for Policy Support (DIA)
dp	parachute drift
DPA	Defense Production Act
DPAS	Defense Priorities and Allocation System
DPC	Defense Planning Committee (NATO)
DPG	Defense Planning Guidance
DPICM	dual purpose improved conventional munitions
DPLSM	dipulse group modem
DPM	dissemination program manager

DPMO	Defense Prisoner of War (POW)/Missing Personnel (MP) Office
DPP	data patch panel; distributed production program
DPPDB	digital point positioning database
DPQ	defense planning questionnaire (NATO)
DPR	display non-nodal routing
DPRB	Defense Planning and Resources Board
DPRE	displaced persons, refugees, and evacuees
DPS	data processing system
DPSC	Defense Personnel Support Center
DPSK	differential phase shift keying
DR	dead reckoning; digital receiver; disaster relief
DRB	Defense Resources Board
DRe	dead reckoning error
DRMD	deployments requirements manning document
DRMO	Defense Reutilization Marketing Office
DRMS	distance root-mean-square
DRT	dead reckoning tracer
DRTC	designated reporting technical control
DS	Directorate for Information Systems and Services (DIA); direct support; doctrine sponsor
DSA	defense special assessment (DIA)
DSAA	Defense Security Assistance Agency
DSAR	Defense Supply Agency regulation
DSB	digital in-band trunk signaling (DIBTS) signaling buffer
DSC	digital selective calling
DSCS	Defense Satellite Communications System
DSCSOC	Defense Satellite Communications System operations center
DSDI	digital simple data interface
DSG	digital signal generator
DSI	defense simulation internet
DSL	display switch locator (SL) routing
DSMAC	digital scene-matching area correlation
DSN	Defense Switched Network
DSNET	Defense Secure Network
DSNET-2	Defense Secure Network-2
DSO	domestic support operation
DSOE	deployment schedule of events
DSP	Defense Satellite Program; Defense Support Program
DSPL	display system programming language
DSPS	Director, Security Plans and Service
DSSCS	Defense Special Security Communications System
DSSO	data system support organization; defense systems support organization
DSTP	Director of Strategic Target Planning
DSTR	destroy
DSVL	doppler sonar velocity log

DSVT	digital subscriber voice terminal
DTE	data terminal equipment; developmental test and evaluation
DTED	digital terrain elevation data
DTG	date-time group; digital trunk group (digital transmission group)
DTMF	dual tone multi-frequency
DTMR	defense traffic management regulation
DTO	division transportation office
DTOC	division tactical operations center
DTR	defense transportation regulation
DTRA	Defense Threat Reduction Agency
DTS	Defense Transportation System; diplomatic telecommunications service
DUSDL	Deputy Under Secretary of Defense for Logistics
DUSDP	Deputy Under Secretary of Defense for Policy
DUSTWUN	duty status-whereabouts unknown
DVA	Department of Veterans Affairs
DVITS	Digital Video Imagery Transmission System
DVOW	digital voice orderwire
DWMCF	double-wide modular causeway ferry
DWRIA	died of wounds received in action
DWT	deadweight tonnage
DWTS	Digital Wideband Transmission System
DX	direct exchange
DZ	drop zone
DZC	drop zone controller
DZCO	drop zone control officer
DZSO	drop zone safety officer
DZST	drop zone support team
DZSTL	drop zone support team leader

E

E	total probable error
E&DCP	evaluation and data collection plan
E&E	evasion and escape
E&M	ear and mouth; special signaling leads
E&R	evasion and recovery
E1	Echelon 1
E2	Echelon 2
E3	Echelon 3; electromagnetic environmental effects
E4	Echelon 4
E5	Echelon 5
E-8C	joint surveillance, target attack radar system (JSTARS) aircraft
EA	electronic attack (previously ECM); emergency action; executive agent; executive assistant

ea	each
EAC	echelons above corps (Army); emergency action; emergency action committee
EAD	earliest arrival date; echelons above division (Army); extended active duty
EAF	expeditionary aerospace forces
EALT	earliest anticipated launch time
EAM	emergency action message
EAP	emergency action plan; emergency action procedures
EAP-CJCS	emergency action procedures of the Chairman of the Joint Chiefs of Staff
EARLY	evasion and recovery supplemental data report
E-ARTS	en route automated radar tracking system
EASTPAC	eastern Pacific Ocean
EBCDIC	extended binary coded decimal interchange code
EC	electronic combat; error control; European Community
ECAC	Electromagnetic Compatibility Analysis Center
ECB	echelons corps and below (Army)
ECC	evacuation control center
ECN	electronic change notice; Minimum Essential Emergency Communications Network
ECOSOC	Economic and Social Council (UN)
ECP	emergency command precedence; engineering change proposal
ECS	expeditionary combat support
ECU	environmental control unit
ED	envelope delay; evaluation directive
EDC	estimated date of completion
EDD	earliest delivery date
EDI	electronic data interchange
EDSS	equipment deployment and storage system
EE	emergency establishment
EEBD	emergency escape breathing device
EECT	end evening civil twilight
EED	electro-explosive device
EEFI	essential elements of friendly information
EELV	evolved expendable launch vehicle
EEO	equal employment opportunity
EEPROM	electronic erasable programmable read-only memory
EER	extended echo ranging
EEZ	exclusive economic zone
EFTO	encrypt for transmission only
EGS	Earth ground station
EHF	extremely high frequency
EI	exercise item
EIA	Electronic Industries Association
EIS	Environmental Impact Statement
ELBA	emergency locator beacon

ELCAS	elevated causeway system
ELCAS(M)	elevated causeway system (modular)
ELCAS(NL)	elevated causeway system (Navy lighterage)
ELD	emitter locating data
ELECTRO-OPTINT	electro-optical intelligence
ELINT	electronic intelligence
ELOS	extended line of sight
ELPP	equal level patch panel
ELR	extra-long-range aircraft
ELSEC	electronics security
ELSET	element set
ELT	emergency locator transmitter
ELV	expendable launch vehicle
ELVA	emergency low visibility approach
EM	executive manager
EMC	electromagnetic compatibility
EMCON	emission control
EMCON orders	emission control orders
EME	electromagnetic environment
EMEDS	Expeditionary Medical Support
EMI	electromagnetic interface; electromagnetic interference
EMP	electromagnetic pulse
EMR hazards	electromagnetic radiation hazards
EMS	emergency medical services
EMSEC	emanations security
EMT	emergency medical technician; emergency medical treatment
EMV	electromagnetic vulnerability
ENDEX	exercise termination
ENL	enlisted
ENSCE	enemy situation correlation element
ENWGS	Enhanced Naval Warfare Gaming System
EO	electro-optical; end office; equal opportunity; executive order; eyes only
EOB	electronic order of battle; enemy order of battle
EOC	early operational capability; emergency operations center
EOD	explosive ordnance disposal
EOI	electro-optic(al) imagery
EO-IR	electro-optical-infrared
EOL	end of link
EOM	end of message
EOP	emergency operating procedures
EOW	engineering orderwire
EP	electronic protection; emergency procedures; execution planning
EPA	Environmental Protection Agency; evasion plan of action
EPBX	electronic private branch exchange
EPF	enhanced palletized load system (PLS) flatrack

EPH	emergency planning handbook
EPIC	El Paso Intelligence Center
EPIRB	emergency position-indicating radio beacon
EPROM	erasable programmable read-only memory
EPW	enemy prisoner of war
EPW/CI	enemy prisoner of war/civilian internee
ERGM	extended range guided munitions
ERO	engine running on or offload
ERSD	estimated return to service date
ERT	emergency response team (FEMA)
ERT-A	emergency response team – advance
ES	electronic warfare support (previously ESM)
ESB	engineer support battalion
ESC	Electronics Systems Center
ESF	Economic Support Fund; emergency support function
ESG	executive steering group
ESGN	electrically suspended gyro navigation
ESI	extremely sensitive information
ESO	embarkation staff officer
EST	emergency service team; emergency support team (FEMA); en route support team
ETA	estimated time of arrival
ETAC	emergency tactical air control; enlisted terminal attack controller
ETD	estimated time of departure
ETI	estimated time of intercept
ETIC	estimated time for completion; estimated time in commission
ETM	electronic transmission
ETPL	endorsed TEMPEST products list
ETR	export traffic release
ETS	European telephone system
ETSS	extended training service specialist
ETX	end of text
EU	European Union
E-UAV	endurance unmanned aerial vehicle
EUB	essential user bypass
EURV	essential user rekeying variable
EUSA	Eighth US Army
EUSCS	effective US controlled ships
EVC	evasion chart
EW	early warning; electronic warfare
EWC	electronic warfare coordinator
EWCS	electronic warfare control ship
EW/GCI	early warning/ground-controlled intercept
EWIR	electronic warfare integrated reprogramming
EWO	electronic warfare officer
EXCIMS	Executive Council for Modeling and Simulations

EXCOM	extended communications search
ExCom	executive committee
EXDIR	Executive Director (CIA)
EXDIR/ICA	Executive Director for Intelligence Community Affairs (USG)
EXER	exercise
EXORD	execute order
EXPLAN	exercise plan
EZ	extraction zone
EZCO	extraction zone control officer

F

F	flare patterns; flash
FA	feasibility assessment; field artillery
FAA	Federal Aviation Administration; Foreign Assistance Act
FAC	forward air controller
FAC(A)	forward air controller (airborne)
FACSFAC	fleet area control and surveillance facility
FAD	feasible arrival date; force activity designator
FAE	fuel air explosive
FAM	functional area manager
FAMP	forward area minefield planning
FAO	Food and Agriculture Organization (UN); foreign area officer
FAPES	Force Augmentation Planning and Execution System
FAR	Federal Aviation Regulation
FARP	forward arming and refueling point
FAS	Foreign Agricultural Service (USDA); frequency assignment subcommittee; fueling at sea; functional account symbol
FASCAM	family of scatterable mines
FAX	facsimile
FBI	Federal Bureau of Investigation
FBIS	Foreign Broadcast Information Service
FC	field circular; floating causeway; floating craft; force commander
FCA	functional configuration audit
FCC	Federal Communications Commission
FCE	forward command element
FCG	foreign clearance guide
FCO	Federal coordinating officer
FCP	fire control party
FCT	firepower control team
FDA	Food and Drug Administration
FDBM	functional database manager
FDC	fire direction center
FDESC	force description

FDL	fast deployment logistics
FDLP	flight deck landing practice
FDM	frequency division multiplexing
FDO	fire direction officer; flexible deterrent option; flight deck officer
FDR/FA	flight data recorder/fault analyzer
FDS	fault detection system
FDSL	fixed directory subscriber list
FDSS	fault detection subsystem
FDSSS	flight deck status and signaling system
FDUL	fixed directory unit list
FDX	full duplex
FEBA	forward edge of the battle area
FEC	forward error correction
FED-STD	federal standard
FEK	frequency exchange keying
FEMA	Federal Emergency Management Agency
FEP	fleet satellite (FLTSAT) extremely high frequency (EHF) package
FEU	forty-foot equivalent unit
FEZ	fighter engagement zone
FF	navy fast frigate
Ff	fatigue correction factor
FFA	free-fire area
FFCC	force fires coordination center
FFE	fire for effect; flame field expedients
FFG	guided missile frigate
FFH	fast frequency hopping
FFH-net	fast-frequency-hopping net
FFHT-net	fast-frequency-hopping training net
FFP	fresh frozen plasma
FFTU	forward freight terminal unit
FG	fighter group
FGMDSS	Future Global Maritime Distress and Safety System
FH	fleet hospital
FHA	Bureau for Food and Humanitarian Assistance; Federal Highway Administration; foreign humanitarian assistance
F-hour	effective time of announcement by the Secretary of Defense to the Military Departments of a decision to mobilize Reserve units
FHWA	Federal Highway Administration
FIA	functional interoperability architecture
FIC	force indicator code
FID	foreign internal defense
FIDAF	foreign internal defense augmentation force
FIE	fly-in echelon
FIFO	first-in-first-out

FIR	first-impressions report; flight information region
FIRCAP	foreign intelligence requirements capabilities and priorities
FIS	flight information service; foreign intelligence service
FISC	fleet and industrial supply center
FISINT	foreign instrumentation signals intelligence
FISS	foreign intelligence security service
FIST	fire support team; fleet imagery support terminal
FIWC	fleet information warfare center
FIXe	navigational fix error
FLAR	forward-looking airborne radar
FLETC	Federal Law Enforcement Training Center
FLIP	flight information publication; flight instruction procedures
FLIR	forward-looking infrared
FLO/FLO	float-on/float-off
FLOLS	fresnel lens optical landing system
FLOT	forward line of own troops
FLP	force level planning
FLS	forward logistic site
FLTCINC	fleet commander in chief
FLTSAT	fleet satellite
FLTSATCOM	fleet satellite communications
FM	field manual (Army); financial management; flare multiunit; force module; frequency modulation; functional manager
FMA-net	frequency management A-net
FMAS	foreign media analysis subsystem
FMC	force movement characteristics; full mission-capable
FMCC	force movement control center (USMC)
FMCH	fleet multichannel
FMCR	Fleet Marine Corps Reserve
FMF	Fleet Marine Force
FMFP	foreign military financing program
FMID	force module identifier
FMO	frequency management office
FMP	force module package
FMS	force module subsystem; foreign military sales
FMSC	frequency management sub-committee
FMT-net	frequency management training net
FNOC	Fleet Numerical Oceanographic Command
FNS	foreign nation support
FO	fiber optic; flash override; forward observer
FOB	forward operating base; forward operations base
FOC	full operational capability
FOD	field operations division; foreign object damage
FOFW	fiber optic field wire
FOI	fault detection isolation
FOIA	Freedom of Information Act
FOIU	fiber optic interface unit

FOL	fiber optic link; forward operating location
FON	freedom of navigation (operations)
FORSCOM	United States Army Forces Command
FORSTAT	force status and identity report
FOS	full operational status
FOT	follow-on operational test
FOUO	for official use only
FOV	field of view
FP	firing point; force protection; frequency panel
FPA	foreign policy advisor
FPC	field press censorship
FPD	foreign post differential
FPM	Federal personnel manual
FPO	Fleet post office
FPTS	forward propagation by tropospheric scatter
FR	final report; frequency response
FRA	Federal Railroad Administration (DOT)
FRAG	fragmentation code
FRAG order	fragmentary order
FRD	formerly restricted data
FREQ	frequency
FRMAC	Federal Radiological Monitoring and Assessment Center (DOE)
FRN	force requirement number
FROG	free rocket over ground
FRP	Federal response plan (USG)
FRRS	frequency resource record system
FS	fighter squadron; file separator; file server; flare single-unit
fs	search radius safety factor
FSA	fire support area
FSB	fire support base; forward staging base; forward support battalion
FSC	fire support coordinator
FSCC	fire support coordination center
FSCL	fire support coordination line
FSCM	fire support coordinating measure
FSCOORD	fire support coordinator
FSE	fire support element
FSK	frequency shift key
FSN	foreign service national
FSO	fire support officer; flight safety officer; foreign service officer
FSS	fast sealift ships; fire support station; flight service station
FSSG	force service support group (USMC)
FSST	forward space support to theater
FST	fleet surgical team
FSU	former Soviet Union; forward support unit
FSW	feet of seawater

ft	feet; foot
ft3	cubic feet
FTC	Federal Trade Commission
FTRG	fleet tactical readiness group
FTS	Federal telecommunications system; Federal telephone service; file transfer service
FTU	field training unit; freight terminal unit
FTX	field training exercise
FUAC	functional area code
F/V	fishing vessel
Fv	aircraft speed correction factor
FVT	Force Validation Tool
FW	fighter wing; weather correction factor
FWD	forward
FWDA	friendly weapon danger area
FWF	former warring factions
FY	fiscal year
FYDP	Future Years Defense Program

G

G-1	Army or Marine Corps component manpower or personnel staff officer (Army division or higher staff, Marine Corps brigade or higher staff)
G-2	Army or Marine Corps component intelligence staff officer (Army division or higher staff, Marine Corps brigade or higher staff)
G-3	Army or Marine Corps component operations staff officer (Army division or higher staff, Marine Corps brigade or higher staff)
G-4	Army or Marine Corps component logistics staff officer (Army division or higher staff, Marine Corps brigade or higher staff)
G/A	ground to air
GA	Tabun, a nerve agent
GAA	general agency agreement
GAMSS	global air mobility support system
GAO	General Accounting Office
GATB	guidance, apportionment, and targeting board
GB	group buffer; Sarin, a nerve agent
GBL	government bill of lading
GBR	ground-based radar
GBS	Global Broadcast Service; Global Broadcast System
GC	Geneva Convention
GC3A	global command, control, and communications assessment
GC4A	global command, control, communications, and computer assessment
GCA	ground controlled approach

GCCS	Global Command and Control System
GCE	ground combat element (MAGTF)
GCI	ground control intercept
GCRI	general collective routing indicator (RI)
GCS	ground control station
GCSS	Global Combat Support System
GD	Soman, a nerve agent
GDIP	General Defense Intelligence Program
GDIPP	General Defense Intelligence Proposed Program
GDP	General Defense Plan (SACEUR)
GDSS	Global Decision Support System
GE	general engineering
GENADMIN	general admin (message)
GENSER	general service (message)
GENTEXT	general text
GEO	geosynchronous earth orbit
GEOCODE	geographic code
GEOFILE	geolocation code file; standard specified geographic location file
GEOLOC	geographic location; geographic location code
GEOREF	geographic reference; world geographic reference system
GF	a nerve agent
GFE	government-furnished equipment
GFI	government-furnished information
GFM	government-furnished material
GFMPL	Graphics Fleet Mission Program Library
GFOAR	global family of operation plans assessment report
GFU	group framing unit
GHz	gigahertz
GI&S	geospatial information and services
GIAC	graphic input aggregate control
GIC	(*gabarit international de chargement)* international loading gauge
GII	global information infrastructure
GLCM	ground launched cruise missile
GLINT	gated laser intensifier
GLO	ground liaison officer
GM	group modem
GMD	group mux and/or demux
GMDSS	Global Maritime Distress and Safety System
GMF	ground mobile forces
GMI	general military intelligence
GMR	graduated mobilization response; ground mobile radar
GMT	Greenwich Mean Time
GMTI	ground moving target indicator
GOCO	government-owned, contractor-operated
GOES	geostationary operational environmental satellite
GOS	grade of service

GOSG	general officer steering group
GOTS	government off-the-shelf
GP	group
GPD	gallons per day
GPEE	general purpose encryption equipment
GPM	gallons per minute
GPMDM	group modem
GPMRC	Global Patient Movement Requirements Center
GPS	global positioning system
GR	graduated response
GRASP	general retrieval and sort processor
GRCA	ground reference coverage area
GRREG	graves registration
GS	general service; general support; ground speed; group separator
GSA	General Services Administration; general support artillery
GSE	ground support equipment
GSI	glide slope indicator
GSM	ground station module
GSO	general services officer
GS-R	general support-reinforcing
GSR	general support-reinforcing; ground surveillance radar
GSSA	general supply support area
GSSC	Global Satellite Communications Support Center
gt	gross ton
GTL	gun-target line
GTM	global transportation management
GTN	Global Transportation Network
GUARD	US National Guard and Air Guard
GUARDS	General Unified Ammunition Reporting Data System
G/VLLD	ground/vehicle laser locator designator
GW	guerrilla warfare
GWC	global weather central
GWEN	Ground Wave Emergency Network

H

H&I	harassing and interdicting
HA	holding area
HAARS	high-altitude airdrop resupply system
HAB	high altitude burst
HAC	helicopter aircraft commander
HACC	humanitarian assistance coordination center
HAHO	high-altitude high-opening parachute technique
HALO	high-altitude low-opening parachute technique
HARM	high-speed antiradiation missile
HARP	high altitude release point
HAST	humanitarian assistance survey team

HATR	hazardous air traffic report
HAZ	hazardous cargo
HAZMAT	hazardous materials
HB	heavy boat
HCA	humanitarian and civic assistance
HCAS	hostile casualty
HCL	hydrochloride
HCO	helicopter control officer
HCP	hardcopy printer
HCS	helicopter combat support (Navy); helicopter control station; helicopter coordination section
HD	a mustard agent; harmonic distortion
HDC	harbor defense commander; helicopter direction center
HDPLX	half duplex
HE	heavy equipment; high explosive
HEAT	helicopter external air transport; high explosive antitank
HEC	helicopter element coordinator
HEFOE	hydraulic electrical fuel oxygen engine
HEL-H	heavy helicopter
HEL-L	light helicopter
HEL-M	medium helicopter
HELO	helicopter
HEMP	high-altitude electromagnetic pulse
HEMTT	heavy expanded mobile tactical truck
HEO	highly elliptical orbit
HERF	hazards of electromagnetic radiation to fuel
HERO	electromagnetic radiation hazards; hazards of electromagnetic radiation to ordnance
HERP	hazards of electromagnetic radiation to personnel
HET	heavy equipment transporter
HF	high frequency
HFDF	high frequency direction-finding
HH	homing pattern
HHD	headquarters and headquarters detachment
H-hour	seaborne assault landing hour; specific time an operation or exercise begins
HICAP	high-capacity firefighting foam station
HIDACZ	high-density airspace control zone
HIDTA	high-intensity drug trafficking area
HIFR	helicopter in-flight refueling
HIMEZ	high-altitude missile engagement zone
HIRSS	hover infrared suppressor subsystem
HIV	human immuno-deficiency virus
HJ	crypto key change
HLPS	heavy-lift pre-position ship
HLZ	helicopter landing zone
HMH	Marine heavy helicopter squadron
HMLA	Marine light/attack helicopter squadron

HMM	Marine medium helicopter squadron
HMMWV	high mobility multipurpose wheeled vehicle
HMW	health, morale, and welfare
HN	host nation
HNS	host-nation support
HOB	height of burst
HOC	human intelligence operations cell; humanitarian operations center
HOD	head of delegation
HOGE	hover out of ground effect
HOIS	hostile intelligence service
HOM	head of mission
HOSTAC	helicopter operations from ships other than aircraft carriers (USN publication)
HOTPHOTOREP	hot photo interpretation report
HPA	high power amplifier
HPMSK	high priority mission support kit
HPT	high-payoff target
HQ	HAVE QUICK; headquarters
HQCOMDT	headquarters commandant
HQDA	Headquarters, Department of the Army
HQFM-net	HAVE QUICK frequency modulation net
HQFMT-net	HAVE QUICK frequency modulation training net
HQMC	Headquarters, Marine Corps
HRJTF	humanitarian relief joint task force
HRS	horizon reference system
HRT	hostage rescue team
HS	helicopter antisubmarine warfare squadron (Navy); homing single-unit
HSB	high speed boat
HSCDM	high speed cable driver modem
HSD	human intelligence support detachment
HSE	headquarters support element; human intelligence support element
HSEP	hospital surgical expansion package (USAF)
HSI	hyperspectral imagery
HSLS	health service logistic support
HSM	humanitarian service medal
HSPR	high speed pulse restorer
HSS	health service support
HSSDB	high speed serial data buffer
HST	helicopter support team
HT	hatch team
HTERRCAS	hostile terrorist casualty
HTG	hard target graphic
HTH	high test hypochlorite
HU	hospital unit
HUD	head-up display

HUMINT	human intelligence; human resources intelligence
HUMRO	humanitarian relief operation
HUS	hardened unique storage
HVAA	high value airborne asset
HVT	high-value target
HWM	high water mark
Hz	hertz

I

I	immediate; individual
I&W	indications and warning
IA	information assurance; initial assessment
IADB	Inter-American Defense Board
IADS	Integrated Air Defense System
IAEA	International Atomic Energy Agency (UN)
IAF	initial approach fix
IAP	international airport
IAR	interoperability assessment report
IASC	Interagency Standing Committee (UN); interim acting service chief
IATA	International Air Transport Association
IATACS	Improved Army Tactical Communications System
IATO	interim authority to operate
IAW	in accordance with
I/B	inboard
IBB	International Broadcasting Bureau
IBES	intelligence budget estimate submission
IBM	International Business Machines
IBS	Integrated Broadcast System
IC	incident commander; intelligence community; intercept
ICAD	individual concern and deficiency
ICAO	International Civil Aviation Organization
ICBM	intercontinental ballistic missile
ICC	information coordination center; intelligence coordination center (USCG)
ICD	International Code of Diseases; International Cooperation and Development Program (USDA)
ICEDEFOR	Iceland Defense Forces
IC/EXCOM	Intelligence Community Executive Committee
ICITAP	International Crime Investigative Training Assistance Program (DOJ)
ICM	improved conventional munitions; integrated collection management
ICN	idle channel noise; interface control net
ICNIA	integrated communications, navigation, and identification avionics
ICOD	intelligence cutoff data

ICON	imagery communications and operations node; intermediate coordination node
ICP	incident control point; intertheater communications security (COMSEC) package; interface change proposal; inventory control point
ICRC	International Committee of the Red Cross
ICRI	interswitch collective routing indicator
ICS	incident command system; internal communications system; inter-Service chaplain support
ICSAR	interagency committee on search and rescue
ICU	intensive care unit; interface control unit
ICVA	International Council of Voluntary Agencies
ID	identification
IDAD	internal defense and development
IDB	integrated database
IDCA	International Development Cooperation Agency
IDDF	intermediate data distribution facility
IDEAS	Intelligence Data Elements Authorized Standards
IDEX	imagery data exploitation system
IDF	intermediate distribution frame
IDHS	intelligence data handling system
IDNDR	International Decade for Natural Disaster Reduction (UN)
IDO	installation deployment officer
IDP	imminent danger pay
IDS	individual deployment site; integrated deployment system; interface design standards; intrusion detection system
IDSS	interoperability decision support system
IDT	inactive duty training
IDZ	inner defense zone
IED	improvised explosive device
IEL	illustrative evaluation scenario
IEMATS	improved emergency message automatic transmission system
IER	information exchange requirement
IES	imagery exploitation system
IESS	imagery exploitation support system
IEW	intelligence and electronic warfare
IF	intermediate frequency
IFC	intelligence fusion center
IFCS	improved fire control system
IFF	identification, friend or foe
IFFN	identification, friend, foe, or neutral
IFF/SIF	identification, friend or foe/selective identification feature
IFR	instrument flight rules
IFRC	International Federation of Red Cross and Red Crescent Societies
IG	inspector general
IGSM	interim ground station module (JSTARS)

IHADSS	integrated helmet and display sight system (Army)
IIB	interagency information bureau
IICL	Institute of International Container Lessors
III	incapacitating illness or injury
IIM	intelligence information management
IIP	interoperability improvement program
IIR	imagery interpretation report; imaging infrared; intelligence information report
IJC3S	initial joint command, control, and communications system; Integrated Joint Command, Control, and Communications System
IL	intermediate location
ILO	International Labor Organization (UN)
ILS	instrument landing system; integrated logistic support
IMA	individual mobilization augmentee
IMC	instrument meteorological conditions; International Medical Corps
IMDG	international maritime dangerous goods (UN)
IMET	international military education and training
IMF	International Monetary Fund (UN)
IMI	international military information
IMINT	imagery intelligence
IMIT	international military information team
IMLTU	intermatrix line termination unit
IMM	integrated materiel management
IMMDELREQ	immediate delivery required
IMO	information management officer; International Maritime Organization
IMOSAR	International Maritime Organization (IMO) search and rescue manual
IMP	implementation; inventory management plan
IMS	information management system; international military staff; international military standardization
IMSU	installation medical support unit
IMU	intermatrix unit; inertial measuring unit
IN	Air Force component intelligence officer (staff); impulse noise; instructor
INCERFA	uncertainty phase (ICAO)
INCNR	increment number
INCSEA	incidents at sea
IND	improvised nuclear device
INF	infantry
INFLIGHTREP	in-flight report (voice only)
INFOSEC	information security
ING	Inactive National Guard
INID	intercept network in dialing
INJILL	injured or ill

INL	International Narcotics and Law Enforcement Affairs (USG)
INM	international narcotics matters
INMARSAT	international maritime satellite
INR	Bureau of Intelligence and Research, Department of State
INREQ	information request
INS	Immigration and Naturalization Service; inertial navigation system; insert code
INSCOM	United States Army Intelligence and Security Command
INTAC	individual terrorism awareness course
INTACS	integrated tactical communications system
INTELSAT	International Telecommunications Satellite Organization
INTELSITSUM	intelligence situation summary
INTERCO	International Code of signals
INTERPOL	International Criminal Police Organization
INTERPOL-USNCB	International Criminal Police Organization, United States National Central Bureau (DOJ)
INTREP	intelligence report
INTSUM	intelligence summary
INU	inertial navigation unit; integration unit
INV	invalid
INVOL	involuntary
I/O	input/output
IO	information objectives; information operations; intelligence oversight; international organization
IOC	Industrial Operations Command; initial operational capability; intelligence operations center; investigations operations center
IOI	injured other than hostilities or illness
IOM	installation, operation, and maintenance; International Organization for Migration
IOP	interface operating procedure
IOU	input/output unit
IP	initial point; initial position; instructor pilot; Internet Protocol
IPA	intelligence production agency
IPB	intelligence preparation of the battlespace
IPBD	intelligence program budget decision
IPC	interagency planning cell
IPDM	intelligence program decision memorandum
IPDP	inland petroleum distribution plan
IPDS	imagery processing and dissemination system; inland petroleum distribution system
IPE	industrial plant equipment
IPIR	initial photo interpretation report
IPL	integrated priority list
IPO	International Program Office
IPOM	intelligence program objective memorandum

IPP	impact point prediction; industrial preparedness program
IPR	intelligence production requirement
IPRG	intelligence program review group
IPS	illustrative planning scenario; interoperability planning system
IPSG	intelligence program support group
IPSP	intelligence priorities for strategic planning
IPT	integrated planning team; integrated process team; Integrated Product Team
I/R	internment/resettlement
IR	incident report; information rate; information requirement; infrared; intelligence requirement
IRAC	interdepartment radio advisory committee
IRC	International Red Cross; International Rescue Committee
IRCCM	infrared counter countermeasures
IRCM	infrared countermeasures
IRDS	infrared detection set
IRF	Immediate Reaction Forces (NATO)
IRINT	infrared intelligence
IRO	international relief organization
IR pointer	infrared pointer
IRR	Individual Ready Reserve; integrated readiness report
IRS	Internal Revenue Service
IRST	infrared search and track
IRSTS	infrared search and track sensor; Infrared Search and Track System
IRT	Initial Response Team
IS	information system; interswitch
ISA	inter-Service agreement
ISB	intermediate staging base
ISDB	integrated satellite communications (SATCOM) database
ISE	intelligence support element
ISMMP	integrated continental United States (CONUS) medical mobilization plan
ISO	International Organization for Standardization; isolation
ISOO	Information Security Oversight Office
ISOPAK	International Organization for Standardization package
ISOPREP	isolated personnel report
ISP	internet service provider
ISR	intelligence, surveillance, and reconnaissance
ISSA	inter-Service support agreement
ISSM	information system security manager
ISSO	information systems security organization
IST	integrated system test; interswitch trunk
ITA	international telegraphic alphabet
ITAC	intelligence and threat analysis center (Army)
ITALD	improved tactical air-launched decoy
ITAR	international traffic in arms regulation (coassembly)

ITF	intelligence task force (DIA)
ITO	installation transportation officer
ITRO	inter-Service training organization
ITU	International Telecommunications Union
ITV	in-transit visibility
ITW/AA	integrated tactical warning and attack assessment
IUWG	inshore undersea warfare group
IV	intravenous
IVSN	Initial Voice Switched Network
IW	information warfare
IW-D	defensive information warfare
IWG	intelligence working group; interagency working group
IWSC	Information Warfare Support Center

J

J-3	operations directorate of a joint staff
J-4	logistics directorate of a joint staff
J-5	plans directorate of a joint staff
J-6	command, control, communications, and computer systems directorate of a joint staff
J-7	Operational Plans and Joint Force Development Directorate, Joint Staff
J-8	Director for Force Structure, Resource, and Assessment, Joint Staff
JAAP	joint airborne advance party
JAAR	joint after-action report
JAARS	joint after-action reporting system
JAAT	joint air attack team
JA/ATT	joint airborne/air transportability training
JAC	joint analysis center
JACC	joint airspace control center
JACCC	joint airlift coordination and control cell
JACC/CP	joint airborne communications center/command post
JADO	joint air defense operations
JAG	Judge Advocate General
JAI	joint administrative instruction; joint airdrop inspection
JAIC	joint air intelligence center
JAIEG	joint atomic information exchange group
JAMPS	Joint Interoperability of Tactical Command and Control Systems (JINTACCS) automated message preparation system
JANAP	Joint Army, Navy, Air Force publication
JAO	joint air operations
JAOC	joint air operations center
JAPO	joint area petroleum office
JAR	joint activity report
JARCC	joint air reconnaissance control center

JARS	joint automated readiness system
JASC	joint action steering committee
JASSM	Joint Air-to-Surface Standoff Missile
JAT	joint acceptance test
JATACS	joint advanced tactical cryptological support
JATF	joint amphibious task force
JBP	Joint Blood Program
JBPO	Joint Blood Program Office
JC2WC	joint command and control warfare center
JCASREP	joint casualty report
JCAT	joint crisis action team
JCC	joint command center; joint course catalog
JCCB	Joint Configuration Control Board
JCCC	joint combat camera center; joint communications control center
JCCP	joint casualty collection point
JCEOI	joint communications-electronics operating instructions
JCET	joint combined exercise for training
JCEWS	joint force commander's electronic warfare staff
JCGRO	joint central graves registration office
JCIDO	Joint Combat Identification Office
JCIOC	joint counterintelligence operations center
JCISB	Joint Counterintelligence Support Branch
JCLL	joint center for lessons learned
JCMA	joint communications security (COMSEC) monitor activity
JCMC	joint crisis management capability
JCMEB	joint civil-military engineering board
JCMEC	joint captured materiel exploitation center
JCMOTF	joint civil-military operations task force
JCMPO	Joint Cruise Missile Project Office
JCMT	joint collection management tools
JCN	joint communications network
JCS	Joint Chiefs of Staff
JCSAN	Joint Chiefs of Staff Alerting Network
JCSAR	joint combat search and rescue
JCSC	joint communications satellite center
JCSE	joint communications support element
JCSM	Joint Chiefs of Staff memorandum
JCSS	joint communications support squadron
JCTN	joint composite track network
JDA	joint duty assignment
JDAL	Joint Duty Assignment List
JDAM	Joint Direct Attack Munition
JDAMIS	Joint Duty Assignment Management Information System
JDC	joint deployment community; Joint Doctrine Center
JDEC	joint document exploitation center
JDIG	Joint Drug Intelligence Group

JDISS	joint deployable intelligence support system
JDN	joint data network
JDSS	Joint Decision Support System
JDSSC	Joint Data Systems Support Center
JDST	joint decision support tools
JDTC	joint deployment training center
JDWP	Joint Doctrine Working Party
JEAP	Joint Electronic Intelligence (ELINT) Analysis Program
JECG	joint exercise control group
JECPO	Joint Electronic Commerce Program Office
JEEP	joint emergency evacuation plan
JEL	Joint Electronic Library
JEM	joint exercise manual
JEMP	joint exercise management package
JEPES	Joint Engineer Planning and Execution System
JET	Joint Operation Planning and Execution System (JOPES) editing tool
JEWC	Joint Electronic Warfare Center
JEZ	joint engagement zone
JFA	joint field activity
JFACC	joint force air component commander
JFAST	joint flow and analysis system for transportation
JFC	joint force commander
JFE	joint fires element
JFIP	Japanese facilities improvement project
JFLCC	joint force land component commander
JFMCC	joint force maritime component commander
JFMO	Joint Frequency Management Office
JFP	joint force package (packaging)
JFRB	Joint Foreign Release Board
JFRG	joint force requirements generator
JFS	joint force surgeon
JFSOCC	joint force special operations component commander
JFTR	joint Federal travel regulations
JFUB	Joint Facilities Utilization Board
JI	joint inspection
JIADS	joint integrated air defense system
JIATF	joint interagency task force (DOD)
JIATF-E	joint interagency task force - East
JIATF-S	joint interagency task force - South
JIATF-W	joint interagency task force - West
JIB	joint information bureau
JIC	joint intelligence center
JICC	joint information coordination center
JICO	joint interface control officer
JICPAC	Joint Intelligence Center, Pacific
JIDC	joint interrogation and debriefing center
JIEO	joint interoperability engineering organization

JIEP	joint intelligence estimate for planning
JIES	joint interoperability evaluation system
JILE	joint intelligence liaison element
JIMP	joint implementation master plan
JIMPP	joint industrial mobilization planning process
JIMS	joint information management system
JINTACCS	Joint Interoperability of Tactical Command and Control Systems
JIOP	joint interface operational procedures
JIOP-MTF	joint interface operating procedures-message text formats
JIPB	joint intelligence preparation of the battlespace
JIPC	joint imagery production complex
JIPTL	joint integrated prioritized target list
JISE	joint intelligence support element
JITC	joint interoperability test command
JLCC	joint lighterage control center
JLNCHREP	joint launch report
JLOTS	joint logistics over-the-shore
JLRC	joint logistics readiness center
JM&S	joint modeling and simulation
JMAARS	joint model after-action review system
JMAO	joint mortuary affairs office; joint mortuary affairs officer
JMAS	joint manpower automation system
JMAT	joint mobility assistance team
JMC	joint military command; joint movement center
JMCG	Joint Mobility Control Group
JMCIS	joint maritime command information system
JMCOMS	joint maritime communications system
JMEM	Joint Munitions Effectiveness Manual
JMEM-SO	Joint Munitions Effectiveness Manual-Special Operations
JMET	joint mission-essential task
JMETL	joint mission-essential task list
JMFU	joint meteorological and oceanographic (METOC) forecast unit
JMIC	Joint Military Intelligence College
JMICS	Joint Worldwide Intelligence Communications System (JWICS) mobile integrated communications system
JMIE	joint maritime information element
JMIP	joint military intelligence program
JMITC	Joint Military Intelligence Training Center
JMMC	Joint Material Management Center
JMNA	joint military net assessment
JMO	joint force meteorological and oceanographic (METOC) officer; joint maritime operations
JMO(AIR)	joint maritime operations (air)
JMP	joint manpower program
JMPA	joint military postal activity; joint military satellite communications (MILSATCOM) panel administrator

JMPAB	Joint Materiel Priorities and Allocation Board
JMRC	joint mobile relay center
JMRO	Joint Medical Regulating Office
JMRR	Joint Monthly Readiness Review
JMSEP	joint modeling and simulation executive panel
JMSWG	Joint Multi-Tactical Digital Information Link (Multi-TADIL) Standards Working Group
JMT	joint military training
JMTCA	joint munitions transportation coordinating activity
JMTCSS	Joint Maritime Tactical Communications Switching System
JMUA	Joint Meritorious Unit Award
JNACC	joint nuclear accident coordinating center
JNOCC	Joint Operation Planning and Execution System (JOPES) Network Operation Control Center
JNPE	joint nuclear planning element
JOA	joint operations area
JOC	joint operations center; joint oversight committee
JOCC	joint operations command center
JOG	joint operations graphic
JOGS	joint operation graphics system
JOPES	Joint Operation Planning and Execution System
JOPESIR	Joint Operation Planning and Execution System Incident Reporting System
JOPESREP	Joint Operation Planning and Execution System Reporting System
JOPP	joint operation planning process
JOR	joint operational requirement
JORD	joint operational requirements document
JOSG	joint operational steering group
JOT&E	joint operational test and evaluation
JOTS	Joint Operational Tactical System
JP	joint publication
JPATS	joint primary aircraft training system
JPC	joint planning cell
JPD	joint planning document
JPEC	joint planning and execution community
JPERSTAT	joint personnel status and casualty report
JPG	joint planning group
JPMRC	joint patient movement requirements center
JPMT	joint patient movement team
JPN	joint planning network
JPO	Joint Petroleum Office; Joint Program Office
JPOC	joint planning orientation course
JPOI	joint program of instruction
JPOM	joint preparation and onward movement
JPO-STC	Joint Program Office for Special Technology Countermeasures

JPOTF	joint psychological operations task force
JPOTG	joint psychological operations task group
JPRC	joint personnel reception center
JPRC	joint personnel receiving center
JPS	joint processing system
JPTTA	joint personnel training and tracking activity
JQR	joint qualification requirements
JRA	joint rear area
JRAC	joint rear area coordinator
JRACO	joint rear area communications officer
JRADS	Joint Resource Assessment Data System
JRAOC	joint rear area operations center
JRB	Joint Requirements Oversight Council (JROC) Review Board
JRC	joint reception center; joint reconnaissance center
JRCC	joint reception coordination center
JRFL	joint restricted frequency list
JRG	joint review group
JRMB	Joint Requirements and Management Board
JROC	Joint Requirements Oversight Council
JRS	joint reporting structure
JRSC	jam-resistant secure communications; joint rescue sub-center
JRSOI	joint reception, staging, onward movement, and integration
JRTC	joint readiness training center
JRTOC	joint rear tactical operations center
JS	joint staff
JSAC	joint strike analysis cell; joint strike analysis center
JSAM	joint security assistance memorandum; Joint Service Achievement Medal; joint standoff surface attack missile
JSAN	Joint Staff automation for the nineties
JSAS	joint strike analysis system
JSC	Joint Spectrum Center
JSCAT	joint staff crisis action team
JSCC	joint Services coordination committee
JSCM	joint Service commendation medal
JSCP	Joint Strategic Capabilities Plan
JSDS	Joint Staff doctrine sponsor
J-SEAD	joint suppression of enemy air defenses
JSIDS	joint Services imagery digitizing system
JSIR	joint spectrum interference resolution
JSISC	Joint Staff Information Service Center
JSMS	joint spectrum management system
JSO	joint specialty officer or joint specialist
JSOA	joint special operations area
JSOACC	joint special operations air component commander
JSOC	joint special operations command
JSOFI	Joint Special Operations Forces Institute

JSOTF	joint special operations task force
JSOW	joint stand-off weapon
JSPD	joint strategic planning document
JSPDSA	joint strategic planning document supporting analyses
JSPS	Joint Strategic Planning System
JSR	joint strategy review
JSRC	joint search and rescue center; joint subregional command (NATO)
JSS	joint surveillance system
JSSA	joint Services survival, evasion, resistance, and escape (SERE) agency
JSSIS	joint staff support information system
JSST	joint space support team
JSTAR	joint system threat assessment report
JSTARS	Joint Surveillance Target Attack Radar System
JSTE	joint system training exercise
JTA	joint table of allowances; joint technical architecture
JTADS	Joint Tactical Air Defense System (Army); Joint Tactical Display System
JTAGS	joint tactical ground station (Army); joint tactical ground system
JTAMD	joint theater air and missile defense
JTAO	joint tactical air operations
JTAR	joint tactical air strike request
JTASC	joint training analysis and simulation center
JTASG	Joint Targeting Automation Steering Group
JTAV	joint total asset visibility
JTB	Joint Transportation Board
JTC	joint technical committee
JTCB	joint targeting coordination board
JTCC	joint transportation corporate information management center
JTDC	joint track data coordinator
JTCG-ME	Joint Technical Coordinating Group for Munitions Effectiveness
JTD	joint table of distribution
JT&E	joint test and evaluation
JTF	joint task force
JTF-6	joint task force-6
JTF-B	joint task force-Bravo
JTFCEM	joint task force contingency engineering management
JTF HQ	joint task force headquarters
JTFP	Joint Tactical Fusion Program
JTIC	joint transportation intelligence center
JTIDS	Joint Tactical Information Distribution System
JTL	joint target list
JTLS	joint theater-level simulation
JTM	joint training manual

JTMD	joint table of mobilization distribution; Joint Terminology Master Database; joint theater missile defense
JTMP	joint training master plan
JTMS	joint training master schedule
JTP	joint training plan
JTR	joint travel regulations
JTRB	joint telecommunication resources board
JTSG	joint targeting steering group
JTSSCCB	Joint Tactical Switched Systems Configuration Control Board
JTSST	joint training system support team
JTT	joint targeting toolbox; joint training team
JTTP	joint tactics, techniques, and procedures
JUH-MTF	Joint User Handbook-Message Text Formats
JUIC	joint unit identification code
JULL	Joint Universal Lessons Learned (report)
JULLS	Joint Universal Lessons Learned System
JUO	joint urban operation
JUSMAG	Joint United States Military Advisory Group
JUWTF	joint unconventional warfare task force
JVB	Joint Visitors Bureau
JVIDS	Joint Visual Integrated Display System
JWAC	joint warfare analysis center
JWARS	Joint Warfare Analysis and Requirements System
JWC	Joint Warfare Center
JWCA	joint warfighting capabilities assessment
JWFC	Joint Warfighting Center
JWG	joint working group
JWICS	Joint Worldwide Intelligence Communications System
JWID	joint warrior interoperability demonstration

K

KAL	key assets list
KAPP	Key Assets Protection Program
kb	kilobit
kbps	kilobits per second
KC-135	Stratotanker
KEK	key encryption key
KG	key generator
kg	kilogram
kHz	kilohertz
KIA	killed in action
km	kilometer
KP	key pulse
kt	kiloton(s); knot (nautical miles per hour)
KVG	key variable generator

kW	kilowatt
KWOC	keyword-out-of-context

L

L	length
l	search subarea length
LA	lead agent; legal adviser; line amplifier; loop key generator (LKG) adapter
LAADS	low altitude air defense system
LAAM	light anti-aircraft missile
LACH	lightweight amphibious container handler
LACV	lighter, air cushioned vehicle
LAD	latest arrival date
LAMPS	Light Airborne Multipurpose System (helicopter)
LAN	local area network
LANDCENT	Allied Land Forces Central Europe (NATO)
LANDSAT	land satellite
LANDSOUTH	Allied Land Forces Southern Europe (NATO)
LANTIRN	low-altitude navigation and targeting infrared for night
LAO	limited attack option
LAPES	low-altitude parachute extraction system
LARC	lighter, amphibious resupply, cargo
LARS	lightweight airborne recovery system
LASH	lighter aboard ship
LASINT	laser intelligence
LAT	latitude
LAV	light armored vehicle
lb	pound
LBR	Laser Beam Rider
LC	lake current
LCAC	landing craft air cushion
LCAP	low combat air patrol
LCB	line of constant bearing
LCC	amphibious command ship; land component commander; launch control center; lighterage control center; link communications circuit
LCCS	landing craft control ship
LCE	logistics capability estimator
LCES	line conditioning equipment scanner
LCM	landing craft, mechanized; life-cycle management
LCO	lighterage control officer
LCP	lighterage control point
LCPL	landing craft personnel (large)
LCU	landing craft, utility; launch correlation unit
LCVP	landing craft, vehicle, personnel
LD	line of departure

LDF	lightweight digital facsimile
LDI	line driver interface
LDO	laser designator operator
LDR	leader; low data rate
LEA	law enforcement agency
LEAP	Light ExoAtmospheric Projectile
LEASAT	leased satellite
LEAU	Law Enforcement Assistance Unit (FAA)
LED	law enforcement desk; light emitting diode
LEDET	Law Enforcement Detachment (USCG)
LEO	low Earth orbit
LEP	linear error probable
LERTCON	alert condition
LES	law enforcement sensitive; leave and earnings statement; Lincoln Laboratories Experimental Satellite
LET	light equipment transport
LF	landing force; low frequency
LFA	lead federal agency
LFORM	landing force operational reserve material
LFSP	landing force support party
LG	deputy chief of staff for logistics
LGB	laser-guided bomb
LGM	laser-guided missile; loop group multiplexer
LGM-30	Minuteman
LGW	laser guided weapon
LHA	amphibious assault ship (general purpose); amphibious assault ship (multi-purpose)
L-hour	specific hour on C-day at which a deployment operation commences or is to commence
LHT	line-haul tractor
LIMDIS	limited distribution
LIMFAC	limiting factor
LIS	logistics information system
LIWA	land information warfare activity
LKA	attack cargo ship
LKG	loop key generator
LKP	last known position
LLLGB	low-level laser-guided bomb
LLLTV	low-light level television
LLTR	low-level transit route
LM	loop modem
LMAV	laser MAVERICK
LMCC	logistic and movement control center
LMF	language media format
LMSR	large, medium speed roll-on/roll-off
LMW	lead mobility wing
LNA	low voice amplifier
LNO	liaison officer

LOA	Lead Operational Authority; letter of assist; letter of offer and acceptance; logistics over-the-shore (LOTS) operation area
LOAC	law of armed conflict
LOAL	lock-on after launch
LOBL	lock-on before launch
LOC	line of communications; logistics operations center
LOC ACC	location accuracy
LOCAP	low combat air patrol
LOCE	Linked Operational Intelligence Centers Europe; Linked Operations-Intelligence Centers Europe
LOG	logistics
LOGAIR	logistics aircraft
LOGAIS	logistics automated information system
LOGCAP	logistics civilian augmentation program
LOGCAT	logistics capability assessment tool
LOGDET	logistics detail
LOGEX	logistics exercise
LOGFAC	Logistics Feasibility Assessment Capability
LOGFOR	logistics force packaging system
LOGMARS	logistics applications of automated marking and reading symbols
LOGMOD	logistics module
LOGPLAN	logistics planning system
LOGSAFE	logistic sustainment analysis and feasibility estimator
LOI	letter of instruction; loss of input
LO/LO	lift-on/lift-off
LOMEZ	low-altitude missile engagement zone
LONG	longitude
LOP	line of position
LORAN	long-range aid to navigation
LO/RO	lift-on/roll-off
LOROP	long range oblique photography
LOS	line of sight
LOTS	logistics over-the-shore
LP	listening post
LPD	amphibious transport dock; low probability of detection
LPH	amphibious assault ship, landing platform helicopter
LPI	low probability of intercept
LPU	line printer unit
LPV	laser-protective visor
LRC	logistics readiness center
LRD	laser range finder-detector
LRG	long-range aircraft
LRM	low rate multiplexer
LRP	load and roll pallet
LRRP	long range reconnaissance patrol
LRS	launch and recovery site

LRST	long-range surveillance team
LRSU	long-range surveillance unit
LSA	logistic support analysis
LSB	landing support battalion; lower sideband
LSCDM	low speed cable driver modem
LSD	landing ship dock; least significant digit
LSE	landing signal enlisted
LSO	landing safety officer; landing signal officer
LSPR	low speed pulse restorer
LSS	local sensor subsystem
LST	landing ship, tank; laser spot tracker; tank landing ship
LSV	logistics support vessel
LT	large tug; local terminal; long ton
L/T	long ton
LTD	laser target designator
LTD/R	laser target designator/ranger
LTG	local timing generator
LTON	long ton
LTS	low-altitude navigation and targeting infrared for night (LANTIRN) targeting system
LTT	loss to theater
LTU	line termination unit
LUA	launch under attack
LUT	local user terminal
LVS	Logistics Vehicle System (USMC)
LW	leeway
LWR	Lutheran World Relief
LZ	landing zone
LZCO	landing zone control officer

M

M&S	modeling and simulation
M88A1	recovery vehicle
MA	master
mA	milliampere(s)
MAAG	military assistance advisory group
MAAP	master air attack plan
MACCS	Marine air command and control system
MACG	Marine air control group
MACOM	major command (Army)
MACP	mortuary affairs collection point
MACSAT	multiple access commercial satellite
MAD	military air distress
MADCP	mortuary affairs decontamination collection point
MAEB	mean area of effectiveness for blast
MAEF	mean area of effectiveness for fragments
MAF	mobility air forces

MAFC	Marine air-ground task force (MAGTF) all-source fusion center
MAG	Marine aircraft group
MAGTF	Marine air-ground task force
MAGTFACE	Marine air-ground task force aviation combat element
MAJCOM	major command (USAF)
MANFOR	manpower force packaging system
MANPADS	man-portable air defense system
MANPER	manpower and personnel module
MAP	Military Assistance Program; missed approach point; missed approach procedure
MARAD	Maritime Administration
MARDIV	Marine division
MARFOR	Marine Corps forces
MARFOREUR	Marine Corps Forces, Europe
MARFORLANT	Marine Corps Forces, Atlantic
MARFORPAC	Marine Corps Forces, Pacific
MARINCEN	Maritime Intelligence Center
MARLO	Marine liaison officer
MAROP	marine operators
MARPOL	International Convention for the Prevention of Pollution from Ships
MARS	Military Affiliate Radio System
MARSA	military assumes responsibility for separation of aircraft
MART	mobile Automatic Digital Network (AUTODIN) remote terminal
MASCAL	mass casualty
MASF	mobile aeromedical staging facility
MASH	mobile Army surgical hospital
MASINT	measurement and signature intelligence
MAST	military assistance to safety and traffic; mobile ashore support terminal
MATCALS	Marine air traffic control and landing system
MAW	Marine aircraft wing
MAX	maximum
MB	medium boat; megabyte
MBA	main battle area
MBBLs	thousands of barrels
MBI	major budget issue
Mbps	megabytes per second
Mbs	megabits per second
MC	military community; mission-capable
MC-130	Combat Talon (I and II)
MCA	maximum calling area; military civic action; mission concept approval; movement control agency
MCAP	maximum calling area precedence
MCAS	Marine Corps air station

MCC	Marine component commander; master control center; military cooperation committee; military coordinating committee; mission control center; mobility control center; movement control center
MCCC	mobile consolidated command center
MCCDC	Marine Corps Combat Development Command
MCCISWG	military command, control, and information systems working group
MCD	medical crew director
MCDA	military and civil defense assets (UN)
MCDS	modular cargo delivery system
MCEB	Military Communications-Electronics Board
MCEWG	Military Communications-Electronics Working Group
MCIA	Marine Corps Intelligence Activity
MCIO	military criminal investigation organization
MCM	Manual for Courts-Martial; military classification manual; mine countermeasures
MCMG	Military Committee Meteorological Group (NATO)
MCMOPS	mine countermeasures operations
M/CM/S	mobility, countermobility, and/or survivability
MCO	Marine Corps order
MCOO	modified combined obstacle overlay
MCS	maneuver control system; mine countermeasures ship; modular causeway system
MCSF	mobile cryptologic support facility
MCT	movement control team
MCU	maintenance communications unit
MCW	modulated carrier wave
M-DARC	military direct access radar channel
M-day	mobilization day; unnamed day on which mobilization of forces begins
MDCI	multidiscipline counterintelligence
MDF	Main Defense Forces (NATO); main distribution frame
MDITDS	migration defense intelligence threat data system
MDR	medium data rate
MDS	mission design series
MDSS II	Marine air-ground task force (MAGTF) Deployment System II
MDSU	mobile diving and salvage unit
MDW	Military District of Washington
MDZ	maritime defense zone
MEA	munitions effect assessment; munitions effectiveness assessment
MEB	Marine expeditionary brigade
MEBU	mission essential backup
MEC	medium endurance cutter
MED	manipulative electronic deception
MEDCAP	medical civic action program

MEDCOM	US Army Medical Command
MEDEVAC	medical evacuation
MEDINT	medical intelligence
MEDLOG	medical logistics (USAF AIS)
MEDLOGCO	medical logistics company
MEDLOG JR	medical logistics, junior (USAF)
MEDMOB	Medical Mobilization Planning and Execution System
MEDNEO	medical noncombatant evacuation operation
MEDREGREP	medical regulating report
MEDRETE	medical readiness training exercise
MEDS	meteorological data system
MEDSOM	medical supply, optical, and maintenance unit
MEDSTAT	medical status
MEF	Marine expeditionary force
MEF (FWD)	Marine expeditionary force (forward)
MEFPAKA	manpower and equipment force packaging
MEL	maintenance expenditure limit; minimum equipment list
MEP	mobile electric power
MEPCOM	military entrance processing command
MEPES	Medical Planning and Execution System
MEPRS	Military Entrance Processing and Reporting System
MERCO	merchant ship reporting and control
MERINT	merchant intelligence
MERSHIPS	merchant ships
MESAR	minimum-essential security assistance requirements
MET	medium equipment transporter
METAR	meteorological airfield report; meteorological aviation report
METARS	routine aviation weather report (roughly translated from French; international standard code format for hourly surface weather observations)
METCON	control of meteorological information (roughly translated from French); meteorological control (Navy)
METL	mission-essential task list
METOC	meteorological and oceanographic
METSAT	meteorological satellite
METT-T	mission, enemy, terrain and weather, troops and support available—time available
MEU	Marine expeditionary unit
MEU(SOC)	Marine expeditionary unit (special operations capable)
MEWSG	Multi-Service Electronic Warfare Support Group (NATO)
MEZ	missile engagement zone
MF	medium frequency; mobile facility; multi-frequency
MFC	meteorological and oceanographic (METOC) forecast center; multinational force commander
MFDS	Modular Fuel Delivery System
MFE	manpower force element
MFFIMS	mass fatality field information management system

MFO	multinational force and observers
MFP	major force program
MFPF	minefield planning folder
MFS	multifunction switch
MGM	master group multiplexer
MGRS	military grid reference system
MGS	mobile ground system
MGT	management
MHC	management headquarters ceiling
MHE	materials handling equipment
MHW	mean high water
MHz	megahertz
MI	military intelligence; movement instructions
MIA	missing in action
MIB	Military Intelligence Board
MICAP	mission capable/mission capability
MICON	mission concept
MICRO-MICS	micro-medical inventory control system
MICRO-SNAP	micro-shipboard non-tactical automated data processing system
MIDB	modernized integrated database; modernized intelligence database
MIF	maritime interception force
MIIDS/IDB	Military Intelligence Integrated Data System/Integrated Database
MIJI	meaconing, interference, jamming, and intrusion
MILCON	military construction
MILDEC	military deception
MILDEP	Military Department
MILGP	military group (assigned to American Embassy in host nation)
MILOB	military observer
MILOC	military oceanography group (NATO)
MILPERS	military personnel
MILSATCOM	military satellite communications
MILSPEC	military performance specification
MILSTAMP	military standard transportation and movement procedures
MILSTAR	military strategic and tactical relay system
MIL-STD	military standard
MILSTRIP	military standard requisitioning and issue procedure
MILTECH	military technician
MILVAN	military van (container)
MIM	maintenance instruction manual
MINEOPS	joint minelaying operations
MIO	maritime interception operations
MIPE	mobile intelligence processing element
MIPR	military interdepartmental purchase request
MIS	maritime intelligence summary

MISCAP	mission capability
MISREP	mission report
MISS	missing
MIST	military information support team
MITASK	mission tasking
MITO	minimum interval takeoff
MITT	mobile integrated tactical terminal
MIUW	mobile inshore undersea warfare
MIUWU	mobile inshore undersea warfare unit
MIW	mine warfare
MJCS	Joint Chiefs of Staff memorandum
MJSOTF	multinational joint special operations task force
MLA	mission load allowance
MLC	Marine Logistics Command
MLE	maritime law enforcement
MLO	military liaison office
MLP	message load plan
MLPP	multilevel precedence and preemption
MLRS	Multiple Launch Rocket System
MLS	microwave landing system; multilevel security
MLW	mean low water
MMC	materiel management center
MMG	DOD Master Mobilization Guide
MMI	man/machine interface
MMLS	mobile microwave landing system
MNF	multinational force
MNFC	multinational force commander
MNP	master navigation plan
MNS	mine neutralization system (USN); mission needs statement
MO	medical officer; month
MOA	memorandum of agreement; military operating area
MOADS	maneuver-oriented ammunition distribution system
MOB	main operations base; mobilization
MOBCON	mobilization control
MOBREP	military manpower mobilization and accession status report
MOCC	mobile operations control center
MOD	Minister (Ministry) of Defense
MODEM	modulator/demodulator
MODLOC	miscellaneous operational details, local operations
MOD T-AGOS	modified tactical auxiliary general ocean surveillance
MOE	measure of effectiveness
MOG	maximum (aircraft) on ground; movement on ground (aircraft); multinational observer group
MOGAS	motor gasoline
MOLE	multichannel operational line evaluator
MOM	military ordinary mail

MOMAT	mobility matting
MOMSS	mode and message selection system
MOOTW	military operations other than war
MOP	measure of performance; memorandum of policy
MOPP	mission-oriented protective posture
MOR	memorandum of record
MOS	military occupational specialty
MOU	memorandum of understanding
MOUT	military operations in urban terrain; military operations on urbanized terrain
MOVREP	movement report
MOW	maintenance orderwire
MP	military police
MPA	maritime patrol aircraft; mission and payload assessment; mission planning agent
MPAT	military patient administration team
MPC	military payment certificate; military personnel center
MPE/S	maritime pre-positioning equipment and supplies
MPF	maritime pre-positioning force
MPG	mensurated point graphic
mph	miles per hour
MPLAN	Marine Corps Mobilization Management Plan
MPM	medical planning module
MPO	military post office
MPS	maritime pre-positioning ship; message processor shelter; Military Postal Service
MPSA	Military Postal Service Agency
MPSRON	maritime pre-positioning ships squadron
MR	milliradian; mobile reserve
MRAALS	Marine remote area approach and landing system
MRCI	maximum rescue coverage intercept
MRE	meal, ready to eat
MRG	movement requirements generator
MRL	minimum-risk level
MRMC	US Army Medical Research and Materiel Command
MRO	medical regulating office
MROC	multicommand required operational capability
MRR	minimum-risk route
MRRR	mobility requirement resource roster
MRS	meteorological radar subsystem; movement report system
MRSA	Materiel Readiness Support Agency
MRT	maintenance recovery team
MRU	mountain rescue unit
MS	message switch
ms	millisecond
MSC	major subordinate command; Military Sealift Command; military staff committee; mission support confirmation
MSCA	military support to civil authorities

MSCD	military support to civil defense
MSCLEA	military support to civilian law enforcement agencies
MSCO	Military Sealift Command Office
MSD	marginal support date; mobile security division
MS-DOS	Microsoft disk operating system
MSDS	mission specific data set
MSE	mission support element; mobile subscriber equipment;
MSECR	HIS 6000 security module
MSEL	master scenario events list
MSF	*Medicins Sans Frontieres* ("Doctors Without Borders"); mission support force; multiplex signal format
MSG	Marine Security Guard; message
MSGID	message identification
MSI	multi-spectral imagery
MSIC	Missile and Space Intelligence Center
MSIS	Marine safety information system
MSK	mission support kit
MSL	master station log; mean sea level
MSNAP	merchant ship naval augmentation program
MSO	marine safety office(r); military satellite communications (MILSATCOM) systems organization; mobilization staff officer
MSP	mission support plan
MSPES	mobilization stationing, planning, and execution system
MSPF	maritime special purpose force
MSPS	mobilization stationing and planning system
MSR	main supply route; mission support request
MSRR	modeling and simulation resource repository
MSRV	message switch rekeying variable
MSS	meteorological satellite subsystem
MSSG	Marine expeditionary unit (MEU) service support group
MST	mission support team
M/T	measurement ton
MT	measurement ton; military technician; ministry team
MTA	military training agreement
MTBF	mean time between failures
MT Bn	motor transport battalion
MT/D	measurement tons per day
MTF	medical treatment facility; message text format
MTG	master timing generator
MTI	moving target indicator
MTL	mission tasking letter
MTMC	Military Traffic Management Command
MTMCTEA	Military Traffic Management Command Transportation Engineering Agency
MTMS	maritime tactical message system
MTO	message to observer; mission type order
MTOE	modified table of organization and equipment

MTON	measurement ton
MTP	mission tasking package
MTS/SOF-IRIS	multifunction system
MTT	magnetic tape transport; mobile training team
MTW	major theater war
MTX	message text format
MU	marry up
MUL	master urgency list (DOD)
MULE	modular universal laser equipment
MUREP	munitions report
MUSARC	major United States Army reserve commands
MUST	medical unit, self-contained, transportable
MUX	multiplex
MV	merchant vessel; motor vessel
mV	millivolt
MWBP	missile warning bypass
MWC	Missile Warning Center (NORAD)
MWD	military working dog
MWDT	military working dog team
MWF	medical working file
MWOD	multiple word-of-day
MWR	missile warning receiver; morale, welfare, and recreation
MWSS	Marine wing support squadron

N

N	number of required track spacings; number of search and rescue units (SRUs)
N-1	Navy component manpower or personnel staff officer
N-2	Navy component intelligence staff officer
N-3	Navy component operations staff officer
N-4	Navy component logistics staff officer
N-5	Navy component plans staff officer
N-6	Navy component communications staff officer
NAAG	North Atlantic Treaty Organization (NATO) Army Armaments Group
NAC	North American Aerospace Defense Command (NORAD) Air Center; North Atlantic Council (NATO)
NACE	National Military Command System (NMCS) Automated Control Executive
NACISA	North Atlantic Treaty Organization (NATO) Communications and Information Systems Agency
NACISC	North Atlantic Treaty Organization (NATO) Communications and Information Systems Committee
NACSEM	National Communications Security/Emanations Security (COMSEC/EMSEC) Information Memorandum
NACSI	national communications security (COMSEC) instruction

NACSIM	national communications security (COMSEC) information memorandum
NADEFCOL	North Atlantic Treaty Organization (NATO) Defense College
NADEP	naval aircraft depot
NAF	naval air facility; nonappropriated funds
NAFAG	North Atlantic Treaty Organization (NATO) Air Force Armaments Group
NAI	named area of interest
NAIC	National Air Intelligence Center
NAK	negative acknowledgement
NALC	naval ammunition logistic code
NALE	naval and amphibious liaison element
NALSS	naval advanced logistic support site
NAMP	North Atlantic Treaty Organization (NATO) Annual Manpower Plan
NAMTO	Navy material transportation office
NAOC	national airborne operations center (E-4B aircraft)
NAPCAP	North Atlantic Treaty Organization (NATO) Allied Pre-Committed Civil Aircraft Program
NAPMA	North Atlantic Treaty Organization (NATO) Airborne Early Warning and Control Program Management Agency
NAR	notice of ammunition reclassification
NARC	non-automatic relay center
NAS	naval air station
NASA	National Aeronautics and Space Administration
NASAR	National Association for Search and Rescue
NAS computer	national airspace system computer
NAT	nonair-transportable (cargo)
NATO	North Atlantic Treaty Organization
NATOPS	Naval Air Training and Operating Procedures Standardization
NAU	Narcotics Assistance Unit
NAVAIDS	navigational aids
NAVAIR	naval air; Naval Air Systems Command
NAVAIRSYSCOM	Naval Air Systems Command (Also called NAVAIR)
NAVATAC	Navy Antiterrorism Analysis Center; Navy Antiterrorist Alert Center
NAVCHAPDET	naval cargo handling and port group detachment
NAVCHAPGRU	Navy cargo handling and port group
NAVCOMSTA	naval communications station
NAVFACENGCOM	naval facilities engineering command
NAVFOR	Navy forces
NAVMAG	naval magazine
NAVMEDCOMINST	Navy medical command instruction
NAVMEDLOGCOM	Navy Medical Logistical Command
NAVMEDP	Navy medical pamphlet

NAVMTO	naval military transportation office
NAVORD	naval ordnance
NAVORDSTA	naval ordnance station
NAVSAFECEN	naval safety center
NAVSAT	navigation satellite
NAVSEA	naval sea
NAVSEAINST	naval sea instruction
NAVSEALOGCEN	naval sea logistics center
NAVSEASYSCOM	Naval Sea Systems Command
NAVSO	United States Navy Forces, Southern Command
NAVSOC	naval special operations command; naval special operations component; naval special warfare special operations component; Navy special operations component
NAVSOF	naval special operations forces
NAVSPACECOM	Naval Space Command
NAVSPECWARCOM	Naval Special Warfare Command
NAVSPOC	Naval Space Operations Center
NAVSUP	naval supply
NAVSUPINST	Navy Support Instruction
NAVSUPSYSCOM	Naval Supply Systems Command
NAWCAD	Naval Air Warfare Center, Aircraft Division
NB	narrowband
NBC	nuclear, biological, and chemical
NBCCS	nuclear, biological, and chemical (NBC) contamination survivability
NBDP	narrow band direct printing
NBG	naval beach group
NBI	nonbattle injury
NBS	National Bureau of Standards
NBST	narrowband secure terminal
NC3A	nuclear command, control, and communications (C3) assessment
NCA	National Command Authorities
NCAA	North Atlantic Treaty Organization (NATO) Civil Airlift Agency
NCB	naval construction brigade
NCC	naval component commander; Navy component command; network control center; North American Aerospace Defense Command (NORAD) Command Center
NCCS	Nuclear Command and Control System
NCD	net control device
NCF	naval construction force
NCFSU	naval construction force support unit
NCHB	Navy cargo handling battalion
NCHF	Navy cargo handling force
NCIC	National Crime Information Center

NCIS	Naval Criminal Investigative Service
NCISRA	Naval Criminal Investigative Service resident agent
NCISRO	Naval Criminal Investigative Service regional office
NCISRU	Naval Criminal Investigative Service resident unit
NCMP	Navy Capabilities and Mobilization Plan
NCO	noncombat operations; noncommissioned officer
NCOIC	noncommissioned officer in charge
NCR	national cryptologic representative; National Security Agency/Central Security Service Representative; naval construction regiment
NCRDEF	national cryptologic representative defense
NCS	National Communications System; naval control of shipping; net control station
NCSC	National Computer Security Center
NCSE	national intelligence support team (NIST) communications support element
NCSORG	naval control of shipping organization
NCT	network control terminal
NCTAMS	naval computer and telecommunications area master station
NCTS	naval computer and telecommunications station
NCW	naval coastal warfare
NCWC	naval coastal warfare commander
NDA	national defense area
NDAA	National Defense Authorization Act
N-day	day an active duty unit is notified for deployment or redeployment
NDB	nondirectional beacon
NDCS	national drug control strategy
NDHQ	National Defence Headquarters, Canada
NDL	national desired ground zero list
NDMC	North Atlantic Treaty Organization (NATO) Defense Manpower Committee
NDMS	National Disaster Medical System
NDOC	National Defense Operations Center
NDP	national disclosure policy
NDPB	National Drug Policy Board
NDPC	National Disclosure Policy Committee
NDRF	National Defense Reserve Fleet
NDSF	National Defense Sealift Fund
NDU	National Defense University
NEA	Northeast Asia
NEAT	naval embarked advisory team
NEMT	National Emergency Management Team
NEO	noncombatant evacuation operation
NEOPACK	noncombatant evacuation operation package
NEREP	Nuclear Execution and Reporting Plan

NESDIS	National Environmental Satellite, Data and Information Service (DOC)
NETS	Nationwide Emergency Telecommunications System
NETT	new equipment training team
NEW	naval expeditionary warfare; net explosive weight
NEWAC	North Atlantic Treaty Organization (NATO) Electronic Warfare Advisory Committee
NFA	no-fire area
NFD	nodal fault diagnostics
NFIB	National Foreign Intelligence Board
NFIP	National Flood Insurance Program (FEMA); National Foreign Intelligence Program
NFLS	naval forward logistic site
NFO	naval flight officer
NG	National Guard
NGB	National Guard Bureau
NGFS	naval gunfire support
NGIC	National Ground Intelligence Center
NGO	nongovernmental organization
NHCS	nonhostile casualty
NI	national identification (number); noted item
NIBRS	National Incident-Based Reporting System
NIC	National Intelligence Council; naval intelligence center
NICI	National Interagency Counternarcotics Institute
NID	naval intelligence database
NIDMS	National Military Command System (NMCS) Information for Decision Makers System
NIDS	National Military Command Center (NMCC) information display system
NIE	national intelligence estimate
NIEX	no-notice interoperability exercise
NIEXPG	No-Notice Interoperability Exercise Planning Group
NII	national information infrastructure
NIMA	National Imagery and Mapping Agency
NIMCAMP	National Information Management and Communications Master Plan
NIPRNET	Non-Secure Internet Protocol Router Network
NIPS	Naval Intelligence Processing System
NISH	noncombatant evacuation operation (NEO) intelligence support handbook
NISP	Nuclear Weapons Intelligence Support Plan
NIST	National Institute of Standards and Technology; national intelligence support team
NITF	national imagery transmission format
NIU	North Atlantic Treaty Organization (NATO) interface unit
NIWA	naval information warfare activity
NL	Navy lighterage
NLO	naval liaison officer

.NL.	not less than
NLT	not later than
NM	network management
nm	nautical mile
NMB	North Atlantic Treaty Organization (NATO) military body
NMCB	naval mobile construction battalion
NMCC	National Military Command Center
NMCM	not mission capable, maintenance
NMCS	National Military Command System; not mission capable, supply
NMD	national missile defense
NMET	naval mobile environmental team
NMFS	National Marine Fisheries Services
NMIST	National Military Intelligence Support Team (DIA)
NMJIC	National Military Joint Intelligence Center
NMOC	network management operations center
NMPS	Navy mobilization processing site
NMR	news media representative
NMS	national military strategy
NMSA	North Atlantic Treaty Organization (NATO) Mutual Support Act
NNAG	North Atlantic Treaty Organization (NATO) Naval Armaments Group
NOAA	National Oceanic and Atmospheric Administration
NOACT	Navy overseas air cargo terminal
NOC	network operations center
NOCONTRACT	not releasable to contractors or consultants
NOEA	nuclear operations emergency action
NOFORN	not releasable to foreign nationals
NOG	Nuclear Operations Group
NOIC	Naval Operational Intelligence Center
NOK	next of kin
NOMS	Nuclear Operations Monitoring System
NOP	nuclear operations
NOPLAN	no operation plan available or prepared
NORAD	North American Aerospace Defense Command
NORM	normal; not operationally ready, maintenance
NORS	not operationally ready, supply
NOSC	Network Operations and Security Center
NOTAM	notice to airmen
NOTMAR	notice to mariners
NPC	Nonproliferation Center
NPES	Nuclear Planning and Execution System
NPG	nonunit personnel generator
NPS	National Park Service; nonprior service; Nuclear Planning System
NPWIC	National Prisoner of War Information Center

NQ	nonquota
NR	North Atlantic Treaty Organization (NATO) restricted; number
NRC	National Response Center; non-unit-related cargo
NRCHB	Naval Reserve cargo handling battalion
NRCHF	Naval Reserve cargo handling force
NRCHTB	Naval Reserve cargo handling training battalion
NRFI	not ready for issue
NRG	notional requirements generator
NRL	nuclear weapons (NUWEP) reconnaissance list
NRO	National Reconnaissance Office
NROC	Northern Regional Operations Center (CARIBROC-CBRN)
NRP	non-unit-related personnel
NRPC	Naval Reserve Personnel Center
NRT	near real time
NRTD	near real time dissemination
NRZ	non-return-to-zero
NS	nuclear survivability
NSA	National Security Agency; national shipping authority; North Atlantic Treaty Organization (NATO) Standardization Agency
NSA/CSS	National Security Agency/Central Security Service
NSC	National Security Council
NSC/DC	Deputies Committee of the National Security Council
NSCID	National Security Council Intelligence Directive
NSC/IWG	National Security Council/Interagency Working Group
NSC/PC	National Security Council/Principals Committee
NSCS	National Security Council System
NSD	National Security Directive
NSDA	non-self deployment aircraft
NSDD	national security decision directive
NSE	national support element; Navy support element
NSEP	national security emergency preparedness
NSF	National Science Foundation
NSFS	naval surface fire support
NSG	north-seeking gyro
NSI	not seriously injured
NSL	no-strike list
NSM	national search and rescue (SAR) manual
NSN	National Stock Number
NSNF	nonstrategic nuclear forces
NSO	non-Single Integrated Operational Plan (SIOP) option
NSOC	National Signals Intelligence (SIGINT) Operations Center; Navy Satellite Operations Center
NSOOC	North Atlantic Treaty Organization (NATO) Staff Officer Orientation Course
NSP	national search and rescue plan

N-Sp/CC	North American Aerospace Defense Command (NORAD)-US Space Command/Command Center
NSRL	national signals intelligence (SIGINT) requirements list
NSS	national security strategy; non-self-sustaining
NSSA	National Security Space Architect
NSSCS	non-self-sustaining containership
NSST	naval space support team
NSTISSC	National Security Telecommunications and Information Systems Security Committee
NSTL	national strategic targets list
NSTS	National Secure Telephone System
NSW	naval special warfare
NSWG	naval special warfare group
NSWTE	naval special warfare task element
NSWTG	naval special warfare task group
NSWTU	naval special warfare task unit
NSWU	naval special warfare unit
NT	nodal terminal
NTACS	Navy tactical air control system
NTAP	National Track Analysis Program
NTB	national target base
NTC	National Training Center
NTCS-A	Navy Tactical Command System Afloat
NTDS	naval tactical data system
NTF	nuclear task force
NTIC	Navy Tactical Intelligence Center
NTISS	National Telecommunications and Information Security System
NTISSI	National Telecommunications and Information Security System (NTISS) Instruction
NTISSP	National Telecommunications and Information Security System (NTISS) Policy
NTMPDE	National Telecommunications Master Plan for Drug Enforcement
NTMS	national telecommunications management structure
NTPS	near-term pre-positioned ships
NTSB	National Transportation Safety Board
NTSS	National Time-Sensitive System
NTU	new threat upgrade
NUC	non-unit-related cargo
NUCINT	nuclear intelligence
NUDET	nuclear detonation
NUDETS	nuclear detonation detection and reporting system
NUP	non-unit-related personnel
NURC	non-unit-related cargo
NUWEP	policy guidance for the employment of nuclear weapons
NVD	night vision device
NVG	night vision goggle

NW	not waiverable
NWARS	National Wargaming System
NWB	normal wideband
NWBLTU	normal wideband line termination unit
NWDC	Navy Warfare Development Command
NWFP	Northwest Frontier Province (Pakistan)
NWP	naval warfare publication
NWREP	nuclear weapons report
NWS	National Weather Service
NWT	normal wideband terminal

O

O	contour pattern
O&M	operation and maintenance
OA	objective area; operating assembly
OADR	originating agency's determination required
OAE	operational area evaluation
OAFME	Office of the Armed Forces Medical Examiner
OAG	operations advisory group
OAJCG	Operation Alliance joint control group
OAR	Chairman of the Joint Chiefs of Staff operation plans assessment report
OAS	offensive air support; Organization of American States
OASD	Office of the Assistant Secretary of Defense
OAU	Organization of African Unity
O/B	outboard
OB	order of battle
OBA	oxygen breathing apparatus
OBFS	offshore bulk fuel system
OBST	obstacle
OCA	offensive counterair; operational control authority
OCC	Operations Computer Center (USCG)
OCCA	Ocean Cargo Clearance Authority
OCD	orderwire clock distributor
OCE	officer conducting the exercise
OCEANCON	control of oceanographic information
OCJCS	Office of the Chairman of the Joint Chiefs of Staff
OCJCS-PA	Office of the Chairman of the Joint Chiefs of Staff-Public Affairs
OCMI	officer in charge, Marine inspection
OCO	offload control officer
OCONUS	outside the continental United States
OCOP	outline contingency operation plan
OCP	operational configuration processing
OCR	Office of Collateral Responsibility
OCU	orderwire control unit (Types I, II, and III)
OCU-1	orderwire control unit-1

OD	operational detachment; other detainee
ODA	operational detachment-Alpha
ODATE	organization date
O-Day	off-load day
ODB	operational detachment-Bravo
ODC	Office of Defense Cooperation
ODCSLOG	Office of the Deputy Chief of Staff for Logistics (Army)
ODCSOPS	Office of the Deputy Chief of Staff for Operations and Plans (Army)
ODCSPER	Office of the Deputy Chief of Staff for Personnel (Army)
ODIN	Operational Digital Network
ODJS	Office of the Director, Joint Staff
ODR	Office of Defense representative
ODZ	outer defense zone
OEBGD	Overseas Environmental Baseline Guidance Document
OEG	operational exposure guide
OER	operational electronic intelligence (ELINT) requirements
OES	office of emergency services
OET	Office of Emergency Transportation (DOT)
OF	officer (NATO)
OFDA	Office of Foreign Disaster Assistance
OFHIS	operational fleet hospital information system
OFOESA	Office of Field Operational and External Support Activities
OGA	other government agency
OGS	overseas ground station
OH	overhead
OI	Office of Intelligence (USCS); operating instruction
OIC	officer in charge
OICC	operational intelligence coordination center
OID	operation order (OPORD) identification
OIR	operational intelligence requirements; other intelligence requirements
OJT	on-the-job training
OLD	on-line tests and diagnostics
OLS	operational linescan system; optical landing system
OM	contour multiunit
OMA	Office of Military Affairs (CIA)
OMB	Office of Management and Budget
OMC	Office of Military Cooperation
OMF	officer master file
OMT	orthogonal mode transducer
OMT/OMTP	operational maintenance test(ing)/test plan
ONDCP	Office of National Drug Control Policy
ONI	Office of Naval Intelligence
OOB	order of battle
OOD	officer of the deck
OODA	observe, orient, decide, act
OOS	out of service

OP	observation post; operational publication (USN); ordnance pamphlet
OP3	overt peacetime psychological operations (PSYOP) program
OPBAT	Operation Bahamas, Turks, and Caicos
OPCEN	operations center (USCG)
OPCOM	operational command (NATO)
OPCON	operational control
OPDEC	operational deception
OPDOC	operational documentation
OPDS	offshore petroleum discharge system
OPELINT	operational electronic intelligence
OPFOR	opposing force; opposition force
OPG	operations planning group
OPLAN	operation plan
OPLAW	operational law
OPM	Office of Personnel Management; operations per minute
OPNAVINST	Chief of Naval Operations Instruction
OPORD	operation order
OPP	off-load preparation party; orderwire patch panel
OPR	office of primary responsibility
OPREP	operational report
OPS	operational project stock; operations
OPSDEPS	Service Operations Deputies
OPSEC	operations security
OPSTK	operational stock
OPSUM	operation summary
OPT	operational planning team
OPTAR	operating target
OPTEMPO	operating tempo
OPTINT	optical intelligence
OPZONE	operation zone
OR	operational readiness; other rank(s) (NATO)
ORBAT	order of battle
ORCON	originator controlled
ORD	Operational Requirements Document
ORG	origin (GEOLOC)
ORIG	origin
OS	operating system
OSA	operational support airlift
OSAT	out-of-service analog test
OSC	on-scene commander; operations support center
OSD	Office of the Secretary of Defense
OSE	on scene endurance
OSG	operational support group
OSI	open system interconnection; operational subsystem interface
OSIA	on-site inspection activity

OSINT	open-source intelligence
OSIS	open-source information system
OSO	operational support office
OSOCC	on-site operations coordination center
OSP	operations support package
OSPG	overseas security policy group
OSRI	originating station routing indicator
OSV	ocean station vessel
OT	operational test
OT&E	operational test and evaluation
OTC	officer in tactical command; over the counter
OTG	operational target graphic
OTH	over the horizon
OTH-B	over-the-horizon backscatter (radar)
OTHT	over-the-horizon targeting
OTS	Officer Training School
OUB	offshore petroleum discharge system (OPDS) utility boat
OUSD	Office of the Under Secretary of Defense
OUSD(AT&L)	Office of the Under Secretary of Defense (Acquisition, Technology, and Logistics)
OUSD(P)	Office of the Under Secretary of Defense (Policy)
OUT	outsize cargo
OVE	on-vehicle equipment
OVER	oversize cargo
OW	orderwire

P

P	parallel pattern; priority
PA	parent relay; probability of arrival; public affairs
PAA	primary aircraft authorization
PABX	private automatic branch exchange (telephone)
PAD	positional adjustment; precision aircraft direction
PADD	person authorized to direct disposition of remains
PADS	position azimuth determining system
PAG	public affairs guidance
PAI	primary aircraft inventory
PAL	permissive action link; personnel allowance list; program assembler language
PALS	precision approach landing system
PAM	pulse amplitude modulation
PANS	procedures for air navigation services
PAO	Public Affairs Office; public affairs officer
PAR	performance assessment report; population at risk; precision approach radar
PARKHILL	high frequency cryptological device
PARPRO	peacetime application of reconnaissance programs
PAS	personnel accounting symbol

PAT	public affairs team
PAWS	phased array warning system
PAX	passengers; public affairs plans
PB	particle beam; patrol boat; President's budget
PBA	production base analysis
PBCR	portable bar code recorder
PBD	program budget decisions
PC	patrol craft; personal computer; pilot in command; principals committee
Pc	cumulative probability of detection
PCC	policy coordinating committee; primary control center
PCF	personnel control facility
PCL	positive control launch
PC-LITE	processor, laptop imagery transmission equipment
PCM	pulse code modulation
PCO	primary control officer
PCRTS	primary casualty receiving and treatment ship
PCS	permanent change of station; personal communications system; primary control ship; processing subsystem; processor controlled strapping
PCT	personnel control team
PCZ	physical control zone
PD	position description; Presidential directive; probability of damage; probability of detection; procedures description; program definition; program director
Pd	drift compensated parallelogram pattern
PDAI	primary development/test aircraft inventory
PDD	Presidential decision directive
PDDA	power driven decontamination apparatus
PDG	positional data graphic
PDM	program decision memorandum
PDOP	position dilution of precision
PDS	position determining system; protected distribution system
PE	peace enforcement; peacetime establishment; personal effects; program element
PEAD	Presidential emergency action document
PEAS	psychological operations (PSYOP) effects analysis subsystem
PEC	program element code
PECK	patient evacuation contingency kit
PEDB	planning and execution database
PEGEO	personnel geographic location
PEI	principal end item
PEM	program element monitor
PEO	peace enforcement operations
PEP	personnel exchange program
PER	personnel
PERE	person eligible to receive effects

PERINTSUM	periodic intelligence summary
PERMREP	permanent representative (NATO)
PERSCO	personnel support for contingency operations
PERSCOM	personnel command (Army)
PERSINS	personnel information system
PES	preparedness evaluation system
PFD	personal flotation device
PFDB	planning factors database
PFIAB	President's Foreign Intelligence Advisory Board
PFID	positive friendly identification
PFP	Partnership for Peace (NATO)
PGM	precision-guided munitions
PHIBCB	amphibious construction battalion
PHIBGRU	amphibious group
PHIBOP	amphibious operation
PHIBRON	amphibious squadron
PHOTINT	photographic intelligence
PHS	Public Health Service
PI	point of impact; probability of incapacitation; procedural item
PIC	parent indicator code; person identification code; pilot in command
PID	plan identification number
PIDD	planned inactivation or discontinued date
PIF	problem identification flag
PII	pre-incident indicators
PIM	pretrained individual manpower
PIN	personnel increment number
PINS	precise integrated navigation system
PIO	press information officer; public information officer
PIPS	plans integration partitioning system
PIR	priority intelligence requirements
PIRAZ	positive identification and radar advisory zone
PIW	person in water
PK	peacekeeping; probability of kill
PKG-POL	packaged petroleum, oils, and lubricants
PKI	public key infrastructure
PKO	peacekeeping operations
PKP	purple k powder
PL	phase line; public law
PLA	plain language address
PLAD	plain language address directory
PLAT	pilot's landing aid television
PLB	personal locator beacon
PLC	power line conditioner
PLGR	precise lightweight global positioning system (GPS) receiver
PLL	phase locked loop

PLL/ASL	prescribed load list/authorized stock level
PLRS	position location reporting system
PLS	palletized load system; personal locator system; pillars of logistic support
PLT	platoon; program library tape
PM	parallel track multiunit; preventive medicine; program manager; provost marshal
PMAI	primary mission aircraft inventory
P/M/C	passengers/mail/cargo
PMC	parallel multiunit circle; partial mission-capable
PMCM	partial mission-capable, maintenance
PMCS	partial mission-capable, supply
PMD	program management directive
PME	professional military education
PMEL	precision measurement equipment laboratory
PMGM	program manager's guidance memorandum
PMI	patient movement item
PMIS	psychological operations (PSYOP) management information subsystem
PMN	parallel track multiunit non-return
PMO	production management office(r); program management office
PMOS	primary military occupational specialty
PMR	parallel track multiunit return; patient movement request
PMRC	patient movement requirements center
PMS	portable meteorological subsystem
PN	pseudonoise
PNID	precedence network in dialing
PNVS	pilot night vision system
P/O	part of
PO	peace operations; petty officer
POADS	psychological operations automated data system
POAI	primary other aircraft inventory
POAT	psychological operations assessment team
POB	persons on board; psychological operations battalion
POC	point of contact
POD	plan of the day; port of debarkation; probability of detection
POE	port of embarkation; port of entry
POG	port operations group; psychological operations group
POI	period of interest; program of instruction
POL	petroleum, oils, and lubricants
POLAD	political advisor
POLCAP	bulk petroleum capabilities report
POM	program objective memorandum
POMCUS	pre-positioning of materiel configured to unit sets
POLMIL	political-military
POMSO	Plans, Operations, and Military Support Office(r) (NG)
POP	performance oriented packaging

POPS	port operational performance simulator
POR	proposed operational requirement
PORTS	portable remote telecommunications system
POS	port of support; position; primary operating stocks; probability of success
POSF	port of support file
POSSUB	possible submarine
POSTMOB	post mobilization
POTF	psychological operations task force
POTG	psychological operations task group
POTUS	President of the United States
POV	privately owned vehicle
POW	prisoner of war
P/P	patch panel
p-p	peak-to-peak
PPA	personnel information system (PERSINS) personnel activity
PPAG	proposed public affairs guidance
PPBS	Planning, Programming, and Budgeting System
PPD	program planning document
PPDB	point positioning database
PPE	personal protective equipment
Pplan	programming plan
ppm	parts per million
PPP	power projection platform; primary patch panel; priority placement program
PPR	prior permission required
PPS	precision positioning service
PR	personnel recovery; primary zone; production requirement; program review
PRA	primary review authority
PRANG	Puerto Rican Air National Guard
PRBS	pseudorandom binary sequence
PRC	populace and resources control
PRCA	Presidential Reserve Callup Authority
PRD	personnel readiness division; Presidential review directive
PRECOM	preliminary communications search
PREMOB	pre-mobilization
PREPO	pre-positioned force, equipment, or supplies; pre-positioning
PRF	personnel resources file; pulse repetition frequency
PRG	program review group
PRI	movement priority for forces having the same latest arrival date (LAD); priority; progressive routing indicator
PRIFLY	primary flight control
Prime BEEF	Prime Base Engineer Emergency Forces
PRM	Presidential review memorandum
PRMFL	perm file

PRN	pseudorandom noise
PROBSUB	probable submarine
PROC	processor; Puerto Rican Operations Center
PROFIS	professional officer filler information system
PROM	programmable read-only memory
PROPIN	caution, proprietary information involved
PROVORG	providing organization
proword	procedure word
PRP	personnel reliability program
PRRIS	Puerto Rican radar integration system
PRSL	primary zone/switch location
PRT	pararescue team
PRU	pararescue unit; primary reporting unit
PS	parallel track single-unit; processing subsystem
PSA	port support activity
PSB	poststrike base
PSC	principal subordinate command
PSD	planning systems division
PSE	peculiar support equipment
PSHD	port security and harbor defense
PSHDGRU	port security and harbor defense group
PSI	personnel security investigation
PSK	phase shift keying
PSL	parallel track single-unit long-range aid to navigation (LORAN)
PSMS	Personnel Status Monitoring System
PSN	packet switching node; public switch network
PSO	post security officer
PSP	perforated steel planking; power support platform
PSPS	psychological operations (PSYOP) studies program subsystem
PSS	parallel single-unit spiral
P-STATIC	precipitation static
PSTN	public switched telephone network
PSU	port security unit
PSV	pseudosynthetic video
PSYOP	psychological operations
PSYWAR	psychological warfare
PTA	position, time, altitude
PTAI	primary training aircraft inventory
PTC	peace through confrontation; primary traffic channel
PTT	postal telephone and telegraph; public telephone and telegraph; push-to-talk
PTTI	precise time and time interval
pub	publication
PUK	packup kit
PUL	parent unit level
PVNTMED	preventive medicine

PVT	positioning, velocity, and timing
PW	prisoner of war
pW	picowatt
PWB	printed wiring board (assembly)
PWD	programmed warhead detonation
PWF	personnel working file
PWIS	Prisoner of War Information System
PWR	pre-positioned wartime reserves
PWRMR	pre-positioned war materiel requirement
PWRMS	pre-positioned war reserve materiel stock
PWRS	pre-positioned war reserve stock
PZ	pickup zone

Q

QA	quality assurance
QAM	quadrature amplitude modulation
QAT	quality assurance team
QC	quality control
QD	quality distance
QDR	quality deficiency report
QEEM	quick erect expandable mast
QM	quartermaster
QPSK	quadrature phase shift keying
QRA	quick reaction antenna
QRE	quick reaction element
QRG	quick response graphic
QRP	quick response posture
QRS	quick reaction strike
QRSA	quick reaction satellite antenna
Q-ship	decoy ship
QSTAG	quadripartite standardization agreement
QTY	quantity
QUADCON	quadruple container

R

R	routine; search radius
R&D	research and development
R&R	rest and recuperation
R2P2	rapid response planning process
RA	risk analysis
RAA	redeployment assembly area
RABFAC	radar beacon forward air controller
RADAREXREP	radar exploitation report
RADAY	radio day
RADC	regional air defense commander
RADF	radarfind

RADHAZ	electromagnetic radiation hazards
RADINT	radar intelligence
RAE	right of assistance entry
RAF	Royal Air Force (UK)
R-AFF	regimental affiliation
RAM	raised angle marker; random access memory
RAMCC	regional air movement coordination center
RAOB	rawindsonde observation
RAOC	rear area operations center; regional air operations center
RAP	Radiological Assistance Program (DOE); rear area protection; Remedial Action Projects Program (JCS)
RAS	recovery activation signal
RAST	recovery assistance, securing, and traversing systems
RASU	random access storage unit
RATT	radio teletype
RB	radar beacon; short-range coastal or river boat
RBC	red blood cell
RBE	remain-behind equipment
RBECS	Revised Battlefield Electronic Communications, Electronics, Intelligence, and Operations (CEIO) System
RBI	RED/BLACK isolator
RB std	rubidium standard
RC	receive clock; regional coordinator; Reserve Component; river current
RCA	residual capabilities assessment; riot control agent
RCAT	regional counterdrug analysis team
RCC	relocation coordination center; rescue coordination center
RCCPDS	Reserve Component common personnel data system
RCEM	regional contingency engineering management
RCHB	reserve cargo handling battalion
RCM	Rules for Courts-Martial
RCMP	Royal Canadian Mounted Police
RC NORTH	Regional Command North (NATO)
RCO	regional coordinating office (DOE)
RCP	resynchronization control panel
RCS	radar cross section
RC SOUTH	Regional Command South (NATO)
RCSP	remote call service position
RCT	rescue coordination team (Navy)
RCU	rate changes unit; remote control unit
RCVR	receiver
RD	receive data; ringdown
RDA	research, development, and acquisition
R-day	redeployment day
RDD	required delivery date
RDF	radio direction finder; rapid deployment force
RDO	request for deployment order
RDT&E	research, development, test and evaluation

REACT	rapid execution and combat targeting
REAC/TS	radiation emergency assistance center/training site (DOE)
READY	resource augmentation duty program
RECA	Residual Capability Assessment
RECAS	residual capability assessment system
RECAT	residual capability assessment team
RECCE	reconnaissance
RECCEXREP	reconnaissance exploitation report
RECMOB	reconstitution-mobilization
RECON	reconnaissance
RED HORSE	Rapid Engineers Deployable Heavy Operations Repair Squadron, Engineers
REF	reference(s)
REGT	regiment
REL	relative
RELCAN	releasable to Canada
REM	roentgen equivalent mammal
REMT	regional emergency management team
REPOL	petroleum damage and deficiency report; reporting emergency petroleum, oils, and lubricants
REPSHIP	report of shipment
REPUNIT	reporting unit
REQCONF	request confirmation
REQSTATASK	air mission request status tasking
RESA	research, evaluation, and system analysis
RESCAP	rescue combat air patrol
RESCORT	rescue escort
RET	retired
RF	radio frequency; reserve force; response force
RFA	radio frequency authorization; restrictive fire area
RFC	response force commander
RF/EMPINT	radio frequency/electromagnetic pulse intelligence
RFI	radio frequency interference; ready for issue; request for information
RFL	restrictive fire line
RFP	request for proposal
RFS	request for service
RFW	request for waiver
RG	reconstitution group
RGR	Rangers
RH	reentry home
Rh	Rhesus factor
RHIB	rigid hull inflatable boat
RI	radiation intensity; Refugees International; routing indicator
RIB	rubberized inflatable boat
RIC	routing indicator code
RIG	recognition identification group

RIMS	registrant information management system
RINT	unintentional radiation intelligence
RIP	register of intelligence publications
RIS	reconnaissance information system
RISOP	red integrated strategic offensive plan
RISTA	reconnaissance, intelligence, surveillance, and target acquisition
RIT	remote imagery transceiver
RJTD	reconstitution joint table of distribution
RLD	ready-to-load date
RLE	rail liaison element
RLG	regional liaison group; ring laser gyro
RLGM	remote loop group multiplexer
RLGM/CD	remote loop group multiplexer/cable driver
RLP	remote line printer
RM	risk management
RMC	remote multiplexer combiner; rescue mission commander; Resource Management Committee (CSIF); returned to military control
RMKS	remarks
RMO	regional Marine officer
RMS	requirements management system; root-mean-square
RMU	receiver matrix unit
RNAV	area navigation
RNP	remote network processor
R/O	receive only
Ro	search radius rounded to next highest whole number
ROA	restricted operations area
ROC	regional operations center; required operational capability
ROCU	remote orderwire control unit
ROE	rules of engagement
ROEX	rules of engagement exercise
ROG	railhead operations group
ROK	Republic of Korea
ROM	read-only memory; rough order of magnitude
RON	remain overnight
RO/RO	roll-on/roll-off
ROS	reduced operational status
ROTC	Reserve Officer Training Corps
ROTHR	relocatable over-the-horizon backscatter radar (USN)
ROWPU	reverse osmosis water purification unit
ROZ	restricted operations zone
RP	reconstitution priority; release point (road); retained personnel
RPT	report
RPTOR	reporting organization
RPV	remotely piloted vehicle
RQMT	requirement

RQT	rapid query tool
RR	reattack recommendations
RRC	regional reporting center
RRDF	roll-on/roll-off (RO/RO) discharge facility
RRF	rapid reaction force; Ready Reserve Fleet; Ready Reserve Force
RRPP	rapid response planning process
RS	rate synthesizer; requirement submission
RSA	retrograde storage area
RSC	red station clock; rescue sub-center
RSG	reference signal generator
RSI	rationalization, standardization, and interoperability
RSL	received signal level
RSO	regional security officer
RSOC	regional signals intelligence (SIGINT) operations center
RSOI	reception, staging, onward movement, and integration
RSP	recognized surface picture; Red Switch Project (DOD)
RSPA	Research and Special Programs Administration
RSS	radio subsystem; remote sensors subsystem; root-sum-squared
RSSC	regional satellite communications (SATCOM) support center; regional satellite support cell; regional signals intelligence (SIGINT) support center (NSA); regional space support center
RSTA	reconnaissance, surveillance, and target acquisition
RSTV	real-time synthetic video
RSU	rapid support unit; rear support unit; remote switching unit
R/T	receiver/transmitter
RT	remote terminal; rough terrain
RTA	residual threat assessment
RTB	return to base
RTCC	rough terrain container crane
RTCH	rough terrain container handler
RTD	returned to duty
RTF	return to force
RTFL	rough terrain forklift
RTLP	receiver test level point
RTM	real-time mode
RTOC	rear tactical operations center
RTS	remote transfer switch
RTTY	radio teletype
RU	release unit; rescue unit
RUIC	Reserve unit identification number
RUSCOM	rapid ultrahigh frequency (UHF) satellite communications
RV	long-range seagoing rescue vessel; reentry vehicle; rekeying variable; rendezvous
RVR	runway visibility recorder
RVT	remote video terminal

RWCM	regional wartime construction manager
RWI	radio and wire integration
RWR	radar warning receiver
RX	receive; receiver
RZ	recovery zone; return-to-zero

S

S&F	store-and-forward
S&M	scheduling and movement
S&T	scientific and technical
S&TI	scientific and technical intelligence
S-2	battalion or brigade intelligence staff officer (Army; Marine Corps battalion or regiment)
S-3	battalion or brigade operations staff officer (Army; Marine Corps battalion or regiment)
S-4	battalion or brigade logistics staff officer (Army; Marine Corps battalion or regiment)
SA	security assistance; selective availability (GPS); senior adviser; situational awareness; staging area; stand-alone switch
SAAFR	standard use Army aircraft flight route
SAAM	special assignment airlift mission
SAB	scientific advisory board (USAF)
SABER	situational awareness beacon with reply
SAC	special agent in charge
SACC	supporting arms coordination center
SACEUR	Supreme Allied Command, Europe
SACLANT	Supreme Allied Command, Atlantic
SACS	secure telephone unit (STU) access control system
SAF	Secretary of the Air Force
SAFE	selected area for evasion
SAFE-CP	selected area for evasion-contact point
SAFER	evasion and recovery selected area for evasion (SAFE) area activation request
SAG	surface action group
SAI	single agency item
SAID	selected area for evasion (SAFE) area intelligence description
SAL	small arms locker
SAL-GP	semiactive laser-guided projectile (USN)
SALM	single-anchor leg mooring
SALT	supporting arms liaison team
SALTS	streamlined automated logistics transfer system; streamlined automated logistics transmission system
SALUTE	size, activity, location, unit, time, and equipment
SAM	space available mail; special airlift mission; surface-to-air missile

SAMM	security assistance management manual
SAMS	School of Advanced Military Studies
SAO	security assistance office/officer; security assistance organization; selected attack option
SAOC	sector air operations center
SAP	special access program
SAPI	special access program for intelligence
SAPO	subarea petroleum office
SAR	satellite access request; search and rescue; special access requirement; synthetic aperture radar;
SARC	surveillance and reconnaissance center
SARIR	search and rescue incident report
SARMC	search and rescue mission coordinator
SARMIS	search and rescue management information system
SARREQ	search and rescue request
SARSAT	search and rescue satellite-aided tracking
SARSIT	search and rescue situation summary report
SARTEL	search and rescue (SAR) telephone (private hotline)
SARTF	search and rescue task force
SAS	sealed authenticator system; special ammunition storage
SASP	special ammunition supply point
SASS	supporting arms special staff
SASSY	supported activities supply systems
SAT	satellite; security alert team
SATCOM	satellite communications
SAW	surface acoustic wave
SAU	search attack unit
SB	standby base
SBIRS	space-based infrared system
SBL	space-based laser
SBPO	Service blood program officer
SBR	special boat squadron
SBRPT	subordinate reporting organization
SBS	senior battle staff; support battle staff
SBSS	science-based stockpile stewardship
SBU	special boat unit
SC	sea current; search and rescue (SAR) coordinator; station clock
SCAS	stability control augment system
SCATANA	security control of air traffic and navigation aids
SC ATLANTIC	Strategic Command, Atlantic (NATO)
SCATMINEWARN	scatterable minefield warning
SCC	security classification code; Space Control Center (USSPACECOM); Standards Coordinating Committee
SCDL	surveillance control data link
SCE	service cryptologic element
SCEUROPE	Strategic Command, Europe (NATO)
SCF(UK)	Save the Children Fund (United Kingdom)

SCF(US)	Save the Children Federation (United States)
SCG	switching controller group
SCI	sensitive compartmented information
SCIF	sensitive compartmented information facility
SCMP	strategic command, control, and communications (C3) master plan
SCNE	self-contained navigation equipment
SCO	state coordinating officer
SCOC	systems control and operations concept
SCONUM	ship control number
SCP	secure conferencing project; system change proposal
SCPT	strategic connectivity performance test
SCRB	software configuration review board
SCT	shipping control teams; single channel transponder
SCTIS	single channel transponder injection system
SCTS	single channel transponder system
SCT-UR	single channel transponder ultrahigh frequency (UHF) receiver
SCUD	surface-to-surface missile system
SDA	Seventh-Day Adventist (ADRA)
S-day	day the President authorizes selective reserve call-up
SDF	self defense force
SDIO	Strategic Defense Initiative Organization
SDLS	satellite data link standards
SDMX	space division matrix
SDN	system development notification
SDNRIU	secure digital net radio interface unit
SDR	system design review
SDSG	space division switching group
SDSM	space division switching matrix
SDV	sea-air-land team (SEAL) delivery vehicle; submerged delivery vehicle
SEABEE	Navy construction engineer; sea barge
SEAD	suppression of enemy air defenses
SEAL	sea-air-land team
SEAVAN	military container moved via ocean
SEC	submarine element coordinator
SECAF	Secretary of the Air Force
SECARMY	Secretary of the Army
SecDef	Secretary of Defense
SECNAV	Secretary of the Navy
SECNAVINST	Secretary of the Navy instruction
SECOMP	secure en route communications package
SECORD	secure cord switchboard
SECRA	secondary radar data only
SECSTATE	Secretary of State
SECTRANS	Secretary of Transportation
SED	signals external data

SEDAS	spurious emission detection acquisition system
SEF	sealift enhancement feature
SEI	specific emitter identification
SEL REL	selective release
SELRES	Selected Reserve
SEMA	special electronic mission aircraft
SEMS	standard embarkation management system
SEO/SEP	special enforcement operation/special enforcement program
SEP	sealift enhancement program; signal entrance panel; spherical error probable
SERE	survival, evasion, resistance, escape
SERER	survival, evasion, resistance, escape, recovery
SETA	system engineering and technical assistance
SEW	shared early warning
S/EWCC	signals intelligence/electronic warfare coordination center
SEWS	satellite early warning system
SF	security forces; single frequency; special forces
SFAF	standard frequency action format
SFCP	shore fire control party
SFG	security forces group; special forces group
SFI	spectral composition
SFMS	special forces medical sergeant
SFOB	special forces operations base
SFOD-A/B/C	special forces operational detachment-A/B/C
SFS	security forces squadron
SG	supergroup; Surgeon General
SGEMP	system-generated electromagnetic pulse
SGSA	squadron group systems advisor
SGSI	stabilized glide slope indicator
SHAPE	Supreme Headquarters Allied Powers, Europe
SHD	special handling designator
SHF	super-high frequency
SHORAD	short-range air defense
SHORADEZ	short-range air defense engagement zone
SI	special intelligence
SIA	station of initial assignment
SIAGL	survey instrument azimuth gyroscope lightweight
SIC	subject identification code
SICR	specific intelligence collection requirement
SID	secondary imagery dissemination; standard instrument departure
SIDAC	single integrated damage analysis capability
SIDL	standard intelligence documents list
SIDS	secondary imagery dissemination system
SIF	selective identification feature
SIG	signal
SIGINT	signals intelligence

SIGSEC	signal security
SII	seriously ill or injured; statement of intelligence interest
SIMLM	single integrated medical logistics management; single integrated medical logistics manager
SINCGARS	single-channel ground and airborne radio system
SINS	ship's inertial navigation system
SIO	senior intelligence officer; special information operations
SIOP	Single Integrated Operational Plan
SIOP-ESI	Single Integrated Operational Plan-Extremely Sensitive Information
SIPRNET	SECRET Internet Protocol Router Network
SIR	serious incident report; specific information requirement
SIRADS	stored imagery repository and dissemination system
SIRMO	senior information resources management official
SIS	special information systems
SIT	special interest target
SITLM	single integrated theater logistic manager
SITREP	situation report
SIV	special interest vessel
SJA	Staff Judge Advocate
SJS	Secretary, Joint Staff
SKE	station-keeping equipment
SL	sea level; switch locator
SLAM	stand-off land attack missile
SLAR	side-looking airborne radar
SLBM	submarine-launched ballistic missile
SLC	satellite laser communications; single line concept
SLCM	sea-launched cruise missile
SLCP	ship lighterage control point; ship's loading characteristics pamphlet
SLD	system link designator
SLEP	service life extension program
SLGR	small, lightweight ground receiver (GPS)
SLIT	serial-lot item tracking
SLO	space liaison officer
SLOC	sea line of communications
SLP	seaward launch point
SLRP	survey, liaison, and reconnaissance party
SLWT	side loadable warping tug
SM	Secretary, Joint Staff, memorandum; staff memorandum; system manager
SMART	special medical augmentation response team
SMC	midpoint compromise track spacing; search and rescue (SAR) mission coordinator; system master catalog
SMCA	single manager for conventional ammunition
SMCC	strategic mobile command center
SMCM	surface mine countermeasures
SMCOO	spectrum management concept of operations

SMCR	Selected Marine Corps Reserve
SMD	strategic missile defense
SMDC	Space & Missile Defense Command (Army)
SMEB	significant military exercise brief
SMEO	small end office
SMFT	semitrailer mounted fabric tank
SMI	security management infrastructure
SMIO	search and rescue (SAR) mission information officer
SMO	senior meteorological and oceanographic officer; strategic mobility office(r); support to military operations
SMP	sub-motor pool
SMPT	School of Military Packaging Technology
SMRI	service message routing indicator
SMTP	simple message transfer protocol
SMU	special mission unit; supported activities supply system (SASSY) management unit
S/N	signal to noise
SNCO	staff noncommissioned officer
SNIE	special national intelligence estimates
SNF	strategic nuclear forces
SNLC	Senior North Atlantic Treaty Organization (NATO) Logisticians Conference
SNM	system notification message
SNOI	signal not of interest
SO	safety observer; special operations
SOA	separate operating agency; special operations aviation; speed of advance; status of action; sustained operations ashore
SOAF	status of action file
SOC	special operations command
SOCA	special operations communications assembly
SOCC	Sector Operations Control Center (NORAD)
SOCCE	special operations command and control element
SOCCENT	Special Operations Component, United States Central Command
SOCCT	special operations combat control team
SOCEUR	Special Operations Component, United States European Command
SOCEX	special operations capable exercise
SOCLANT	Special Operations Component, United States Atlantic Command
SOCOORD	special operations coordination element
SOCP	special operations communication package
SOCPAC	Special Operations Component, United States Pacific Command
SOCRATES	Special Operations Command, Research, Analysis, and Threat Evaluation System

SOCSOUTH	Special Operations Component, United States Southern Command
SOD	special operations division; strategy and options decision (Planning, Programming, and Budgeting System)
SODARS	special operations debrief and retrieval system
SOE	special operations executive
SOF	special operations forces; supervisor of flying
SOFA	status-of-forces agreement
SOFAR	sound fixing and ranging
SOFME	special operations forces medical element
SOG	special operations group
SOI	signal of interest; signal operating instructions; space object identification
SOIC	senior officer of the intelligence community
SOLAS	safety of life at sea
SOLE	special operations liaison element
SOLL	special operations low-level
SOM	start of message
SOMA	status of mission agreement
SOMPF	special operations mission planning folder
SONMET	special operations naval mobile environment team
SOP	standard operating procedure; standing operating procedure
SOPA	senior officer present afloat (USN)
SO-peculiar	special operations-peculiar
SOR	statement of requirement
SORTIEALOT	sortie allotment message
SORTS	Status of Resources and Training System
SOS	special operations squadron
SOSB	special operations support battalion
SOSC	special operations support command (theater army)
SOSE	special operations staff element
SOSG	station operations support group
SOSR	suppress, obscure, secure, and reduce
SOTA	signals intelligence (SIGINT) operational tasking authority
SOTAC	special operations terminal attack controller
SOUTHAF	Southern Command Air Forces
SOUTHROC	Southern Region Operational Center (USSOUTHCOM)
SOW	special operations wing; standoff weapon; statement of work
SOWT	special operations weather team
SOWT/TE	special operations weather team/tactical element
SP	security police
SPA	special psychological operations (PSYOP) assessment; submarine patrol area
SPCC	ships parts control center (USN)
SPEAR	strike protection evaluation and antiair research
SPEC	specified

SPECAT	special category
SPECWAR	special warfare
SPINS	special instructions
SPINTCOMM	special intelligence communications handling system
SPIREP	specialist intelligence report; spot intelligence report
SPLX	simplex
SPM	single point mooring; single port manager
SPMAGTF	special purpose Marine air-ground task force
SPO	system program office
SPOC	search and rescue (SAR) points of contact; space command operations center; Space Operations Center (USSPACECOM)
SPOD	seaport of debarkation
SPOE	seaport of embarkation
SPOTREP	spot report
SPP	shared production program
SPR	software problem report
SPRINT	special psychiatric rapid intervention team
SPS	special psychological operations (PSYOP) study; standard positioning system
SPSC	system planning and system control
SPTD CMD	supported command
SPTG CMD	supporting command
sqft	square feet
SR	special reconnaissance
SRA	specialized-repair activity
SRAM	short-range air-to-surface attack missile; system replacement and modernization
SRB	software release bulletin; system review board (JOPES)
SRBM	short-range ballistic missile
SRC	Single Integrated Operational Plan (SIOP) response cell; standard requirements code; survival recovery center
SRCC	service reserve coordination center
SRF	secure Reserve force
SRG	short-range aircraft
SRI	surveillance, reconnaissance, and intelligence (Marine Corps)
SRIG	surveillance, reconnaissance, and intelligence group (USMC)
SROC	Senior Readiness Oversight Council; Southern Region Operational Center, United States Southern Command
SROE	standing rules of engagement
SRP	Sealift Readiness Program; sealift reserve program; seaward recovery point; Single Integrated Operational Plan (SIOP) reconnaissance plan
SRP/PDS	stabilization reference package/position determining system
SRR	search and rescue region

SRS	search and rescue sector
SRSG	special representative to the Secretary-General
SRT	scheduled return time; special reaction team; standard remote terminal; strategic relocatable target
SRTD	signals research and target development
SRU	search and rescue unit
SR-UAV	short-range unmanned aerial vehicle
SRWBR	short range wide band radio
S/S	steamship
SS	submarine
SSA	software support activity; special support activity (NSA); strapdown sensor assembly; supply support activity; supply support area
SSB	single side band; surveillance support branch
SSBN	fleet ballistic missile submarine
SSB-SC	single sideband-suppressed carrier
SSC	small scale contingency; surveillance support center
SSCO	shipper's service control office
SSCRA	Soldiers and Sailors Civil Relief Act
SSF	software support facility
SSI	standing signal instruction
SSM	surface-to-surface missile
SSMI	special sensor microwave imager
SSMS	single shelter message switch
SSN	attack submarine, nuclear; Social Security number; space surveillance network
SS (number)	sea state (number)
SSO	special security officer; spot security office
SSP	signals intelligence (SIGINT) support plan
SSPO	strategic systems program office
SSS	Selective Service System; shelter subsystem
SSSC	surface, subsurface search surveillance coordination
SST	space support team
ST	short ton; small tug; strike team
S/T	short ton
ST&E	security test and evaluation
STA	system tape A
STAB	space tactical awareness brief
STA clk	station clock
STAMMIS	standard Army multi-command management information system
STANAG	standardization agreement (NATO)
STANAVFORLANT	Standing Naval Forces, Atlantic (NATO)
STAR	scheduled theater airlift route; standard attribute reference; standard terminal arrival route; surface-to-air recovery; system threat assessment report
STARC	state area coordinators
START	Strategic Arms Reduction Treaty

STARTEX	start of exercise
STB	super tropical bleach
STC	secondary traffic channel
STD	standard
STDM	synchronous time division multiplexer
STEL STU III	Standford telecommunications (secure telephone)
STEP	software test and evaluation program; standard tactical entry point; standard tool for employment planning
STICS	scalable transportable intelligence communications system
STO	special technical operations
STOC	special technical operations coordinator
STOD	special technical operations division
STOL	short takeoff and landing
STOMPS	stand-alone tactical operational message processing system
STON	short ton
STOVL	short takeoff and vertical landing aircraft
STP	security technical procedure
STR	strength
STRATOPS	strategic operations division
STREAM	standard tensioned replenishment alongside method
STS	special tactics squadron
STT	special tactics team
STU	secure telephone unit
STU-III	secure telephone unit III
STW	strike warfare
STWC	strike warfare commander
STX	start of text
SU	search unit
SUBJ	subject
sub-JIB	subordinate joint information bureau
SUBOPAUTH	submarine operating authority
SUBROC	submarine rocket
SUC	surf current
SUIC	service unit identification code
SUPE	supervisory commands program
SURG	surgeon
SUROBS	surf observation
SURPIC	surface picture
SUW	surface warfare
SUWC	surface warfare commander
S/V	sailboat
SVC	Service
SVIP	secure voice improvement program
SVLTU	service line termination unit
SVR	surface vessel radar
SVS	secure voice system
Sw	switch
SWA	Southwest Asia

SWAT	special weapons and tactics
SWBD	switchboard
SWC	strike warfare commander; swell/wave current
SWO	staff weather officer
SWORD	submarine warfare operations research division
SWSOCC	Southwest Sector Operation Control Center North American Aerospace Defense Command (NORAD)
SYDP	six year defense plan
SYG	Secretary General (UN)
SYNC	synchronization
SYS	system
SYSCON	systems control

T

T	search time available; short ton; trackline pattern
T&DE	test and diagnostic equipment
T&E	test and evaluation
T2	technology transfer
TA	target acquisition; theater Army
TAA	tactical assembly area
TAACOM	Theater Army Area Command
TAADS	The Army Authorization Document System
TAB	tactical air base
TAC	tactical advanced computer; terminal access controller
TAC(A)	tactical air coordinator (airborne)
TACAIR	tactical air
TACAMO	take charge and move out (EA-6B aircraft)
TACAN	tactical air navigation
TACC	tactical air command center (USMC); tactical air control center (USN); tanker/airlift control center (USAF)
TAC-D	tactical deception
TACDAR	tactical detection and reporting
TACINTEL	tactical intelligence
TACLOG	tactical-logistical
TACM	tactical air command manual
TACON	tactical control
TACOPDAT	tactical operational data
TA/CP	technology assessment/control plan
TACP	tactical air control party
TACRON	tactical air control squadron
T-ACS	tactical auxiliary crane ship
TACS	tactical air control system; theater air control system
TACSAT	tactical satellite
TACSIM	tactical simulation
TACSTANS	tactical standards
TACT	tactical aviation control team
TACTRAGRULANT	Tactical Training Group, Atlantic

TAD	tactical air direction; temporary additional duty (non-unit-related personnel); theater air defense; time available for delivery
TADC	tactical air direction center
TADCS	tactical airborne digital camera system
TADIL	tactical digital information link
TADS	Tactical Air Defense System
TAES	theater aeromedical evacuation system
TAF	tactical air force
TAFDS	tactical airfield fuel dispensing system
TAFIM	technical architecture framework for information management
TAFS	tactical aerodrome forecasts
TAFT	technical assistance field team
TAG	technical assessment group; The Adjutant General; Tomahawk land-attack missile aimpoint graphic
T-AGOS	tactical auxiliary general ocean surveillance
T-AH	hospital ship
TAI	International Atomic Time; target area of interest; total active inventory
TAIS	transportation automated information systems
TAK	cargo ship
T-AKR	fast logistics ship
TALCE	tanker airlift control element
TALD	tactical air-launched decoy
TALO	theater airlift liaison officer
TAMCA	theater Army movement control agency
TAMCO	theater Army movement control center
TAMMC	theater army material management command
TAMMIS	theater Army medical management and information system
tanalt	tangent altitude
TAO	tactical actions officer
TAOC	tactical air operations center (USMC)
TAP	troopship
TAR	Training and Administration of the Reserve
TARBUL	target bulletin
TARE	tactical record evaluation
TAREX	target plans and operations
TARS	tethered aerostat radar system
TAS	tactical atmospheric summary; true air speed
T-ASA	Television Audio Support Agency
TASCID	tactical Automatic Digital Network (AUTODIN) satellite compensation interface device
TASCO	tactical automatic switch control officer
TASIP	tailored analytic intelligence support to individual electronic warfare and command and control warfare projects
TASMO	tactical air support for maritime operations

TASOSC	theater Army special operations support command
TASS	tactical automated security system; tactical automated switch system
TAT	tactical analysis team; technical assistance team
TATC	tactical air traffic control
TAV	total asset visibility
T-AVB	aviation logistics support ship
TAW	tactical airlift wing
TBD	to be determined
TBM	tactical ballistic missile; theater ballistic missile
TBMCS	theater battle management core system
TBMD	theater ballistic missile defense
TBP	to be published
TBSL	to be supplied later
TBTC	transportable blood transshipment center
TC	tidal current; transmit clock and/or telemetry combiner; Transportation Corps (Army)
TCA	terminal control area; time of closest approach; traditional CINC activity
TC-ACCIS	Transportation Coordinator's Automated Command and Control Information System
TC-AIMS	Transportation Coordinator's Automated Information for Movement System
TCAIMS II	Transportation Coordinator's Automated Information for Movement System II
TCC	transmission control code; transportation component command
TCCF	tactical communications control facility
TCEM	theater contingency engineering management
TCF	tactical combat force; technical control facility
TCM	theater construction manager
TCMD	transportation control and movement document
TCN	third country national; transportation control number
TCSEC	trusted computer system evaluation criteria
TD	timing distributor; total drift; transmit data
TDA	Table of Distribution and Allowance
TDAD	Table of Distribution and Allowance (TDA) designation
T-day	effective day coincident with Presidential declaration of a National Emergency and authorization of partial mobilization
TDBM	technical database management
TDD	target desired ground zero (DGZ) designator
TDF	tactical digital facsimile
TDI	target data inventory
TDIC	time division interface controller
TDIG	time division interface group
TDIM	time division interface module
TDM	time division multiplexed

TDMA	time division multiple access
TDMF	time division matrix function
TDMM	time division memory module
TDMX	time division matrix
TDSG	time division switching group
TDSGM	time division switching group modified
TDT	theater display terminal
TDY	temporary duty
TE	transaction editor
TEA	Transportation Engineering Agency
tech	technical
TECHCON	technical control
TECHDOC	technical documentation
TECHEVAL	technical evaluation
TECHINT	technical intelligence
TECHOPDAT	technical operational data
TECS II	Treasury Enforcement Communications System
TED	trunk encryption device
TEL	transporter-erector-launcher (missile platform)
TELEX	teletype
TELINT	telemetry intelligence
TELNET	telecommunication network
TENCAP	tactical exploitation of national capabilities
TEOB	tactical electronic order of battle
TEP	test and evaluation plan; theater engagement plan
TERCOM	terrain contour matching
TERF	terrain flight
TERPES	tactical electronic reconnaissance processing and evaluation system
TERPROM	terrain profile matching
TERS	tactical event reporting system
TES	tactical event system; theater event system
TEU	twenty-foot equivalent unit
TF	task force
TFCICA	task force counterintelligence coordinating authority
TFE	tactical field exchange; transportation feasibility estimator
TFR	temporary flight restriction
TFS	tactical fighter squadron
TG	task group
TGC	trunk group cluster
TGEN	table generate
TGM	trunk group multiplexer
TGMOW	transmission group module and/or orderwire
TGT	target
TGTINFOREP	target information report
TGU	trunk compatibility unit
THREATCON	terrorist threat condition
TI	threat identification; training instructor

TIAP	theater intelligence architecture program
TIARA	tactical intelligence and related activities
TIBS	tactical information broadcast service
TIC	target information center
TIDP	technical interface design plan
TIDS	tactical imagery dissemination system
TIFF	tagged image file format
TII	total inactive inventory
TIO	target intelligence officer
TIP	target intelligence package
TIPG	telephone interface planning guide
TIPI	tactical information processing and interpretation system; tactical information processing interpretation
TIPS	tactical optical surveillance system (TOSS) imagery processing system
TIS	technical interface specification; thermal imaging system
TISG	technical interoperability standards group
TISS	thermal imaging sensor system
TL	team leader
TLAM	Tomahawk land-attack missile
TLAM/N	Tomahawk land attack missile/nuclear
TLC	traffic load control
TLCF	teleconference
TLP	transmission level point
TLR	trailer
TLX	teletype
TM	tactical missile; target materials; team member; technical manual; theater missile; TROPO modem
TMAO	theater mortuary affairs officer
TMD	tactical munitions dispenser; theater missile defense
TMEP	theater mortuary evacuation point
TMG	timing
TMIS	theater medical information system
TML	terminal
TMMMC	theater medical materiel management center
TMN	trackline multiunit non-return
TMO	traffic management office; transportation management office
TMP	target materials program; telecommunications management program; theater manpower forces
TMR	trackline multiunit return
T/M/S	type, model, and/or series (also as TMS)
TNAPS	tactical network analysis and planning system
TNAPS+	tactical network analysis and planning system plus
T-net	training net
TNF	theater nuclear force
T/O	table of organization
TO	technical order; theater of operations

TO&E	table of organization and equipment
TOA	table of allowance
TOAI	total overall aircraft inventory
TOC	tactical operations center
TOCU	tropospheric scatter (TROPO) orderwire control unit
TOD	time of day
TOE	table of organization and equipment
TOFC	trailer on flatcar
TOH	top of hill
TOI	track of interest
TOPINT	technical operational intelligence
TOR	term of reference; time of receipt
TOS	time on station
TOSS	tactical optical surveillance system
TOT	time on target
TP	technical publication; turn point
TPB	tactical psychological operations battalion
TPC	two person control
TPC/PC	tactical pilotage chart and/or pilotage chart
TPED	tasking, processing, exploitation, and dissemination
TPERS	type personnel element
TPFDD	time-phased force and deployment data
TPFDL	time-phased force and deployment list
TPL	technical publications list; telephone private line
TPMRC	theater patient movement requirements center
TPT	tactical petroleum terminal
TPTRL	time-phased transportation requirements list
TPU	tank pump unit
TRA	technical review authority
TRAC2ES	transportation command regulating and command and control evacuation system
TRACON	terminal radar approach control facility
TRADOC	United States Army Training and Doctrine Command
TRAM	target recognition attack multisensor
TRANSEC	transmission security
TRAP	tactical recovery of aircraft and personnel (Marine Corps); tactical related applications; tanks, racks, adapters, and pylons; terrorism research and analysis program
TRC	tactical radio communication; transmission release code
TRCC	tactical record communications center
TRE	tactical receive equipment
TREAS	Department of the Treasury
TREE	transient radiation effects on electronics
TRICON	triple container
TRI-TAC	Tri-Service Tactical Communications Program
TRK	truck; trunk
TRNG	training
TRO	training and readiness oversight

TROPO	troposphere; tropospheric scatter
TRS	tactical reconnaissance squadron
TS	terminal service; top secret
TSA	theater storage area; travel security advisory
TSB	trunk signaling buffer
TSCM	technical surveillance countermeasures
TSCO	target selection confusion of the operator; top secret control officer
TSE	tactical support element
TSEC	transmission security
TSG	test signal generator
TSGCE	tri-Service group on communications and electronics
TSGCEE	tri-Service group on communications and electronic equipment (NATO)
TSM	trunk signaling message
TSN	trackline single-unit non-return; track supervision net
TSO	technical standard order; telecommunications service order
TSP	telecommunications service priority
TSR	telecommunications service request; trackline single-unit return
TSS	tactical shelter system; timesharing system; time signal set; traffic service station
TSSP	tactical satellite signal processor
TSSR	tropospheric scatter (TROPO)-satellite support radio
TST	tactical support team; theater support team
TSWA	temporary secure working areas
TT	terminal transfer
TT&C	telemetry, tracking, and commanding
TTB	transportation terminal battalion
TTD	technical task directive
TTL	transistor-transistor logic
TTM	threat training manual
TTP	tactics, techniques, and procedures
TTR	tactical training range
TTT	time to target
TTU	transportation terminal unit
TTY	teletype
TUBA	transition unit box assembly
TUCHA	type unit characteristics file
TUCHAREP	type unit characteristics report
TUDET	type unit equipment detail file
TV	television
TVA	Tennessee Valley Authority
TW/AA	tactical warning and attack assessment
TWC	total water current
TWCF	transportation working Capital Fund
TWCM	theater wartime construction manager
TWD	transnational warfare counterdrug analysis

TWDS	tactical water distribution system
TWPL	teletypewriter private line
TWX	teletypewriter exchange
TX	transmitter; transmit
TYCOM	type commander

U

U	wind speed
UARM	unconventional assisted recovery mechanism
UAV	unmanned aerial vehicle
U/C	unit cost; upconverter
UCFF	Unit Type Code Consumption Factors File
UCMJ	Uniform Code of Military Justice
UCP	Unified Command Plan
UCT	underwater construction team
UDAC	unauthorized disclosure analysis center
UDC	unit descriptor code
UDESC	unit description
UDL	unit designation list
UDP	unit deployment program
UDT	underwater demolition team
UE	unit equipment
UFO	ultrahigh frequency (UHF) follow-on satellite system
UGM-84A	Harpoon
UGM-96A	Trident I
UHF	ultrahigh frequency
UAOBS	upper air observation
UAR	unconventional assisted recovery
UHV	Upper Huallaga Valley
UIC	unit identification code
UICIO	unit identification code information officer
UIRV	unique interswitch rekeying variable
UIS	unit identification system
UJTL	Universal Joint Task List
UK	United Kingdom
UK(I)	United Kingdom and Ireland
ULC	unit level code
ULF	ultra low frequency
ULN	unit line number
UMCC	unit movement control center
UMD	unit manning document; unit movement data
UMIB	urgent marine information broadcast
UMMIPS	uniform material movement and issue priority system
UMPR	unit manpower personnel record
UMT	unit ministry team
UN	United Nations
UNAAF	Unified Action Armed Forces

UNC	United Nations Command
UNCTAD	United Nations Conference on Trade and Development
UND	urgency of need designator
UNDHA	United Nations Department of Humanitarian Affairs
UN-DMT	United Nations disaster management team
UNDP	United Nations development programme
UNDPKO	United Nations Department for Peacekeeping Operations
UNEF	United Nations emergency force
UNEP	United Nations environment program
UNESCO	United Nations Educational, Scientific, and Cultural Organization
UNHCHR	United Nations High Commissioner for Human Rights
UNHCR	United Nations Office of the High Commissioner for Refugees
UNICEF	United Nations Children's Fund
UNIFIL	United Nations Interim Force in Lebanon
UNITAF	unified task force
UNITAR	United Nations Institute for Training and Research
UNITREP	unit status and identity report
UNLOC	United Nations logistic course
UNMIH	United Nations Mission in Haiti
UNMILPOC	United Nations military police course
UNMOC	United Nations military observers course
UNMOVCC	United Nations movement control course
UNO	unit number
UNODIR	unless otherwise directed
UNPA	United Nations Participation Act
UNPROFOR	United Nations protection force
UNREP	underway replenishment
UNREP CONSOL	underway replenishment consolidation
UNRWA	United Nations Relief and Works Agency for Palestine Refugees in the Near East
UNSC	United Nations Security Council
UNSOC	United Nations staff officers course
UNSCR	United Nations Security Council resolution
UNTAC	United Nations Transition Authority in Cambodia
UNTSO	United Nations Truce and Supervision Organization
UP&TT	unit personnel and tonnage table
UPU	Universal Postal Union
URDB	user requirements data base
USA	United States Army
USAB	United States Army barracks
USACCSA	United States Army Command and Control Support Agency
USACFSC	United States Army Community and Family Support Center
USACIDC	United States Army Criminal Investigations Command
USAF	United States Air Force

USAFE	United States Air Forces in Europe
USAFEP	United States Air Force, Europe pamphlet
USAFLANT	United States Air Force, Atlantic Command
USAFR	United States Air Force Reserve
USAFSOC	United States Air Force, Special Operations Command
USAFSOF	United States Air Force, Special Operations Forces
USAID	United States Agency for International Development
USAITAC	United States Army Intelligence Threat Analysis Center
USAJFKSWC	United States Army John F. Kennedy Special Warfare Center
USAMMA	United States Army Medical Materiel Agency
USAMPS	United States Army Military Police School
USAO	United States Attorney Office
USAR	United States Army Reserve
USARCENT	United States Army, Central Command
USAREUR	United States Army, European Command
USARJ	United States Army, Japan
USARLANT	United States Army, Atlantic Command
USARPAC	United States Army, Pacific Command
USARSO	United States Army, Southern Command
USASOC	United States Army, Special Operations Command
USB	upper side band
USBP	United States Border Patrol
USC	United States Code
USCENTAF	United States Central Command Air Forces
USCENTCOM	United States Central Command
USCG	United States Coast Guard
USCINCACOM	Commander in Chief, United States Atlantic Command
USCINCCENT	Commander in Chief, United States Central Command
USCINCEUR	United States Commander in Chief, Europe
USCINCPAC	Commander in Chief, United States Pacific Command
USCINCSO	Commander in Chief, United States Southern Command
USCINCSOC	Commander in Chief, United States Special Operations Command
USCINCSPACE	Commander in Chief, United States Space Command
USCINCSTRAT	Commander in Chief, United States Strategic Command
USCINCTRANS	Commander in Chief, United States Transportation Command
USCS	United States Cryptologic System; United States Customs Service
USDA	United States Department of Agriculture
USD(A&T)	Under Secretary of Defense for Acquisition and Technology
USDAO	United States Defense Attaché Office
USDELMC	United States Delegation to the NATO Military Committee
USD(P)	Under Secretary of Defense for Policy
USDR	United States Defense Representative
USD(R&E)	Under Secretary of Defense for Research and Engineering

USELEMNORAD	United States Element, North American Aerospace Defense Command
USERID	user identification
USEUCOM	United States European Command
USFJ	United States Forces, Japan
USFK	United States Forces, Korea
USFORAZORES	United States Forces, Azores
USFS	United States Forest Service
USFWS	United States Fish and Wildlife Service
USG	United States Government
USIA	United States Information Agency
USIC	United States interdiction coordinator
USIS	United States Information Service
USJFCOM	United States Joint Forces Command
USLANTFLT	United States Atlantic Fleet
USLO	United States liaison officer
USMARFORCENT	United States Marine Component, Central Command
USMARFORLANT	United States Marine Component, Atlantic Command
USMARFORPAC	United States Marine Component, Pacific Command
USMARFORSOUTH	United States Marine Component, Southern Command
USMC	United States Marine Corps
USMCEB	United States Military Communications-Electronics Board
USMCR	United States Marine Corps Reserve
USMER	United States merchant ship vessel locator reporting system
USMILGP	United States military group
USMILREP	United States military representative
USMOG-W	United States Military Observer Group - Washington
USMS	United States Marshals Service
USMTF	United States message text format
USMTM	United States military training mission
USN	United States Navy
USNAVCENT	United States Naval Forces, Central Command
USNAVEUR	United States Naval Forces, Europe
USNCB	United States National Central Bureau (INTERPOL)
USNMR	United States National Military representative
USNMTG	United States North Atlantic Treaty Organization (NATO) Military Terminology Group
USNR	United States Navy Reserve
USNS	United States Naval Ship
USPACAF	United States Air Forces, Pacific Command
USPACFLT	United States Pacific Fleet
USPACOM	United States Pacific Command
USPHS	United States Public Health Service
USPS	United States Postal Service
USREPMC	United States representative to the military committee (NATO)
USSOCOM	United States Special Operations Command
USSOUTHAF	United States Air Force, Southern Command

USSOUTHCOM	United States Southern Command
USSPACECOM	United States Space Command
USSS	United States Secret Service (TREAS); United States Signals Intelligence (SIGINT) System
USSTRATCOM	United States Strategic Command
USTRANSCOM	United States Transportation Command
USUN	United States Mission to the United Nations
USW	undersea warfare
USW/USWC	undersea warfare and/or undersea warfare commander
USYG	Under Secretary General
UT1	unit trainer; Universal Time
UTC	Coordinated Universal Time; unit type code
UTM	universal transverse mercator
UTO	unit table of organization
UTR	underwater tracking range
UUV	unmanned underwater vehicle
UVEPROM	ultraviolet erasable programmable read-only memory
UW	unconventional warfare
UWOA	unconventional warfare operating area
UXO	unexploded explosive ordnance; unexploded ordnance

V

V	search and rescue unit ground speed; sector pattern; volt
v	velocity of target drift
VA	Veterans Administration; vulnerability assessment
VAC	volts, alternating current
VARVAL	vessel arrival data, list of vessels available to marine safety offices and captains of the port
VAT B	(weather) visibility (in miles), amount (of clouds, in eighths), (height of cloud) top (in thousands of feet), (height of cloud) base (in thousands of feet)
VBS	visit, board, search
VBSS	visit, board, search, and seizure
VCC	voice communications circuit
VCNOG	Vice Chairman, Nuclear Operations Group
VCO	voltage controlled oscillator
VCOPG	Vice Chairman, Operations Planners Group
VCR	violent crime report
VCXO	voltage controlled crystal oscillator; voltage controlled oscillator
VDC	volts, direct current
VDR	voice digitization rate
VDS	video subsystem
VDSD	visual distress signaling device
VDU	visual display unit
VDUC	visual display unit controller
VEH	vehicle; vehicular cargo

VERTREP	vertical replenishment
VF	voice frequency
VFR	visual flight rules
VFTG	voice frequency telegraph
VHF	very high frequency
VI	visual information
VICE	advice
VID	visual identification information display
VIDOC	visual information documentation
VINSON	encrypted ultrahigh frequency communications system
VIP	very important person; visual information processor
VIRS	verbally initiated release system
VIS	visual imaging system
VISA	Voluntary Intermodel Sealift Agreement; Voluntary Intermodal Shipping Agreement
VISOBS	visual observer
VIXS	video information exchange system
VLA	vertical line array; visual landing aid
VLF	very low frequency
VLR	very-long-range aircraft
VLZ	vertical landing zone
VMC	visual meteorological conditions
VMF	variable message format
VNTK	target vulnerability indicator designating degree of hardness; susceptibility of blast; and K-factor
VO	validation office
VOCODER	voice encoder
VOCU	voice orderwire control unit
VOD	vertical onboard delivery
VOL	volunteer
vol	volume
VOLS	vertical optical landing system
VOR	very high frequency omnidirectional range station
VORTAC	very high frequency omnidirectional range station and/or tactical air navigation
VOX	voice actuation (keying)
VP	video processor
VPB	version planning board
VPD	version planning document
VS	sector single-unit
VS&PT	vehicle summary and priority table
VSAT	very small aperture terminal
VSII	very seriously ill or injured
VSP	voice selection panel
VSR	sector single-unit radar
V/STOL	vertical and/or short takeoff and landing aircraft
VTA	voluntary tanker agreement
VTC	video teleconferencing

VTOL	vertical takeoff and landing
VTOL-UAV	vertical takeoff and landing unmanned aerial vehicle
VTS	vessel traffic service
VTT	video teletraining
VU	volume unit
VV&A	verification, validation, and accreditation
VV&C	verification, validation, and certification

W

W	sweep width
w	search subarea width
WAAR	Wartime Aircraft Activity Report
WACBE	World Area Code Basic Encyclopedia
WAGB	icebreaker (USCG)
WAN	wide-area network
WARM	wartime reserve mode
WARMAPS	wartime manpower planning system
WAS	wide area surveillance
WASP	war air service program
WATCHCON	watch condition
WB	wideband
WC	wind current
WCA	water clearance authority
WCCS	Wing Command and Control System
WCDO	War Consumables Distribution Objective
W-day	declared by the NCA, W-day is associated with an adversary decision to prepare for war
WDT	warning and display terminal
WEAX	weather
WES	weapon engagement status
WETM	weather team
WEU	Western European Union
WEZ	weapon engagement zone
WFE	warfighting environment
WFP	World Food Programme (UN)
WGS	World Geodetic System
WGS-84	World Geodetic System 1984
WH	wounded due to hostilities
WHEC	high-endurance cutter (USCG)
WHNRS	wartime host-nation religious support
WHNS	wartime host-nation support
WHNSIMS	Wartime Host Nation Support Information Management System
WHO	World Health Organization
WIA	wounded in action

WISDIM	Warfighting and Intelligence Systems Dictionary for Information Management; WIS Dictionary for Information Management
WISP	Wartime Information Security Program
WLG	Washington liaison group
WMD	weapons of mass destruction
WMEC	Coast Guard medium-endurance cutter
WMO	World Meteorological Organization
WMP	Air Force War and Mobilization Plan; War and Mobilization Plan
WOC	wing operations center (USAF)
WOD	word-of-day
WORM	write once read many
WPA	water jet propulsion assembly
WPAL	wartime personnel allowance list
WPARR	War Plans Additive Requirements Roster
WPB	Coast Guard patrol boat
WPC	Washington Planning Center
WPM	words per minute
WPN	weapon
WPS	Worldwide Port System
WR	war reserve; weapon radius
WRC	World Radiocommunication Conference
WRL	weapons release line
WRM	war reserve materiel
WRMS	war reserve materiel stock
WRR	weapons response range (as well as wpns release rg)
WRS	war reserve stock
WRSK	war readiness spares kit; war reserve spares kit
WSC	Worldwide Military Command and Control System (WWMCCS) Intercomputer Network (WIN) site coordinator
WSE	weapon support equipment
WSES	surface effect ship (USCG)
WSESRB	Weapon System Explosive Safety Review Board
WSR	weapon system reliability
WT	gross weight; warping tug; weight
WTCA	water terminal clearance authority
WTLO	water terminal logistic office
Wu	uncorrected sweep width
WVRD	World Vision Relief and Development, Inc.
WWABNCP	worldwide airborne command post
WWII	World War II
WWSVCS	Worldwide Secure Voice Conferencing System
WX	weather

X

X	initial position error
XCVR	transceiver
XO	executive officer
XSB	barrier single unit

Y

Y	search and rescue unit (SRU) error
YR	year

Z

Z	zulu
z	effort
Zt	total available effort
ZULU	time zone indicator for Universal Time

Intentionally Blank

APPENDIX B
TERMINOLOGY POINTS OF CONTACT

1. US NATO Military Terminology Group

a. **Office, US NATO Military Terminology Group**, Operational Plans and Joint Force Development Directorate, J-7, JDETD, ATTN: Chairman, US NATO Military Terminology Group, 7000 Joint Staff, Pentagon, Washington, DC 20318-7000; Tel (703) 695-6292, DSN 225-6292; Fax (703) 693-8897, DSN 223-8897

b. **Military Service Terminology Representatives**

- **Army:** US Army HQDA, ODSCSOPS (DAMO-SSP), Washington, DC 20310-0460; Tel (703) 697-6949, DSN 227-6949; Fax (703) 614-2896, DSN 224-2896

- **Navy:** Navy Warfare Development Command (N5T), Sims Hall, 686 Cushing Road, Newport, RI 02841-1207; Tel (401) 841-2717; DSN 948-2717; Fax (401) 841-3286; Fax DSN 948-3286

- **Air Force:** HQ AFDC/DL, 1480 Air Force Pentagon, Washington, DC 20330-1480; Tel (703) 693-7932, DSN 223-7932; Fax (703) 695-8245, DSN 225-8245

- **Marine Corps:** Doctrine Division (C427), Marine Corps Combat Development Command Quantico, VA 22134-5021; Tel (703) 784-6227, DSN 278-6227; Fax (703) 784-2917, DSN 278-2917

- **Coast Guard:** US Coast Guard Headquarters (OPD1), 2100 2nd St SW, Washington, DC 20593- 0001; Tel (202) 267-0583; Fax (202) 297-4278

2. DOD Terminology Points of Contact

a. **Office of the Secretary of Defense**, WHS Directives and Records Division, Rm. 2A286, Pentagon, Washington, DC 20301-1155; Tel (703) 697-4111, DSN 227-4111; Fax (703) 614-8532, DSN 224-8532

b. **Joint Staff Manpower and Personnel Directorate** (J-1) Military Secretariat, 1000 Joint Staff, Pentagon, Washington, DC 20318-1000; Tel (703) 697-9644, DSN 227-9644; Fax (703) 693-1596, DSN 223-1596

- **Operations Directorate** (J-3) Office of the Military Secretariat, 3000 Joint Staff, Pentagon, Washington, DC 20318-3000; Tel (703) 695-4705, DSN 225-4705; Fax (703) 614-1755, DSN 224-1755

- **Logistics Directorate** (J-4) Logistics Planning Division, 4000 Joint Staff, Pentagon, Washington, DC 20318-4000; Tel (703) 697-0595, DSN 227-0595; Fax (703) 697-0566, DSN 227-0566

- **Strategic Plans and Policy Directorate** (J-5) Policy Division, 5000 Joint Staff, Pentagon, Washington, DC 20318-5000; Tel (703) 614-8715, DSN 224-8715; Fax (703) 697-1337, DSN 227-1337

- **Command, Control, Communications, and Computer Systems Directorate** (J-6) C4 Architecture and Integration Division, 6000 Joint Staff, Pentagon, Washington, DC 20318-6000; Tel (703) 693-5332, DSN 223-5332; Fax (703) 697-6610, DSN 227-6610

- **Operational Plans and Joint Force Development Directorate** (J-7) Joint Doctrine Division, ATTN: Chairman, US NATO Military Terminology Group, 7000 Joint Staff, Pentagon, Washington, DC 20318-7000; Tel (703) 695-6292, DSN 225-6292; Fax (703) 693-8897, DSN 223-8897

- **Force Structure, Resources, and Assessment Directorate** (J-8) Forces Division, 8000 Joint Staff, Pentagon, Washington, DC 20318-8000; Tel (703) 697-0799, DSN 227-0799; Fax (703) 614-6601, DSN 224-6601

- **US Military Communications-Electronics Board** (USMCEB), Washington, DC 20318-6100; Tel (703) 614-7924, DSN 224-7924; Fax (703) 693-3322, DSN 223-3322

c. **Defense Agencies**

- **Defense Information Systems Agency** (DISA), JIEO, Center for Standards, 10701 Parkridge Boulevard, Reston, VA 22091-4398; Tel (703) 735-3532, DSN 364-3532; Fax (703) 735-3256, DSN 364-3256

- **Defense Intelligence Agency** (DIA) ATTN: J2J, Pentagon, Washington, DC 20340-5037; Tel (703) 695-1032, DSN 225-1032; Fax (703) 697-9650, DSN 227-9650

- **Defense Logistics Agency** (DLA) ATTN: DASC-DD, 8725 Kingman Road, Fort Belvoir, VA 22060-6220; Tel (703) 767-1268, DSN 427-1268; Fax (703) 767-5559, DSN 427-5559

- **National Imagery and Mapping Agency** (NIMA) ATTN: PCO/DFJ, Mail Stop P-37, 12310 Sunrise Valley Drive, Reston, VA 20191-3449; Tel (703) 263-3148, DSN 570-3148; Fax (703) 264-3139, DSN 570-3139

- **Defense Special Weapons Agency** (DSWA) ATTN: OPOE, 6801 Telegraph Road, Alexandria, VA 22310-3398; Tel (703) 325-6844, DSN 221-6844; Fax (703) 325-6226, DSN 221-6226

- **National Security Agency** (NSA) Central Security Service, ATTN: N-51, Rm. 2A256, Pentagon, Washington, DC 20301-1155; Tel (301) 688-7819, DSN 923-7819; Fax (301) 497-2844, DSN 923-2844

d. **Combatant Commands**

- **US Central Command** (USCENTCOM) ATTN: CCJ5-O, 7115 S Boundary Blvd, MacDill AFB, FL 33621-5101; Tel (813) 828-6447, DSN 968-6447; Fax (813) 828-5917, DSN 968-5917

- **US European Command** (USEUCOM) ATTN: CHF, ECJ5-D Unit 30400, Box 1000, APO AE 09128-4209; Tel 011-49-711-680-5277, DSN 314-430-5277; Fax 011-49-711-680-7338, DSN 314-430-7338

- **US Joint Forces Command** (USJFCOM) ATTN: JWFC Code JW100, 116 Lake View Parkway, Suffolk, VA 23435-2697

- **US Pacific Command** (USPACOM) ATTN: J383 Box 64013, Camp H. M. Smith, HI 96861-4013; Tel (808) 477-8268, DSN 477-1164; Fax (808) 477-8280, DSN 477-8280

- **US Southern Command** (USSOUTHCOM) ATTN: SCJ5-PS 3511 NW 91st Ave, Miami, FL 33172-1217; Tel (305) 437-1511, DSN 312-567-1511, Fax (305) 437-1854, DSN 312-567-1854

- **US Space Command** (USSPACECOM) ATTN: J5X, 250 S Peterson Blvd, Suite 116, Peterson AFB, CO 80914-3130; Tel (719) 554-3164, DSN 692-3164; Fax (719) 554-5493, DSN 692-5493

- **US Special Operations Command** (USSOCOM) ATTN: SOOP-PJ-D, 7701 Tampa Point Boulevard, MacDill AFB, FL 33608-6001; Tel (813) 828-7548/3114, DSN 299-7548/3114; Fax (813) 828-9805, DSN 299-9805

- **US Strategic Command** (USSTRATCOM) ATTN: J512, 901 SAC Boulevard, Ste 2E-18, Offutt AFB, NE 68113-6500; Tel (402) 294-2080, DSN 271-2080; Fax (402) 294-1035, DSN 271-1035

- **US Transportation Command** (USTRANSCOM) ATTN: TCJ5-SR, 508 Scott Drive, Scott AFB, IL 62225-7001; Tel (618) 256-5103, DSN 576-5103; Fax (618) 256-7957, DSN 576-7957

Intentionally Blank

APPENDIX C
ADMINISTRATIVE INSTRUCTIONS

1. User Comments

Users are highly encouraged to submit comments on this publication to the Operational Plans and Joint Force Development Directorate, J-7, Joint Doctrine Division, ATTN: Chairman, US NATO Military Terminology Group, 7000 Joint Staff, Pentagon, Washington, DC 20318-7000; Tel (703) 695-6292, DSN 225-6292; Fax (703) 693-8897, DSN 223-8897. These comments should address content (accuracy, usefulness, consistency, and organization), writing, and appearance.

2. Authorship

The lead agent and the Joint Staff doctrine sponsor for this publication is the Director for Operational Plans and Joint Force Development (J-7).

3. Supersession

This publication supersedes JP 1-02, 23 March 1994, *Department of Defense Dictionary of Military and Associated Terms*.

4. Change Recommendations

a. Recommendations for urgent changes to this publication should be submitted:

TO: JOINT STAFF WASHINGTON DC//J7-JDETD//

Routine changes should be submitted to the Director for Operational Plans and Joint Force Development (J-7), JDETD, ATTN: Chairman, US NATO Military Terminology Group, 7000 Joint Staff, Pentagon, Washington, DC 20318-7000 via the designated terminology point of contact found in Appendix B.

b. Sample Format for Submission of Proposed Changes:

To: (Designated Terminology Point of Contact – See Appendix B.)

Subject: Recommendation for Changes to the "DOD Dictionary of Military and Associated Terms (JP 1-02)"

1. The following changes are recommended for JP 1-02:

a. Change: **grazing fire** – Fire ~~which is~~ approximately parallel to the ground ~~and~~ where the center of the cone of fire does not rise above ~~the height of a man standing~~ one meter from the ground. See also fire.

Reason: (State the reason for the recommended change, and include appropriate reference to source material, if applicable.)

NOTE: For proposed changes, added words are underlined, deleted words are line through.

b. Add: **lines of communications** – All the routes, land, water, and air, which connect an operating military force with a base of operations and along which supplies and military forces move.

Reason: (State the reason for the recommended addition and include appropriate reference to source material, if applicable.)

c. Delete: **frustrated cargo**.

Reason: (State the reason for deleting the term and include appropriate reference to source material, if applicable.)

2. Point of contact for this action is: ________________________
(name and rank or title)

Tel. XXX-XXXX
Address of Submitting Office:

____
____
____

c. When a Joint Staff directorate submits a proposal to the Chairman of the Joint Chiefs of Staff that would change terminology reflected in this publication, that directorate will include a proposed change to this publication as an enclosure to its proposal.

d. Record of Changes:

15 October 2001: JMTGM-50-01 (JP 3-0), JMTGM-051-01 (JPs 4-07, 0-2, 3-18, 4-02, and 3-07.6), and JMTGM-052-01 (JP 3-02) added.

19 December 2001: JPs 4-04, 4-09, and 3-05.1 added.

23 January 2002: JPs 3-60 and 5-00.1 added.

5. Distribution

a. JP 1-02 may be purchased from the Government Printing Office (GPO). Tel. (202) 512-0312, GPO Bookstore, or write to: Superintendent of Documents, US Government Printing Office, Washington, DC 20402.

b.

Army:	US Army AG Publication Center SL 1655 Woodson Road Attn: Joint Publications St. Louis, MO 63114-6181
Air Force:	Air Force Publications Distribution Center 2800 Eastern Boulevard Baltimore, MD 21220-2896
Navy:	CO, Naval Inventory Control Point 700 Robbins Avenue Bldg 1, Customer Service Philadelphia, PA 19111-5099
Marine Corps:	Commander (Attn: Publications) 814 Radford Blvd, Suite 20321 Albany, GA 31704-0321
Coast Guard:	Commandant (G-OPD), US Coast Guard 2100 2nd Street, SW Washington, DC 20593-0001
	Commander USJFCOM JWFC Code JW2102 Doctrine Division (Publication Distribution) 116 Lake View Parkway Suffolk, VA 23435-2697

c. Local reproduction is authorized and access to unclassified publications is unrestricted. However, access to and reproduction authorization for classified joint publications must be in accordance with DOD Regulation 5200.1-R, *Information Security Program*.

Intentionally Blank

JOINT CHIEFS OF STAFF

Made in United States
North Haven, CT
04 November 2026

21774412R00339